D0163546

PASSIVE AND ACTIVE FILTERS

THEORY AND IMPLEMENTATIONS

PASSIVE AND ACTIVE FILTERS

THEORY AND IMPLEMENTATIONS

Wai-Kai Chen
University of Illinois at Chicago

John Wiley & Sons
New York
Chichester
Brisbane
Toronto
Singapore

Library of Congress Cataloging in Publication Data:

Chen, Wai-Kai, 1936–
Passive and active filters.

Includes index.
1. Electric filters, Active. 2. Electric filters,
Passive. 3. Analog electronic systems. I. Title.
TK7872.F5C469 1986 621.3815'324 85-9497
ISBN 0-471-82352-X

Printed in the United States of America

10 9 8 7

To Shiao-Ling and
Jerome and Melissa

PREFACE

Electrical filters permeate modern technology so much that it is difficult to find any electronic system that does not employ a filter in one form or another. Thus, the topic of filters is important for study by electrical engineering students preparing to enter the profession and by practicing engineers wishing to extend their skills. The purpose of the book is to fill such needs.

The book is designed to provide the basic material for an introductory senior or first-year graduate course in the theory and design of modern analog filters, as opposed to digital filters. It is intended for use as a text for a two-quarter or one-semester course. It should be preceded by a course on linear networks and systems analysis in which Laplace transform and state-variable techniques are introduced. Some familiarity of the theory of a complex variable and matrix algebra is necessary for the last three chapters.

With the rapid advance of solid-state technology and the wide use of digital computers, few curricula in electrical engineering can afford more than a one-semester or two-quarter course in analog filter design. The trend is toward shortening the passive filter synthesis and expanding the active design. This is to recognize the tremendous advances made in integrated-circuit processing techniques, which frequently render active elements much cheaper to produce than some of the passive ones. As a result, modern synthesis techniques lean heavily toward the use of active elements, which, together with resistors and capacitors, virtually eliminate the need to use inductors in many frequency ranges. Such active *RC* or inductorless filters are attractive in that they usually weigh less and require less space than their passive counterparts, and they can be fabricated in microminiature form using integrated-circuit technology, thereby making inexpensive mass production possible. Thus, they are taking over an ever-increasing share of the total filter production and applications. However, active *RC* filters have the finite bandwidth of the active devices, which places a limit on the high-frequency performance. Most of the active *RC* filters are used up to approximately 30 kHz. This is quite adequate for use in voice and data communication systems. The passive filters, on the other hand, do not have such a limitation, and they can be employed up to approximately 500 MHz, the limitation being the parasitics associated with the passive elements. Furthermore, passive filters generally have a lower sensitivity than the active ones.

Based on these observations, the decision was made to concentrate on inductorless filters in which the active element is the operational amplifier. The passive filter synthesis is included to provide background material for other topics. For example, active filters are designed from passive resistively terminated *LC* ladder prototypes. In addition, other basic passive synthesis techniques are included to provide the students with a solid background in analog filters. By putting together theories, techniques, and procedures that can be used to analyze, design, and implement analog filters, it is hoped that after the completion of this book the students will be able to do some simple filter design work and will possess enough background for advanced study.

A by-product of this choice of topics is the stress of the importance of the operational amplifier. Two developments that have profoundly affected the modern practice of electrical engineering are the microprocessor for digital systems and the operational amplifier for the analog systems. It is imperative that the students have experience with both. This book stresses the usefulness of the operational amplifier.

The scope of this book should be quite clear from a glance at the table of contents. Chapter 1 introduces the fundamentals of analog filter design by covering the basic active building blocks and properties of network functions. Chapter 2 gives a fairly complete exposition of the subject of approximation, which includes the usual Butterworth, Chebyshev, inverse Chebyshev, and Bessel–Thomson responses. Frequency transformation is also covered. With the concept of positive-real functions established in Chapter 1, the properties of the driving-point immittance functions of the *LC* and *RC* one-port networks and their realization techniques are examined in Chapter 3. These techniques are extended to realize various classes of transfer functions in the form of *LC* or *RC* ladders or parallel ladders. Chapter 4 discusses the synthesis of resistively terminated lossless two-port networks and presents explicit design formulas for the doubly terminated lossless Butterworth and Chebyshev ladder networks. These ladder structures also serve as the prototypes for the synthesis of the coupled active filters described in Chapter 8.

The next four chapters deal with active filter synthesis. Chapter 5 introduces two basic approaches and two network configurations. The advantages and disadvantages of each approach are examined. The subject of sensitivity is covered in Chapter 6. In addition to discussing the various types of sensitivity and their interrelations, general relations of network function sensitivities are given. Chapter 7 begins the discussion of active filters, which include the single-amplifier general biquads and the multiple-amplifier general biquads. The study of simulated ladder networks is undertaken in Chapter 8. Simulation of the passive ladder is accomplished in three ways: the use of simulated inductors, the use of frequency-dependent negative resistors, and the simulation of the block-diagram representation of the ladder.

The last three chapters contain material on advanced passive filter synthesis. Chapter 9 is concerned with the design of broadband matching networks, which are used to equalize a resistive generator to a frequency-dependent load and to achieve a preassigned transducer power-gain characteristic. The theory of passive cascade synthesis is presented in Chapter 10. The students will soon discover that active synthesis is much simpler than its passive counterpart, because with the use of the operational amplifier as a buffer, individual sections can be designed independently.

Finally, the general problem of compatibility of two frequency-dependent impedances is taken up in Chapter 11. It is shown that the problem essentially reduces to that of the existence of a certain type of all-pass function.

The intent of the book is to provide a unified and modern treatment of filter design techniques. The arrangement of topics is such that they reinforce one another. The book stresses basic concepts, modern design techniques, and implementation procedures. The theory is supplemented and illustrated by numerous practical examples. The prerequisite knowledge is a typical undergraduate mathematics background of calculus, complex variables, and simple matrix algebra plus a working knowledge in Laplace transform technique. When the level of the material is beyond that assumed, references are given to the relevant literature. As a result, the book can also be used as a guide on filters to the practicing engineers who desire a good solid introduction to the field.

The material presented in the book has been classroom-tested at the University of Illinois at Chicago for the past several years. There is little difficulty in fitting the book into a one-semester or two-quarter course on analog filter design. For example, the first eight chapters contain material suitable for a one-semester course. The entire book is designed for a two-quarter course. For this reason, the material of the last three chapters on passive synthesis, which could have been made to follow Chapter 4, is relegated to the end, so that the first eight chapters would fit nicely into a one-semester course.

A special feature of the book is that it contains material on the design of broadband matching networks and on compatible impedances. These topics are normally excluded from undergraduate curricula, but recent advances makes these important filters available to undergraduates. The serious students will find the perusal of these chapters to be a gratifying and stimulating experience.

A variety of problems are included at the end of each chapter to enhance and extend the presentation. Most of these problems have been class-tested to ensure that their levels of difficulty and their degrees of complexity are consistent with the intent of the book. Also, the triangle symbol is used to represent the operational amplifier, because it appears to be the most commonly accepted usage in the literature.

I am indebted to many of my students over the years who participated in testing the material of this book, and to my colleagues at the University of Illinois at Chicago for providing a stimulating milieu for discussions. A personal note of appreciation goes to Puszka Łuszka for inspirational discussions. Special thanks are due to Yi-Sheng Zhu, Joseph Chiang and Eishi Yasui, who gave the complete manuscript a careful and critical reading and assisted me in preparing many of the tables, and to Jing-Liang Wan of Dalian Institute of Technology for proofreading the complete manuscript. Yi-Sheng Zhu and Eishi Yasui assisted me in preparing the index. Dr. Carlos Lisboa also proofread the first three chapters with some detailed comments. Finally, I express my appreciation to my wife, Shiao-Ling, and children, Jerome and Melissa, for their patience and understanding during the preparation of the book.

Wai-Kai Chen

Chicago, Illinois

CONTENTS

PASSIVE AND ACTIVE FILTERS

THEORY AND IMPLEMENTATIONS

chapter 1
FUNDAMENTALS OF NETWORK SYNTHESIS

Design is the primary concern for the engineer as opposed to analysis for the scientist. It has traditionally implied the use of cut-and-try methods, the know-how from experience, and the use of handbooks and charts. It was not until after the Second World War that a scientific basis for design fully emerged, distinguished by the name *synthesis.* Synthesis has since become one of the most important subjects in modern electrical engineering. The contrast between analysis and synthesis is illustrated schematically in Figure 1.1. If the network and the excitation are given and the response is to be determined, the solution process is called *analysis.* When the excitation and the response are given and it is required to determine a network with no trial and error, it is known as *synthesis.* In this sense, analysis and synthesis are opposites. The word analysis comes from the Greek *lysis* and *ana.* It means the loosening up of a complex system. Synthesis, on the other hand, means putting together a complex system from its parts or components. In network synthesis, we are primarily concerned with the design of networks to meet prescribed excitation–response characteristics.

An electrical *filter* is a device designed to separate, pass, or suppress a group of signals from a mixture of signals. On a larger scale, televisions and radios are typical examples of electrical filters. When a television is tuned to a particular channel, it will only pass those signals transmitted by that channel and will block all other signals.

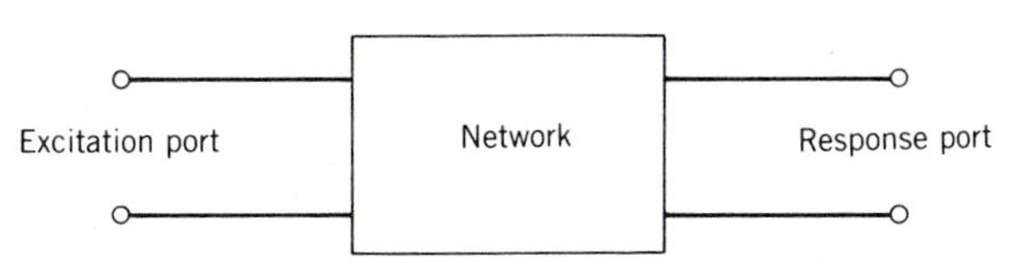

Figure 1.1

On a smaller scale, filters are basic electronic components used in the design of communication systems such as telephone, television, radio, radar, and computer. In fact, electrical filters permeate modern technology so much that it is difficult to find any electronic system that does not employ a filter in one form or another. The purpose of this book is to introduce some fundamental concepts and methods of modern filter design, and no electrical engineering graduate is equipped to understand the literature or advanced practice without a knowledge of these fundamentals.

1.1 NETWORK SYNTHESIS PROBLEM

There are important differences between analysis and synthesis. First of all, in analysis there is normally a unique solution. By contrast, in synthesis solutions are not unique and there may exist no solution at all. For example, in Figure 1.2, the generator with internal resistance $R_1 = 10\,\Omega$ is capable of supplying a maximum power of $|V_g|^2/4R_1 = 100/40 = 2.5$ W; yet the required output power is $|V_2|^2/R_2 = 36/10 = 3.6$ W. It is impossible to design a passive network to achieve this. As another example, consider the impedance function

$$Z(s) = \frac{4(s^2+7s+10)}{s^2+5s+4} \tag{1.1}$$

This impedance can be realized as the driving-point impedance of at least four different RC networks shown in Figure 1.3. Thus, by specifying the driving-point impedance there are many networks that will yield the same input impedance.

Second, network analysis uses a few basic methods such as nodal, loop, or state-variable techniques. Synthesis, on the other hand, makes use of a variety of methods and involves approximation. Figure 1.4 is the ideal low-pass filter transmission characteristic, where the amplitude of the desired transfer function is constant from $\omega = 0$ to $\omega = \omega_c$ and zero for all ω greater than ω_c. Such niceties cannot be achieved with a finite number of network elements. What then can be done to obtain a desired transmission characteristic? Instead of seeking overly idealistic performance criteria, we specify the maximum permissible loss or attenuation over a given frequency band and the minimum allowable loss over another frequency band. The shaded area of Figure 1.5 represents such a compromise, where maximum deviations are specified from 0 to ω_c and from ω_s to ∞. This is known as the *approximation problem*, and will be discussed in greater detail in Chapter 2.

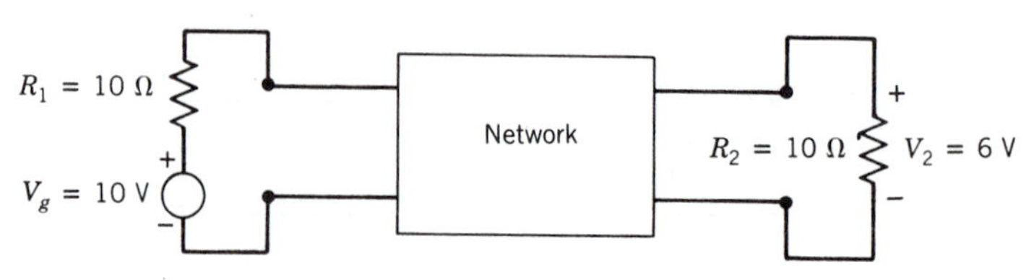

Figure 1.2

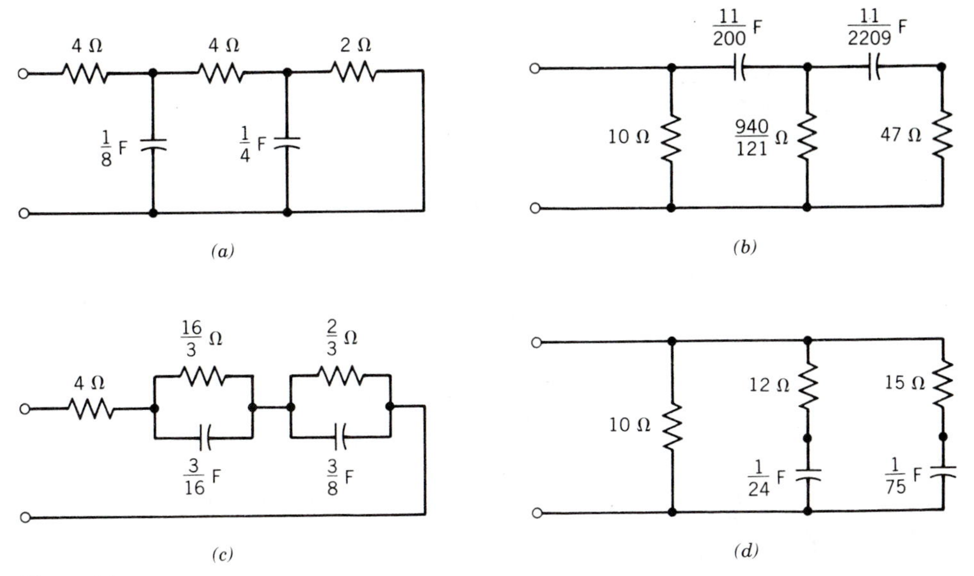

Figure 1.3

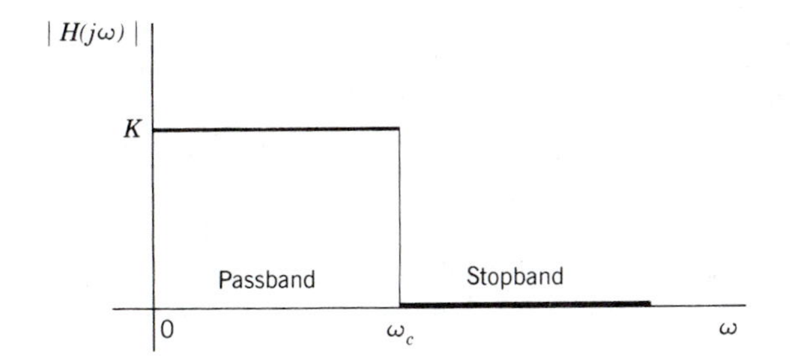

Figure 1.4

How is the synthesis of a network accomplished? The steps we follow are usually as follows: The first step is the determination of a suitable model to represent the system and the setting of specifications. For example, if one wishes to design a network to couple a voltage generator to the input of an operational amplifier, and to achieve the ideal low-pass transmission characteristic of Figure 1.4, the network model of Figure 1.6 and the transfer voltage-ratio function

$$|H(j\omega)| = \frac{|V_2(j\omega)|}{|V_g(j\omega)|} \tag{1.2}$$

are most appropriate. As mentioned above, with a finite number of elements, the characteristic of Figure 1.4 is not realizable. To circumvent this difficulty, we must alter our requirements. The second step is, therefore, to determine the allowable tolerances from the specifications. One such compromise is shown in Figure 1.5. The next

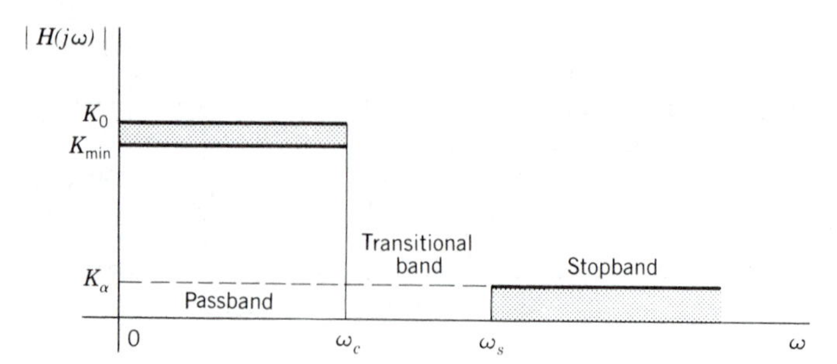

Figure 1.5

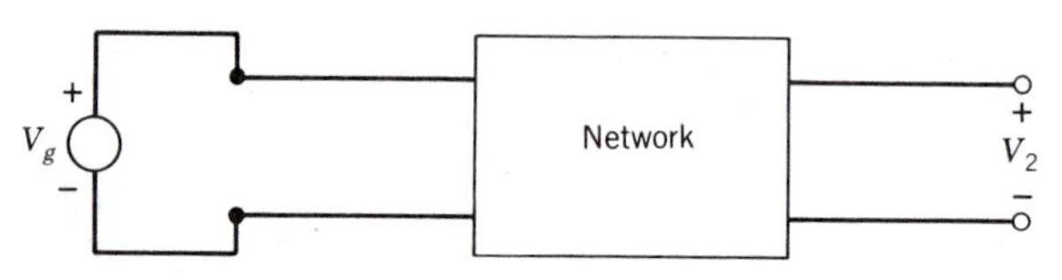

Figure 1.6

step is the determination of methods for expressing or approximating the filter characteristic so as to facilitate its realization. Once this is done, we proceed to the synthesis of physical networks that meet our specifications. In general, there are many possible solutions with different network configurations and different element values. The final step is then to select what appears to be the best solution based on criteria involving costs, performance, sensitivity, convenience, and engineering judgment. The selection of a solution from a number of alternatives is an integral part of engineering that manifests itself in synthesis much more than in analysis.

1.2 CLASSIFICATION OF FILTERS

Electrical filters may be classified in a number of ways. An *analog filter* is a filter used to process analog or continuous-time signals, whereas a *digital filter* is used to process discrete-time or digital signals. Analog filters may further be divided into *passive* or *active* filters, depending on the type of elements used in their realizations.

Filters are also classified according to the functions they perform. A *passband* is a frequency band in which the attenuation of the filter transmission characteristic is small, whereas in the *stopband* the opposite is true. The patterns of passband and stopband give rise to the four most common filter names whose ideal characteristics are depicted in Figure 1.7. An *ideal low-pass* characteristic is shown in (*a*) with passband extending from $\omega=0$ to $\omega=\omega_c$ and stopband from ω_c to infinity, where ω_c is called the *angular* or *radian cutoff frequency*, or simply *cutoff frequency*. An *ideal high-pass* characteristic is the one shown in (*b*) with passband extending from ω_c to infinity and stopband from 0 to ω_c. Figure 1.7(*c*) is the characteristic for an *ideal*

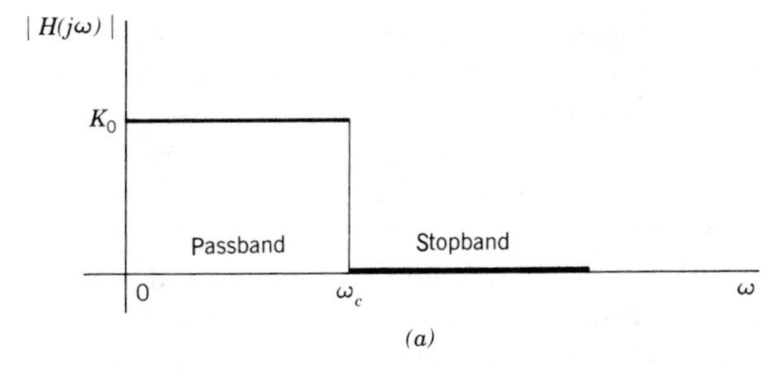

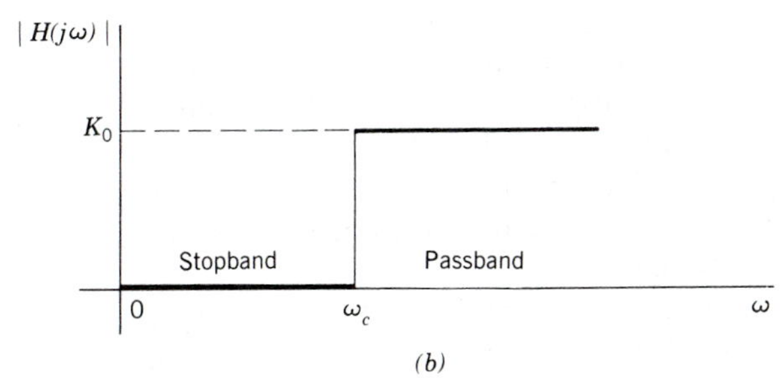

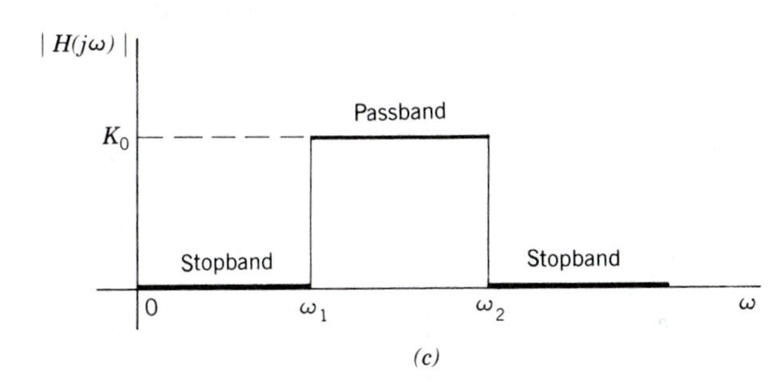

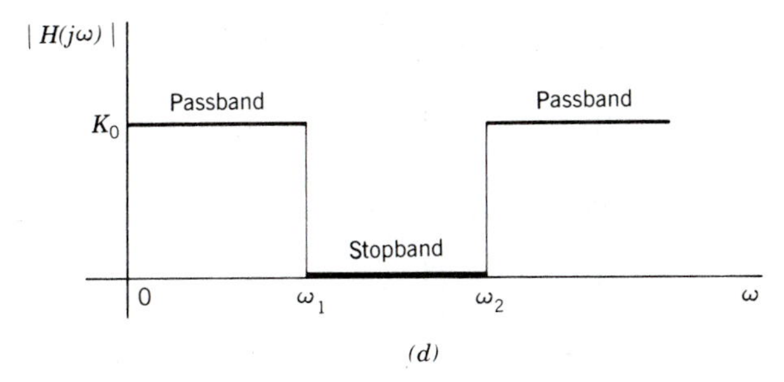

Figure 1.7

band-pass filter, in which radian frequencies extending from ω_1 to ω_2 are passed, while all other frequencies are stopped. Finally, the *ideal band-elimination* characteristic is shown in (*d*), where the radian frequencies from ω_1 to ω_2 are stopped and all others are passed. The band-elimination filters are also known as *notch* filters.

A familiar example that makes use of low-pass, high-pass, and band-pass filters is in the detection of signals generated by a telephone set with push buttons as in TOUCH-TONE dialing. In TOUCH-TONE dialing, the 10 decimal digits from 0 to 9, together with two extra buttons * and # used for special purposes, need to be identified. By using 8 signal frequencies in the frequency band from 697 to 1633 Hz, and arranging them in two groups, 4 low-band frequencies and 4 high-band frequencies, as depicted in Figure 1.8, 16 distinct signals can be identified. Each signal is represented by a pair of tones, one from the low band and one from the high band. Pressing a push button is therefore identified by a unique pair of signal frequencies. As the telephone number is pushed a set of signals is transmitted and then converted to suitable dc signals that are used by the switching system to connect the caller to the party being called. To detect the proper number to be called, it is necessary to identify the individual tones in the respective groups. This is accomplished by the 8 band-pass filters as shown in Figure 1.9. Each of these band-pass filters passes one tone and rejects all others, and is followed by a detector that is energized when its input voltage exceeds a certain threshold. When a detector is energized, its output provides the required dc switching signal.

As another example, consider the transmission of a low-frequency signal such as a voice signal over a distance. To do this it is necessary to modulate a high-frequency signal carrier with this low-frequency signal before transmission. At the receiver,

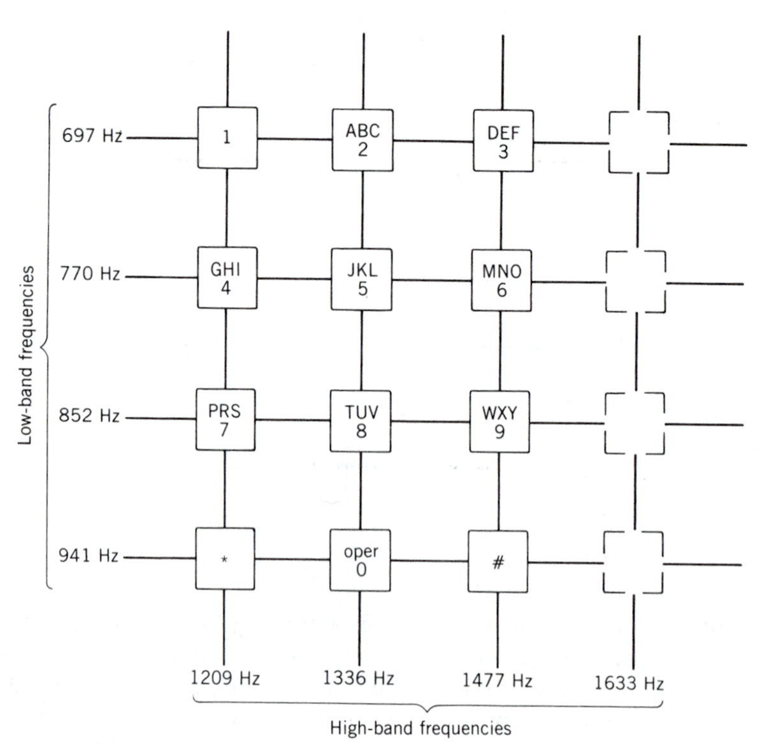

Figure 1.8

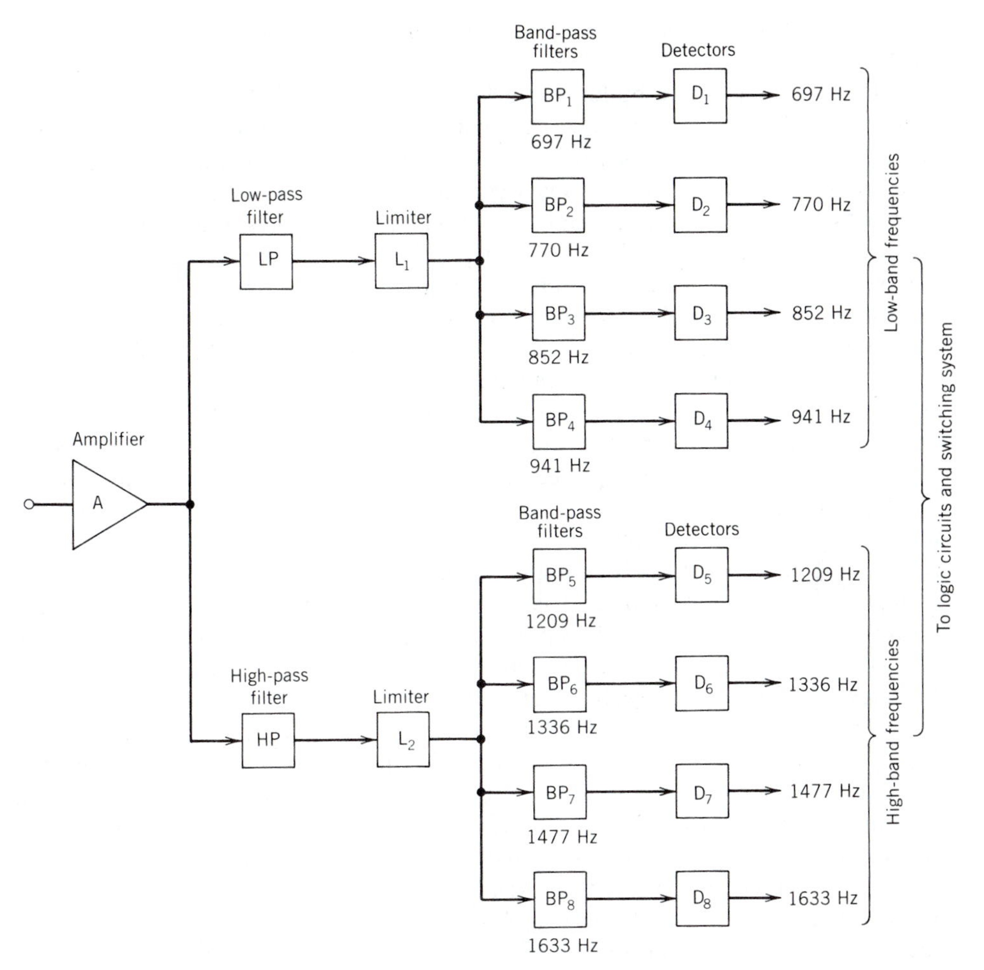

Figure 1.9

the transmitted signal goes through a mixer where it is multiplied by a signal at the modulating frequency. The output of the mixer is passed through a low-pass filter to recover the desired low-frequency signal.

1.3 BUILDING BLOCKS

The networks that we will design in this book consist of interconnections of *elements*. These elements are idealizations of physical devices in that they are assumed to be *lumped*, *linear*, *finite*, and *time-invariant*. In general, they can be classified into two broad categories: *basic building blocks* and *secondary building blocks*. This classification is based on the observation that every secondary building block can be constructed by interconnecting elements in the basic building block categories. In this sense, the secondary building blocks are not as fundamental as the basic building blocks.

Four familiar basic building blocks are the *resistor*, the *capacitor*, the *self-inductor*, and the *mutual inductor*, as shown in Figure 1.10. They are represented by the resistance R, capacitance C, self-inductance L, and mutual inductance M. In addition, with the development of inexpensive operational amplifiers, the use of active filters built with resistors, capacitors, and operational amplifiers has been extensive. Such filters are attractive in that they usually weigh less and require less space than their passive counterparts, and they can be fabricated in microminiature form using integrated-circuit technology. As a result, they can be mass-produced inexpensively. For this reason, we consider the operational amplifier as a basic building block.

The *operational amplifier*, abbreviated as the *op amp*, is one of the most versatile building blocks. Basically, it is a direct-coupled differential amplifier with extremely high gain used with external feedback for gain–bandwidth control. A simplified schematic diagram of the integrated-circuit realization of the popular, inexpensive 741-type op amp with internal frequency compensation is shown in Figure 1.11(*a*), with pin and external connections indicated in (*b*) and (*c*). For our purposes, we consider the op amp as a basic building block defined only by its behavior observed at its input and output ports. Some basic understanding of the internal circuitry of the op amp is very useful, but one does not need to know the intricate details of its design and configuration to use it as an element in active filters. We refer readers to the many good textbooks on this subject for details.†

The op amp is a five-terminal device, two inputs, one output, and two power supply terminals, as depicted by the standard symbol shown in Figure 1.12(*a*). The power supply terminals are usually omitted in the representation and only the output and input terminals are exhibited as in Figure 1.12(*b*). The terminal marked V^+ is called the *noninverting terminal* and that marked V^- is the *inverting terminal*. The equivalent circuit of the op amp is shown in Figure 1.13, where r_i is the input resistance, r_o the output resistance, and A the amplifier gain. The open circuit output voltage of the op amp is related to the difference of the input terminal voltages by

$$V_o = A(V^+ - V^-) \tag{1.3a}$$

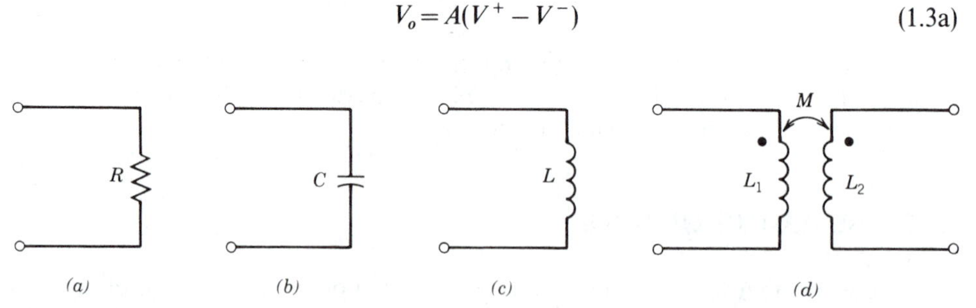

Figure 1.10

†J. K. Roberge 1975, *Operational Amplifiers: Theory and Practices.* New York: John Wiley.
R. G. Irvine 1981, *Operational Amplifier Characteristics and Applications.* Englewood Cliffs, NJ: Prentice-Hall.

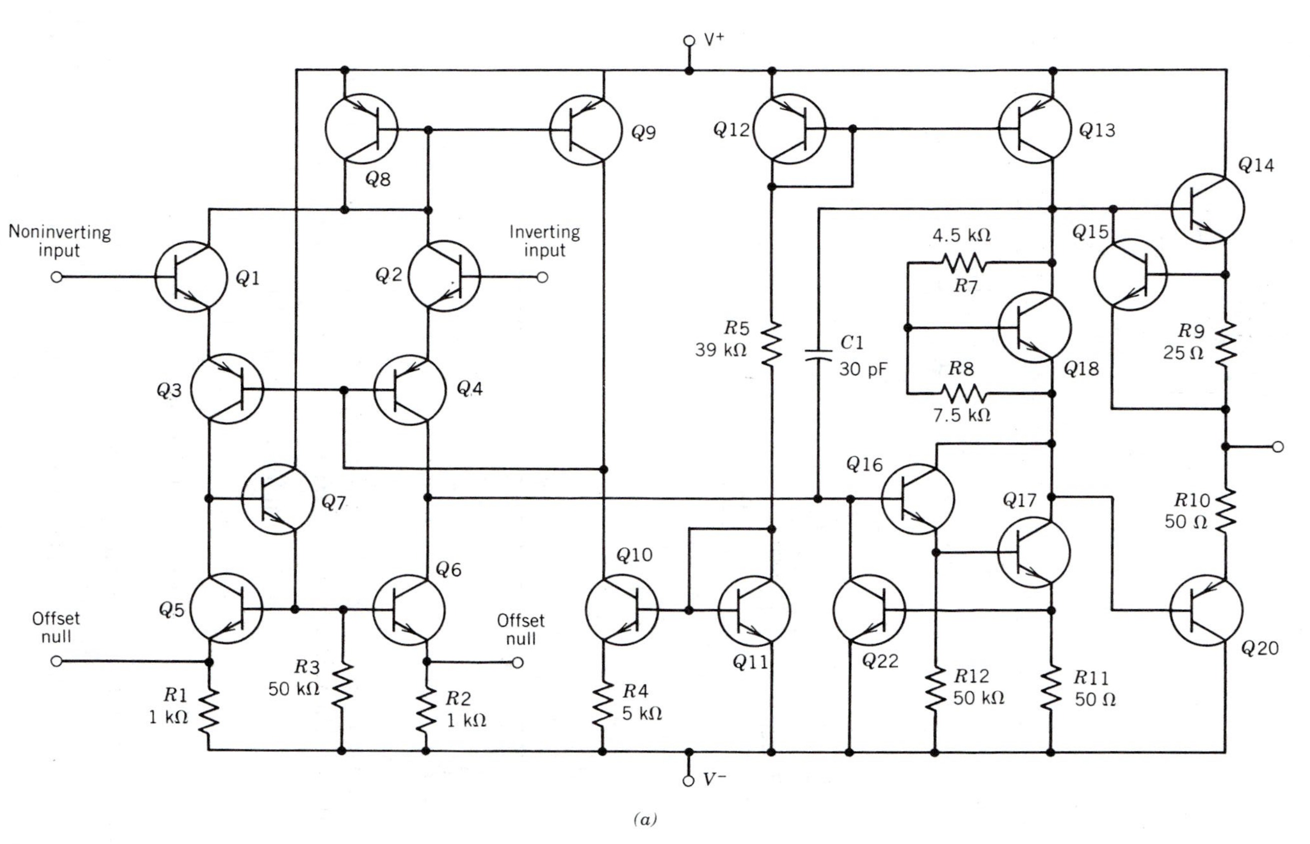

V+
V−
Noninverting input
Inverting input
Offset null
Offset null
Q1
Q2
Q3
Q4
Q5
Q6
Q7
Q8
Q9
Q10
Q11
Q12
Q13
Q14
Q15
Q16
Q17
Q18
Q20
Q22
R1 1 kΩ
R2 1 kΩ
R3 50 kΩ
R4 5 kΩ
R5 39 kΩ
C1 30 pF
4.5 kΩ R7
R8 7.5 kΩ
R9 25 Ω
R10 50 Ω
R11 50 Ω
R12 50 kΩ

(a)

Figure 1.11

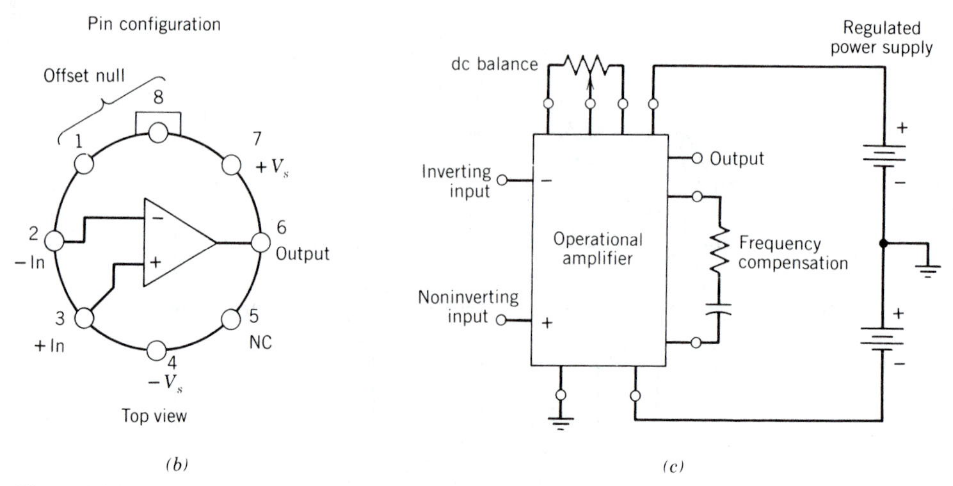

Figure 1.11

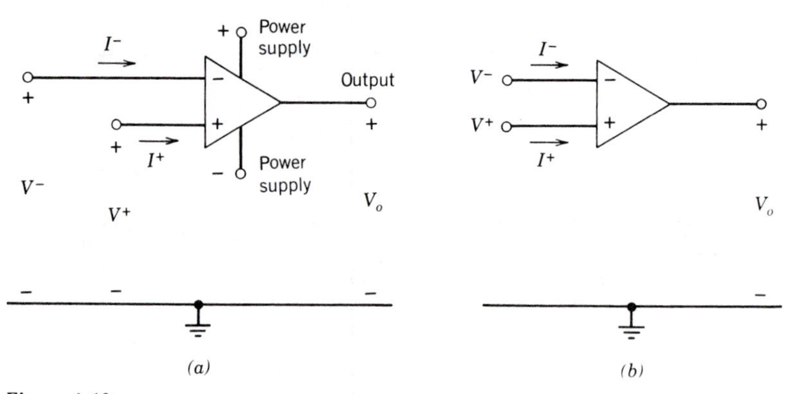

Figure 1.12

Usually, one of the input terminals is grounded, and nearly all op amps have only one output terminal. The values of the parameters of a typical op amp are listed as follows:

$$A \geqq 10{,}000 \qquad \text{at frequencies } f \leqq 10 \text{ kHz}$$

$$r_i = 500 \text{ k}\Omega, \qquad r_o = 300\ \Omega$$

$$|V^+ - V^-| < 1 \text{ mV} \qquad \text{(typically a few microvolts)}$$

Figure 1.14 shows a typical op amp gain versus frequency characteristic.† Idealized input–output characteristics of an op amp are sketched in Figure 1.15, where the output saturation voltage takes place when it reaches the supply voltage, which is

†Most op amps are internally and/or externally compensated to achieve a single-pole roll-off known as the *one-pole roll-off model*. In this case, the pole is typically located between 10 and 100 Hz. For the *two-pole roll-off model* shown in Figure 1.14, the poles at 1 and 100 kHz are due to stray capacitors.

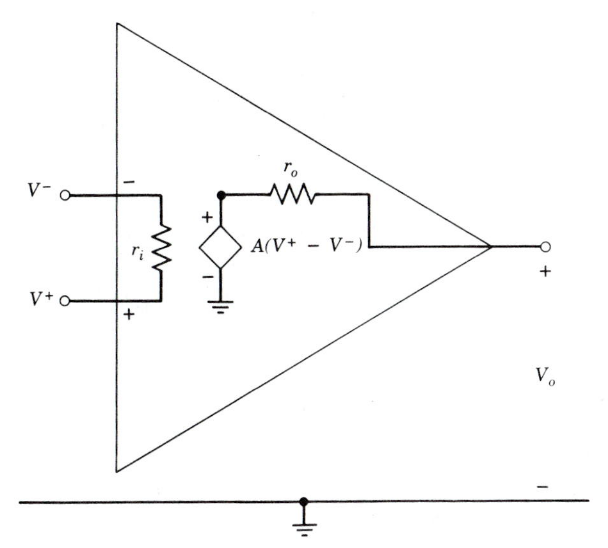

Figure 1.13

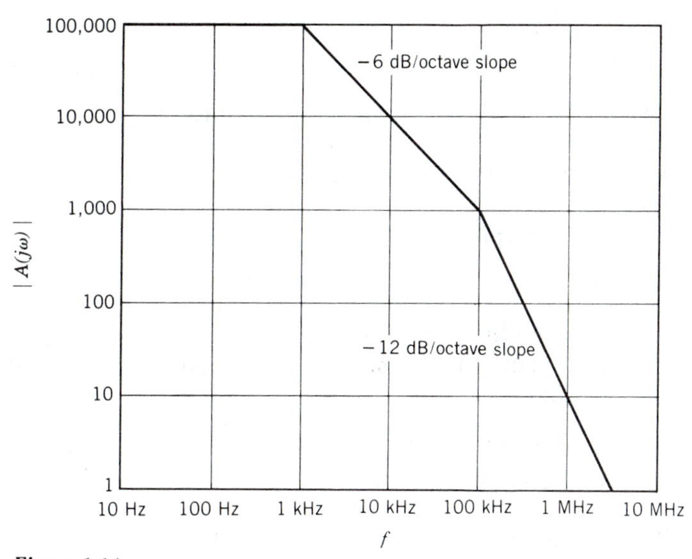

Figure 1.14

usually less than 15 V. This shows that the range of input voltage for linear operation is very small, typically a few microvolts, and a good op amp can be approximated by the so-called ideal op amp.

The *ideal op amp* is the idealization of the real-world op amp with the following characteristics:

1 infinite input impedance, $r_i = \infty$.

2 Zero output impedance, $r_o = 0$.

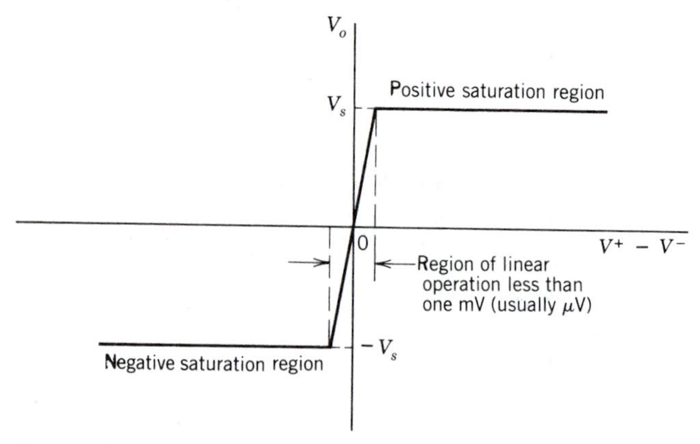

Figure 1.15

3 Infinite voltage gain, $A = \infty$.

4 Infinite bandwidth.

5 Zero output signal when the input voltage is zero: $V_o = 0$ for $V^+ = V^-$. This is known as the *zero offset*.

6 Characteristics that do not drift with environmental factors such as temperature.

In practice, the ideal op amp is a good approximation to the physical op amp, and can be used with considerable simplification in the design and analysis of various networks where op amps are employed. Since the output of the ideal op amp must be finite, (1.3a) implies that the potential difference between the input terminals of the amplifier must be zero as A approaches infinity:

$$V^+ - V^- = V_o/A \to 0 \qquad \text{as } A \to \infty \tag{1.3b}$$

Moreover, because the input impedance is infinite, the input current must be identically zero. Thus, for the ideal op amp we should apply the following rules in the computation of network functions:

Rule 1. The potential difference between the input terminals is zero.

Rule 2. The current at each of the input terminals is zero.

Thus, from Figure 1.12 these assumptions will imply that

$$V^+ = V^- \qquad \text{and} \qquad I^+ = I^- = 0 \tag{1.4}$$

If, for example, the positive terminal of an op amp is grounded, the negative terminal is nearly at the ground potential at all times, and we say a *virtual ground* exists at the negative terminal. We remark that the triangular symbol for an op amp as depicted in Figure 1.12 signifies the unilateral nature of the device: the difference of input voltages determines the output voltage, but the output voltage does not influence the input.

To see how the above rules help in computation, we analyze several secondary building blocks constructed from the basic building blocks.

1.3.1 **Inverting Amplifier**

Consider the RC amplifier of Figure 1.16, the equivalent network of which is shown in Figure 1.17. The loop equations of the amplifier are found to be

$$\begin{bmatrix} Z_1+r_i & -r_i \\ -r_i-Ar_i & r_i+r_o+Z_2+Ar_i \end{bmatrix}\begin{bmatrix} I_{m1} \\ I_{m2} \end{bmatrix}=\begin{bmatrix} V_{\text{in}} \\ 0 \end{bmatrix} \tag{1.5}$$

Solving for I_{m1} and I_{m2} by Cramer's rule, we obtain

$$\begin{aligned} V_o &= Ar_i(I_{m2}-I_{m1})+r_oI_{m2}=(Ar_i+r_o)I_{m2}-Ar_iI_{m1} \\ &= \frac{(Ar_i+r_o)(r_i+Ar_i)-Ar_i(r_i+r_o+Ar_i+Z_2)}{Z_1(r_i+r_o+Ar_i+Z_2)+r_i(r_o+Z_2)}V_{\text{in}} \end{aligned} \tag{1.6}$$

or

$$\frac{V_o}{V_{\text{in}}}=\frac{r_i(r_o-AZ_2)}{Z_1(r_i+r_o+Ar_i+Z_2)+r_i(r_o+Z_2)} \tag{1.7}$$

As $r_i\to\infty$ and $r_o\to 0$, the amplifier gain reduces to

$$\frac{V_o}{V_{\text{in}}}=-\frac{AZ_2}{(1+A)Z_1+Z_2} \tag{1.8}$$

For the ideal op amp, we let $A\to\infty$, obtaining

$$\frac{V_o}{V_{\text{in}}}=-\frac{Z_2}{Z_1} \tag{1.9}$$

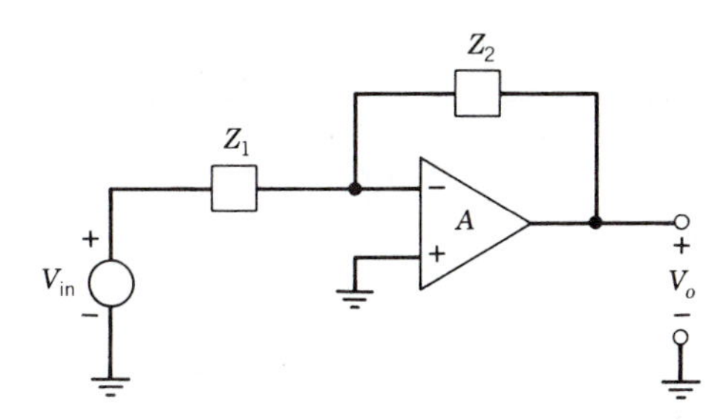

Figure 1.16

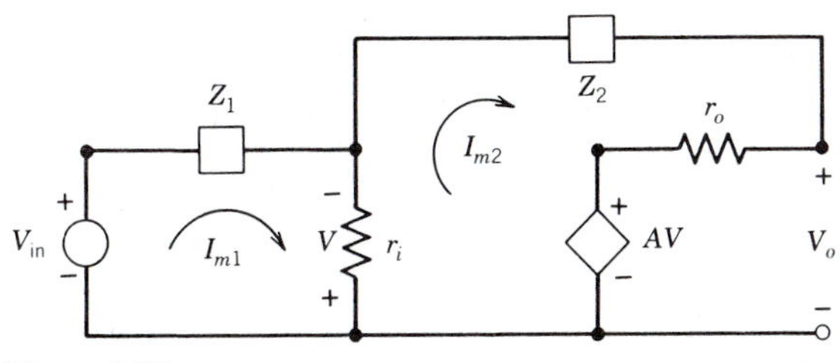

Figure 1.17

For constant Z_1 and Z_2, (1.9) represents an ideal voltage-controlled voltage source (VCVS). Because the polarity of output voltage is the negative of the input, the amplifier is known as an *inverting amplifier*. Thus, an inverting amplifier is a good approximation to the VCVS. The VCVS is an important secondary building block in active filter design.

We now make the approximation at the outset and represent the op amp with the virtual ground model as shown in Figure 1.18. From this network, we can write down the equations

$$I_1 = I_2 \tag{1.10a}$$

$$V_{\text{in}} = I_1 Z_1 \tag{1.10b}$$

$$V_o = -I_2 Z_2 \tag{1.10c}$$

from which we immediately obtain the gain formula (1.9).

In the inverting amplifier of Figure 1.16, let $Z_1 = R$ and $Z_2 = 1/sC$. The resulting network is shown in Figure 1.19 with gain function

$$\frac{V_o}{V_{\text{in}}} = -\frac{1}{RCs} \tag{1.11}$$

This function is readily recognized to be the integration operation. We may write (1.11) in the time domain as

$$v_o(t) = -\frac{1}{RC}\int_{0-}^{t} v_{\text{in}}(\tau)\,d\tau \tag{1.12}$$

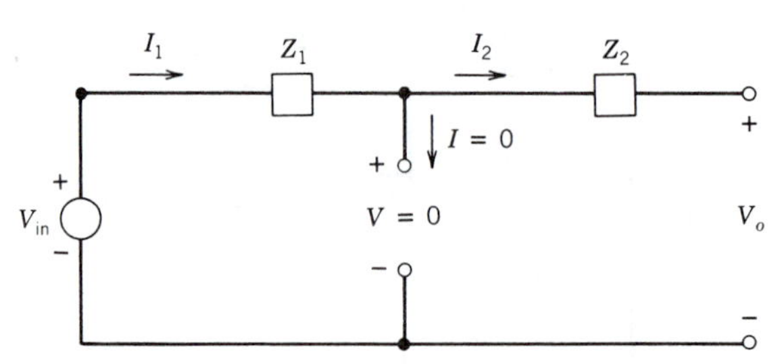

Figure 1.18

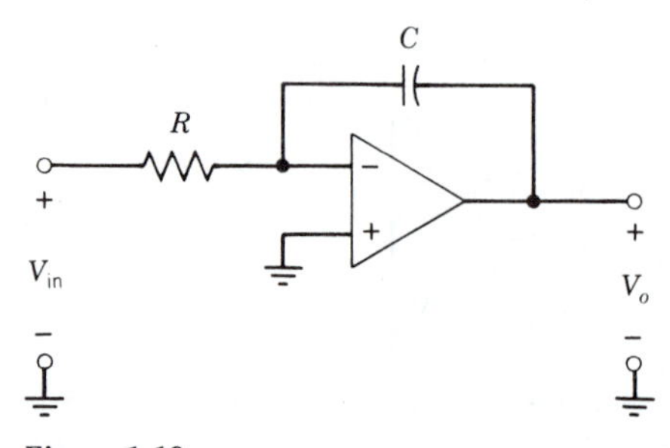

Figure 1.19

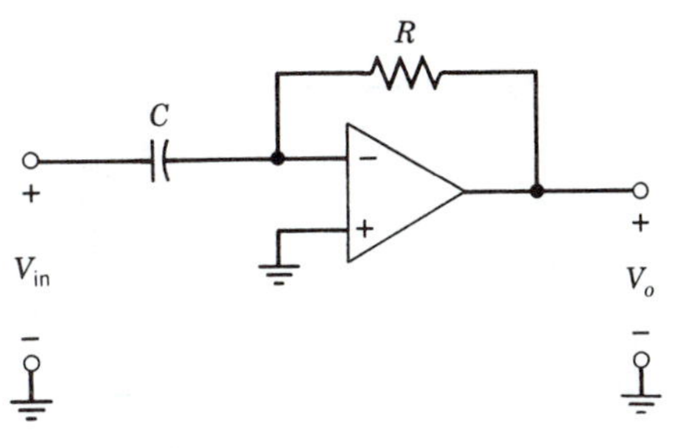

Figure 1.20

On the other hand, when $Z_1 = 1/sC$ and $Z_2 = R$ as shown in Figure 1.20, the amplifier gain function becomes

$$\frac{V_o}{V_{\text{in}}} = -RCs \tag{1.13}$$

or in time domain

$$v_o(t) = -RC\frac{dv_{\text{in}}(t)}{dt} \tag{1.14}$$

The network of Figure 1.19 is known as an *inverting integrator* and that of Figure 1.20 is called an *inverting differentiator*. These are again secondary building blocks used in analog computation. The differentiator, however, is rarely seen in active circuits because it tends to reduce circuit performance.

1.3.2 **Noninverting Amplifier**

A noninverting amplifier using a single op amp is shown in Figure 1.21. The amplifier can be represented by the equivalent network of Figure 1.22, from which the loop equations are found to be

$$\begin{bmatrix} r_i + Z_1 & -Z_1 \\ Ar_i - Z_1 & Z_1 + Z_2 + r_o \end{bmatrix} \begin{bmatrix} I_{m1} \\ I_{m2} \end{bmatrix} = \begin{bmatrix} V_{\text{in}} \\ 0 \end{bmatrix} \tag{1.15}$$

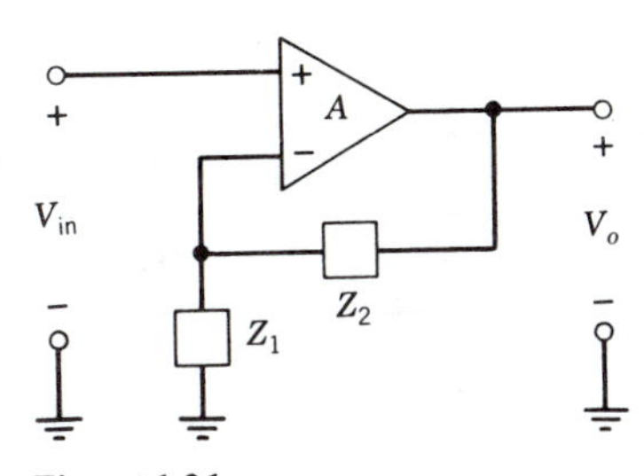

Figure 1.21

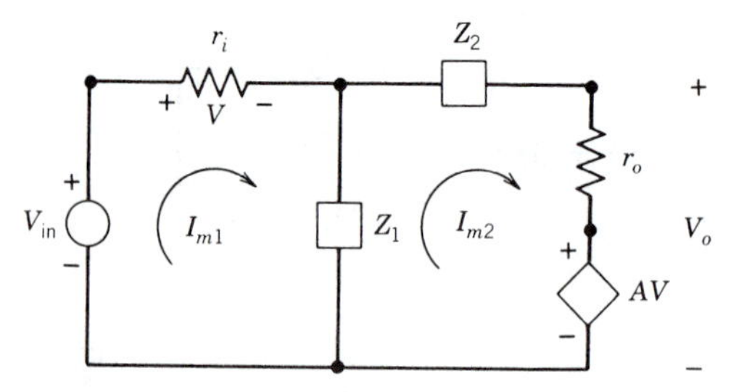

Figure 1.22

Solving for the loop currents I_{m1} and I_{m2} yields the output voltage as

$$V_o = r_o I_{m2} + A r_i I_{m1}$$
$$= \frac{r_o(Z_1 - Ar_i) + Ar_i(r_o + Z_1 + Z_2)}{r_i(r_o + Z_1 + Z_2) + Z_1(Z_2 + r_o) + Ar_i Z_1} V_{in} \quad (1.16)$$

The voltage gain function becomes

$$\frac{V_o}{V_{in}} = \frac{r_o Z_1 + Ar_i(Z_1 + Z_2)}{r_i(r_o + Z_1 + Z_2) + Z_1(Z_2 + r_o) + Ar_i Z_1} \quad (1.17)$$

In the limit, as $A \to \infty$, we have

$$\frac{V_o}{V_{in}} = 1 + \frac{Z_2}{Z_1} \quad (1.18)$$

Thus, for the ideal op amp and constant Z_1 and Z_2, the amplifier of Figure 1.21 represents the ideal VCVS. It is noninverting because there is no reversal of polarities between the input and output voltages. In the case of $Z_2 = 0$, (1.18) becomes unity and the corresponding network degenerates to that of a "follower" circuit, that is, the output voltage follows the input. Under this situation, the impedance Z_1 is redundant and can be removed. The resulting network is shown in Figure 1.23, which has a closed-loop voltage gain of unity.

Instead of applying the limiting process, the gain formula (1.18) can easily be obtained by representing the op amp with the virtual ground model shown in Figure 1.24, from which we have the following equations:

$$I_1 = -I_2 \quad (1.19a)$$

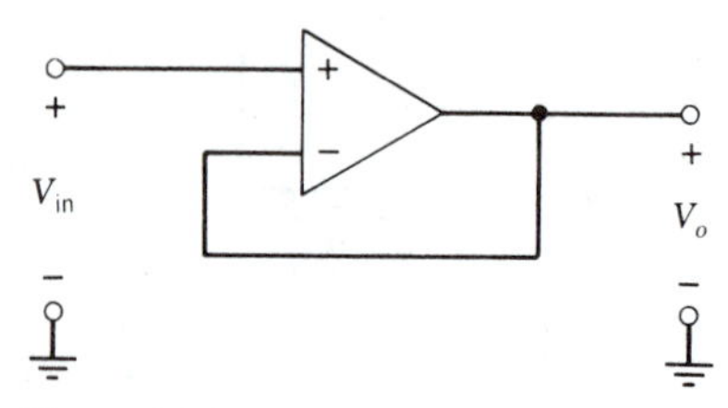

Figure 1.23

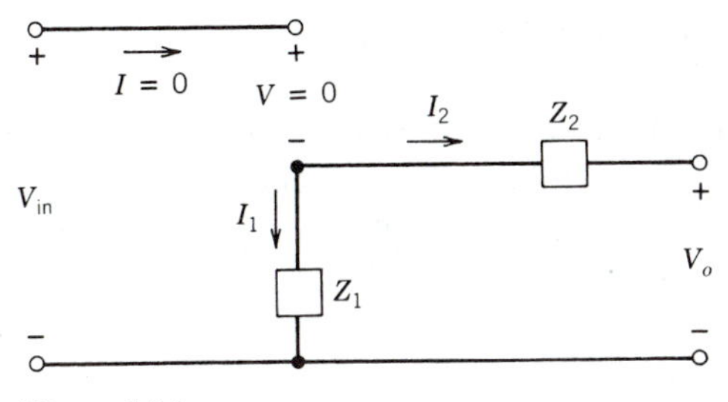

Figure 1.24

$$V_{in} = I_1 Z_1 \tag{1.19b}$$

$$Z_1 I_1 - Z_2 I_2 = V_o \tag{1.19c}$$

Combining these equations gives the gain formula (1.18).

1.3.3 **Gyrator**

Another useful secondary building block is the *gyrator* represented by the symbols shown in Figure 1.25. The gyrator is a device whose terminal voltages and currents are related by the equations

$$V_1 = -rI_2 \tag{1.20a}$$

$$V_2 = rI_1 \tag{1.20b}$$

for Figure 1.25(*a*), and

$$V_1 = rI_2 \tag{1.21a}$$

$$V_2 = -rI_1 \tag{1.21b}$$

for Figure 1.25(*b*). Thus, it is a two-port device characterized by a single parameter r, called the *gyration resistance*. The arrow to the right or the left in Figure 1.25 shows the *direction of gyration*. We remark that the gyrator is not a reciprocal device. For if we first short-circuit the right-hand side and apply a voltage $V_1 = V$ to the left-hand side, and if we next short-circuit the left-hand side and apply the same voltage ($V_2 = V$) to

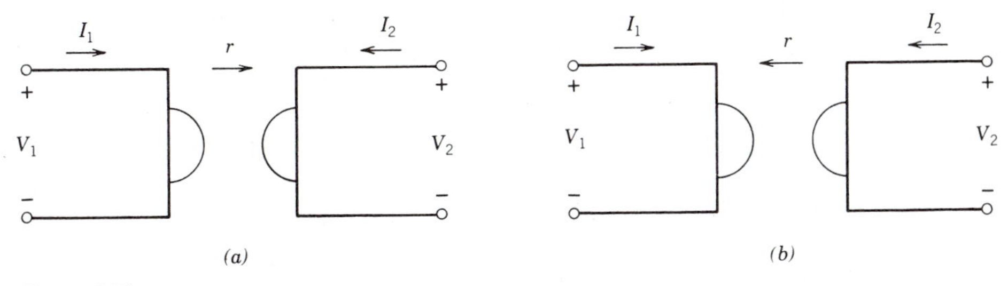

Figure 1.25

the right-hand side, then it will be found that $I_2 = -I_1$, confirming that the gyrator is not reciprocal. In fact, it is *antireciprocal*, and neither stores nor dissipates energy. In this respect, it is similar to the ideal transformer, being a lossless two-port device.

A realization of the gyrator with two op amps and seven resistors is shown in Figure 1.26. To see this, we represent each op amp by its virtual ground model, and obtain the following equations:

$$V_1 = RI_c \tag{1.22a}$$

$$V_d = 2RI_c \tag{1.22b}$$

$$V_d + RI_b = V_1 \tag{1.22c}$$

$$-V_2 - RI_d + V_d = 0 \tag{1.22d}$$

$$V_e - V_d + 2RI_d = 0 \tag{1.22e}$$

$$V_e - V_2 + RI_f = 0 \tag{1.22f}$$

$$V_1 - V_2 = RI_a \tag{1.22g}$$

$$I_1 = I_a + I_b \tag{1.22h}$$

$$I_2 = -I_a + I_f \tag{1.22i}$$

Eliminating all the voltage and current variables except I_1, I_2, V_1, and V_2 gives

$$V_1 = RI_2 \tag{1.23a}$$

$$V_2 = -RI_1 \tag{1.23b}$$

which realizes a gyrator of Figure 1.25(*b*) with gyration resistance R.

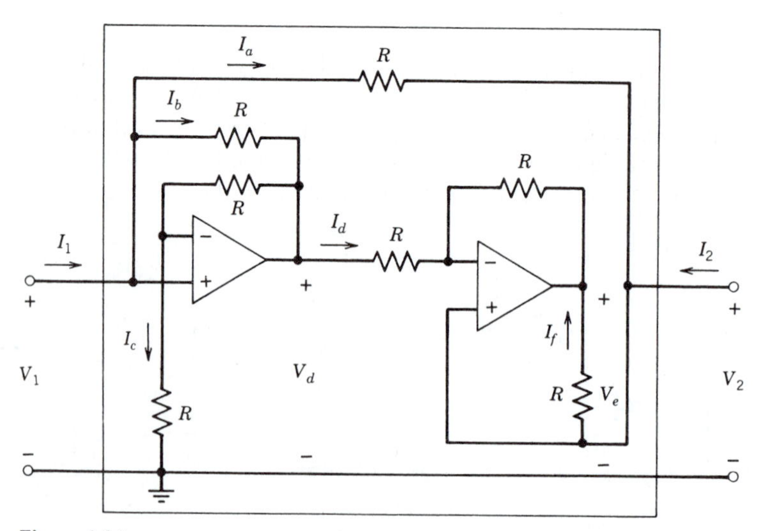

Figure 1.26

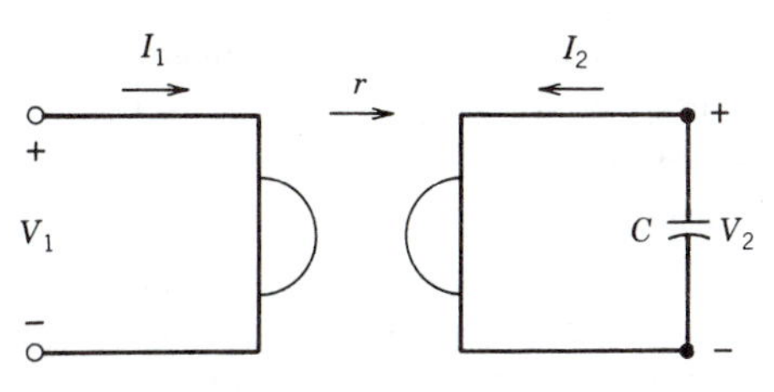

Figure 1.27

An important application of the gyrator in active filter design is in the realization of inductors. The network of Figure 1.27 is a schematic realization of an inductor. We know that $I_2 = -sCV_2$. Therefore, upon inserting $v-i$ relations associated with the gyrator given in (1.20), we observe that

$$V_1 = -rI_2 = sCrV_2 = sCr^2I_1 \tag{1.24}$$

The input impedance of the terminated gyrator becomes

$$Z_{\text{in}} = V_1/I_1 = (Cr^2)s = Ls \tag{1.25}$$

where $L = Cr^2$, which is equivalent to a self-inductor of inductance L. Likewise, it can be shown that the input impedance of an inductor-terminated gyrator is that of a capacitor.

1.4 TRANSFER FUNCTIONS

It is well known that a one-port network can be represented by its Thévenin or Norton equivalent. The equivalence is maintained only at the terminal pair, not inside the original network. For the transfer functions, we are dealing with two ports, an input port and an output port, such as shown in Figure 1.28. The network is called a *two-port network* or simply a *two-port*. Fundamental to the concept of a port is the assumption that the instantaneous current entering one terminal of the port is always equal to the instantaneous current leaving the other terminal of the port. This assumption is crucial in subsequent derivations and the resulting conclusions. If it is violated, the terminal pair does not constitute a port. Figure 1.28 is a general representation of a two-port that is electrically and magnetically isolated except at the two ports. By focusing attention on the two ports, we are interested in the behavior of the network

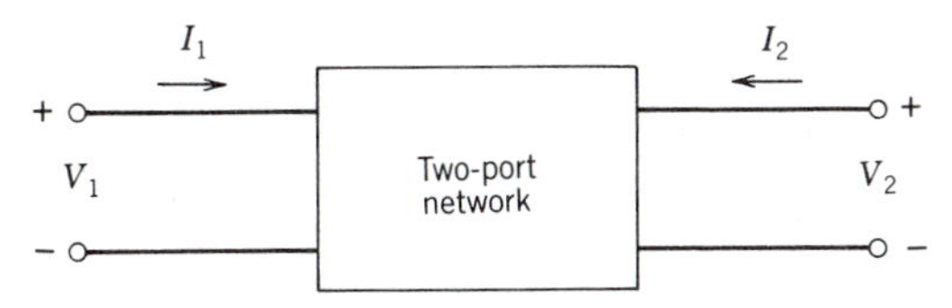

Figure 1.28

only at the two ports. Our discussion is entirely based on the assumption that the two-port is devoid of independent sources inside and has zero initial conditions. To proceed, we establish the sign convention for the references of port voltages and currents as indicated in Figure 1.28.

An example of the use of the two-port concepts is given by the inverting amplifier of Figure 1.16, where the input port variable is V_{in} and the output port variable is V_o. The gyrator circuit of Figure 1.26 is another example of the two-port network. Observe that there are four variables V_1, I_1, V_2, and I_2 associated with a two-port. We can specify any two of the four variables as being independent; the other two become dependent. The dependence of two of the four variables on the other two is described in a number of ways depending on the choice of the independent variables.

Suppose that we choose V_1 and V_2 as the independent variables. Then the port currents I_1 and I_2 and the port voltages V_1 and V_2 are related by the matrix equation

$$\begin{bmatrix} I_1 \\ I_2 \end{bmatrix} \triangleq \begin{bmatrix} y_{11} & y_{12} \\ y_{21} & y_{22} \end{bmatrix} \begin{bmatrix} V_1 \\ V_2 \end{bmatrix} \tag{1.26}$$

The four parameters y_{ij} $(i, j = 1, 2)$ are known as the *short-circuit admittance parameters* or simply the *y-parameters*. The coefficient matrix is called the *short-circuit admittance matrix*. Observe that if either V_1 or V_2 is zero, the four parameters y_{ij} may be defined as

$$y_{11} = \left.\frac{I_1}{V_1}\right|_{V_2=0}, \qquad y_{12} = \left.\frac{I_1}{V_2}\right|_{V_1=0} \tag{1.27a}$$

$$y_{21} = \left.\frac{I_2}{V_1}\right|_{V_2=0}, \qquad y_{22} = \left.\frac{I_2}{V_2}\right|_{V_1=0} \tag{1.27b}$$

The name "short-circuit" becomes obvious.

To express V_1 and V_2 in terms of I_1 and I_2, we may begin by computing the inverse of the short-circuit admittance matrix in (1.26). For this we let $\Delta_y \triangleq y_{11}y_{22} - y_{12}y_{21}$ and obtain

$$\begin{bmatrix} V_1 \\ V_2 \end{bmatrix} = \frac{1}{\Delta_y} \begin{bmatrix} y_{22} & -y_{12} \\ -y_{21} & y_{11} \end{bmatrix} \begin{bmatrix} I_1 \\ I_2 \end{bmatrix} \triangleq \begin{bmatrix} z_{11} & z_{12} \\ z_{21} & z_{22} \end{bmatrix} \begin{bmatrix} I_1 \\ I_2 \end{bmatrix} \tag{1.28}$$

The last step defined a new matrix called the *open-circuit impedance matrix*, the elements of which are the *open-circuit impedance parameters* or simply the *z-parameters*. The reason for the name again becomes apparent by noting from (1.28) that

$$z_{11} = \left.\frac{V_1}{I_1}\right|_{I_2=0}, \qquad z_{12} = \left.\frac{V_1}{I_2}\right|_{I_1=0} \tag{1.29a}$$

$$z_{21} = \left.\frac{V_2}{I_1}\right|_{I_2=0}, \qquad z_{22} = \left.\frac{V_2}{I_2}\right|_{I_1=0} \tag{1.29b}$$

The condition $I_2 = 0$ or $I_1 = 0$ is met by having the corresponding pair of terminals open.

In synthesis, specifications are often given in terms of the transfer functions. However, realization is usually accomplished in terms of the y- or z-parameters,

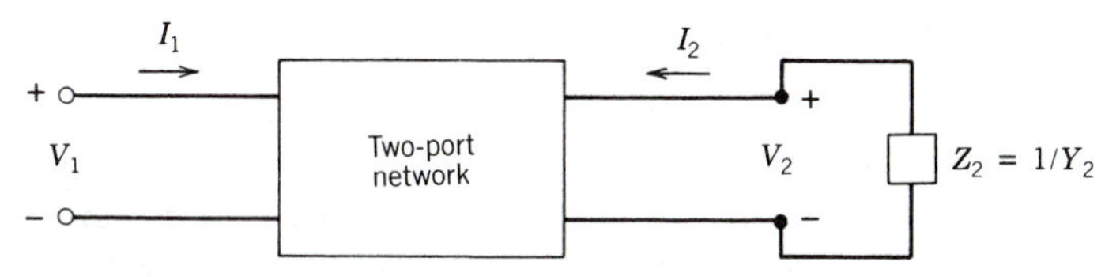

Figure 1.29

pointing to a need to express one set of functions in terms of the other. Figure 1.29 shows a two-port network terminated in a load impedance $Z_2 = 1/Y_2$. For this termination, the output voltage and current are related by

$$V_2 = -Z_2 I_2 \tag{1.30}$$

Substituting this in (1.26) yields

$$\begin{bmatrix} I_1 \\ 0 \end{bmatrix} = \begin{bmatrix} y_{11} & y_{12} \\ y_{21} & y_{22} + Y_2 \end{bmatrix} \begin{bmatrix} V_1 \\ V_2 \end{bmatrix} \tag{1.31}$$

Solving for V_1 and V_2 and taking the ratio, we obtain the *transfer voltage-ratio* or *voltage-gain function* as

$$G_{12} \triangleq \frac{V_2}{V_1} = -\frac{y_{21}}{y_{22} + Y_2} \tag{1.32}$$

For the *transfer admittance function*, we substitute (1.30) in (1.32) and obtain

$$Y_{12} \triangleq \frac{I_2}{V_1} = \frac{y_{21} Y_2}{y_{22} + Y_2} \tag{1.33}$$

Likewise, if we substitute (1.30) in (1.28), the resulting equation is

$$\begin{bmatrix} V_1 \\ 0 \end{bmatrix} = \begin{bmatrix} z_{11} & z_{12} \\ z_{21} & z_{22} + Z_2 \end{bmatrix} \begin{bmatrix} I_1 \\ I_2 \end{bmatrix} \tag{1.34}$$

Solving this for I_1 and I_2 and computing their ratio gives the *transfer current-ratio* or *current-gain function* as

$$\alpha_{12} \triangleq \frac{I_2}{I_1} = -\frac{z_{21}}{z_{22} + Z_2} \tag{1.35}$$

Finally, combining (1.30) with (1.35) results in the *transfer impedance function*

$$Z_{12} \triangleq \frac{V_2}{I_1} = \frac{z_{21} Z_2}{z_{22} + Z_2} \tag{1.36}$$

In the special situation where the output port is open-circuited, the transfer impedance and voltage-ratio functions reduce to

$$Z_{12} = z_{21}, \qquad G_{12} = -y_{21}/y_{22} \tag{1.37a}$$

When the output port is short-circuited, (1.33) and (1.35) become

$$Y_{12} = y_{21}, \qquad \alpha_{12} = -z_{21}/z_{22} \tag{1.37b}$$

1.4.1 Transducer Power Gain

The most useful measure of power flow is called the *transducer power gain* G, which is defined as the ratio of average power delivered to the load to the maximum available average power at the source. It is a function of the two-port parameters, the load and the source impedances. The transducer power gain is the most meaningful description of the power transfer capabilities of a two-port network as it compares the power delivered to the load with the power that the source is capable of supplying under optimum conditions. To illustrate this definition, we derive an expression for G in terms of the impedance parameters of the two-port and the source and load impedances.

Refer to Figure 1.30. The input impedance Z_{11} of the two-port when the output is terminated in Z_2 can be determined from (1.34) as

$$Z_{11} \triangleq \frac{V_1}{I_1} = z_{11} - \frac{z_{12}z_{21}}{z_{22}+Z_2} \tag{1.38}$$

The average power P_2 delivered to the load is given by

$$P_2 = |I_2(j\omega)|^2 \operatorname{Re} Z_2(j\omega) \tag{1.39}$$

where Re denotes "the real part of." The maximum available average power P_{avail} from the source occurs when the input impedance Z_{11} is conjugately matched with the source impedance, that is, $Z_{11}(j\omega) = \bar{Z}_1(j\omega)$, where the bar denotes the complex conjugate. Under this condition, the average power delivered by the source to the load $\bar{Z}_1(j\omega)$ is

$$P_{\text{avail}} = \frac{|V_g(j\omega)|^2}{|Z_1(j\omega)+\bar{Z}_1(j\omega)|^2} \operatorname{Re} Z_1(j\omega) = \frac{|V_g(j\omega)|^2}{4 \operatorname{Re} Z_1(j\omega)} \tag{1.40}$$

Combining (1.39) and (1.40) yields the transducer power gain

$$G \triangleq \frac{P_2}{P_{\text{avail}}} = \frac{4|z_{21}|^2(\operatorname{Re} Z_1)(\operatorname{Re} Z_2)}{|(z_{11}+Z_1)(z_{22}+Z_2) - z_{12}z_{21}|^2} \tag{1.41}$$

the variable $s = j\omega$ being dropped in the expression, for simplicity. Likewise, in terms of the admittance parameters, G becomes

$$G = \frac{4|y_{21}|^2(\operatorname{Re} Y_1)(\operatorname{Re} Y_2)}{|(y_{11}+Y_1)(y_{22}+Y_2) - y_{12}y_{21}|^2} \tag{1.42}$$

where $Y_1 = 1/Z_1$ and $Y_2 = 1/Z_2$.

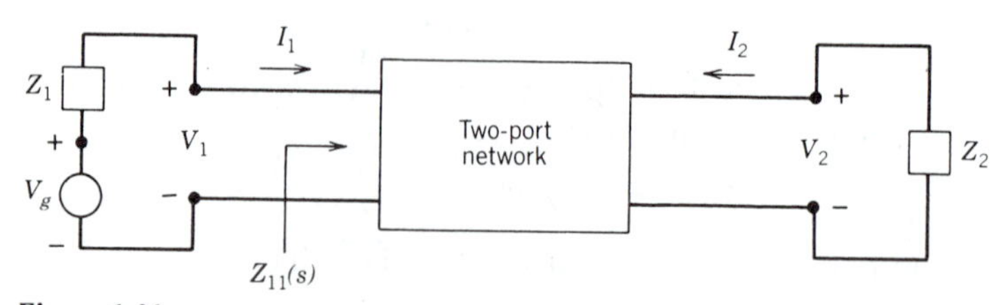

Figure 1.30

1.4.2 Series and Parallel Connection of Two-Port Networks

Simple two-ports are interconnected to yield more complicated and practical two-port networks. In this section, we study two useful and common connections of two-port networks used in synthesis: the series and parallel connections.

Two two-ports are said to be connected in *series* if they are connected as shown in Figure 1.31. This connection forces the equality of the terminal currents of the two-ports. From Figure 1.31 we have

$$\begin{bmatrix} V_1 \\ V_2 \end{bmatrix} = \begin{bmatrix} V_{1a} \\ V_{2a} \end{bmatrix} + \begin{bmatrix} V_{1b} \\ V_{2b} \end{bmatrix} = \begin{bmatrix} z_{11a} & z_{12a} \\ z_{21a} & z_{22a} \end{bmatrix} \begin{bmatrix} I_1 \\ I_2 \end{bmatrix} + \begin{bmatrix} z_{11b} & z_{12b} \\ z_{21b} & z_{22b} \end{bmatrix} \begin{bmatrix} I_1 \\ I_2 \end{bmatrix}$$

$$= \begin{bmatrix} z_{11a} + z_{11b} & z_{12a} + z_{12b} \\ z_{21a} + z_{21b} & z_{22a} + z_{22b} \end{bmatrix} \begin{bmatrix} I_1 \\ I_2 \end{bmatrix} \tag{1.43}$$

where subscripts a and b are used to distinguish the z-parameters of component two-ports N_a and N_b. Equation (1.43) shows that the open-circuit impedance matrix of the composite two-port is simply the sum of the open-circuit impedance matrices of the individual two-ports. The ideal transformer in Figure 1.31 can be removed if *Brune's test* as shown in Figure 1.32 is satisfied:† the voltage marked V is zero.

Another useful connection, called a *parallel connection*, is depicted in Figure 1.33. By carrying out a similar argument as in the series case, we can show that the short-circuit admittance matrix of the composite two-port network is the sum of the short-circuit admittance matrices of the individual two-ports. We remark that the ideal transformers used in Figures 1.31 and 1.33 are necessary to make sure that the nature of the ports is not altered after the connection, recalling the requirement of a port as stated at the beginning of Section 1.4. Like the series case, the ideal transformer in the

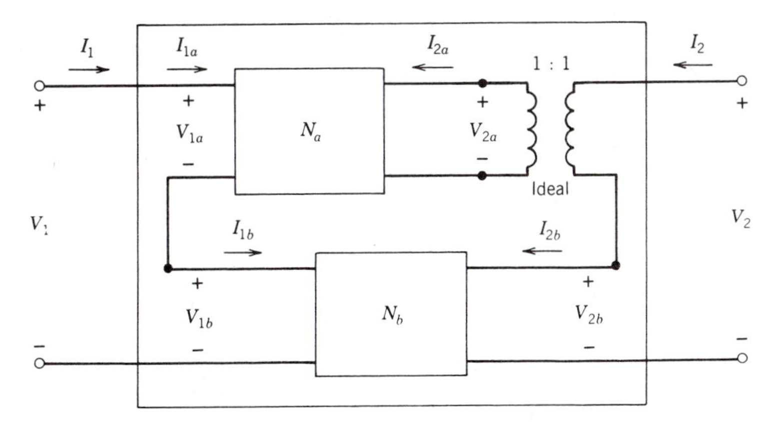

Figure 1.31

†See, for example, L. Weinberg 1962, *Network Analysis and Synthesis*. New York: McGraw-Hill. The test is named after Otto Brune, who is now Principal Research Officer of the National Research Laboratories, Pretoria, South Africa.

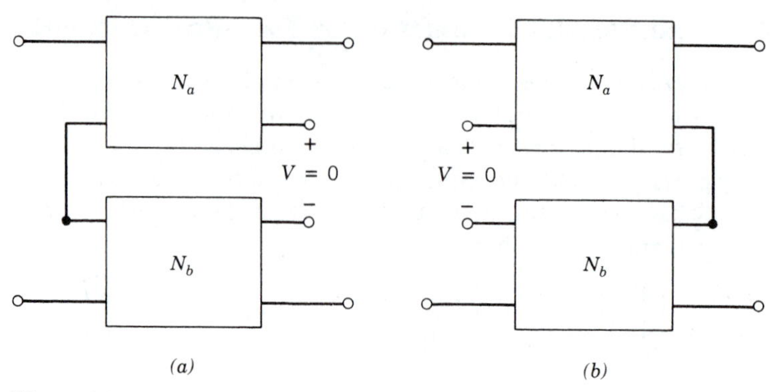

Figure 1.32

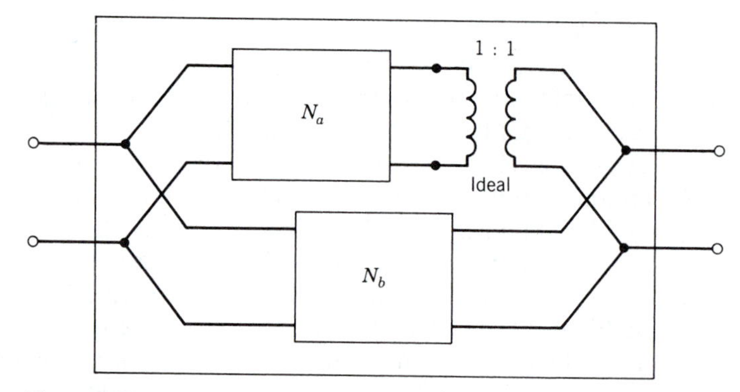

Figure 1.33

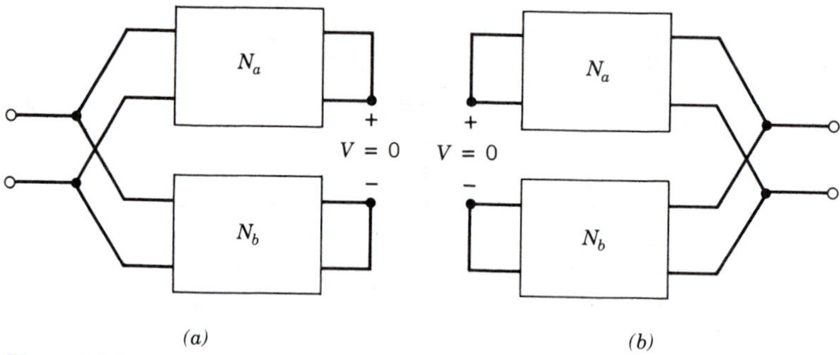

Figure 1.34

parallel connection of Figure 1.33 can be removed if *Brune's test* as shown in Figure 1.34 is satisfied:† the voltage marked V is zero.

The bridged-T network of Figure 1.35 can be viewed as the parallel connection of two two-ports as shown in Figure 1.36. It is easy to verify that Brune's test is satisfied and the short-circuit admittance matrices can be added to give the overall

†See footnote on page 23.

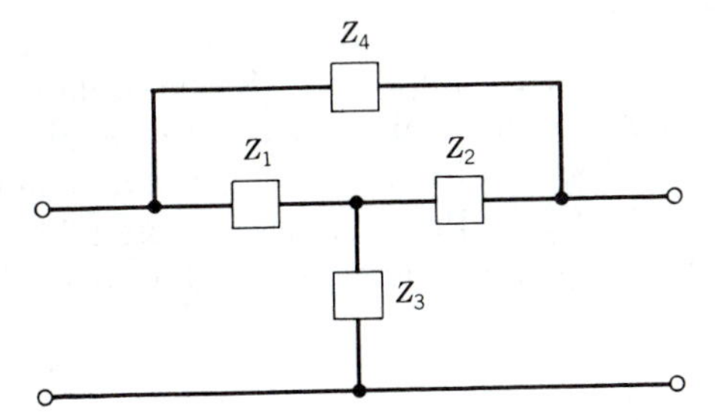

Figure 1.35

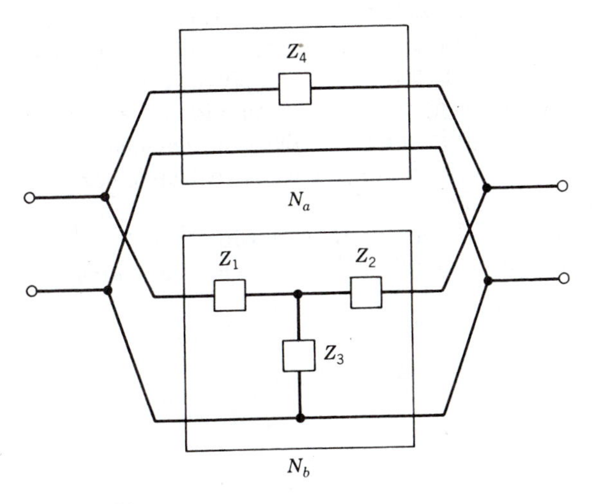

Figure 1.36

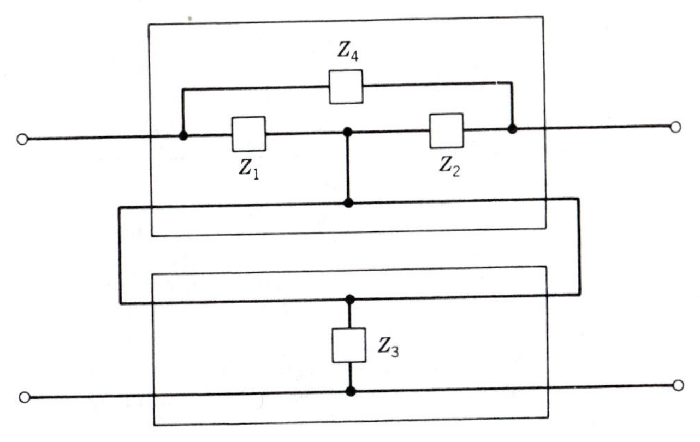

Figure 1.37

admittance matrix. On the other hand, if the bottom conductor in the component two-port N_a is replaced by a nonzero impedance, Brune's condition will not be satisfied in the resulting two-port and an ideal transformer is required. Alternatively, the two-port network of Figure 1.35 can be viewed as the series connection of two two-ports as shown in Figure 1.37. Again, Brune's test of Figure 1.32 is satisfied. Thus, the open-circuit impedance matrices can be added to yield the overall impedance matrix of the bridged-T network of Figure 1.35.

1.5 TELLEGEN'S THEOREM

Kirchhoff's current and voltage laws are algebraic constraints arising from the interconnection of network elements and are independent of the characteristics of the elements. In this section, we show that they imply the conservation of energy. In other words, conservation of energy need not be an added postulate for the discipline of network theory. This result is known as *Tellegen's theorem*,† and will help us understand the fundamental properties of physically realizable impedance functions to be discussed later in this chapter.

Consider the network of Figure 1.38 with branch voltages v_k and currents i_k $(k=1, 2, 3, 4, 5)$ as labeled. Kirchhoff's current law equations are found to be

$$i_1+i_2=0 \tag{1.44a}$$

$$-i_2+i_3+i_4=0 \tag{1.44b}$$

$$-i_4-i_5=0 \tag{1.44c}$$

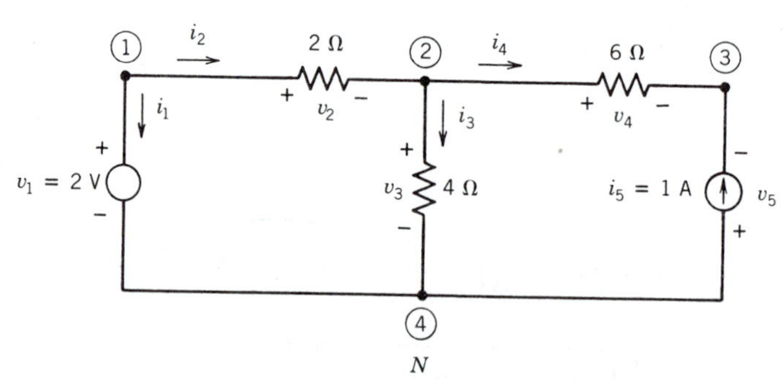

Figure 1.38

†B. D. H. Tellegen, "A general network theorem, with applications," *Proc. Inst. Radio Engrs. Australia*, vol. 14, 1953, pp. 265–270. Bernard D. H. Tellegen (1900–) is with Philips Research Laboratories, Netherlands, and Technological University, Delft.

which can be put in matrix form as

$$\begin{bmatrix} 1 & 1 & 0 & 0 & 0 \\ 0 & -1 & 1 & 1 & 0 \\ 0 & 0 & 0 & -1 & -1 \end{bmatrix} \begin{bmatrix} i_1 \\ i_2 \\ i_3 \\ i_4 \\ i_5 \end{bmatrix} = \begin{bmatrix} 0 \\ 0 \\ 0 \end{bmatrix} \tag{1.45a}$$

or more compactly

$$\mathbf{Ai}(t) = \mathbf{0} \tag{1.45b}$$

where the coefficient matrix $\mathbf{A}$ is called the *incidence matrix* and $\mathbf{i}(t)$ the *branch-current vector*. If we write

$$\mathbf{A} = [a_{kj}] \tag{1.46}$$

the elements of $\mathbf{A}$ are determined by

$$a_{kj} = \begin{cases} +1, & \text{if branch current } i_j \text{ leaves node } k \\ -1, & \text{if branch current } i_j \text{ enters node } k \\ 0, & \text{otherwise.} \end{cases} \tag{1.47}$$

Note that only $n-1$ current equations are required, where n is the number of nodes of the network. As usual, the nodal equation corresponding to the potential-reference point is ignored in (1.44).

The branch voltages v_k can be expressed in terms of the nodal voltages v_{nj} ($j = 1, 2, 3$) from nodes j to the common reference node 4, as follows:

$$v_1 = v_{n1} \tag{1.48a}$$

$$v_2 = v_{n1} - v_{n2} \tag{1.48b}$$

$$v_3 = v_{n2} \tag{1.48c}$$

$$v_4 = v_{n2} - v_{n3} \tag{1.48d}$$

$$v_5 = -v_{n3} \tag{1.48e}$$

or, in matrix notation,

$$\begin{bmatrix} v_1 \\ v_2 \\ v_3 \\ v_4 \\ v_5 \end{bmatrix} = \begin{bmatrix} 1 & 0 & 0 \\ 1 & -1 & 0 \\ 0 & 1 & 0 \\ 0 & 1 & -1 \\ 0 & 0 & -1 \end{bmatrix} \begin{bmatrix} v_{n1} \\ v_{n2} \\ v_{n3} \end{bmatrix} \tag{1.49a}$$

Using the incidence matrix $\mathbf{A}$ defined in (1.45), the above equation can be compactly written as

$$\mathbf{v}(t) = \mathbf{A}'\mathbf{v}_n(t) \tag{1.49b}$$

where the prime denotes the matrix transpose, $\mathbf{v}(t)$ the *branch-voltage vector*, and $\mathbf{v}_n(t)$ the *nodal voltage vector.*

The instantaneous power delivered to branch k is $v_k i_k$. Let us sum the branch instantaneous powers $v_k i_k$ for all b branches of the network. Then

$$\sum_{k=1}^{b} v_k i_k = v_1 i_1 + v_2 i_2 + \cdots + v_k i_k = \mathbf{v}'\mathbf{i} = \mathbf{i}'\mathbf{v} \tag{1.50}$$

where the time variable t is omitted, for simplicity. Substituting (1.49b) in (1.50) and invoking (1.45b), we obtain

$$\sum_{k=1}^{b} v_k i_k = (\mathbf{A}'\mathbf{v}_n)'\mathbf{i} = \mathbf{v}_n'(\mathbf{A}\mathbf{i}) = \mathbf{v}_n'\mathbf{0} = 0 \tag{1.51}$$

This equation shows that the sum of instantaneous powers delivered to all the elements is equal to zero. If we integrate (1.51) between any two limits t_0 to t, the total energy stored in the network from t_0 to t is

$$w(t) = \sum_{k=1}^{b} \int_{t_0}^{t} v_k(\tau) i_k(\tau)\, d\tau = \text{constant} \tag{1.52}$$

meaning that conservation of energy is a direct consequence of Kirchhoff's current and voltage laws.

Since Kirchhoff's laws are independent of the characteristics of the network elements, an even more general result can be stated. Consider another network $\hat{N}$ of Figure 1.39, which has the same topological configuration, the same references for branch currents and voltages, and the same numbering for the branches as the network N of Figure 1.38. As a result, both networks have the same incidence matrix $\mathbf{A}$, and all the equations (1.44) to (1.49) remain valid for the network $\hat{N}$ as well. Let $\hat{\mathbf{i}}(t)$ and $\hat{\mathbf{v}}(t)$ denote the branch-current vector and branch-voltage vector of $\hat{N}$. Then

$$\mathbf{A}\hat{\mathbf{i}}(t) = \mathbf{0} \tag{1.53}$$

$$\hat{\mathbf{v}}(t) = \mathbf{A}'\hat{\mathbf{v}}_n(t) \tag{1.54}$$

hold, where $\hat{\mathbf{v}}_n(t)$ is the nodal voltage vector for $\hat{N}$.

Refer to Figure 1.39. The branch currents and voltages are represented by

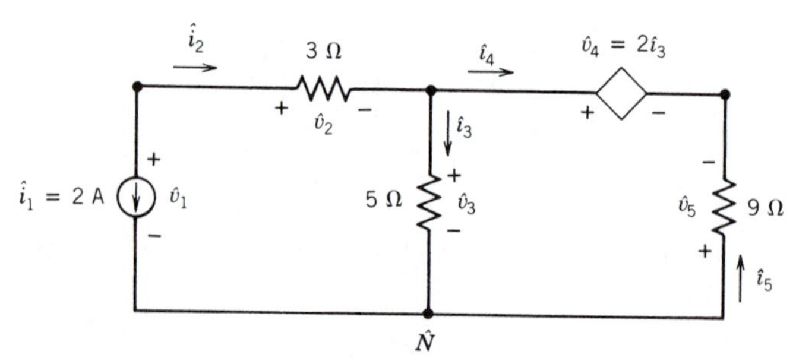

Figure 1.39

$\hat{i}_k$ and $\hat{v}_k$, respectively. We next compute the sum of products

$$\sum_{k=1}^{b} v_k\hat{i}_k = \mathbf{v}'\hat{\mathbf{i}} = (\mathbf{A}'\mathbf{v}_n)'\hat{\mathbf{i}} = \mathbf{v}_n'(\mathbf{A}\hat{\mathbf{i}}) = \mathbf{v}_n'\mathbf{0} = 0 \tag{1.55}$$

where (1.49b) and (1.53) were used. A similar derivation gives

$$\sum_{k=1}^{b} \hat{v}_k i_k = \hat{\mathbf{v}}'\mathbf{i} = 0 \tag{1.56}$$

By taking the transposes of these equations and observing that the transpose of a scalar—namely, a 1×1 matrix—is itself, we obtain several variations of (1.55) and (1.56) as follows:

$$\mathbf{v}'(t)\hat{\mathbf{i}}(t) = \hat{\mathbf{v}}'(t)\mathbf{i}(t) = \hat{\mathbf{i}}'(t)\mathbf{v}(t) = \mathbf{i}'(t)\hat{\mathbf{v}}(t) = 0 \tag{1.57}$$

Equation (1.57) is referred to as *Tellegen's theorem.* Note that the entity $\mathbf{v}'(t)\hat{\mathbf{i}}(t)$ does not have physical significance as the sum of instantaneous powers because $\mathbf{v}(t)$ and $\hat{\mathbf{i}}(t)$ belong to two different networks. In the special case where they belong to the same network, $\mathbf{v}'(t)\mathbf{i}(t)$ is the total instantaneous power delivered to all the elements of the network. Tellegen's theorem is valid whether the networks are linear or nonlinear and time-invariant or time-varying. In addition to its fundamental importance, the theorem provides a way to check the accuracy of the numerical results of a computer solution. After all the voltages and currents are computed, the sum of the products of the corresponding terms must be zero for every value of t.

Example 1.1 We use the networks of Figures 1.38 and 1.39 to verify Tellegen's theorem. For the network N of Figure 1.38, the branch-current vector and the branch-voltage vector are found to be

$$\mathbf{i} = \begin{bmatrix} i_1 \\ i_2 \\ i_3 \\ i_4 \\ i_5 \end{bmatrix} = \begin{bmatrix} 1/3 \\ -1/3 \\ 2/3 \\ -1 \\ 1 \end{bmatrix}, \qquad \mathbf{v} = \begin{bmatrix} v_1 \\ v_2 \\ v_3 \\ v_4 \\ v_5 \end{bmatrix} = \begin{bmatrix} 2 \\ -2/3 \\ 8/3 \\ -6 \\ -26/3 \end{bmatrix} \tag{1.58}$$

For the network $\hat{N}$ of Figure 1.39, they are given by

$$\hat{\mathbf{i}} = \begin{bmatrix} i_1 \\ i_2 \\ i_3 \\ i_4 \\ i_5 \end{bmatrix} = \begin{bmatrix} 2 \\ -2 \\ -3/2 \\ -1/2 \\ 1/2 \end{bmatrix}, \qquad \hat{\mathbf{v}} = \begin{bmatrix} v_1 \\ v_2 \\ v_3 \\ v_4 \\ v_5 \end{bmatrix} = \begin{bmatrix} -27/2 \\ -6 \\ -15/2 \\ -3 \\ 9/2 \end{bmatrix} \tag{1.59}$$

It is straightforward to verify that

$$\mathbf{v}'\hat{\mathbf{i}} = \hat{\mathbf{i}}'\mathbf{v} = 2 \times 2 + (-2) \times \left(-\frac{2}{3}\right) + \left(-\frac{3}{2}\right) \times \left(\frac{8}{3}\right) + \left(-\frac{1}{2}\right) \times (-6) + \left(\frac{1}{2}\right) \times \left(-\frac{26}{3}\right) = 0 \tag{1.60a}$$

$$\hat{\mathbf{v}}'\mathbf{i}=\mathbf{i}'\hat{\mathbf{v}}=\left(\frac{1}{3}\right)\times\left(-\frac{27}{2}\right)+\left(-\frac{1}{3}\right)\times(-6)+\left(\frac{2}{3}\right)\times\left(-\frac{15}{2}\right)+(-1)\times(-3)+1\times\left(\frac{9}{2}\right)=0 \tag{1.60b}$$

$$\mathbf{v}'\mathbf{i}=\mathbf{i}'\mathbf{v}=\left(\frac{1}{3}\right)\times 2+\left(-\frac{1}{3}\right)\times\left(-\frac{2}{3}\right)+\left(\frac{2}{3}\right)\times\left(\frac{8}{3}\right)+(-1)\times(-6)+1\times\left(-\frac{26}{3}\right)=0 \tag{1.61a}$$

$$\hat{\mathbf{v}}'\hat{\mathbf{i}}=\hat{\mathbf{i}}'\hat{\mathbf{v}}=2\times\left(-\frac{27}{2}\right)+(-2)\times(-6)+\left(-\frac{3}{2}\right)\times\left(-\frac{15}{2}\right)+\left(-\frac{1}{2}\right)\times(-3)+\left(\frac{1}{2}\right)\times\left(\frac{9}{2}\right)=0 \tag{1.61b}$$

1.6 PASSIVE IMMITTANCES

In the early chapters of this book, we shall be concerned with the design of one-ports composed exclusively of passive elements such as resistors, inductors, capacitors, and coupled inductors. The one-ports are specified in terms of their driving-point *immittances*, *im*pedances, or ad*mittances.* A basic question that must be answered before design can proceed is the following: Given an immittance function, is it possible to find a one-port composed only of R, L, C, and M elements that realizes the given immittance function? This is known as the *realizability problem.* The answer to this question was first given by Brune.† and will be briefly discussed below.

Consider a linear *RLCM* one-port network of Figure 1.40, which is excited by a voltage source v_1. Assume that the overall network has b branches. For our purposes, the branch corresponding to the voltage source v_1 is numbered branch 1 and all other branches are numbered from 2 to b. Using Laplace transformation, Kirchhoff's

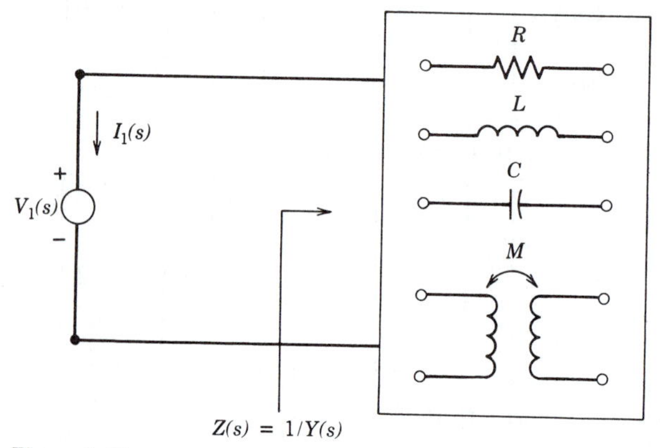

Figure 1.40

†See footnote on p. 34.

current law equations (1.45b) for the network of Figure 1.40 can be written as

$$\mathbf{A}\mathbf{I}(s)=\mathbf{0} \tag{1.62}$$

and from (1.49b)

$$\mathbf{V}(s)=\mathbf{A}'\mathbf{V}_n(s) \tag{1.63}$$

where $\mathbf{I}(s)$, $\mathbf{V}(s)$, and $\mathbf{V}_n(s)$ are the Laplace transforms of $\mathbf{i}(t)$, $\mathbf{v}(t)$, and $\mathbf{v}_n(t)$, respectively. Taking the complex conjugate of (1.62) and combining this with (1.63) gives

$$\mathbf{V}'(s)\bar{\mathbf{I}}(s)=\mathbf{V}_n'(s)\mathbf{A}\bar{\mathbf{I}}(s)=\mathbf{V}_n'(s)\mathbf{0}=0 \tag{1.64}$$

or

$$\sum_{k=1}^{b} V_k(s)\bar{I}_k(s)=0 \tag{1.65}$$

where $V_k(s)$ and $I_k(s)$ denote the Laplace transforms of the branch voltage $v_k(t)$ and the branch current $i_k(t)$, respectively.

In accordance with the reference directions shown in Figure 1.40, the driving-point impedance of the one-port is defined as the ratio of $V_1(s)$ to $-I_1(s)$, or

$$Z(s)\triangleq\frac{V_1(s)}{-I_1(s)}=\frac{V_1(s)\bar{I}_1(s)}{-I_1(s)\bar{I}_1(s)}=-\frac{V_1(s)\bar{I}_1(s)}{|I_1(s)|^2} \tag{1.66}$$

Equation (1.65) can be rewritten as

$$-V_1(s)\bar{I}_1(s)=\sum_{k=2}^{b} V_k(s)\bar{I}_k(s) \tag{1.67}$$

Substituting this in (1.66) leads to

$$Z(s)=\frac{1}{|I_1(s)|^2}\sum_{k=2}^{b} V_k(s)\bar{I}_k(s) \tag{1.68}$$

Likewise, we can show that the dual relation for the input admittance

$$Y(s)\triangleq\frac{-I_1(s)}{V_1(s)}=\frac{1}{|V_1(s)|^2}\sum_{k=2}^{b} \bar{V}_k(s)I_k(s) \tag{1.69}$$

holds for the one-port network of Figure 1.40.

We now consider the specific types of elements inside the one-port. If branch k is a resistor of resistance R_k, then

$$V_k=R_kI_k \tag{1.70}$$

If it is a capacitor of capacitance C_k, then

$$V_k=\frac{1}{sC_k}I_k \tag{1.71}$$

Finally, if branch k is an inductor of self-inductance L_k and mutual inductances M_{kj}, then

$$V_k = sL_k I_k + \sum_{\substack{\text{all } j \\ j \neq k}} sM_{kj} I_j \tag{1.72}$$

Substituting these in (1.68) and rewriting the summation as sums over all resistors R, all capacitors C, and all inductors LM, we obtain

$$\begin{aligned} Z(s) &= \frac{1}{|I_1(s)|^2}\left[\sum_R R_k|I_k|^2 + \sum_C \frac{1}{sC_k}|I_k|^2 + \sum_{LM}\left(sL_k|I_k|^2 + \sum_{\substack{\text{all } j \\ j \neq k}} sM_{kj} I_j \bar{I}_k\right)\right] \\ &= \frac{1}{|I_1(s)|^2}\left[F_0(s) + \frac{1}{s}V_0(s) + sM_0(s)\right] \end{aligned} \tag{1.73}$$

where

$$F_0(s) \triangleq \sum_R R_k|I_k(s)|^2 \geqq 0 \tag{1.74}$$

$$V_0(s) \triangleq \sum_C \frac{1}{C_k}|I_k(s)|^2 \geqq 0 \tag{1.75}$$

$$M_0(s) \triangleq \sum_{LM}\left[L_k|I_k(s)|^2 + \sum_{\substack{\text{all } j \\ j \neq k}} M_{kj} I_j(s)\bar{I}_k(s)\right] \tag{1.76}$$

The quantities F_0, V_0, and M_0 are closely related to the average power and stored energies of the one-port under steady-state sinusoidal conditions. The average power dissipated in the resistors is

$$P_{\text{ave}} = \frac{1}{2}\sum_R R_k|I_k(j\omega)|^2 = \frac{1}{2}F_0(j\omega) \tag{1.77}$$

Thus, $F_0(j\omega)$ represents twice the average power dissipated in the resistors of the one-port network. The average electric energy stored in all the capacitors is

$$E_C = \frac{1}{4\omega^2}\sum_C \frac{1}{C_k}|I_k(j\omega)|^2 = \frac{1}{4\omega^2}V_0(j\omega) \tag{1.78}$$

$V_0(j\omega)$ represents $4\omega^2$ times the average electric energy stored in the capacitors. Likewise, the average magnetic energy stored in the inductors of the one-port is given by

$$E_M = \frac{1}{4}\sum_{LM}\left[L_k|I_k(j\omega)|^2 + \sum_{\substack{\text{all } q \\ q \neq k}} M_{kq} I_q(j\omega)\bar{I}_k(j\omega)\right] = \frac{1}{4}M_0(j\omega) \tag{1.79}$$

showing that $M_0(j\omega)$ represents four times the average magnetic energy stored in the inductors. With these physical interpretations, we conclude that all the three quantities F_0, V_0, and M_0 are real and nonnegative. Thus, we rewrite (1.73) as

$$Z(s) = \frac{1}{|I_1(s)|^2}\left(F_0 + \frac{1}{s}V_0 + sM_0\right) \tag{1.80}$$

An equation similar to (1.80) can be written for (1.69) in terms of F_0, V_0, and M_0. The result is given by

$$Y(s)=\frac{1}{|V_1(s)|^2}\left(F_0+\frac{1}{\bar{s}}V_0+\bar{s}M_0\right) \tag{1.81}$$

This equation is not needed to carry on the subsequent discussion, just as the nodal system of equations itself is really superfluous. However, just as the nodal equations provide helpful insight and often simplify computations, so is the dual approach.

Up to this point, we have prepared ourselves for the major effort of establishing analytic properties of passive one-port immittances. We are now ready to embark on this task. As a first step, we set $s=\sigma+j\omega$ and compute the real part and imaginary part of $Z(s)$. The results are

$$\text{Re } Z(s)=\frac{1}{|I_1(s)|^2}\left(F_0+\frac{\sigma}{\sigma^2+\omega^2}V_0+\sigma M_0\right) \tag{1.82}$$

$$\text{Im } Z(s)=\frac{\omega}{|I_1(s)|^2}\left(M_0-\frac{1}{\sigma^2+\omega^2}V_0\right) \tag{1.83}$$

where Im stands for "the imaginary part of."

Observe that these equations apply no matter what the value of s may be except at the zeros of $I_1(s)$. The equations are important in that many analytic properties of passive impedances can be obtained from them. However, before we do this, we mention that the imaginary part ω of the complex frequency $s=\sigma+j\omega$ is called the *real frequency*, whereas the real part of s, misleading as it may be, is called the *imaginary frequency*, and was in general use before 1930. Another convention is to name ω the *radian frequency* and σ the *neper frequency*, thus avoiding the near metaphysical names. But, no matter what we call them, the two components of frequency add together to give the complex frequency. For the present, we shall use the term real frequency for ω. When we speak of the real-frequency axis, we mean the $j\omega$-axis of the complex-frequency plane.

We now state some of the consequences of (1.82) and (1.83) as a theorem for later reference.

Theorem 1.1 If $Z(s)$ is the driving-point impedance of a linear, passive, lumped, reciprocal, and time-invariant one-port network N, then

(a) Whenever $\sigma \geqq 0$, Re $Z(s) \geqq 0$.

(b) If N contains no resistors, then

$$\sigma>0 \text{ implies Re } Z(s)>0$$
$$\sigma=0 \text{ implies Re } Z(s)=0$$
$$\sigma<0 \text{ implies Re } Z(s)<0$$

(c) If N contains no capacitors, then

$$\omega>0 \text{ implies Im } Z(s)>0$$
$$\omega=0 \text{ implies Im } Z(s)=0$$
$$\omega<0 \text{ implies Im } Z(s)<0$$

(d) If N contains no self- and mutual inductors, then

$$\omega > 0 \text{ implies Im } Z(s) < 0$$
$$\omega = 0 \text{ implies Im } Z(s) = 0$$
$$\omega < 0 \text{ implies Im } Z(s) > 0$$

Similar results can be stated for the admittance function $Y(s)$. The resulting theorem is obtained simply by replacing $Z(s)$, Re $Z(s)$, and Im $Z(s)$ in Theorem 1.1 by $Y(s)$, Re $Y(s)$, and $-$Im $Y(s)$, respectively; everything else being the same.

1.7 POSITIVE-REAL FUNCTIONS

Theorem 1.1 demonstrates that the driving-point impedance $Z(s)$ of a passive *LMC*, *RLM*, or *RC* one-port network maps different regions of the complex-frequency s-plane into various regions of the Z-plane. In this section, we assert that the driving-point immittance of a passive one-port is a positive-real function. Conversely, every positive-real function can be realized as the driving-point immittance function of a passive one-port network containing only passive elements such as positive R's, L's, C's, ideal transformers, and coupled coils. The concept of positive-real function is introduced below.

Definition 1.1 : ***Positive – real function.*** A *positive-real function* $F(s)$, abbreviated as a *PR* function, is an analytic function of the complex variable $s = \sigma + j\omega$ satisfying the following three conditions:

(a) $F(s)$ is analytic in the open RHS (*r*ight-*h*alf of the s-plane), i.e., $\sigma > 0$.

(b) $F(\bar{s}) = \bar{F}(s)$ for all s in the open RHS.

(c) Re $F(s) \geqq 0$ whenever Re $s \geqq 0$.

The above definition establishes a new class of mathematical functions. Our objective is to show that positive realness is a necessary and sufficient condition for a passive one-port immittance. By making a study of this class of functions, we can deduce many properties that we could not establish from physical reasoning alone. The concept of a positive-real function, as well as many of its properties, is due to Otto Brune.†

The above definition holds for both rational and transcendental functions. A *rational function* is defined as a ratio of two polynomials. Network functions associated with any linear lumped system, which we deal with exclusively in this book, are rational. For a rational function $F(s)$, the analyticity requirement (a) in Definition 1.1 is redundant, since, as will be shown shortly, the other two conditions would require that $F(s)$ be devoid of poles in the open RHS. The second condition (b) is equivalent to

†O. Brune, "Synthesis of a finite two-terminal network whose driving-point impedance is a prescribed function of frequency," *J. Math. Phys.*, vol. 10, 1931, pp. 191–236. This is the same as his M.I.T. Sc.D. dissertation, 1930.

stating that $F(s)$ is real when s is real, and for a rational $F(s)$ it is always satisfied if all the coefficients of the polynomials are real.

To show that the first condition (a) is implied by the other two for rational functions $F(s)$, we consider a RHS pole s_0 of order n of $F(s)$. In the neighborhood of this pole, the Laurent series expansion of $F(s)$ takes the form

$$F(s)=\frac{a_{-n}}{(s-s_0)^n}+\frac{a_{-n+1}}{(s-s_0)^{n-1}}+\cdots+\frac{a_{-1}}{s-s_0}+\sum_{k=0}^{\infty} a_k(s-s_0)^k \tag{1.84}$$

If sufficiently small neighborhood of s_0 is chosen, the first term of the Laurent series expansion can be made much larger in magnitude than the rest. Write

$$a_{-n}=me^{j\theta} \tag{1.85}$$

$$s-s_0=re^{j\phi} \tag{1.86}$$

Then in the neighborhood of s_0, Re $F(s)$ can be approximated by

$$\text{Re } F(s) \approx \text{Re}\left[\frac{a_{-n}}{(s-s_0)^n}\right]=\frac{m}{r^n}\cos(\theta-n\phi) \tag{1.87}$$

Since θ is a fixed angle and ϕ can vary from 0 to 2π in this neighborhood, Re $F(s)$ changes sign $2n$ times as ϕ varies from 0 to 2π. The case $n=3$ is illustrated in Figure 1.41. As we can see, the real part of a rational function in the neighborhood of a pole of order 3 alters signs six times, contradicting condition (c) of Definition 1.1 for positive realness. As a result, no such poles can exist and $F(s)$ is analytic in the open RHS. We conclude that a rational function $F(s)$ is positive real if the following two conditions are satisfied:

(i) $F(\sigma)$ is real.

(ii) Re $F(s) \geqq 0$ whenever Re $s \geqq 0$.

We now interpret the definition of a positive-real function $F(s)$ as a mapping. Refer to Figure 1.42. A positive-real function maps the σ-axis of the s-plane into the

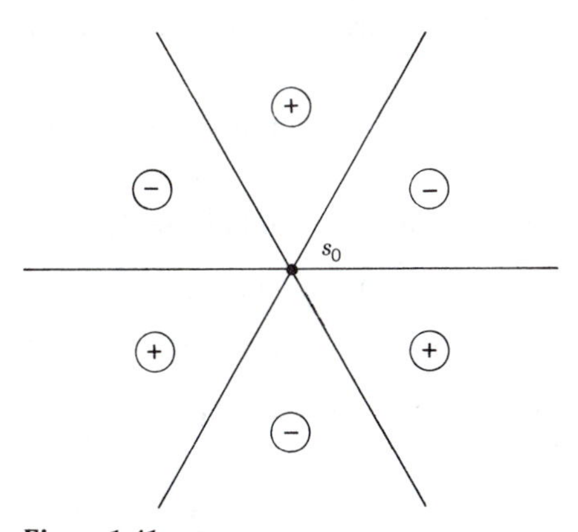

Figure 1.41

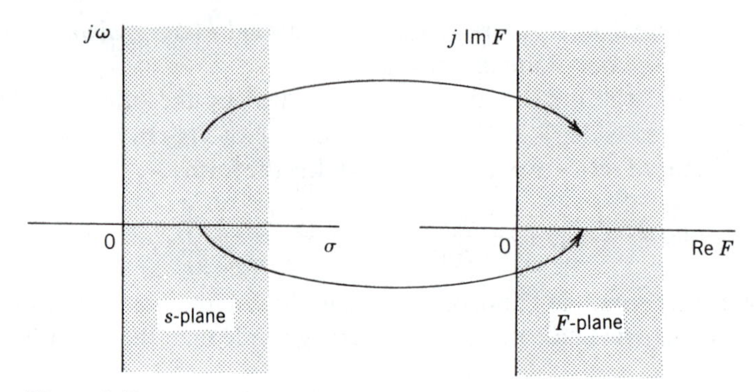

Figure 1.42

Re F axis, and maps the RHS into the right-half of the F-plane. An immediate consequence of the interpretation is that a positive-real function of a positive-real function is itself positive real, or

Property 1.1 If $F_1(s)$ and $F_2(s)$ are positive real, so is $F_1[F_2(s)]$.

This is a useful result. We can use it to obtain other properties. For example, let $F_1(s)=1/s$. Then $F_1(s)$ is positive real, and by Property 1.1 the function defined by the equation

$$F(s)=F_1[F_2(s)]=1/F_2(s) \tag{1.88}$$

is positive real if $F_2(s)$ is positive real. Likewise, letting $F_2(s)=1/s$ shows that

$$F(s)=F_1[F_2(s)]=F_1(1/s) \tag{1.89}$$

is positive real if $F_1(s)$ is.

Property 1.2 If $F(s)$ is positive real, so are $1/F(s)$ and $F(1/s)$.

Property 1.2 states that the reciprocal of a positive-real function is positive real. It follows that a positive-real function cannot have any zeros in the open RHS. For if it did, then its reciprocal, being positive real, would have poles in the open RHS, which is impossible.

Property 1.3 A positive-real function is devoid of poles and zeros in the open RHS.

We illustrate the above results by the following example.

Example 1.2 Consider the rational function

$$F(s)=1+s/(s^2+1) \tag{1.90}$$

To ascertain its positive realness, we compute its real part by substituting $s=\sigma+j\omega$ and obtain

$$F(s)=F(\sigma+j\omega)=1+\frac{\sigma(\sigma^2+\omega^2+1)-j(\sigma^2+\omega^2-1)\omega}{(\sigma^2-\omega^2+1)^2+4\sigma^2\omega^2} \tag{1.91}$$

Since

$$\text{Re } F(s) = 1 + \frac{\sigma(\sigma^2 + \omega^2 + 1)}{(\sigma^2 - \omega^2 + 1)^2 + 4\sigma^2\omega^2} \geqq 0, \qquad \sigma \geqq 0 \tag{1.92}$$

and the fact that all the coefficients of $F(s)$ are real, $F(s)$ is a positive-real function, so are the functions defined by

$$\frac{1}{F(s)} = \frac{s^2+1}{s^2+s+1} \tag{1.93}$$

$$F(1/s) = 1 + \frac{1/s}{(1/s)^2+1} = 1 + \frac{s}{s^2+1} \tag{1.94}$$

$F(1/s)$ turns out to be the same as $F(s)$. This, of course, is not true in general. Also observe that $F(s)$ is devoid of poles and zeros in the open RHS. It has a pair of $j\omega$-axis poles at $s = \pm j1$ and a pair of zeros at $s = -\frac{1}{2} \pm j(\sqrt{3}/2)$.

In (1.84) we considered the Laurent series expansion of $F(s)$ about a pole s_0 of order n. The constant a_{-1} is called the *residue* of the function $F(s)$ evaluated at the pole s_0. The other coefficients $a_{-2}, a_{-3}, \ldots, a_{-n}$, and a_k do not have specific names. We showed that if $F(s)$ is positive real, s_0 cannot be located in the open RHS. Can poles be located on the $j\omega$-axis and the function still be positive real? Figure 1.43 shows such a pole $s_0 = j\omega_0$ with a circle of radius r around the pole. Following (1.85) and (1.86), we see that (1.87) remains valid or

$$\text{Re } F(s) \approx (m/r^n) \cos(\theta - n\phi) \tag{1.95}$$

However, in this case, ϕ can vary from $-\frac{1}{2}\pi$ to $\frac{1}{2}\pi$. For Re $Z(s)$ to remain nonnegative for values of s on the circle and in the right-half plane, it is necessary that $\theta = 0$ and $n = 1$. The condition $n = 1$ implies that the pole on the $j\omega$-axis must be simple. This together with the condition $\theta = 0$ implies that the residue

$$a_{-1} = me^{j\theta} = m \tag{1.96}$$

at the $j\omega$-axis pole $j\omega_0$ must be real and positive. Therefore, we permit poles on the $j\omega$-axis provided they are simple and their residues are real and positive.

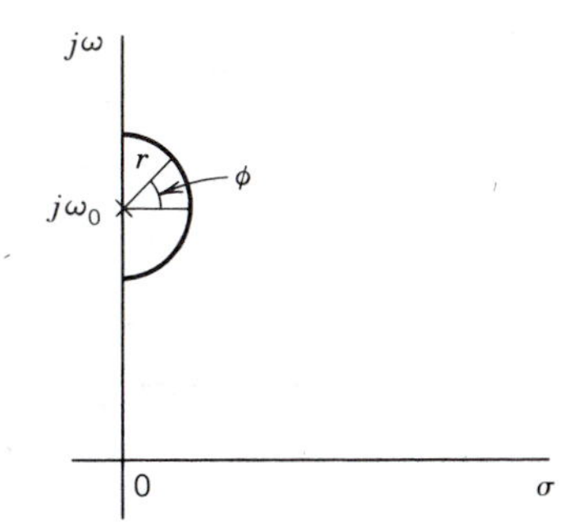

Figure 1.43

Property 1.4 If a positive-real function has any poles or zeros on the $j\omega$-axis (0 and ∞ included), such poles or zeros must be simple. At a simple pole on the $j\omega$-axis, the residue is real and positive.

We now have established a number of properties of the positive-real functions. The real significance of the positive-real functions is their use in the characterization of the passive one-port immittances. This characterization is one of the most penetrating results in network theory, and is stated as

Theorem 1.2 A real rational function is the driving-point immittance of a linear, passive, lumped, reciprocal, and time-invariant one-port network if and only if the function is positive real.

The necessity of the theorem follows directly from (1.82). The sufficiency was first established by Brune in 1930.† He showed that any given positive-real rational function can be realized as the driving-point immittance of a passive one-port network using only the passive elements such as resistors, capacitors, and self- and mutual inductors. Formal constructive proofs of sufficiency will be presented in detail in Chapters 4 and 10.

Example 1.3 Consider the passive one-port of Figure 1.44, the driving-point impedance of which is found to be

$$Z(s)=\frac{3s^4+4s^3+7s^2+3s+3}{2s^4+s^3+3s^2+s+1} \tag{1.97}$$

Since the one-port is passive, by Theorem 1.2 the impedance function $Z(s)$ is positive real. On the other hand, if the function $Z(s)$ is given, the task of ascertaining positive realness of $Z(s)$ is difficult if the conditions of Definition 1.1 are employed for checking. Hence, it is desirable to have alternate but equivalent conditions for checking the positive realness of a function. This will be the topic of the following section.

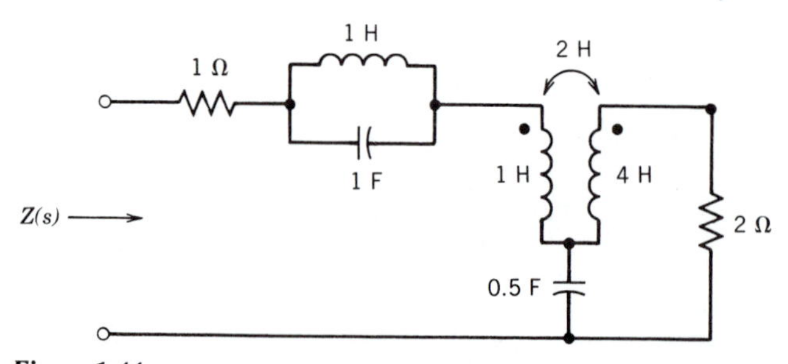

Figure 1.44

†See footnote on p. 34.

1.8 EQUIVALENT POSITIVE-REAL CONDITIONS

The definition of positive realness is admirably elegant and concise, but from a practical viewpoint, condition (c) is almost impossible to use for any nontrivial functions. For this reason, we introduce equivalent conditions that are relatively easy to apply.

Theorem 1.3 A rational function $F(s)$ is positive real if and only if the following four conditions are satisfied:

(i) $F(s)$ is real when s is real.

(ii) $F(s)$ has no poles in the open RHS.

(iii) Poles of $F(s)$ on the $j\omega$-axis, if they exist, are simple, and residues evaluated at these poles are real and positive.

(iv) Re $F(j\omega) \geqq 0$ for all ω, except at the poles.

Proof From the definition of a PR function and previous discussion, we see immediately that all the conditions are necessary. Therefore, only sufficiency needs to be verified. For this we show that if a function $F(s)$ satisfies these four conditions, then it must be PR.

Let $F(s)$ be expanded in a partial fraction, and write

$$F(s) = \left(k_\infty s + \frac{k_0}{s} + \sum_x \frac{k_x}{s + j\omega_x} + \frac{k_x}{s - j\omega_x} \right) + F_1(s)$$

$$= \left(k_\infty s + \frac{k_0}{s} + \sum_x \frac{2k_x s}{s^2 + \omega_x^2} \right) + F_1(s) \tag{1.98}$$

where k_∞, k_0, and k_x are residues at the $j\omega$-axis poles $j\infty$, 0, and $j\omega_x$, respectively, and are real and positive. $F_1(s)$ is the function formed by the terms corresponding to the open left-half s-plane poles of $F(s)$. Therefore, $F_1(s)$ is analytic in the RHS and the entire $j\omega$-axis including the point infinity. For such a function, the minimum value of the real part throughout the region where the function is analytic lies on the boundary,† namely, the $j\omega$-axis. This shows that the minimum value of Re $F_1(s)$ for all Re $s \geqq 0$ occurs on the $j\omega$-axis. But according to (1.98) this value is nonnegative:

$$\text{Re } F_1(j\omega) = \text{Re } F(j\omega) \geqq 0 \tag{1.99}$$

Thus, the real part of $F_1(s)$ must be nonnegative everywhere in the closed RHS, i.e., Re $s \geqq 0$; or

$$\text{Re } F_1(s) \geqq 0 \qquad \text{for Re } s \geqq 0 \tag{1.100}$$

This together with the fact that $F_1(s)$ is real whenever s is real shows that $F_1(s)$ is positive real.

It is easy to verify that each term inside the parentheses of (1.98) is positive real. Since the sum of two or more positive-real functions is positive real, $F(s)$ is positive real. This completes the proof of the sufficiency of the stated conditions.

†See, for example, R. V. Churchill 1960, *Introduction to Complex Variables and Applications.* New York: McGraw-Hill.

In testing a given function for positive realness, it may not always be necessary to use all the conditions of the theorem. In some situations, we may eliminate some functions from consideration by inspection because they violate certain simple necessary conditions. A function is not PR if it has a pole or zero in the open RHS. Another simple test for positive realness is that the highest powers of s in numerator and denominator not differ by more than unity; and this is similarly valid for the lowest powers of s. This follows directly from the fact that a PR function can have at most a simple pole or a simple zero at the origin or infinity, both of which lie on the $j\omega$-axis.

A *Hurwitz polynomial* is a polynomial that has no zeros in the open RHS. This definition permits zeros on the $j\omega$-axis. To distinguish such a polynomial from one that has zeros neither in the open RHS nor on the $j\omega$-axis, the latter is referred to as a *strictly Hurwitz polynomial*. With these terminologies, we see that a PR function is the ratio of two Hurwitz polynominals.

For computational purposes, Theorem 1.3 can be reformulated and put in a much more convenient form. The result is stated as†

Theorem 1.4 A rational function represented in the form

$$F(s)=\frac{P(s)}{Q(s)}=\frac{m_1(s)+n_1(s)}{m_2(s)+n_2(s)} \tag{1.101}$$

where $m_1(s)$, $m_2(s)$ and $n_1(s)$, $n_2(s)$ are the even and odd parts of the polynomials $P(s)$ and $Q(s)$, respectively, is positive real if and only if the following conditions are satisfied:

(i) $F(s)$ is real when s is real.
(ii) $P(s)+Q(s)$ is strictly Hurwitz.
(iii) $m_1(j\omega)m_2(j\omega)-n_1(j\omega)n_2(j\omega)\geqq 0$ for all ω. (1.102)

The testing of the second condition can easily be accomplished by means of the Hurwitz test,† which states that a real polynomial is strictly Hurwitz if and only if the continued-fraction expansion of the ratio of the even part to the odd part or the odd part to the even part of the polynomial yields only real and positive coefficients, and does not terminate prematurely. For $P(s)+Q(s)$ to be strictly Hurwitz, it is necessary and sufficient that the continued-fraction expansion

$$\left[\frac{m_1(s)+m_2(s)}{n_1(s)+n_2(s)}\right]^{\pm 1}=\alpha_1 s+\cfrac{1}{\alpha_2 s+\cfrac{1}{\ddots+\cfrac{1}{\alpha_k s}}} \tag{1.103}$$

yields only real and positive α's, and does not terminate prematurely, i.e., k must equal the degree of $m_1(s)+m_2(s)$ or $n_1(s)+n_2(s)$, whichever is larger. It can be shown that the third condition (1.102) is satisfied if and only if its left-hand-side polynomial does not have real positive roots of odd multiplicity. This may be determined by factoring

†See, for example, M. E. Van Valkenburg 1960, *Introduction to Modern Network Synthesis*, Chapter 4. New York: John Wiley.

it or by the use of Sturm's theorem, which can be found in most texts on elementary theory of equations. We illustrate these results by the following examples.

Example 1.4 Test the following function to see if it is PR:

$$F(s)=\frac{2s^3+2s^2+3s+2}{s^2+1} \tag{1.104}$$

For illustrative purposes, we follow the three steps outlined in Theorem 1.4, as follows:

$$P(s)=m_1(s)+n_1(s)=(2s^2+2)+(2s^3+3s) \tag{1.105a}$$

$$Q(s)=m_2(s)+n_2(s)=(s^2+1)+0 \tag{1.105b}$$

$$P(s)+Q(s)=(3s^2+3)+(2s^3+3s)=m(s)+n(s) \tag{1.105c}$$

Condition (i) is clearly satisfied. To test condition (ii), we perform the Hurwitz test, which gives

$$\frac{n(s)}{m(s)}=\frac{2s^3+3s}{3s^2+3}=2s/3+\cfrac{1}{3s+\cfrac{1}{s/3}} \tag{1.106}$$

Since the continued-fraction expansion does not terminate prematurely and since all of its coefficients are real and positive, we conclude that the polynomial $P(s)+Q(s)$ is strictly Hurwitz. Thus, condition (ii) is satisfied.

To test condition (iii), we first compute

$$m_1(j\omega)m_2(j\omega)-n_1(j\omega)n_2(j\omega)=2(\omega^2-1)^2\geqq 0 \tag{1.107}$$

which is nonnegative for all ω. Therefore, $F(s)$ is positive real.

Example 1.5 Test the following function to see if it is PR:

$$F(s)=\frac{6s^3+32s^2+39s+99}{6s^2+2s+24} \tag{1.108}$$

We apply Theorem 1.4. Condition (i) is clearly satisfied. To test condition (ii), we perform the Hurwitz test, which is

$$\frac{6s^3+41s}{38s^2+123}=\frac{3}{19}s+\cfrac{1}{\cfrac{361}{205}s+\cfrac{1}{\cfrac{410}{2337}s}} \tag{1.109}$$

The expansion yields only real and positive coefficients and does not terminate prematurely. Thus, the polynomial $P(s)+Q(s)$ is strictly Hurwitz and condition (ii) is satisfied. For condition (iii), we compute

$$m_1(j\omega)m_2(j\omega)-n_1(j\omega)n_2(j\omega)=180\omega^4-1284\omega^2+2376 \tag{1.110}$$

This condition is also satisfied, because (1.110) does not possess any real and positive roots of odd multiplicity, meaning that it is nonnegative for all real ω. Thus, the function $F(s)$ is positive real.

1.9 SCALING NETWORK FUNCTIONS

The computations required in synthesis are greatly simplified by choosing convenient, easy-to-handle values of network elements. After obtaining the nominal design, magnitude and frequency scalings are used to change the element values of the network to make the network practically realizable. The way in which these objectives are accomplished is the subject of this discussion.

An impedance $|Z(j\omega)|$ is said to be *magnitude-scaled* by a factor of b if it is multiplied by a real positive constant b. It is *scaled up* if $b>1$, and *down* if $b<1$. Magnitude-scaling by a factor of b is equivalent to moving up or down the impedance characteristic $|Z(j\omega)|$ in the $|Z(j\omega)|$ versus ω plot by the same factor b. To achieve this impedance change, we alter the magnitude of the impedance for every element in the network. The impedance magnitudes for the R, L, and C elements are

$$Z_R=R, \qquad |Z_L(j\omega)|=\omega L, \qquad |Z_C(j\omega)|=1/\omega C \tag{1.111}$$

Multiplying each of these by b yields

$$bZ_R=bR, \qquad b|Z_L(j\omega)|=b\omega L, \qquad b|Z_C(j\omega)|=1/(\omega C/b) \tag{1.112}$$

Thus, if we designate the elements after scaling as "new," then the equations for obtaining the new element values are given by

$$R_{\text{new}}=bR \tag{1.113a}$$
$$L_{\text{new}}=bL \tag{1.113b}$$
$$C_{\text{new}}=C/b \tag{1.113c}$$

In other words, if the elements in the old network are changed according to these equations, all impedances in that network will be scaled in magnitude by a factor of b.

We next consider the scaling of the frequency without affecting the magnitude. An impedance function $Z(s)$ is said to be *frequency-scaled* by a factor of a if s is replaced by s/a, where a is a real positive constant. It is *scaled up* if $a>1$, and *down* if $a<1$. To affect this change, the impedances for the R, L, and C elements are altered as

$$Z_R=R, \qquad Z_L(s/a)=sL/a, \qquad Z_C(s/a)=a/sC \tag{1.114}$$

From these equations, we see that new element values may be expressed in terms of old values by the equations

$$R_{\text{new}}=R \tag{1.115a}$$
$$L_{\text{new}}=L/a \tag{1.115b}$$
$$C_{\text{new}}=C/a \tag{1.115c}$$

So far we have scaled the network functions in magnitude and in frequency separately. We can do both at once by combining (1.113) and (1.115):

$$R_{\text{new}}=bR \tag{1.116a}$$
$$L_{\text{new}}=bL/a \tag{1.116b}$$
$$C_{\text{new}}=C/ab \tag{1.116c}$$

Frequency-scaling is often used in designing filters using normalized frequency. One such example is in the design of low-pass filters where the cutoff frequency ω_c is conveniently chosen to be 1 rad/s. After realization, the desired network with cutoff frequency ω_c is obtained by a frequency-scaling by a factor of ω_c.

We remark that magnitude-scaling discussed above is valid for all network functions expressible as a ratio of determinants of different orders such as occurred in the computations of driving-point immittance functions. Network functions that may be written as the quotient of determinants of the same orders are not affected by magnitude-scaling. Thus, the transfer voltage-ratio function is *unaffected* by magnitude-scaling. All network functions, however, are affected by frequency-scaling. We illustrate these by the following examples.

Example 1.6 The one-port network of Figure 1.44 realizes the driving-point impedance $Z(s)$ of (1.97). It is clear that the element values are not in the practical range, but this will be a characteristic of unscaled networks. Suppose that we wish to increase the impedance level by 100 so that the network will be terminated in a 200-Ω resistor, and at the same time scale the frequency such that the behavior of the unscaled network at 1 rad/s takes place at 10^5 rad/s in the scaled network. Thus, we choose

$$a=10^5 \qquad \text{and} \qquad b=100 \tag{1.117}$$

Making use of (1.116) gives the scaled network as shown in Figure 1.45 with element values as indicated.

Example 1.7 Consider the *RC* amplifier of Figure 1.16, the equivalent network of which is shown in Figure 1.17. The transfer voltage-ratio function was computed earlier in (1.7), and is repeated below as

$$\frac{V_o}{V_{\text{in}}}=\frac{r_i[r_o-AZ_2(s)]}{Z_1(s)[r_i+r_o+Ar_i+Z_2(s)]+r_i[r_o+Z_2(s)]} \tag{1.118}$$

Suppose that we magnitude-scale the network by a factor of b and frequency-scale the network by a factor of a. This is equivalent to replacing r_i, r_o, $Z_1(s)$, and $Z_2(s)$ by br_i, br_o, $bZ_1(s/a)$, and $bZ_2(s/a)$, respectively. Substituting these in (1.118)

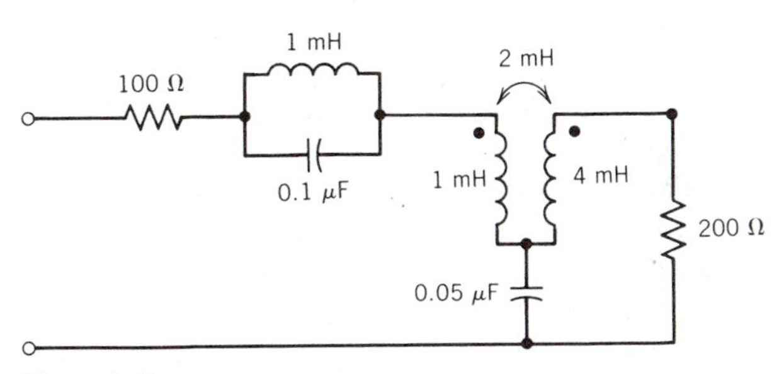

Figure 1.45

results in

$$\frac{V_o}{V_{in}} = \frac{br_i[br_o - AbZ_2(s/a)]}{bZ_1(s/a)[br_i + br_o + Abr_i + bZ_2(s/a)] + br_i[br_o + bZ_2(s/a)]}$$

$$= \frac{r_i[r_o - AZ_2(s/a)]}{Z_1(s/a)[r_i + r_o + Ar_i + Z_2(s/a)] + r_i[r_0 + Z_2(s/a)]} \tag{1.119}$$

showing that the transfer voltage-ratio function is unaffected by magnitude-scaling, and is affected only by the frequency-scaling. This is an important result, and is used as the basis for designing strategies to give practical element values.

1.10 SUMMARY AND SUGGESTED READINGS

We began this chapter by pointing out the important differences between analysis and synthesis, and classified electrical filters in a number of ways. Based on their functions, filters can be classified as *low-pass*, *high-pass*, *band-pass*, and *band-elimination*. They are constructed by using two broad categories of devices: *basic building blocks* and *secondary building blocks*. The basic building blocks consist of the *resistors*, *capacitors*, *self-* and *mutual inductors*, and the *operational amplifiers*. Using these, the secondary building blocks such as the *inverting amplifiers*, the *noninverting amplifiers*, and the *gyrators* are realized.

A one-port network can be represented by its driving-point immittance function. For a two-port network, it can be characterized by a set of four parameters. For our purposes, we introduced two sets of two-port parameters: the *open-circuit impedance parameters* and the *short-circuit admittance parameters*. We expressed the *transducer power gain*, which is defined as the ratio of average power delivered to the load to the maximum available average power at the source, in terms of these parameters and the source and load impedances. Since simple two-ports are interconnected to yield more complicated and practical two-port networks, we studied two useful and common connections of two-port networks used in synthesis: the *series connection* and the *parallel connection*. We showed that when two two-ports are connected in series, the open-circuit impedance matrix of the composite two-port is simply the sum of the open-circuit impedance matrices of the individual two-ports. Similarly, when two-ports are connected in parallel, their short-circuit admittance matrices add.

Kirchhoff's laws are algebraic constraints arising from the interconnection of network elements and are independent of the characteristics of the elements. As a result, they imply the conservation of energy. In other words, conservation of energy need not be an added postulate for the discipline of network theory. This result is known as *Tellegen's theorem*, and would help us understand the fundamental properties of physically realizable impedance functions. Tellegen's theorem is valid whether the networks are linear or nonlinear and time-invariant or time-varying. In addition, it is applicable to two different but topologically identical networks.

Next we were concerned with the realizability of one-ports composed exclusively of passive elements such as resistors, inductors, capacitors, and coupled inductors. We showed that a real rational function is the driving-point immittance of a linear, passive, lumped, reciprocal, and time-invariant one-port network if and only if the function is positive real. This characterization is one of the most penetrating results in network theory, and was first established by Otto Brune in 1930. For this reason, we studied the class of mathematical functions called the *positive-real functions*, and deduced many of their properties. The definition of a positive-real function is admirably elegant and concise, but from a practical viewpoint, it is almost impossible to use for any nontrivial functions. To test a given function for positive realness, we introduced equivalent conditions that are relatively easy to apply.

Finally, to simplify the computations required in synthesis, we discussed the notions of *magnitude-* and *frequency-scalings*. The objective is either to choose convenient, easy-to-handle values of network elements during the design process or to change the element values of a nominal design to make the network practically realizable. We found that if each inductance and each capacitance is multiplied by $1/a$, then the network is frequency-scaled by a factor of a. If every resistance and inductance is multiplied by b and every capacitance is divided by the same b, the network is magnitude-scaled by a factor of b. The magnitude-scaling is valid for all network functions expressible as a ratio of determinants of different orders, and has no effect for network functions that may be written as the quotient of determinants of the same orders. Thus, the transfer voltage-ratio function is unaffected by magnitude-scaling. On the other hand, all network functions are affected by frequency-scaling.

A more detailed discussion on two-port networks is given by Chen [1]. An excellent introduction to various aspects of positive-real functions and their passive one-port realizations can be found in Van Valkenburg [9] and Temes and LaPatra [7]. The subject of scaling is also treated in Van Valkenburg [9].

Before we turn our attention to other topics on network synthesis, we mention that although the conservation of energy is implied by Kirchhoff's two laws, we may regard the conservation of energy and the conservation of charge (Kirchhoff's current law) as two basic postulates for the discipline of network theory. These two postulates would then imply Kirchhoff's voltage law. In fact, it can be shown that any two of the three constraints (Kirchhoff's current law, Kirchhoff's voltage law, and the energy conservation law) would imply the third. These are different ways of stating the same thing.

REFERENCES

1. W. K. Chen 1983, *Linear Networks and Systems*, Chapter 6. Monterey, CA: Brooks/Cole.
2. W. K. Chen 1976, *Theory and Design of Broadband Matching Networks*, Chapter 1. Cambridge, England: Pergamon Press.
3. G. Daryanani 1976, *Principles of Active Network Synthesis and Design*, Chapters 2 and 3. New York: John Wiley.
4. M. S. Ghausi and K. R. Laker 1981, *Modern Filter Design*, Chapters 1 and 2. Englewood Cliffs, NJ: Prentice–Hall.

5. L. P. Huelsman and P. E. Allen 1980, *Introduction to the Theory and Design of Active Filters*, Chapter 1. New York: McGraw–Hill.
6. H. Y-F. Lam 1979, *Analog and Digital Filers: Design and Realization*, Chapters 2–4. Englewood Cliffs, NJ: Prentice–Hall.
7. G. C. Temes and J. W. LaPatra 1977, *Introduction to Circuit Synthesis and Design*, Chapters 2 and 4. New York: McGraw–Hill.
8. M. E. Van Valkenburg 1982, *Analog Filter Design*, Chapters 1 and 2. New York: Holt, Rinehart and Winston.
9. M. E. Van Valkenburg 1960, *Introduction to Modern Network Synthesis*, Chapters 3, 4, and 7. New York: John Wiley.

PROBLEMS

1.1 Verify that the four one-port networks of Figure 1.3 have the same driving-point impedance as given by (1.1).

1.2 Scale the networks of Figure 1.3 so that all elements will have practical values with capacitance in μF and resistance in kΩ.

1.3 The network of Figure 1.44 is to be scaled by increasing the level of impedance by 500 and scaling the frequency such that 1 rad/s becomes 4×10^5 rad/s. Find the element values in the scaled network.

1.4 Figure 1.46 is the hybrid-pi equivalent network of a transistor. Compute the transfer voltage-ratio function V_o/V_{in} for the network.

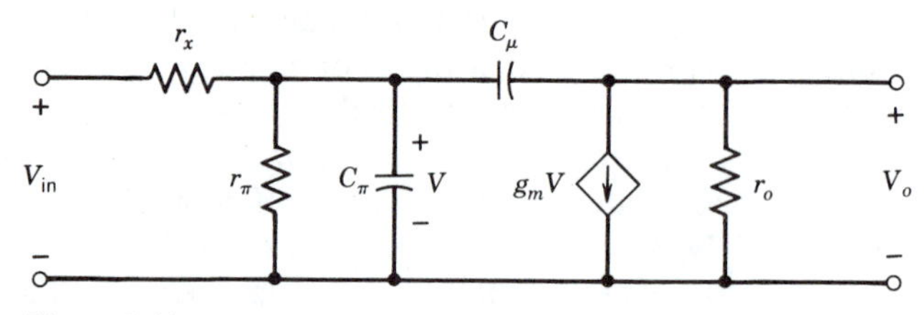

Figure 1.46

1.5 Consider the noninverting amplifier of Figure 1.21 with $Z_1 = R_1$ and $Z_2 = R_2$. Assuming that the op amp gain is modeled as

$$A(s) = 2\pi \times 10^6/s \tag{1.120}$$

$r_i = \infty$ and $r_o = 0$, compute the magnitude of the transfer voltage-ratio function $|V_o(j\omega)/V_{in}(j\omega)|$ at 1 kHz, 10 kHz, 100 kHz, 1 MHz, and 10 MHz. Sketch this ratio as a function of frequency.

1.6 Synthesize the transfer voltage-ratio function

$$\frac{V_o}{V_{in}} = -\frac{s+8}{s+12} \tag{1.121}$$

using the inverting amplifier of Figure 1.16.

1.7 Compute the transfer voltage-ratio function V_o/V_{in} for the op amp network of Figure 1.47, assuming that the op amp is ideal.

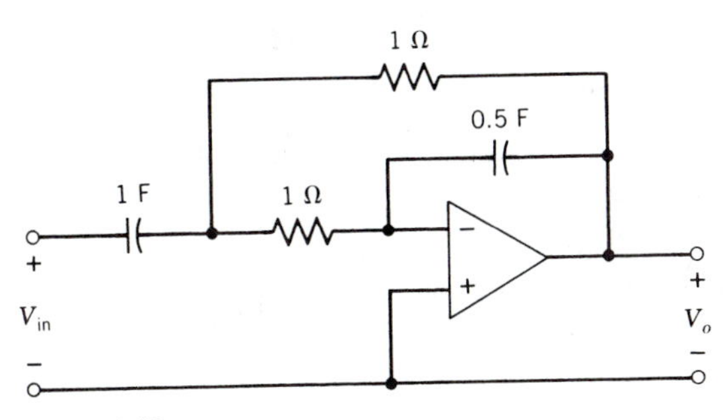

Figure 1.47

1.8 Scale the network of Figure 1.47 so that all elements will have practical values with capacitance in μF and resistance in kΩ.

1.9 Show that the input admittance of the network shown in Figure 1.48 is given by

$$Y(s) \triangleq I/V = (1 + R_2/R_1)Cs \tag{1.122}$$

where the op amp is assumed to be ideal. This network is known as a *capacitance multiplier*.

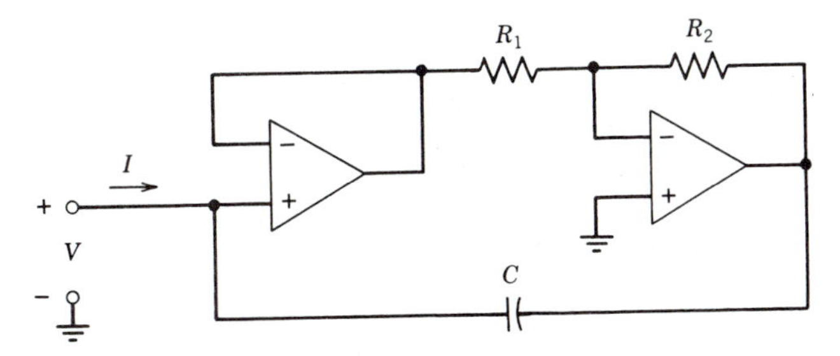

Figure 1.48

1.10 In the network of Figure 1.38, replace the 1-A current source by a 7-Ω resistor. Using the resulting network and the network of Figure 1.39, verify Tellegen's theorem by showing that (1.57) remains valid.

1.11 Test the following function to see if it is positive real:

$$F(s) = \frac{2s^4 + 4s^3 + 5s^2 + 5s + 2}{s^3 + s^2 + s + 1} \tag{1.123}$$

1.12 Determine if the following function is positive real:

$$F(s) = \frac{2s^2 + s + 6}{2s^2 + s + 2} \tag{1.124}$$

1.13 Find the range of "a" for which the following function is positive real:

$$F(s) = \frac{s^2 + 2s + a}{s^2 + 3s + 2} \tag{1.125}$$

1.14 Test the following function to see if it is positive real:

$$F(s)=\frac{14s^2+9s+4}{14s^3+9s^2+6s+1} \tag{1.126}$$

1.15 Determine if the polynomial

$$P(s)=7s^4+8s^3+9s^2+8s+2 \tag{1.127}$$

is Hurwitz or strictly Hurwitz.

1.16 Let $F_1(s)$ and $F_2(s)$ be two positive-real functions. Consider the following combinations of these functions. Are these combinations of themselves positive real, or conditionally positive real? If they are conditionally positive real, what are the conditions? Justify the statements.

(a) $F_1(s)F_2(s)$

(b) $F_1(s)/F_2(s)$

(c) $F_1(s)-F_2(s)$

(d) $[F_1(s)/F_2(s)]^{1/2}$

1.17 Test the following function to see if it is positive real:

$$F(s)=\frac{2s^3+s^2+4s+1}{s^3+3s^2+3s+1} \tag{1.128}$$

1.18 A network function is known to have poles at $s=-2$ and $s=-6$ and zeros at $s=-1$ and $s=-5$. It is also known that the function has value 8/7 at $s=1$. Determine the network function.

1.19 Test the following function to see if it is positive real:

$$F(s)=\frac{6s^3+32s^2+39s+99}{6s^2+2s+24} \tag{1.129}$$

1.20 Verify that the impedance function

$$Z(s)=\frac{6s^2+19s+18}{s^2+s+10} \tag{1.130}$$

is positive real and that the minimum value of its real part Re $Z(j\omega)$ occurs at $s=j2$.

1.21 Verify that the impedance function

$$Z(s)=\frac{2s^3+s^2+4s+1}{s^3+3s^2+3s+1} \tag{1.131}$$

can be realized by the one-port network of Figure 1.49.

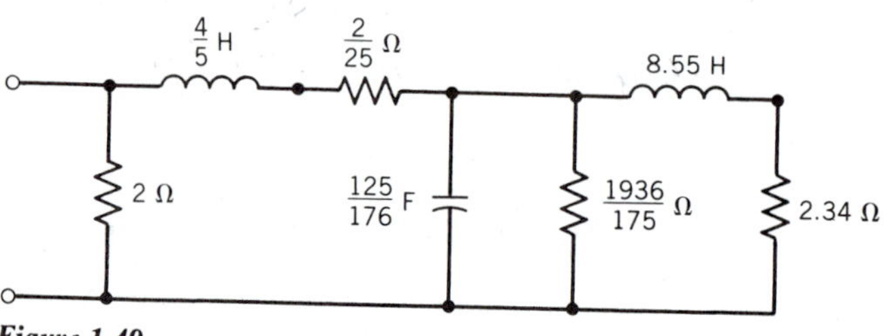

Figure 1.49

1.22 Show that the positive-real impedance

$$Z(s)=\frac{3s^2+s+1}{2s^2+s+3} \tag{1.132}$$

can be realized by the one-port network of Figure 1.50.

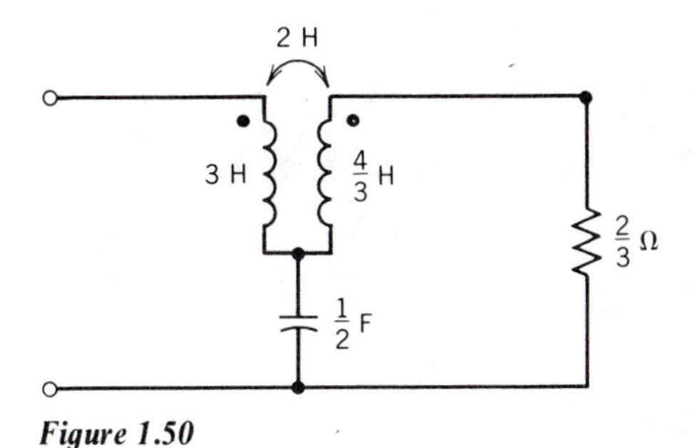

Figure 1.50

1.23 Determine the minimum value of the real positive constant a for which the following function is positive real:

$$F(s)=\frac{2s^2+s+2}{2s^2+(1+2a)s+a} \tag{1.133}$$

1.24 Show that the one-port network of Figure 1.51 is equivalent to a grounded inductor.

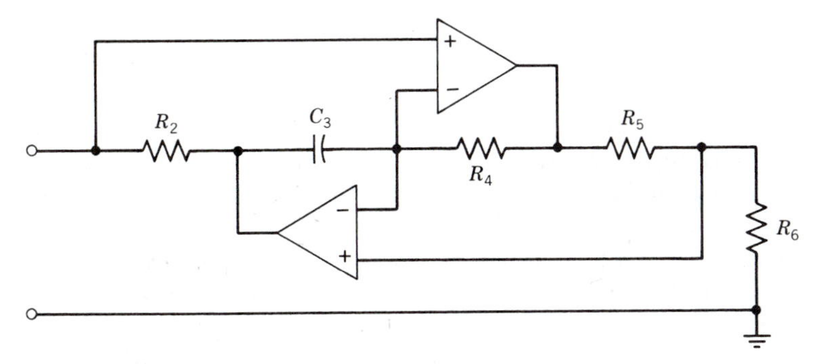

Figure 1.51

1.25 Show that the function

$$F(s)=\frac{s^2+a_1s+a_0}{s^2+b_1s+b_0} \tag{1.134}$$

is positive real if

$$a_1b_1 \geqq (\sqrt{a_0}-\sqrt{b_0})^2 \tag{1.135}$$

1.26 It is claimed that the function

$$F(s)=\left(\frac{s+1}{s+2}\right)^n \tag{1.136}$$

is not positive real for $n>N_0$. Verify this claim and determine the value of N_0 for the function.

chapter 2

FILTER APPROXIMATION AND FREQUENCY TRANSFORMATIONS

In this chapter we consider techniques for approximating ideal filter transmission characteristics. Here we confine our discussion to the low-pass situation, and later we show how the techniques developed for the low-pass case can be applied to other filter types.

The ideal low-pass filter transmission characteristic we seek is that shown in Figure 2.1, where the amplitude of the desired transfer function is constant from $\omega=0$ to $\omega=\omega_c$ and zero for all ω greater than ω_c. Because of its shape, this characteristic is called a *brick-wall* type of response. However, such overly stringent require-

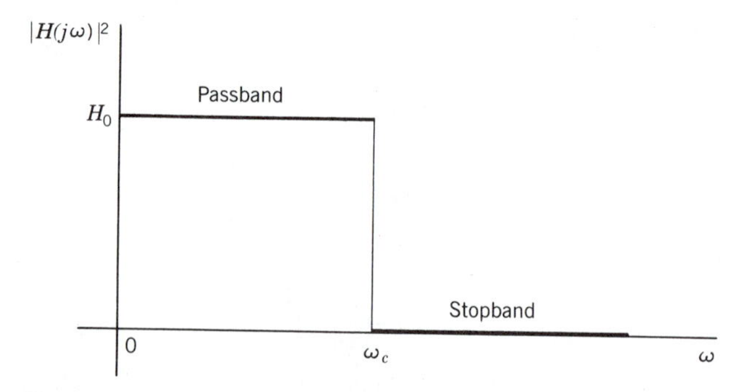

Figure 2.1

ments cannot be realized with finite network elements. The only alternative is to approximate the transmission characteristic. Instead of seeking an overly idealistic performance criterion, we specify the maximum permissible loss or attenuation over a given frequency band of interest called the *passband*, the minimum allowable loss over another frequency band called the *stopband*, and a statement about the selectivity or the tolerable interval between these two bands called the *transitional band*. We then seek a rational function that meets all the specifications and at the same time it must be realizable for the class of networks desired. This is known as the *approximation problem*.

To introduce this concept along with a discussion of the approximating functions, we consider the ideal low-pass brick-wall type of response of Figure 2.1, and show how it can be approximated by four popular rational function approximation schemes: the Butterworth response, the Chebyshev response, the inverse Chebyshev response, and the Bessel–Thomson response.

Confining attention to the low-pass filter characteristic is not to be deemed as restrictive as it may appear. We shall demonstrate this by considering frequency transformations that permit low-pass characteristic to be converted to a high-pass, band-pass, or band-elimination characteristic. This will be presented at the end of this chapter.

2.1 THE BUTTERWORTH RESPONSE

The response function

$$|H(j\omega)|^2 = \frac{H_0}{1+(\omega/\omega_c)^{2n}} \tag{2.1}$$

is known as the nth-order *Butterworth* or *maximally-flat* low-pass response, and was first suggested by Butterworth.† The constant H_0 is usually chosen so that $H^2(0)=H_0$. The nature of this approximation function is seen from the following observations.

First of all, we show that the first $2n-1$ derivatives of $|H(\omega)|^2$ are zero at $\omega=0$. To see this we apply the binomial series expansion

$$(1+x)^{-1} = 1-x+x^2-x^3+x^4-x^5+\dots, \qquad x^2<1 \tag{2.2}$$

to (2.1) and obtain

$$|H(j\omega)|^2 = H_0[1-(\omega/\omega_c)^{2n}+(\omega/\omega_c)^{4n}-(\omega/\omega_c)^{6n}+\cdots], \qquad \omega<\omega_c \tag{2.3}$$

From this expression, it is clear that the first $2n-1$ derivatives are zero at $\omega=0$. Equation (2.3) is known as the *Maclaurin* series expansion of $|H(j\omega)|^2$. If we replace ω by $1/\omega$ in (2.1) and then derive the Maclaurin series expansion for the new function, we obtain a series that begins with a term $H_0\omega_c^{2n}\omega^{2n}$, indicating again that the first $2n-1$ derivatives of the new function are zero at $\omega=0$. This shows that the Butter-

†Named after British engineer S. Butterworth. The term maximally-flat was coined by V. D. Landon.

worth response not only gives a maximally-flat characteristic at dc but also exhibits this characteristic in the stopband at infinity.

Figure 2.2 illustrates the Butterworth response of several orders as well as the ideal brick-wall type of response, which corresponds to the limiting case as n approaches infinity. Observe that all the curves intersect the line $|H(j\omega)|=\sqrt{H_0/2}$ at $\omega=\omega_c$, showing a 3-dB attentuation at ω_c. The radian frequency ω_c is called the *radian cutoff frequency*. For frequencies far above ω_c, the response becomes

$$|H(j\omega)|^2 \approx H_0 \frac{\omega_c^{2n}}{\omega^{2n}} \tag{2.4a}$$

In terms of decibels, we have

$$\alpha(\omega)=10 \log |H(j\omega)|^2 \approx 10 \log H_0 - 20n \log (\omega/\omega_c) \tag{2.4b}$$

yielding an asymptotic slope

$$\frac{d\alpha(\omega)}{d \log (\omega/\omega_c)} = -20n \text{ dB/decade} \tag{2.4c}$$

This means that $\alpha(\omega)$ drops off at a rate of $20n$ dB per decade. (A *decade* is a frequency change by a factor of 10, and an *octave* is a frequency change by a factor of 2.) Thus, a drop off rate of $20n$ dB/decade is equivalent to $6n$ dB/octave.

Apart from the dc value H_0, the Butterworth response is specified by a single parameter n. The manner in which specifications for a filter will be given to the engineer is illustrated by the plot of Figure 2.3. In the passband, which extends from $\omega=0$ to $\omega=\omega_p$, the attenuation should not exceed a preassigned value α_{max} expressed in dB. From ω_p to ω_s we have the transitional band. For the stopband extending from $\omega=\omega_s$ to $\omega=\infty$, the attenuation should not be less than α_{min}. From this information, we can determine n from which the design can proceed. We illustrate this by the following example.

Example 2.1 Suppose that we are required to realize the following specifications with a Butterworth response of minimum order:

(i) The 3-dB radian cutoff frequency is at 1000 rad/s.

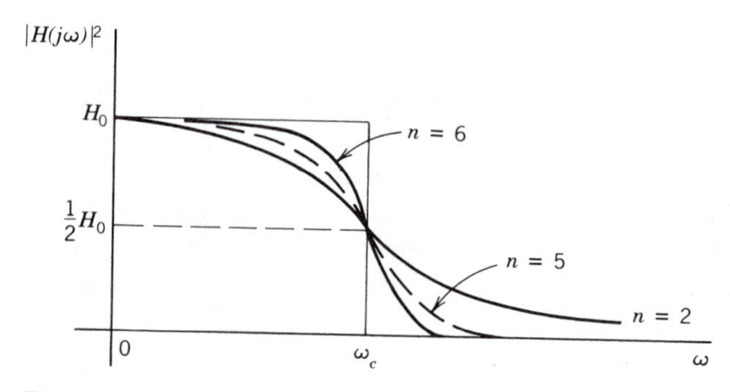

Figure 2.2

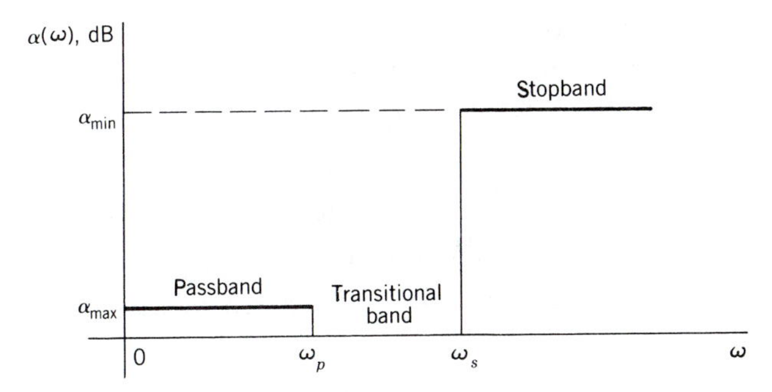

Figure 2.3

(ii) For the passband extending from $\omega=0$ to $\omega=250$ rad/s, the attenuation should not exceed 0.1 dB.

(iii) For the stopband extending from $\omega=2000$ rad/s to $\omega=\infty$, the attenuation should not be less than 60 dB.

From the above specifications, we have

$$\alpha_{\max}=0.1 \text{ dB}, \qquad \alpha_{\min}=60 \text{ dB}, \qquad \omega_c=1000 \text{ rad/s}$$
$$\omega_p=250 \text{ rad/s}, \qquad \omega_s=2000 \text{ rad/s}$$

Using these in (2.1), the passband requirement becomes

$$10 \log\left[1+\left(\frac{250}{1000}\right)^{2n}\right] \leqq \alpha_{\max}=0.1 \tag{2.5a}$$

yielding $n \geqq 1.36$. To satisfy the stopband specifications, we require that

$$10 \log\left[1+\left(\frac{2000}{1000}\right)^{2n}\right] \geqq \alpha_{\min}=60 \tag{2.5b}$$

obtaining $n \geqq 9.97$. Thus, to fulfill both requirements, we choose $n=10$.

The above example illustrates the manner in which specifications for a filter will be given to the engineer. For the passband extending from $\omega=0$ to $\omega=\omega_p$, the attenuation should not exceed $\alpha_{\max}$. For the stopband from $\omega=\omega_s$ to $\omega=\infty$, the attenuation should not be less than $\alpha_{\min}$. From ω_p to ω_s we have a transitional band. These specifications are as indicated in Figure 2.3. To determine the minimum order n required in the approximation, we use (2.1) and obtain the following constraints:

$$10 \log\left[1+\left(\frac{\omega_p}{\omega_c}\right)^{2n}\right]=\alpha_{\max} \tag{2.6a}$$

$$10 \log\left[1+\left(\frac{\omega_s}{\omega_c}\right)^{2n}\right]=\alpha_{\min} \tag{2.6b}$$

Solving these equations for n gives (Problem 2.1)

$$n=\frac{\log[(10^{\alpha_{\min}/10}-1)/(10^{\alpha_{\max}/10}-1)]}{2 \log(\omega_s/\omega_p)} \tag{2.7}$$

Recall that the Butterworth response (2.1) has a 3-dB attenuation at $\omega=\omega_c$. To satisfy specifications calling for less than 3-dB loss at $\omega=\omega_c$, we modify the response function (2.1) so that it appears in the form

$$|H(j\omega)|^2=\frac{H_0}{1+\varepsilon^2(\omega/\omega_c)^{2n}} \tag{2.8}$$

where ε is a nonnegative real constant. This function is again maximally-flat because its first $2n-1$ derivatives are zero at $\omega=0$. At the passband frequency $\omega=\omega_p=\omega_c$, the attenuation in dB is

$$\alpha_{\max}=-10\log\left[\frac{|H(j\omega_p)|^2}{H_0}\right]=10\log(1+\varepsilon^2) \tag{2.9a}$$

where $\alpha_{\max}$ is also the maximum attenuation in the passband. Therefore, the parameter ε is related to the passband loss requirement $\alpha_{\max}$ by

$$\varepsilon=\sqrt{10^{\alpha_{\max}/10}-1} \tag{2.9b}$$

2.1.1 Poles of the Butterworth Function

Once the order of the Butterworth response is determined, our next task is to find the location of the poles. To this end, we resort to the theorem on the uniqueness of analytic continuation in the theory of analytic functions of a complex variable by substituting ω by $-js$ in (2.1),† and obtain

$$|H(j\omega)|^2\Big|_{\omega=-js}=H(j\omega)H(-j\omega)\Big|_{\omega=-js}=H(s)H(-s)$$

$$=\frac{H_0}{1+(-1)^n(s/\omega_c)^{2n}} \tag{2.10a}$$

or

$$H(y)H(-y)=\frac{H_0}{1+(-1)^n y^{2n}} \tag{2.10b}$$

where $y=s/\omega_c$. The poles of this function are the roots of the equation

$$1+(-1)^n y^{2n}=0 \tag{2.11}$$

which can be solved to yield

$$y^{2n}=(-1)^{n+1}=e^{j(n+1)\pi+j2k\pi} \tag{2.12}$$

for all integers k, zero included. Even though there are infinitely many choices of k,

†By a simple change of variable, (2.8) can be put into the form of (2.1): Let $\omega_c'=\varepsilon^{-1/n}\omega_c$. Then $1+\varepsilon^2(\omega/\omega_c)^{2n}=1+(\omega/\omega_c')^{2n}$.

only $2n$ distinct roots are possible. They are located at

$$s_k = \omega_c \exp\left[\frac{j(2k+n-1)\pi}{2n}\right], \quad k=1, 2, \ldots, 2n \tag{2.13}$$

on a circle of radius ω_c. For $n=5$ and $n=6$, the pole locations of (2.10) are shown in Figure 2.4. It is seen that these poles are located symmetrically with respect to both the real and the imaginary axes with quadrantal symmetry. For n odd, a pair of poles are located on the real axis, but they never lie on the imaginary axis for any n. This follows directly from (2.13) and the fact that the poles are separated by π/n radians.

Since we are interested only in networks that are stable, $H(s)$ is devoid of poles in the right-half of the s-plane (RHS). Thus, we assign all the left-half of the s-plane (LHS) poles of (2.10) to $H(s)$, and all the RHS poles to $H(-s)$. To this end we decompose $1+(-1)^n y^{2n}$ into the form

$$1+(-1)^n y^{2n} = q(y)q(-y) \tag{2.14}$$

where

$$q(y) = a_0 + a_1 y + \cdots + a_{n-1}y^{n-1} + a_n y^n = \sum_{m=0}^{n} a_m y^m \tag{2.15}$$

with $a_0 = a_n = 1$, is the Hurwitz polynomial of degree n formed by the LHS poles of (2.10), which are located at

$$y_k = \frac{s_k}{\omega_c} = \exp\left[\frac{j(2k+n-1)\pi}{2n}\right], \qquad k=1, 2, \ldots, n \tag{2.16}$$

Here, a *Hurwitz polynomial* is a polynomial whose zeros all have negative real parts. In accordance with our definition as stated in Section 1.8, $q(y)$ is a strictly Hurwitz polynomial. The polynomial $q(y)$, as defined in (2.15), is called the *Butterworth polynomial*. Finally, the transfer function $H(s)$ with an nth-order Butterworth response can be written as

$$H(y) = \frac{H_0^{1/2}}{q(y)} = \frac{H_0^{1/2}}{y^n + a_{n-1}y^{n-1} + \cdots + a_1 y + a_0} \tag{2.17}$$

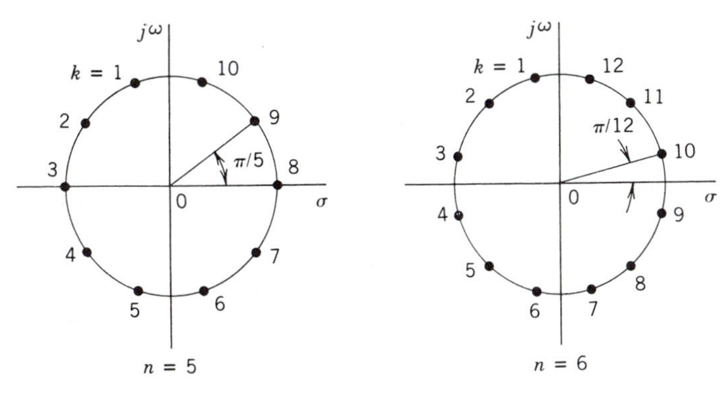

Figure 2.4

Although the pole locations and the Butterworth polynomials can easily be computed from (2.16), it is useful to have them readily available. Table 2.1 gives the pole locations for $n=2$ to $n=10$, Table 2.2 lists the coefficients of the Butterworth polynomials, and Table 2.3 expresses the Butterworth polynomials in factored form. In fact, the coefficients a_k can be calculated explicitly by the recursion formula†

$$\frac{a_k}{a_{k-1}}=\frac{\cos[(k-1)\pi/2n]}{\sin(k\pi/2n)}, \qquad k=1,2,\ldots,n \tag{2.18}$$

with $a_n=1$. Thus, we can calculate the coefficients starting from a_n or a_0, which is known to be unity. Forming the product of the terms given in (2.18) yields an explicit formula for the coefficients a_k as

$$a_k=\prod_{m=1}^{k}\frac{\cos[(m-1)\pi/2n]}{\sin(m\pi/2n)}, \qquad k=1,2,\ldots,n \tag{2.19}$$

with $a_0=1$. Since

$$\cos(m\pi/2n)=\sin[(n-m)\pi/2n] \tag{2.20}$$

it follows that

$$a_k=a_{n-k} \tag{2.21}$$

This indicates that the coefficients of the Butterworth polynomial are symmetric from its two ends, as can be seen from Table 2.2.

Example 2.2 Consider the fifth-order Butterworth response. The transfer function $H(s)$ can be written as

$$H(s)=\frac{H_0^{1/2}}{s^5+a_4s^4+a_3s^3+a_2s^2+a_1s+1} \tag{2.22}$$

The coefficients a_k can be computed recurrently by the formula (2.18), as follows:

$$a_1=a_0(\cos 0)/(\sin 18^\circ)=3.23606798=a_{5-1}=a_4$$
$$a_2=a_1(\cos 18^\circ)/(\sin 36^\circ)=5.23606798=a_{5-2}=a_3$$
$$a_3=a_2(\cos 36^\circ)/(\sin 54^\circ)=5.23606798=a_{5-3}=a_2$$
$$a_4=a_3(\cos 54^\circ)/(\sin 72^\circ)=3.23606798=a_{5-4}=a_1$$
$$a_5=a_4(\cos 72^\circ)/(\sin 90^\circ)=1.00000000=a_{5-5}=a_0$$

which confirm the numbers given in Table 2.2. Alternatively, we can calculate these coefficients from the other end, as follows:

$$a_4=a_5(\sin 90^\circ)/(\cos 72^\circ)=3.23606798=a_{5-4}=a_1$$
$$a_3=a_4(\sin 72^\circ)/(\cos 54^\circ)=5.23606798=a_{5-3}=a_2$$
$$a_2=a_3(\sin 54^\circ)/(\cos 36^\circ)=5.23606798=a_{5-2}=a_3$$

†For a derivation of the formula, see W. K. Chen 1976, *Theory and Design of Broadband Matching Networks*, pp. 120–121. Cambridge, England: Pergamon Press.

Table 2.1 Pole Locations for Normalized Butterworth Response (2.10)

n					
2	$-0.70710678 \pm j0.70710678$				
3	$-0.50000000 \pm j0.86602540$	-1.00000000			
4	$-0.38268343 \pm j0.92387953$	$-0.92387953 \pm j0.38268343$			
5	$-0.30901699 \pm j0.95105652$	$-0.80901699 \pm j0.58778525$	-1.00000000		
6	$-0.25881905 \pm j0.96592583$	$-0.70710678 \pm j0.70710678$	$-0.96592583 \pm j0.25881905$		
7	$-0.22252093 \pm j0.97492791$	$-0.62348980 \pm j0.78183148$	$-0.90096887 \pm j0.43388374$	-1.00000000	
8	$-0.19509032 \pm j0.98078528$	$-0.55557023 \pm j0.83146961$	$-0.83146961 \pm j0.55557023$	$-0.98078528 \pm j0.19509032$	
9	$-0.17364818 \pm j0.98480775$	$-0.50000000 \pm j0.86602540$	$-0.76604444 \pm j0.64278761$	$-0.93969262 \pm j0.34202014$	-1.00000000
10	$-0.15643447 \pm j0.98768834$	$-0.45399050 \pm j0.89100652$	$-0.70710678 \pm j0.70710678$	$-0.89100652 \pm j0.45399050$	$-0.98768834 \pm j0.15643447$

Table 2.2 Coefficients of the Butterworth Polynomial $q(s)=s^n+a_{n-1}s^{n-1}+\cdots+a_1s+1$

n	a_1	a_2	a_3	a_4	a_5	a_6	a_7	a_8	a_9
2	1.41421356								
3	2.00000000	2.00000000							
4	2.61312593	3.41421356	2.61312593						
5	3.23606798	5.23606798	5.23606798	3.23606798					
6	3.86370331	7.46410162	9.14162017	7.46410162	3.86370331				
7	4.49395921	10.09783468	14.59179389	14.59179389	10.09783468	4.49395921			
8	5.12583090	13.13707118	21.84615097	25.68835593	21.84615097	13.13707118	5.12583090		
9	5.75877048	16.58171874	31.16343748	41.98638573	41.98638573	31.16343748	16.58171874	5.75877048	
10	6.39245322	20.43172909	42.80206107	64.88239627	74.23342926	64.88239627	42.80206107	20.43172909	6.39245322

Table 2.3 Factors of Butterworth Polynomials

n	Butterworth Polynomial
1	$s+1$
2	$s^2+1.41421356s+1$
3	$(s+1)(s^2+s+1)$
4	$(s^2+0.76536686s+1)(s^2+1.84775907s+1)$
5	$(s+1)(s^2+0.61803399s+1)(s^2+1.61803399s+1)$
6	$(s^2+0.51763809s+1)(s^2+1.41421356s+1)(s^2+1.93185165s+1)$
7	$(s+1)(s^2+0.44504187s+1)(s^2+1.24697960s+1)(s^2+1.80193774s+1)$
8	$(s^2+0.39018064s+1)(s^2+1.11114047s+1)(s^2+1.66293922s+1)(s^2+1.96157056s+1)$
9	$(s+1)(s^2+0.34729636s+1)(s^2+s+1)(s^2+1.53208889s+1)(s^2+1.87938524s+1)$
10	$(s^2+0.31286893s+1)(s^2+0.90798100s+1)(s^2+1.41421356s+1)(s^2+1.78201305s+1)$ $(s^2+1.97537668s+1)$

$$a_1=a_2(\sin 36°)/(\cos 18°)=3.23606798=a_{5-1}=a_4$$
$$a_0=a_1(\sin 18°)/(\cos 0)=1.00000000=a_{5-0}=a_5$$

This also confirms the symmetric property (2.21).

2.2 THE CHEBYSHEV RESPONSE

Another useful approximation of the ideal brick-wall response is given by

$$|H(j\omega)|^2=\frac{H_0}{1+\varepsilon^2C_n^2(\omega/\omega_c)} \tag{2.23}$$

where $C_n(\omega)$ is the nth-order Chebyshev polynomial of the first kind and $\varepsilon^2\leqq 1$ and H_0 is a constant. The response (2.23) is called the *nth-order Chebyshev*† or *equiripple response.* We begin our discussion by considering the properties of the Chebyshev polynomial.

2.2.1 Chebyshev Polynomials

The *Chebyshev polynomial of order n* is defined by the equations

$$C_n(\omega)=\cos(n\cos^{-1}\omega),\qquad 0\leqq\omega\leqq 1 \tag{2.24a}$$
$$=\cosh(n\cosh^{-1}\omega),\qquad \omega>1 \tag{2.24b}$$

In fact, these two expressions are completely equivalent, each being valid for all ω. To show that the transcendental function (2.24) is indeed a polynomial, it is sufficient

†Named after P. L. Chebyshev (1821–1894). The German transliteration is Tschebyscheff.

to consider (2.24a) and let

$$w = \cos^{-1}\omega \tag{2.25}$$

Substituting it in (2.24a) gives

$$C_n(\omega) = \cos nw \tag{2.26}$$

Using the trigonometric identity

$$\begin{aligned}\cos(n+1)w &= \cos nw \cos w - \sin nw \sin w \\ &= \cos nw \cos w + \tfrac{1}{2}\cos(n+1)w - \tfrac{1}{2}\cos(n-1)w\end{aligned} \tag{2.27}$$

we obtain

$$\cos(n+1)w = 2\cos nw \cos w - \cos(n-1)w \tag{2.28}$$

From (2.24a) and (2.25), we have the desired recurrence formula

$$C_{n+1}(\omega) = 2\omega C_n(\omega) - C_{n-1}(\omega) \tag{2.29}$$

Because the lower-order Chebyshev polynomials are known, i.e.,

$$C_0(\omega) = 1 \qquad \text{and} \qquad C_1(\omega) = \omega \tag{2.30}$$

the higher-order polynomials can be computed recurrently by means of (2.29), as follows:

$$\begin{aligned}
C_2(\omega) &= 2\omega^2 - 1 \\
C_3(\omega) &= 4\omega^3 - 3\omega \\
C_4(\omega) &= 8\omega^4 - 8\omega^2 + 1 \\
C_5(\omega) &= 16\omega^5 - 20\omega^3 + 5\omega \\
C_6(\omega) &= 32\omega^6 - 48\omega^4 + 18\omega^2 - 1 \\
C_7(\omega) &= 64\omega^7 - 112\omega^5 + 56\omega^3 - 7\omega \\
C_8(\omega) &= 128\omega^8 - 256\omega^6 + 160\omega^4 - 32\omega^2 + 1 \\
C_9(\omega) &= 256\omega^9 - 576\omega^7 + 432\omega^5 - 120\omega^3 + 9\omega \\
C_{10}(\omega) &= 512\omega^{10} - 1280\omega^8 + 1120\omega^6 - 400\omega^4 + 50\omega^2 - 1
\end{aligned} \tag{2.31}$$

From the above discussion, we now list below some of the properties of the Chebyshev polynomials:

(i) $C_n(\omega)$ is either an even or an odd function depending on whether n is even or odd. Thus, we can write

$$C_n(-\omega) = C_n(\omega), \qquad n \text{ even} \tag{2.32a}$$

$$C_n(-\omega) = -C_n(\omega), \qquad n \text{ odd} \tag{2.32b}$$

(ii) Every coefficient of $C_n(\omega)$ is an integer, and the one associated with ω_n is 2^{n-1}. Thus, in the limit as ω approaches infinity

$$C_n(\omega) \rightarrow 2^{n-1}\omega^n \tag{2.33}$$

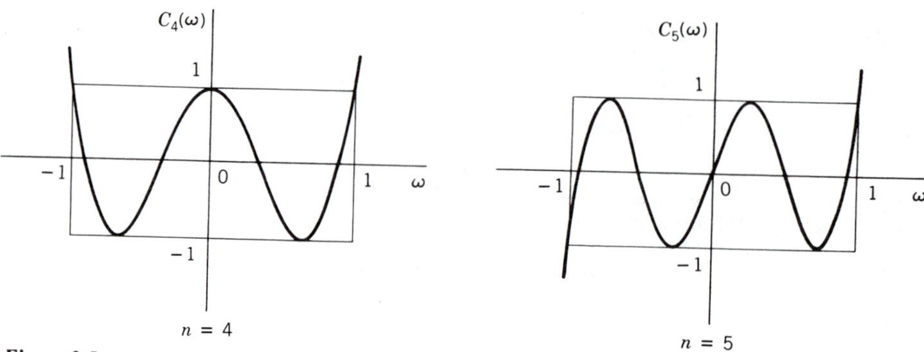

Figure 2.5

(iii) In the range $-1 \leqslant \omega \leqslant 1$, all of the Chebyshev polynomials have the equal-ripple property, varying between a maximum of 1 and a minimum of -1. Outside of this interval, their magnitude increases monotonically as ω is increased, and approaches infinity in accordance with (2.33). Sketches of the polynomials for $n=4$ and $n=5$ are shown in Figure 2.5.

(iv) As indicated in Figure 2.5, the Chebyshev polynomials possess special values at $\omega=0$, 1, or -1:

$$C_n(0)=(-1)^{n/2}, \qquad n \text{ even} \tag{2.34a}$$

$$=0, \qquad n \text{ odd} \tag{2.34b}$$

$$C_n(\pm 1)=1, \qquad n \text{ even} \tag{2.35a}$$

$$=\pm 1, \qquad n \text{ odd} \tag{2.35b}$$

2.2.2 Equiripple Characteristic

We now examine the manner in which the Chebyshev polynomials approximate the ideal brick-wall response. Apart from the constant H_0, (2.23) dictates that we square $C_n(\omega/\omega_c)$, multiply it by the constant ε^2, not greater than unity, add unity, and form the reciprocal of $1+\varepsilon^2 C_n^2(\omega/\omega_c)$. If we carry out all these steps as depicted in Figure 2.6 for $n=4$ and $n=5$, the responses that result in (*d*) have equal maxima and equal minima in the passband. From these plots, it is clear that the total number of troughs and peaks for positive ω is equal to n, all lying within the passband, and outside the band the magnitude decreases monotonically. At the edge of the passband $\omega=\omega_c$, the response goes through a value that is equal to the earlier relative minimum points in the passband. This is in contrast to the Butterworth response, where at the radian cutoff frequency, the response is attenuated by 3 dB from its maximum value at $\omega=0$. Because of the equal-ripple property in the passband, the Chebyshev response is also known as the *equiripple response*. Plots showing equiripple response for $n=4$ and $n=5$ are presented in Figure 2.7.

From (2.23) it is clear that the maximum value of the Chebyshev response occurs

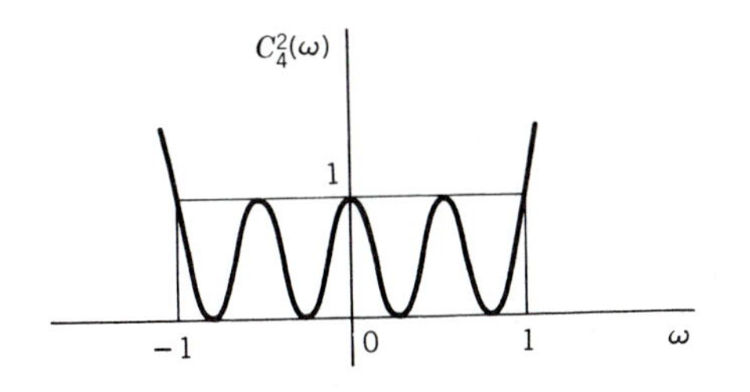

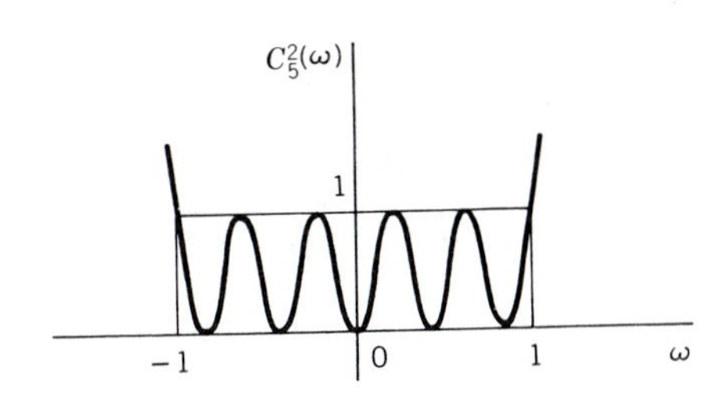

(a)

$\epsilon^2 C_4^2(\omega)$

ϵ^2

−1 0 1 ω

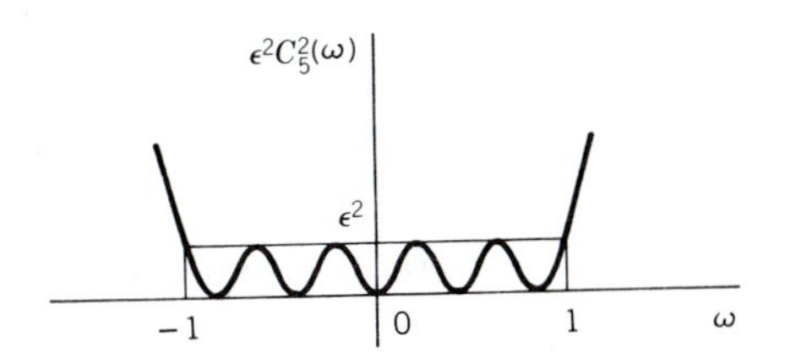

(b)

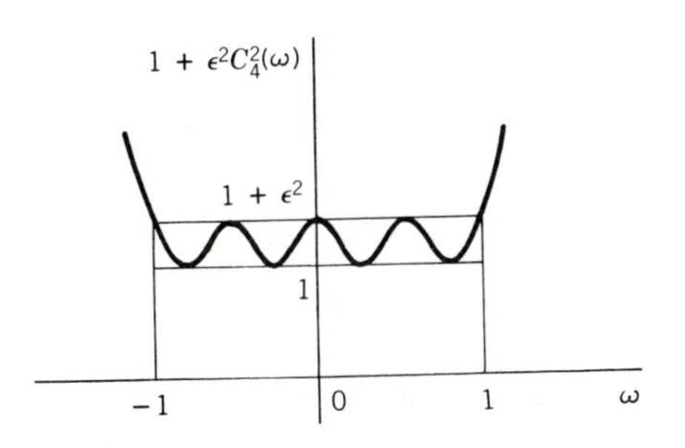

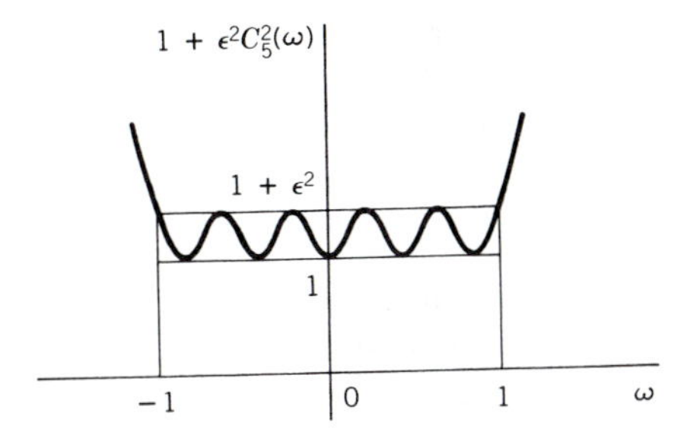

(c)

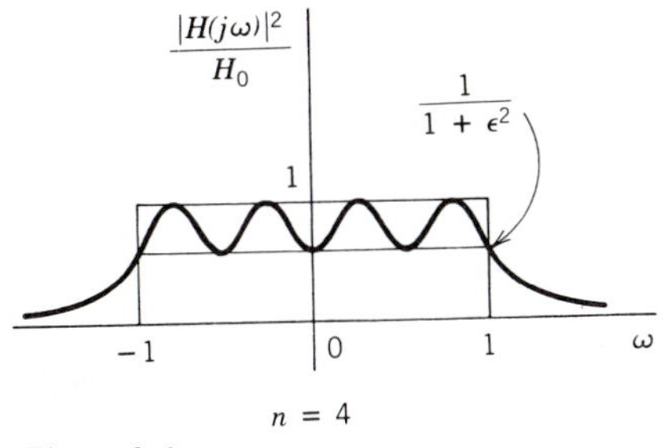

$n = 4$

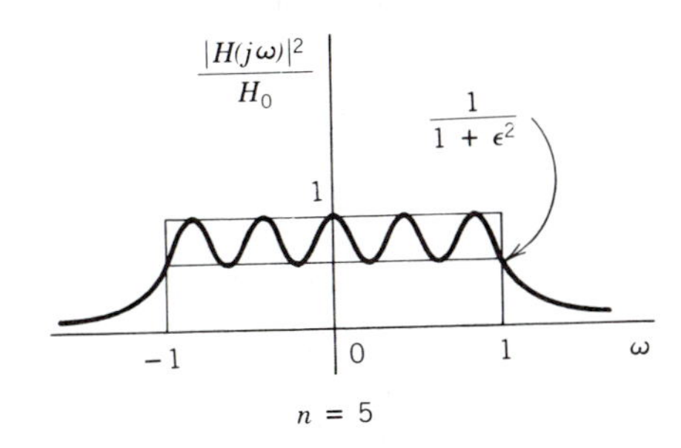

$n = 5$

(d)

Figure 2.6

at the points of ω where $C_n(\omega)$ vanishes or at the zeros of $C_n(\omega)$ with

$$|H(j\omega)|^2_{\max} = H_0 \tag{2.36}$$

and that the minimum value in the passband is

$$|H(j\omega)|^2_{\min} = \frac{H_0}{1+\varepsilon^2} \tag{2.37}$$

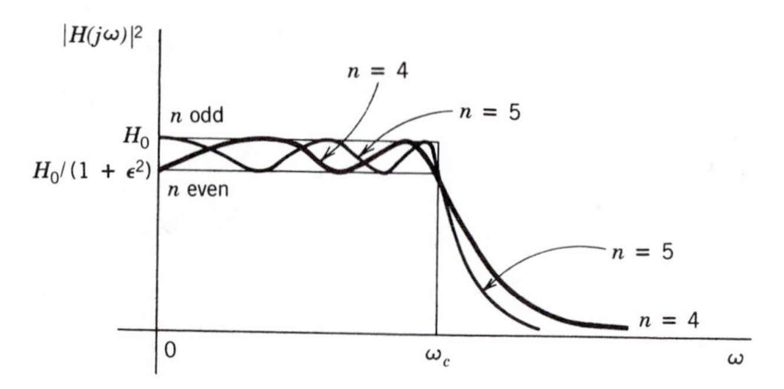

Figure 2.7

occurring at the points of ω where $C_n(\omega)=\pm 1$. Thus, the dc value is

$$H^2(0)=H_0, \qquad n \text{ odd} \tag{2.38a}$$

$$=\frac{H_0}{1+\varepsilon^2}, \qquad n \text{ even} \tag{2.38b}$$

At the edge of the passband, the value is

$$|H(j\omega_c)|^2=\frac{H_0}{1+\varepsilon^2} \tag{2.39}$$

for all n, as illustrated in Figure 2.7. Observe that the quantity ε plays an important role in determining the maxima and the minima of the ripple, and is called the *ripple factor*. For a fixed H_0, the peak-to-peak ripple in the passband, usually stated in terms of decibels, is determined by the ripple factor alone. Let $\alpha_{\max}$ be the peak-to-peak ripple in dB in the passband. Then from (2.36) and (2.37) we have

$$\alpha_{\max}=10 \log(1+\varepsilon^2) \tag{2.40}$$

yielding

$$\varepsilon=\sqrt{10^{\alpha\max/10}-1} \tag{2.41}$$

For frequencies far above ω_c, the response approaches

$$|H(j\omega)|^2 \rightarrow \frac{H_0}{2^{2n-2}\varepsilon^2(\omega/\omega_c)^{2n}} \tag{2.42}$$

In terms of decibels, we have the attenuation

$$\alpha(\omega)=10 \log H_0-6(n-1)-20 \log \varepsilon-20n \log(\omega/\omega_c) \tag{2.43}$$

showing an asymptotic slope of $-20n$ dB/decade or $-6n$ dB/octave. This drop rate is the same as that given for the Butterworth response except that it is offset by a value depending on both n and ε. Consideration of two limiting cases of the ripple factor is helpful in interpreting the difference, and the results are sketched in Figure 2.8. Observe that for $\varepsilon=1$, the radian cutoff frequency ω_c corresponds to a point having

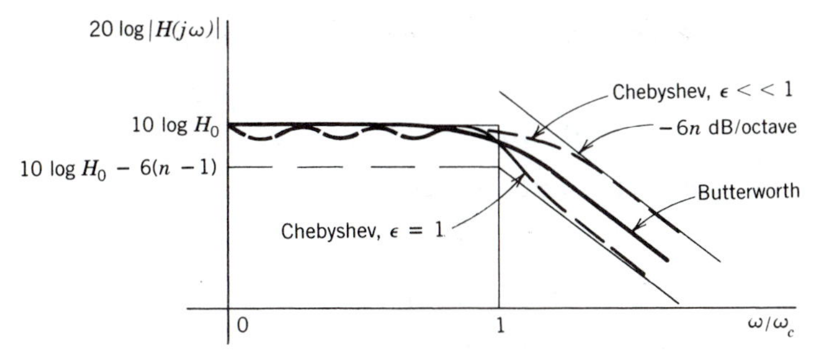

Figure 2.8

3-dB attenuation from its maximum value, a convenient value for comparison with all the Butterworth response. Because of the equal-ripple property, this also gives a 3-dB ripple in the passband.

Apart from the constant H_0, the Chebyshev response is determined by two parameters: the ripple factor ε and the order of the Chebyshev polynomial n. The ripple factor is fixed by the permissible ripple in the passband by formula (2.41). The order of the Chebyshev response is found by the rate of attenuation in the stopband. If the stopband extends from $\omega=\omega_s$ to $\omega=\infty$, the minimum attenuation in the stopband occurs at $\omega=\omega_s$. If the minimum attenuation in the stopband is required to be $\alpha_{\min}$, expressed in dB, then from (2.23)

$$\alpha_{\min}=-10\log[|H(j\omega_s)|^2/H_0]=10\log[1+\varepsilon^2C_n^2(\omega_s/\omega_c)] \tag{2.44a}$$

which can be written as

$$C_n(\omega_s/\omega_c)=\cosh\left(n\cosh^{-1}\frac{\omega_s}{\omega_c}\right)$$

$$=\frac{1}{\varepsilon}\sqrt{10^{\alpha_{\min}/10}-1} \tag{2.44b}$$

Solving this equation for n gives the desired formula

$$n=\frac{\cosh^{-1}[\varepsilon^{-1}(10^{\alpha_{\min}/10}-1)^{1/2}]}{\cosh^{-1}(\omega_s/\omega_c)} \tag{2.45}$$

We illustrate the above by the following example.

Example 2.3 Determine the order of the Chebyshev response that realizes the following specifications:

(i) The passband extends from $\omega=0$ to $\omega=250$ rad/s.
(ii) The attenuation within the passband should not exceed 0.1 dB.
(iii) The response must give at least 60-dB attenuation at the frequency $\omega=2000$ rad/s or higher.

From the above specifications, we can make the following identifications:

$$\omega_c = 250 \text{ rad/s}, \qquad \omega_s = 2000 \text{ rad/s}$$
$$\alpha_{\max} = 0.1 \text{ dB}, \qquad \alpha_{\min} = 60 \text{ dB}$$

To calculate the ripple factor, we use (2.41)

$$\varepsilon = \sqrt{10^{0.01} - 1} = 0.1526 \tag{2.46a}$$

The required value of n is found from (2.45)

$$n = \frac{\cosh^{-1}[6.5531(10^6 - 1)^{1/2}]}{\cosh^{-1} 8} = 3.424 \tag{2.46b}$$

which is rounded to $n = 4$. The desired magnitude function becomes

$$|H(j\hat{\omega})|^2 = \frac{H_0}{1 + 0.0233(8\hat{\omega}^4 - 8\hat{\omega}^2 + 1)^2} \tag{2.47}$$

where $\hat{\omega} = \omega/\omega_c$ is the normalized frequency.

2.2.3 Poles of the Chebyshev Function

Like the Butterworth case, we next determine the location of the poles of the Chebyshev response. To this end, we again appeal to the theory of analytic continuation by replacing ω by $-js$ in (2.23) and obtain

$$H(s)H(-s) = \frac{H_0}{1 + \varepsilon^2 C_n^2(-jy)} \tag{2.48}$$

where $y = s/\omega_c$, as before in (2.10), is the normalized complex frequency. The poles of this function are given by the roots of the equation

$$1 + \varepsilon^2 C_n^2(-jy) = 0 \tag{2.49}$$

with the *generalized Chebyshev polynomial* defined by the relation

$$C_n(-jy) = \cosh[n \cosh^{-1}(-jy)] \tag{2.50}$$

To put this in a more convenient form, write

$$\cosh^{-1}(-jy) = u + jv \tag{2.51}$$

where u and v are real. Substituting these in (2.49) and expanding the resulting hyperbolic cosine yields

$$C_n(-jy) = \cosh nu \cosh jnv - \sinh nu \sinh jnv = \pm j/\varepsilon \tag{2.52}$$

Applying the relations $\cosh ju = \cos u$ and $\sinh ju = j \sin u$ to (2.52) and equating the real and imaginary parts on both sides results in

$$\cosh nu \cos nv = 0 \tag{2.53a}$$

$$\sinh nu \sin nv = \pm 1/\varepsilon \tag{2.53b}$$

Since $\cosh nu \neq 0$, (2.53a) is satisfied only if $\cos nv = 0$ or

$$v_k = (2k-1)\pi/2n, \qquad k = 1, 2, \ldots, 2n \tag{2.54}$$

giving $2n$ distinct solutions. At these values of v, $\sin nv = \pm 1$, so that from (2.53b) we obtain

$$u_k = \pm(1/n)\sinh^{-1}(1/\varepsilon) \tag{2.55}$$

Substituting u_k and v_k in (2.51) and taking the hyperbolic cosine on both sides gives

$$-jy_k = \cosh(u_k + jv_k) = \cosh u_k \cos v_k - j \sinh u_k \sin v_k \tag{2.56}$$

The $2n$ distinct poles of (2.48) are, therefore, given by

$$y_k = s_k/\omega_c = \sigma_k + j\omega_k, \qquad k = 1, 2, \ldots, 2n \tag{2.57}$$

where

$$\sigma_k = -\sinh a \sin \frac{(2k-1)\pi}{2n} \tag{2.58a}$$

$$\omega_k = \cosh a \cos \frac{(2k-1)\pi}{2n} \tag{2.58b}$$

$$a = \frac{1}{n} \sinh^{-1} \frac{1}{\varepsilon} \tag{2.58c}$$

To find the locus of these poles, we square the real and imaginary parts of y_k and add. This leads to

$$\frac{\sigma_k^2}{\sinh^2 a} + \frac{\omega_k^2}{\cosh^2 a} = 1 \tag{2.59}$$

Equation (2.59) is the locus of an ellipse, the major semi-axis of which is $\cosh a$ and the minor semi-axis of which is $\sinh a$. For $n=5$ and $n=6$, pole locations are presented in Figure 2.9. Observe that these poles, like the Butterworth response, also possess quadrantal symmetry, being symmetric with respect to both the real and the imaginary axes. Hence, the left-hand side of (2.49) can be decomposed into the form

$$1 + \varepsilon^2 C_n^2(-jy) = \varepsilon^2 2^{2n-2} p(y)p(-y) \tag{2.60}$$

where

$$p(y) = b_0 + b_1 y + \cdots + b_{n-1} y^{n-1} + b_n y^n = \sum_{m=0}^{n} b_m y^m \tag{2.61}$$

with $b_n = 1$, is the Hurwitz polynomial of degree n formed by the LHS roots of (2.49), which are given by

$$y_k = s_k/\omega_c = -\sinh a \sin \frac{(2k-1)\pi}{2n} + j \cosh a \cos \frac{(2k-1)\pi}{2n}, \qquad k = 1, 2, \ldots, n \tag{2.62}$$

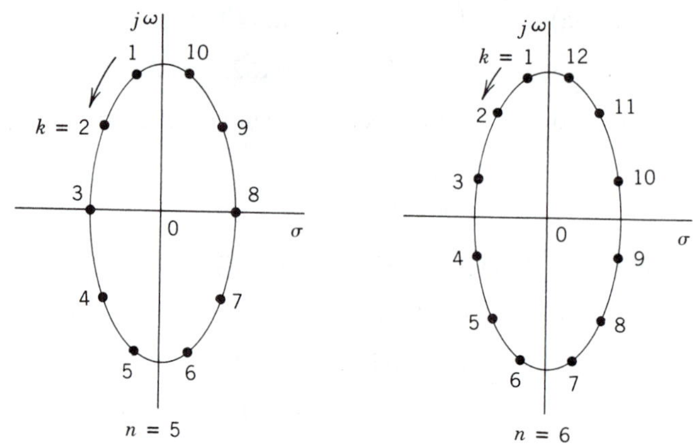

Figure 2.9

To avoid the necessity of computing these Hurwitz polynomials $p(y)$, they have been tabulated. Table 2.4 lists the locations of the LHS poles of the Chebyshev response (2.48) for various ripple factors and for $n=1, 2, \ldots, 10$. Table 2.5 gives the coefficients of the corresponding Hurwitz polynomials $p(y)$. Table 2.6 expresses these polynomials $p(y)$ in factored form.

Comparing (2.62) with (2.16) shows that the real part of a Chebyshev pole is sinh a times the real part of the corresponding Butterworth pole and the imaginary part of a Chebyshev pole is cosh a times the imaginary part of the corresponding Butterworth pole. Thus, by normalizing the major semi-axis to unity, which is equivalent to dividing y_k by cosh a, the Chebyshev and Butterworth poles have the same imaginary part while their real parts are related by the factor tanh a. To obtain the real part of the Chebyshev pole, we simply shift the real part of the corresponding Butterworth pole horizontally from a unit circle to an ellipse with a major semi-axis of unity and a minor semi-axis of tanh a. This is illustrated in Figure 2.10 with $n=4$.

Example 2.4 Determine the normalized transfer function that realizes the fourth-order Chebyshev response with 0.1-dB ripple in the passband.

From (2.41) the 0.1-dB ripple in the passband corresponds to the ripple factor $\varepsilon=0.15262042$. We carry eight significant digits to verify the results given in Tables 2.4 and 2.5. From (2.62) the pole locations for the normalized Chebyshev response are found as follows:

$$y_1 = -\sinh a \sin\frac{\pi}{8} + j\cosh a \cos\frac{\pi}{8} = -0.26415637 + j1.12260981$$

$$y_2 = -\sinh a \sin\frac{3\pi}{8} + j\cosh a \cos\frac{3\pi}{8} = -0.63772988 + j0.46500021$$

$$y_3 = -\sinh a \sin\frac{5\pi}{8} + j\cosh a \cos\frac{5\pi}{8} = -0.63772988 - j0.46500021$$

$$y_4 = -\sinh a \sin\frac{7\pi}{8} + j\cosh a \cos\frac{7\pi}{8} = -0.26415637 - j1.12260981$$

Table 2.4 Pole Locations for Normalized Chebyshev Response (2.48)

n	0.1-dB Ripple ($\varepsilon = 0.15262042$)				
1	-6.55220322				
2	$-1.18617812 \pm j1.38094842$				
3	$-0.48470285 \pm j1.20615528$	-0.96940571			
4	$-0.26415637 \pm j1.12260981$	$-0.63772988 \pm j0.46500021$			
5	$-0.16653368 \pm j1.08037201$	$-0.43599085 \pm j0.66770662$	-0.53891432		
6	$-0.11469337 \pm j1.05651891$	$-0.31334811 \pm j0.77342552$	$-0.42804148 \pm j0.28309339$		
7	$-0.08384097 \pm j1.04183333$	$-0.23491716 \pm j0.83548546$	$-0.33946514 \pm j0.46365945$	-0.37677788	
8	$-0.06398012 \pm j1.03218136$	$-0.18219998 \pm j0.87504111$	$-0.27268154 \pm j0.58468377$	$-0.32164981 \pm j0.20531364$	
9	$-0.05043805 \pm j1.02550963$	$-0.14523059 \pm j0.90181804$	$-0.22250617 \pm j0.66935388$	$-0.27294423 \pm j0.35615576$	-0.29046118
10	$-0.04078867 \pm j1.02071040$	$-0.11837334 \pm j0.92079615$	$-0.18437079 \pm j0.73074797$	$-0.23232075 \pm j0.46916907$	$-0.25752954 \pm j0.16166465$

n	0.5-dB Ripple ($\varepsilon = 0.34931140$)				
1	-2.86277516				
2	$-0.71281226 \pm j1.00404249$				
3	$-0.31322824 \pm j1.02192749$	-0.62645649			
4	$-0.17535307 \pm j1.01625289$	$-0.42333976 \pm j0.42094573$			
5	$-0.11196292 \pm j1.01155737$	$-0.29312273 \pm j0.62517684$	-0.36231962		
6	$-0.07765008 \pm j1.00846085$	$-0.21214395 \pm j0.73824458$	$-0.28979403 \pm j0.27021627$		
7	$-0.05700319 \pm j1.00640854$	$-0.15971939 \pm j0.80707698$	$-0.23080120 \pm j0.44789394$	-0.25617001	
8	$-0.04362008 \pm j1.00500207$	$-0.12421947 \pm j0.85199961$	$-0.18590757 \pm j0.56928794$	$-0.21929293 \pm j0.19990734$	
9	$-0.03445272 \pm j1.00400397$	$-0.09920264 \pm j0.88290628$	$-0.15198727 \pm j0.65531705$	$-0.18643998 \pm j0.34868692$	-0.19840529
10	$-0.02789941 \pm j1.00327317$	$-0.08096724 \pm j0.90506581$	$-0.12610944 \pm j0.71826429$	$-0.15890716 \pm j0.46115406$	$-0.17614995 \pm j0.15890286$

Table 2.4 (continued)

n	1-dB Ripple ($\varepsilon = 0.50884714$)				
1	-1.96522673				
2	$-0.54886716 \pm j0.89512857$				
3	$-0.24708530 \pm j0.96599867$	-0.49417060			
4	$-0.13953600 \pm j0.98337916$	$-0.33686969 \pm j0.40732899$			
5	$-0.08945836 \pm j0.99010711$	$-0.23420503 \pm j0.61191985$	-0.28949334		
6	$-0.06218102 \pm j0.99341120$	$-0.16988172 \pm j0.72722747$	$-0.23206274 \pm j0.26618373$		
7	$-0.04570898 \pm j0.99528396$	$-0.12807372 \pm j0.79815576$	$-0.18507189 \pm j0.44294303$	-0.20541430	
8	$-0.03500823 \pm j0.99645128$	$-0.09969501 \pm j0.84475061$	$-0.14920413 \pm j0.56444431$	$-0.17599827 \pm j0.19820648$	
9	$-0.02766745 \pm j0.99722967$	$-0.07966524 \pm j0.87694906$	$-0.12205422 \pm j0.65089544$	$-0.14972167 \pm j0.34633423$	-0.15933047
10	$-0.02241445 \pm j0.99777551$	$-0.06504927 \pm j0.90010629$	$-0.10131662 \pm j0.71432840$	$-0.12766638 \pm j0.45862706$	$-0.14151928 \pm j0.15803212$

n	1.5-dB Ripple ($\varepsilon = 0.64229086$)				
1	-1.55692704				
2	$-0.46108873 \pm j0.84415805$				
3	$-0.21005618 \pm j0.93934594$	-0.42011237			
4	$-0.11913070 \pm j0.96761105$	$-0.28760695 \pm j0.40079762$			
5	$-0.07652815 \pm j0.97978702$	$-0.20035330 \pm j0.60554168$	-0.24765030		
6	$-0.05325112 \pm j0.98615853$	$-0.14548476 \pm j0.72191815$	$-0.19873588 \pm j0.26424038$		
7	$-0.03917029 \pm j0.98991746$	$-0.10975272 \pm j0.79385217$	$-0.15859728 \pm j0.44055472$	-0.17602970	
8	$-0.03001306 \pm j0.99232369$	$-0.08546998 \pm j0.84125141$	$-0.12791486 \pm j0.56210622$	$-0.15088586 \pm j0.19738545$	
9	$-0.02372663 \pm j0.99395816$	$-0.06831811 \pm j0.87407213$	$-0.10466942 \pm j0.64876011$	$-0.12839605 \pm j0.34519804$	-0.13663622
10	$-0.01922587 \pm j0.99511966$	$-0.05579564 \pm j0.89771042$	$-0.08690374 \pm j0.71242702$	$-0.10950510 \pm j0.45740630$	$-0.12138734 \pm j0.15761147$

Table 2.4 (continued)

n	2-dB Ripple ($\varepsilon = 0.76478310$)				
1	-1.30756027				
2	$-0.40190822 \pm j0.81334508$				
3	$-0.18445539 \pm j0.92307712$	-0.36891079			
4	$-0.10488725 \pm j0.95795296$	$-0.25322023 \pm j0.39679711$			
5	$-0.06746098 \pm j0.97345572$	$-0.17661514 \pm j0.60162872$	-0.21830832		
6	$-0.04697322 \pm j0.98170517$	$-0.12833321 \pm j0.71865806$	$-0.17530643 \pm j0.26304711$		
7	$-0.03456636 \pm j0.98662052$	$-0.09685278 \pm j0.79120823$	$-0.13995632 \pm j0.43908744$	-0.15533980	
8	$-0.02649238 \pm j0.98978701$	$-0.07544391 \pm j0.83910091$	$-0.11290980 \pm j0.56066930$	$-0.13318619 \pm j0.19688088$	
9	$-0.02094714 \pm j0.99194711$	$-0.06031490 \pm j0.87230365$	$-0.09240778 \pm j0.64744750$	$-0.11335493 \pm j0.34449962$	-0.12062980
10	$-0.01697581 \pm j0.99348681$	$-0.04926573 \pm j0.89623740$	$-0.07673317 \pm j0.71125803$	$-0.09668943 \pm j0.45665576$	$-0.10718106 \pm j0.15735285$

n	3-dB Ripple ($\varepsilon = 0.99762835$)				
1	-1.00237729				
2	$-0.32244983 \pm j0.77715757$				
3	$-0.14931010 \pm j0.90381443$	-0.29862021			
4	$-0.08517040 \pm j0.94648443$	$-0.20561953 \pm j0.39204669$			
5	$-0.05485987 \pm j0.96592748$	$-0.14362501 \pm j0.59697601$	-0.17753027		
6	$-0.03822951 \pm j0.97640602$	$-0.10444497 \pm j0.71477881$	$-0.14267448 \pm j0.26162720$		
7	$-0.02814564 \pm j0.98269568$	$-0.07886234 \pm j0.78806075$	$-0.11395938 \pm j0.43734072$	-0.12648537	
8	$-0.02157816 \pm j0.98676635$	$-0.06144939 \pm j0.83654012$	$-0.09196552 \pm j0.55895824$	$-0.10848072 \pm j0.19628003$	
9	$-0.01706520 \pm j0.98955191$	$-0.04913728 \pm j0.87019734$	$-0.07528269 \pm j0.64588414$	$-0.09234789 \pm j0.34366777$	-0.09827457
10	$-0.01383196 \pm j0.99154176$	$-0.04014192 \pm j0.89448274$	$-0.06252250 \pm j0.70986552$	$-0.07878295 \pm j0.45576172$	$-0.08733157 \pm j0.15704479$

Table 2.5 Coefficients of the Polynomial $p(s) = s^n + b_{n-1}s^{n-1} + \cdots + b_1 s + b_0$

0.1-dB Ripple ($\varepsilon = 0.15262042$)

n	b_0	b_1	b_2	b_3	b_4	b_5	b_6	b_7	b_8	b_9
1	6.55220322									
2	3.31403708	2.37235625								
3	1.63805080	2.62949486	1.93881142							
4	0.82850927	2.02550052	2.62679762	1.80377250						
5	0.40951270	1.43555791	2.39695895	2.77070415	1.74396339					
6	0.20712732	0.90176006	2.04784060	2.77905025	2.96575608	1.71216592				
7	0.10237818	0.56178554	1.48293374	2.70514436	3.16924598	3.18350446	1.69322441			
8	0.05178183	0.32643144	1.06662645	2.15924064	3.41845152	3.56476973	3.41291899	1.68102289		
9	0.02559454	0.19176027	0.69421123	1.73411961	2.93387298	4.19161066	3.96384487	3.64896144	1.67269928	
10	0.01294546	0.10703398	0.45721609	1.22966377	2.57903464	3.80850443	5.02617707	4.36536964	3.88905473	1.66676617

0.5-dB Ripple ($\varepsilon = 0.34931140$)

n	b_0	b_1	b_2	b_3	b_4	b_5	b_6	b_7	b_8	b_9
1	2.86277516									
2	1.51620263	1.42562451								
3	0.71569379	1.53489546	1.25291297							
4	0.37905066	1.02545528	1.71686621	1.19738566						
5	0.17892345	0.75251811	1.30957474	1.93736749	1.17249093					
6	0.09476266	0.43236692	1.17186133	1.58976350	2.17184462	1.15917611				
7	0.04473086	0.28207223	0.75565110	1.64790293	1.86940791	2.41265096	1.15121758			
8	0.02369067	0.15254444	0.57356040	1.14858937	2.18401538	2.14921726	2.65674981	1.14608011		
9	0.01118272	0.09411978	0.34081930	0.98361988	1.61138805	2.78149904	2.42932969	2.90273369	1.14257051	
10	0.00592267	0.04928548	0.23726885	0.62696891	1.52743068	2.14423722	3.44092676	2.70974148	3.14987570	1.14006640

Table 2.5 (continued)

				1-dB Ripple ($\varepsilon=0.50884714$)						
n	b_0	b_1	b_2	b_3	b_4	b_5	b_6	b_7	b_8	b_9
1	1.96522673									
2	1.10251033	1.09773433								
3	0.49130668	1.23840917	0.98834121							
4	0.27562758	0.74261937	1.45392476	0.95281138						
5	0.12282667	0.58053415	0.97439607	1.68881598	0.93682013					
6	0.06890690	0.30708064	0.93934553	1.20214039	1.93082492	0.92825096				
7	0.03070667	0.21367139	0.54861981	1.35754480	1.42879431	2.17607847	0.92312347			
8	0.01722672	0.10734473	0.44782572	0.84682432	1.83690238	1.65515567	2.42302642	0.91981131		
9	0.00767667	0.07060479	0.24418637	0.78631094	1.20160717	2.37811881	1.88147976	2.67094683	0.91754763	
10	0.00430668	0.03449708	0.18245121	0.45538923	1.24449142	1.61298557	2.98150939	2.10785235	2.91946571	0.91593199

				1.5-dB Ripple ($\varepsilon=0.64229086$)						
n	b_0	b_1	b_2	b_3	b_4	b_5	b_6	b_7	b_8	b_9
1	1.55692704									
2	0.92520563	0.92217745								
3	0.38923176	1.10298881	0.84022474							
4	0.23130141	0.60470214	1.33087103	0.81347530						
5	0.09730794	0.50419031	0.80441337	1.57113155	0.80141319					
6	0.05782535	0.24758513	0.83401695	1.00055677	1.81596761	0.79494354				
7	0.02432698	0.18365019	0.44733249	1.22429494	1.19561450	2.06289611	0.79107030			
8	0.01445634	0.08613897	0.39173725	0.69590812	1.67617801	1.39030856	2.31091937	0.78856753		
9	0.00608175	0.06034495	0.19776813	0.69725472	0.99316759	2.19012821	1.58489077	2.55957170	0.78685666	
10	0.00361408	0.02760645	0.15821236	0.37141240	1.11564949	1.33908548	2.76635551	1.77945744	2.80861148	0.78563538

Table 2.5 (continued)

					2-dB Ripple ($\varepsilon = 0.76478310$)					
n	b_0	b_1	b_2	b_3	b_4	b_5	b_6	b_7	b_8	b_9
1	1.30756027									
2	0.82306043	0.80381643								
3	0.32689007	1.02219034	0.73782158							
4	0.20576511	0.51679810	1.25648193	0.71621496						
5	0.08172252	0.45934912	0.69347696	1.49954327	0.70646057					
6	0.05144128	0.21027056	0.77146177	0.86701492	1.74585875	0.70122571				
7	0.02043063	0.16612635	0.38263808	1.14459657	1.03954580	1.99366532	0.69809071			
8	0.01286032	0.07293732	0.35870428	0.59822139	1.57958072	1.21171208	2.24225293	0.69606455		
9	0.00510766	0.05437558	0.16844729	0.64446774	0.85686481	2.07674793	1.38374646	2.49128967	0.69467931	
10	0.00321508	0.02333474	0.14400571	0.31775596	1.03891044	1.15852866	2.63625070	1.55574245	2.74060318	0.69369039

					3-dB Ripple ($\varepsilon = 0.99762835$)					
n	b_0	b_1	b_2	b_3	b_4	b_5	b_6	b_7	b_8	b_9
1	1.00237729									
2	0.70794778	0.64489965								
3	0.25059432	0.92834806	0.59724042							
4	0.17698695	0.40476795	1.16911757	0.58157986						
5	0.06264858	0.40796631	0.54893711	1.41502514	0.57450003					
6	0.04424674	0.16342991	0.69909774	0.69060980	1.66284806	0.57069793				
7	0.01566215	0.14615300	0.30001666	1.05184481	0.83144115	1.91155070	0.56842010			
8	0.01106168	0.05648135	0.32076457	0.47189898	1.46669900	0.97194732	2.16071478	0.56694758		
9	0.00391554	0.04759081	0.13138977	0.58350569	0.67893051	1.94386024	1.11232209	2.41014443	0.56594069	
10	0.00276542	0.01803133	0.12775604	0.24920426	0.94992084	0.92106589	2.48342053	1.25264670	2.65973784	0.56522179

Table 2.6 Factors of Chebyshev Polynomial $p(s)$

n	0.1-dB Ripple ($\varepsilon = 0.15262042$)
1	$s+6.55220322$
2	$s^2+2.37235625s+3.31403708$
3	$(s+0.96940571)(s^2+0.96940571s+1.68974743)$
4	$(s^2+0.52831273s+1.33003138)(s^2+1.27545977s+0.62292460)$
5	$(s+0.53891432)(s^2+0.33306737s+1.19493715)(s^2+0.87198169s+0.63592015)$
6	$(s^2+0.22938674s+1.12938678)(s^2+0.62669622s+0.69637408$ $(s^2+0.85608296s+0.26336138)$
7	$(s+37677788)(s^2+0.16768193s+1.09244600)(s^2+0.46983433s+0.75322204)$ $(s^2+0.67893028s+0.33021667)$
8	$(s^2+0.12796025s+1.06949182)(s^2+0.36439996s+0.79889377)$ $(s^2+0.54536308s+0.41621034)(s^2+0.64329961s+0.14561229)$
9	$(s+0.29046118)(s^2+0.10087611s+1.05421401)(s^2+0.29046118s+0.83436770)$ $(s^2+0.44501235s+0.49754361)(s^2+0.54588846s+0.20134548)$
10	$(s^2+0.08157734s+1.04351344)(s^2+0.23674667s+0.86187780)$ $(s^2+0.36874158s+0.56798518)(s^2+0.46464150s+0.27409255)$ $(s^2+0.51505907s+0.09245692)$

n	0.5-dB Ripple ($\varepsilon = 0.34931140$)
1	$s+2.86277516$
2	$s^2+1.42562451s+1.51620263$
3	$(s+0.62645649)(s^2+0.62645649s+1.14244773)$
4	$(s^2+0.35070614s+1.06351864)(s^2+0.84667952s+0.35641186)$
5	$(s+0.36231962)(s^2+0.22392584s+1.03578401)(s^2+0.58624547s+0.47676701)$
6	$(s^2+0.15530015s+1.02302281)(s^2+0.42428790s+0.59001011)$ $(s^2+0.57958805s+0.15699741)$
7	$(s+0.25617001)(s^2+0.11400638s+1.01610751)(s^2+0.31943878+0.67688354)$ $(s^2+0.46160241s+0.25387817)$
8	$(s^2+0.08724015s+1.01193187)(s^2+0.24843894s+0.74133382)$ $(s^2+0.37181515s+0.35865039)(s^2+0.438587s+0.08805234)$
9	$(s+0.19840529)(s^2+0.06890543s+1.00921097)(s^2+0.19840529s+0.78936466)$ $(s^2+0.30397454s+0.45254057)(s^2+0.37287997s+0.15634244)$
10	$(s^2+0.05579882s+1.00733544)(s^2+0.16193449s+0.82569981)$ $(s^2+0.25221888s+0.53180718)(s^2+0.31781432s+0.23791455)$ $(s^2+0.35229989s+0.05627892)$

n	1-dB Ripple ($\varepsilon = 0.50884714$)
1	$s+1.96522673$
2	$s^2+1.09773433s+1.10251033$
3	$(s+0.49417060)(s^2+0.49417060s+0.99420459)$
4	$(s^2+0.27907199s+0.98650488)(s^2+0.67373939s+0.27939809)$
5	$(s+0.28949334)(s^2+0.17891672s+0.98831489)(s^2+0.46841007s+0.42929790$
6	$(s^2+0.12436205s+0.99073230)(s^2+0.33976343s+0.55771960)$ $(s^2+0.46412548s+0.12470689)$

Table 2.6 (continued)

n	1-dB Ripple ($\varepsilon = 0.50884714$)
7	$(s+0.20541430)(s^2+0.09141796s+0.99267947)(s^2+0.25614744s+0.65345550)$
	$(s^2+0.37014377s+0.23045013)$
8	$(s^2+0.07001647s+0.99414074)(s^2+0.19939003s+0.72354268)$
	$(s^2+0.29840826s+0.34085925)(s^2+0.35199655s+0.07026120)$
9	$(s+0.15933047)(s^2+0.05533489s+0.99523251)(s^2+0.15933047s+0.77538620)$
	$(s^2+0.24410845s+0.43856211)(s^2+0.29944334s+0.14236398)$
10	$(s^2+0.04482890s+0.99605837)(s^2+0.13009854s+0.81442274)$
	$(s^2+0.20263323s+0.52053011)(s^2+0.25533277s+0.22663749)$
	$(s^2+0.28303855s+0.04500185)$

n	1.5-dB Ripple ($\varepsilon = 0.64229086$)
1	$s+1.55692704$
2	$s^2+0.92217745s+0.92520563$
3	$(s+0.42011237)(s^2+0.42011237s+0.92649440)$
4	$(s^2+0.23826140s+0.95046327)(s^2+0.57521390s+0.24335649)$
5	$(s+0.24765030)(s^2+0.15305630s+0.96583917)(s^2+0.40070660s+0.40682217)$
6	$(s^2+0.10650224s+0.97534434)(s^2+0.29096953s+0.54233163)$
	$(s^2+0.39747177s+0.10931893)$
7	$(s+0.17602970)(s^2+0.07834059s+0.98147089)(s^2+0.21950545+0.64224692)$
	$(s^2+0.31719456s+0.21924156)$
8	$(s^2+0.06002613s+0.98560709)(s^2+0.17093995s+0.71500904)$
	$(s^2+0.25582972s+0.33232561)(s^2+0.30177173s+0.06172756)$
9	$(s+0.13663622)(s^2+0.04745326s+0.98851577)(s^2+0.13663622s+0.76866946)$
	$(s^2+0.20933884s+0.43184537)(s^2+0.25679210s+0.13564724)$
10	$(s^2+0.03845173s+0.99063278)(s^2+0.11159127s+0.80899715)$
	$(s^2+0.17380748s+0.51510452)(s^2+0.21901021s+0.22121189)$
	$(s^2+0.24277468s+0.03957626)$

n	2-dB Ripple ($\varepsilon = 0.76478310$)
1	$s+1.30756027$
2	$s^2+0.80381643s+0.82306043$
3	$(s+0.36891079)(s^2+0.36891079s+0.88609517)$
4	$(s^2+0.20977450s+0.92867521)(s^2+0.50644045s+0.22156843)$
5	$(s+0.21830832)(s^2+0.13492196s+0.95216702)(s^2+0.35323028s+0.39315003)$
6	$(s^2+0.09394643s+0.96595153)(s^2+0.25666642s+0.53293883)$
	$(s^2+0.35061285s+0.09992612)$
7	$(s+0.15533980)(s^2+0.06913271s+0.97461489)(s^2+0.19370556s+0.63539092)$
	$(s^2+0.27991264s+0.21238555)$
8	$(s^2+0.05298476s+0.98038017)(s^2+0.15088783s+0.70978212)$
	$(s^2+0.22581959s+0.32709869)(s^2+0.26637237s+0.05650064)$
9	$(s+0.12062980)(s^2+0.04189429s+0.98439786)(s^2+0.12062980s+0.76455155)$
	$(s^2+0.18481557s+0.42772746)(s^2+0.22670986s+0.13152933)$

Table 2.6 (continued)

n	2-dB Ripple ($\varepsilon=0.76478310$)
10	$(s^2+0.03395162s+0.98730422)(s^2+0.09853145s+0.80566858)$ $(s^2+0.15346633s+0.51177596)(s^2+0.19337886s+0.21788333)$ $(s^2+0.21436212s+0.03624770)$

n	3-dB Ripple ($\varepsilon=0.99762835$)
1	$s+1.00237729$
2	$s^2+0.64489965s+0.70794778$
3	$(s+0.29862021)(s^2+0.29862021s+0.83917403)$
4	$(s^2+0.17034080s+0.90308678)(s^2+0.41123906s+0.19598000)$
5	$(s+0.17753027)(s^2+0.10971974s+0.93602549)(s^2+0.28725001s+0.37700850)$
6	$(s^2+0.07645903s+0.95483021)(s^2+0.20888994s+0.52181750)$ $(s^2+0.28534897s+0.08880480)$
7	$(s+0.12648537)(s^2+0.05629129s+0.96648298)(s^2+0.15772468s+0.62725902$ $(s^2+0.22791876s+0.20425365)$
8	$(s^2+0.04315631s+0.97417345)(s^2+0.12289879s+0.70357540)$ $(s^2+0.18393103s+0.32089197)(s^2+0.21696145s+0.05029392)$
9	$(s+0.09827457)(s^2+0.03413040s+0.97950420)(s^2+0.09827457s+0.75965789)$ $(s^2+0.15056538s+0.42283380)(s^2+0.18469578s+0.12663567)$
10	$(s^2+0.02766392s+0.98334638)(s^2+0.08028383s+0.80171075)$ $(s^2+0.12504500s+0.50781813)(s^2+0.15756589s+0.21392550)$ $(s^2+0.17466314s+0.03228987)$

confirming Table 2.4 for $n=4$. The desired transfer function becomes

$$H(y)=\frac{\hat{H}_0^{1/2}}{(y-y_1)(y-y_2)(y-y_3)(y-y_4)}$$

$$=\frac{0.81902540}{(y^2+0.52831273y+1.33003138)(y^2+1.27545977y+0.62292460)}$$

$$=\frac{0.81902540}{y^4+1.80377250y^3+2.62679762y^2+2.02550052y+0.82850927} \tag{2.63}$$

The constant $\hat{H}_0^{1/2}$ is chosen to be 0.81902540 so that $H(0)=1/\sqrt{1+\varepsilon^2}=0.98855309$. The coefficients in the denominator of (2.63) confirm those given in Table 2.5 for $n=4$ with 0.1-dB ripple.

2.3 THE INVERSE CHEBYSHEV RESPONSE

A filter whose transfer function has no finite zeros is called an *all-pole filter*. A filter with a Butterworth or Chebyshev response is an all-pole filter. In this section, we

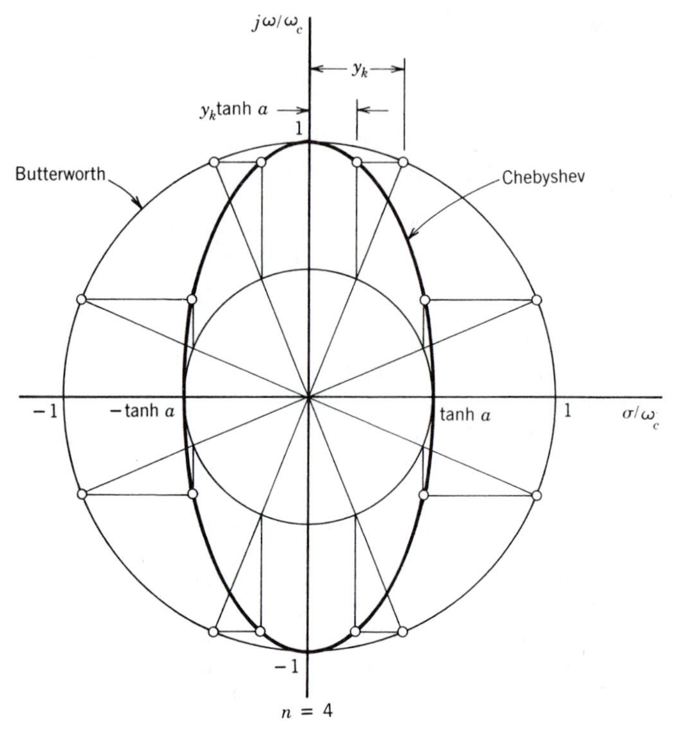

Figure 2.10

introduce a filter whose transfer function has finite zeros and that exhibits a maximally-flat passband response and an equal-ripple stopband response. The associated approximation is known as the inverse Chebyshev response.

The *inverse Chebyshev response of order n* is described by the equation

$$|H(j\omega)|^2 = \frac{H_0 \varepsilon^2 C_n^2(\omega_c/\omega)}{1+\varepsilon^2 C_n^2(\omega_c/\omega)} \tag{2.64}$$

where the parameters H_0, ω_c, ε, and $C_n(\omega)$ are defined the same as in (2.23) for the Chebyshev response. To see how the inverse Chebyshev response will give a maximally-flat response in the passband and an equal-ripple response in the stopband, we return to the Chebyshev response

$$|\hat{H}(j\omega)|^2 = \frac{H_0}{1+\varepsilon^2 C_n^2(\omega/\omega_c)} \tag{2.65}$$

which is plotted in Figure 2.11(a). In (b) we subtract this function from H_0:

$$H_0 - |\hat{H}(j\omega)|^2 = \frac{H_0 \varepsilon^2 C_n^2(\omega/\omega_c)}{1+\varepsilon^2 C_n^2(\omega/\omega_c)} \tag{2.66}$$

Finally, we invert frequency by replacing ω by ω_c^2/ω, yielding the inverse Chebyshev response (2.64) as shown in (c).

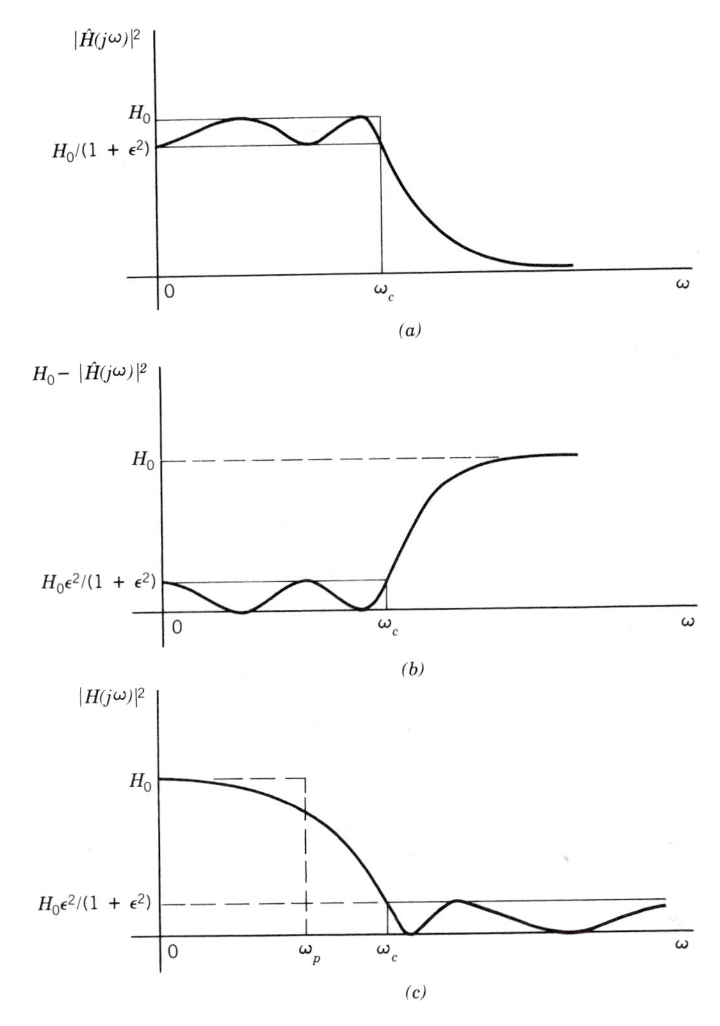

Figure 2.11

As in the Chebyshev response, apart from the constant H_0 the inverse Chebyshev response is determined by two parameters n and ε. At $\omega = \omega_c$, $C_n(1) = 1$ for all n. Then from (2.64)

$$|H(j\omega_c)|^2 = \frac{H_0 \varepsilon^2}{1 + \varepsilon^2} \tag{2.67}$$

As can be seen from Figure 2.11(*c*), the ripple band extends from $\omega = \omega_c$ to $\omega = \infty$ and the ripples oscillate between 0 and $|H(j\omega_c)|^2$. In terms of decibels, the minimum value of the attenuation in the stopband is, from (2.67),

$$\alpha_{\min} = -10 \log[|H(j\omega_c)|^2 / H_0] = 10 \log\left(1 + \frac{1}{\varepsilon^2}\right) \text{ dB} \tag{2.68}$$

Solving this equation for ε gives

$$\varepsilon = (10^{\alpha_{\min}/10} - 1)^{-1/2} \tag{2.69}$$

Thus, with minimum attenuation $\alpha_{\min}$ specified in the stopband, (2.69) fixes the value for ε.

To observe the maximally-flat characteristic in the passband, we note from (2.33) that the asymptotic behavior of the Chebyshev polynomial for large ω is $C_n(\omega) \approx 2^{n-1}\omega^n$, $\omega \gg 1$. Hence, we have

$$C_n(\omega_c/\omega) \approx 2^{n-1}(\omega_c/\omega)^n, \qquad \omega \ll \omega_c \tag{2.70}$$

If we substitute this approximation into (2.64), we obtain

$$|H(j\omega)|^2 \approx \frac{H_0\varepsilon^2 2^{2n-2}(\omega_c/\omega)^{2n}}{1+\varepsilon^2 2^{2n-2}(\omega_c/\omega)^{2n}}$$

$$= \frac{H_0}{1+\omega^{2n}/(2^{n-1}\varepsilon\omega_c^n)^2}, \qquad \omega \ll \omega_c \tag{2.71}$$

Equation (2.71) has the same form as that for the Butterworth response (2.1). Therefore, for small ω the inverse Chebyshev response (2.64) is maximally flat. If ω_p as indicated in Figure 2.11(*c*) is the end of the passband, where the attenuation is largest for $\omega \leqq \omega_p$, then from (2.64) the maximum attenuation $\alpha_{\max}$ in the passband is given by

$$\alpha_{\max} = -10\log[|H(j\omega_p)|^2/H_0] = 10\log\left[1+\frac{1}{\varepsilon^2 C_n^2(\omega_c/\omega_p)}\right] \tag{2.72}$$

which can be rewritten as

$$C_n(\omega_c/\omega_p) = \cosh\left(n\cosh^{-1}\frac{\omega_c}{\omega_p}\right) = \frac{1}{\varepsilon\sqrt{10^{\alpha_{\max}/10}-1}} \tag{2.73}$$

Solving this equation for n gives

$$n = \frac{\cosh^{-1}[\varepsilon^{-1}(10^{\alpha_{\max}/10}-1)^{-1/2}]}{\cosh^{-1}(\omega_c/\omega_p)} \tag{2.74}$$

Example 2.5 Determine the order of the inverse Chebyshev response that realizes the following specifications:

(i) The passband extends from $\omega = 0$ to $\omega = 250$ rad/s.
(ii) The passband attenuation should not exceed 0.1 dB.
(iii) For frequencies $\omega \geqq 2000$ rad/s, the response must be attenuated by at least 60 dB.

From the above specifications, we can make the following identifications:

$$\omega_p = 250 \text{ rad/s}, \qquad \omega_c = 2000 \text{ rad/s}, \qquad \alpha_{\max} = 0.1 \text{ dB}, \qquad \alpha_{\min} = 60 \text{ dB}$$

To determine ε, we use (2.69), yielding

$$\varepsilon = 0.001$$

The required value of n is found using (2.74):

$$n=\frac{\cosh^{-1}[10^3(10^{0.01}-1)^{-1/2}]}{\cosh^{-1}8}=3.424 \tag{2.75}$$

which is rounded up to $n=4$. This number is to be compared to that found in (2.46b) by the Chebyshev response. They are seen to be identical. There is no surprise since the two problems have the same specifications. If we substitute (2.41) in (2.45) and (2.69) in (2.74), there result two identical equations except that one of the denominators is $\cosh^{-1}\omega_s/\omega_c$ and the other is $\cosh^{-1}\omega_c/\omega_p$. In the Chebyshev response, the ratio of the stopband edge frequency ω_s to the passband end frequency ω_c is ω_s/ω_c, while in the inverse Chebyshev response this ratio is ω_c/ω_p. Therefore, they represent the same thing. Our conclusion is that the Chebyshev and the inverse Chebyshev responses require the same value of n to satisfy the general specifications α_{max}, α_{min}, ω_p, and ω_s, as depicted in Figure 2.3.

2.3.1 Zeros and Poles of the Inverse Chebyshev Function

Having found the values of n and ε that will satisfy the specifications, we now proceed to determine the location of the zeros and poles of the inverse Chebyshev function. As before, we replace ω by $-js$ in (2.64) and obtain

$$H(y)H(-y)=\frac{H_0\varepsilon^2C_n^2(j/y)}{1+\varepsilon^2C_n^2(j/y)} \tag{2.76}$$

where $y=s/\omega_c$. The zeros of this function are defined by the roots of the equation

$$C_n(j/y_{kz})=0 \tag{2.77}$$

or

$$\cos\left(n\cos^{-1}\frac{j}{y_{kz}}\right)=0 \tag{2.78}$$

This requires that

$$\cos^{-1}\frac{j}{y_{kz}}=\frac{k\pi}{2n}, \qquad k \text{ odd} \tag{2.79}$$

or

$$y_{kz}=j\sec\frac{k\pi}{2n}, \qquad k=1, 3, \ldots, 2n-1 \tag{2.80}$$

The poles of the inverse Chebyshev response are given by the roots of the equation

$$1+\varepsilon^2C_n^2(j/y)=0 \tag{2.81}$$

Equation (2.81) has the same form as (2.49), except that $1/y$ has replaced $-y$. Since we have already determined the location of the Chebyshev poles in Section 2.2.3, the poles of the inverse Chebyshev response may be determined by first finding the

Chebyshev poles using (2.57), and then replacing y_k by $y_{kp} = -1/y_k$ to obtain the inverse Chebyshev poles y_{kp}. However, the poles of the inverse Chebyshev response will not lie on an ellipse. To see this, let $y_k = r_k e^{j\theta_k}$. The equation for an ellipse in polar coordinates for the Chebyshev poles is obtained from (2.59) as

$$\frac{\cos^2 \theta_k}{\sinh^2 a} + \frac{\sin^2 \theta_k}{\cosh^2 a} = \frac{1}{r_k^2} \tag{2.82}$$

where $\sigma_k = r_k \cos \theta_k$ and $\omega_k = r_k \sin \theta_k$. Thus when $y_k \to -1/y_k$ we have $r_k \to 1/r_k$ and $\theta_k \to \pi - \theta_k$ so that the poles for the inverse Chebyshev fall on the curve

$$\frac{\cos^2 \theta_k}{\sinh^2 a} + \frac{\sin^2 \theta_k}{\cosh^2 a} = r_k^2 \tag{2.83}$$

which is not the equation for an ellipse. Nevertheless, the curve for this equation does look similar to an ellipse.

To obtain the normalized transfer function $H(y)$, we may identify the numerator and denominator of (2.76) as

$$H(y)H(-y) = H_0 \frac{f(y)f(-y)}{g(y)g(-y)} \tag{2.84}$$

where from (2.80) and (2.62) with $y_{kp} = 1/y_k$ $(k = 1, 2, \ldots, n)$

$$f(y) = (y - y_{1z})(y - y_{3z}) \ldots (y - y_{(2n-1)z}) \tag{2.85}$$

$$g(y) = (y - y_{1p})(y - y_{2p}) \ldots (y - y_{np}) \tag{2.86}$$

giving

$$H(y) = H_0^{1/2} \frac{f(y)}{g(y)} \tag{2.87}$$

Example 2.6 Determine the normalized transfer function that realizes the fourth-order inverse Chebyshev response with at least 60-dB attenuation in the stopband.

From (2.80) the zeros of the normalized transfer function $H(y)$ are

$$y_{1z} = j \sec \pi/8 = j1.08239 \tag{2.88a}$$

$$y_{3z} = j \sec 3\pi/8 = j2.61313 \tag{2.88b}$$

$$y_{5z} = j \sec 5\pi/8 = -j2.61313 \tag{2.88c}$$

$$y_{7z} = j \sec 7\pi/8 = -j1.08239 \tag{2.88d}$$

giving

$$f(y) = (y^2 + 1.17157)(y^2 + 6.82845) = y^4 + 8y^2 + 8 \tag{2.89}$$

The minimum 60-dB attenuation in the stopband corresponds to, using (2.69),

$$\varepsilon = (10^6 - 1)^{-1/2} = 0.001 \tag{2.90}$$

We first compute the Chebyshev poles y_k using (2.62), as follows:

$$y_1 = -\sinh a \sin \pi/8 + j \cosh a \cos \pi/8 = -1.25097 + j3.15825 \tag{2.91a}$$

$$y_2 = -\sinh a \sin 3\pi/8 + j \cosh a \cos 3\pi/8 = -3.02010 + j1.30819 \tag{2.91b}$$

$$y_3 = -\sinh a \sin 5\pi/8 + j \cosh a \cos 5\pi/8 = -3.02010 - j1.30819 \tag{2.91c}$$

$$y_4 = -\sinh a \sin 7\pi/8 + j \cosh a \cos 7\pi/8 = -1.25097 - j3.15825 \tag{2.91d}$$

where $a = 1.90023$. For the inverse Chebyshev poles y_{kp}, Re $y_{kp} < 0$, we set

$$y_{1p} = 1/y_1 = -0.10841 - j0.27369 = 1/\bar{y}_4 = \bar{y}_{4p} \tag{2.92a}$$

$$y_{2p} = 1/y_2 = -0.27880 - j0.12077 = 1/\bar{y}_3 = \bar{y}_{3p} \tag{2.92b}$$

where the bar denotes the complex conjugate, obtaining

$$\begin{aligned} g(y) &= (y - y_{1p})(y - y_{2p})(y - y_{3p})(y - y_{4p}) \\ &= (y^2 + 0.21682y + 0.08666)(y^2 + 0.55760y + 0.09231) \\ &= y^4 + 0.77442y^3 + 0.29987y^2 + 0.06834y + 0.00800 \end{aligned} \tag{2.93}$$

The desired transfer function is given by

$$H(y) = H_0^{1/2} \frac{y^4 + 8y^2 + 8}{y^4 + 0.77442y^3 + 0.29987y^2 + 0.06834y + 0.00800} \tag{2.94}$$

2.3.2 **Comparisons of Inverse Chebyshev Response with Other Responses**

In this section, we compare the inverse Chebyshev response with the Butterworth and Chebyshev responses. For convenience in comparing the formulas for the value of n required for a preassigned attenuation in the passband and stopband, we choose the parameters in these responses such that at $\omega = \omega_c$ their value all becomes $1/(1+\varepsilon^2)$, as follows:

$$|H(j\omega)|^2 = \frac{H_0}{1 + \varepsilon^2(\omega/\omega_c)^{2n}} \tag{2.95}$$

$$|H(j\omega)|^2 = \frac{H_0}{1 + \varepsilon^2 C_n^2(\omega/\omega_c)} \tag{2.96}$$

$$|H(j\omega)|^2 = \frac{\dfrac{H_0}{\varepsilon^2 C_n^2(\omega_s/\omega_c)} C_n^2(\omega_s/\omega)}{1 + \dfrac{1}{\varepsilon^2 C_n^2(\omega_s/\omega_c)} C_n^2(\omega_s/\omega)} \tag{2.97}$$

For a fixed value of n, sketches of these functions are presented in Figure 2.12. From these sketches it is clear that all the responses have the same passband extending from $\omega = 0$ to $\omega = \omega_c$. Within the passband, the magnitude varies between H_0 and

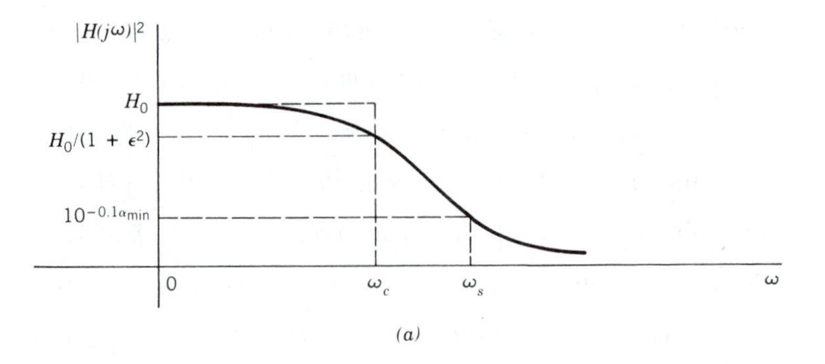

(a)

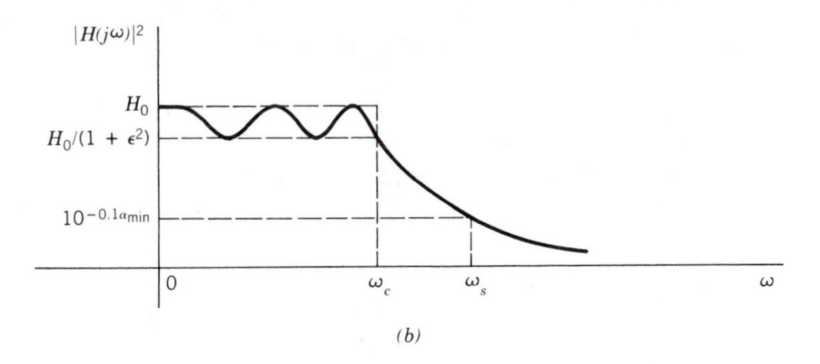

(b)

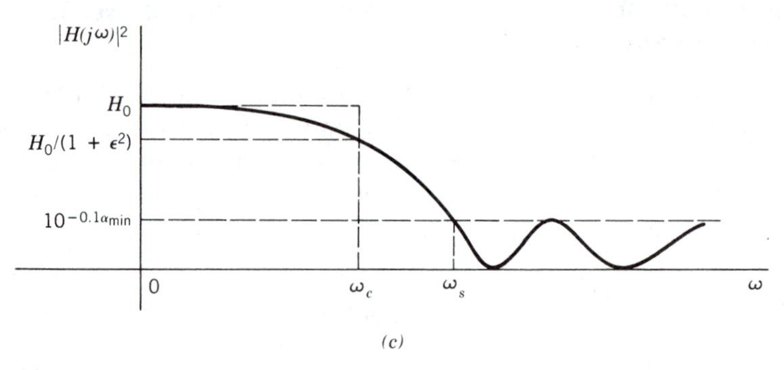

(c)

Figure 2.12

$H_0/(1+\varepsilon^2)$. Therefore, the maximum permissible variation in the passband is determined by

$$\alpha_{\max}=10\log(1+\varepsilon^2)\ \text{dB} \tag{2.98}$$

The stopband extends from $\omega=\omega_s$ to $\omega=\infty$. If $\alpha_{\min}$ is the minimum attenuation acceptable in the stopband, we see that for the Butterworth response (2.95) the required order n is

$$\alpha_{\min}=10\log[1+\varepsilon^2(\omega_s/\omega_c)^{2n}] \tag{2.99}$$

Solving for n yields

$$n=\frac{\log(10^{\alpha_{\min}/10}-1)^{1/2}-\log\varepsilon}{\log(\omega_s/\omega_c)}\approx\frac{\alpha_{\min}/20-\log\varepsilon}{\log(\omega_s/\omega_c)} \tag{2.100}$$

For the Chebyshev response, the value of n is given by (2.45) and is repeated below:

$$n=\frac{\cosh^{-1}[\varepsilon^{-1}(10^{\alpha_{\min}/10}-1)^{1/2}]}{\cosh^{-1}(\omega_s/\omega_c)}\approx\frac{\cosh^{-1}(10^{\alpha_{\min}/20}/\varepsilon)}{\cosh^{-1}(\omega_s/\omega_c)} \tag{2.101}$$

Comparison of (2.100) with (2.101) shows that for the same specifications $\alpha_{\max}$, $\alpha_{\min}$, ω_c, and ω_s, the value of n for the Chebyshev response is considerably less than that for the Butterworth.

Now considering the inverse Chebyshev response, we set $\omega=\omega_s$ in (2.97) and obtain

$$\alpha_{\min}=-10\log[|H(j\omega_s)|^2/H_0]=10\log[1+\varepsilon^2C_n^2(\omega_s/\omega_c)]$$

Solving this for n yields the desired formula

$$n=\frac{\cosh^{-1}[\varepsilon^{-1}(10^{\alpha_{\min}/10}-1)^{1/2}]}{\cosh^{-1}(\omega_s/\omega_c)}\approx\frac{\cosh^{-1}(10^{\alpha_{\min}/20}/\varepsilon)}{\cosh^{-1}(\omega_s/\omega_c)} \tag{2.102}$$

Observe that the above equation is identical with (2.101), a fact that was pointed out in Example 2.5. Even though the Chebyshev and the inverse Chebyshev responses require the same value of n to satisfy the general specifications, the transfer function with an inverse Chebyshev response has finite zeros on the $j\omega$-axis. As a result, the network realization of an inverse Chebyshev response requires more elements, and therefore is more complicated. Nevertheless, the inverse Chebyshev filter is more economical in elements than a Butterworth filter for the same specifications, and is desirable in situations where attenuation is to be distributed over the entire stopband.

Example 2.7 Determine the orders of the Butterworth, Chebyshev, and inverse Chebyshev responses to satisfy the following specifications:

$$\alpha_{\max}=0.1\text{ dB},\qquad \alpha_{\min}=60\text{ dB},\qquad \omega_c=250\text{ rad/s},\qquad \omega_s=2000\text{ rad/s}$$

For the 0.1-dB maximum permissible variation in the passband, the value of ε is found from (2.98) as

$$\varepsilon=\sqrt{10^{\alpha_{\max}/10}-1}=0.15262 \tag{2.103}$$

The order of the Butterworth response is determined from (2.100)

$$n=\frac{\log(10^6-1)^{1/2}-\log 0.15262}{\log(2000/250)}=4.226 \tag{2.104}$$

which is rounded up to $n=5$. For the Chebyshev and inverse Chebyshev responses, we use (2.101) or (2.102), obtaining

$$n=\frac{\cosh^{-1}[(10^6-1)^{1/2}/0.15262]}{\cosh^{-1}(2000/250)}=3.424 \tag{2.105}$$

Thus, we choose $n=4$.

Suppose that we lower the edge of the stopband from 2000 rad/s to 500 rad/s, everything else being the same. Then for the Butterworth response, we have

$$n = 12.678 \tag{2.106}$$

rounded up to $n=13$. For the Chebyshev or the inverse Chebyshev response, the value of n is found to be from (2.101)

$$n = 7.199 \tag{2.107}$$

which is rounded up to $n=8$. Clearly, the latter is more economical in elements than the Butterworth for the same specifications.

2.4 THE BESSEL–THOMSON RESPONSE

Unlike the previous three functions, which approximate the magnitude of a transfer function, in this section we study an approximating function that will give maximally-flat time delay. Delay filters are frequently encountered in the design of communication systems such as in the transmission of a signal through a coaxial cable or an optical fiber, especially in digital transmission, where delay, being insensitive to human ear, plays a vital role in performance.

Let $v_1(t)$ be the input signal to a network N, and let $v_2(t)$ be the output, as depicted in Figure 2.13. If the transmission network is ideal, the output $v_2(t)$ is a delayed replica of the input as shown in (*c*). Mathematically, this requires that

$$v_2(t) = v_1(t - T) \tag{2.108}$$

where T is the delay in seconds. Taking Laplace transform on both sides gives

$$V_2(s) = V_1(s)e^{-sT} \tag{2.109}$$

where $V_1(s)$ and $V_2(s)$ denote the Laplace transforms of $v_1(t)$ and $v_2(t)$, respectively. Then

$$V_2(s)/V_1(s) = H(s) = e^{-sT} \tag{2.110}$$

When $s=j\omega$, the magnitude and phase characteristics become

$$|H(j\omega)| = 1 \tag{2.111}$$

$$\operatorname{Arg} H(j\omega) = -\omega T \tag{2.112}$$

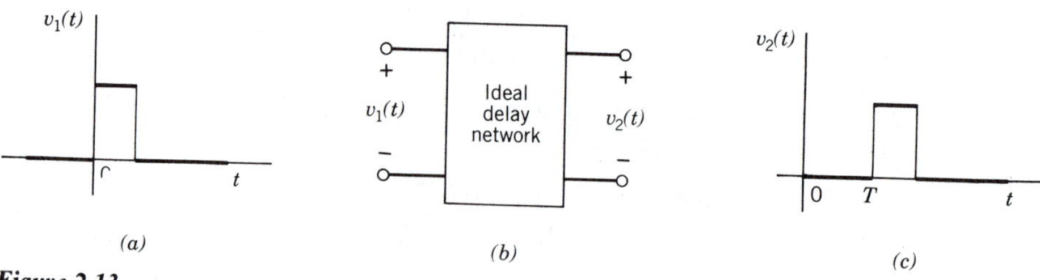

Figure 2.13

If we define the *delay* to be the negative of the derivative of the phase,

$$\text{delay} = -\frac{d \arg H(j\omega)}{d\omega} \tag{2.113}$$

the delay for the ideal transmission system is from (2.112)

$$\text{delay} = T \tag{2.114}$$

Equation (2.113) is also known as the *group delay* or sometimes as *signal delay* or *envelope delay*. In our discussion we will refer to it simply as delay. For the ideal transmission system, the magnitude of its transfer function is constant, being equal to unity for all ω, while the phase is a linear function of ω or equivalently the delay is constant. These are sketched in Figure 2.14.

2.4.1 Maximally-Flat Delay Characteristic

Our objective is to find a function that will approximate constant time delay for as large a range of ω as possible. It turns out that the coefficients of the polynomials used in the transfer function $H(s)$ are closely related to Bessel polynomials, and Thomson† was one of the first to use these polynomials in the approximation. For this reason we shall call the response that results the *Bessel–Thomson response*, although it is frequently known as either Bessel or Thomson response in the literature.

To simplify our notation, we first normalize frequency by letting

$$y = sT \tag{2.115}$$

in (2.110) and then observe that

$$H(y) = e^{-y} = \frac{1}{e^{y}} = \frac{1}{\cosh y + \sinh y} \tag{2.116}$$

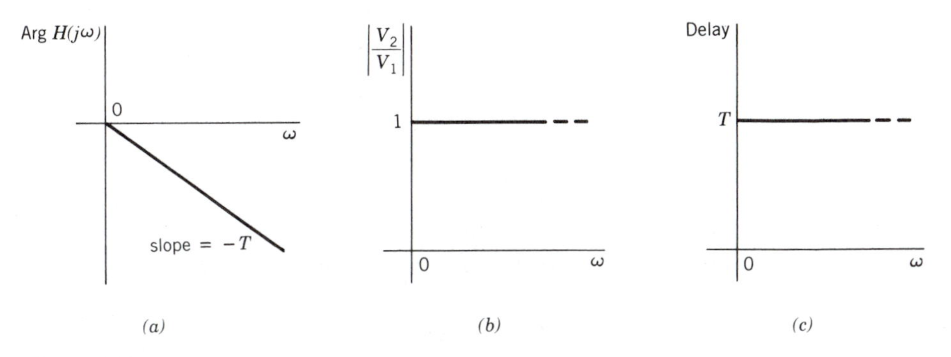

Figure 2.14

†W. E. Thomson, "Delay networks having maximally flat frequency characteristics," *Proc. IEE*, pt. 3, vol. 96, 1949, pp. 487–490.

There are many ways this function can be approximated. One clever method is to rearrange the last equation into the form

$$H(y)=\frac{1/\sinh y}{1+\cosh y/\sinh y}=\frac{1/\sinh y}{1+\coth y} \tag{2.117}$$

The series expansions of the hyperbolic functions are given by

$$\cosh y=1+\frac{y^2}{2!}+\frac{y^4}{4!}+\frac{y^6}{6!}+\cdots \tag{2.118}$$

$$\sinh y=y+\frac{y^3}{3!}+\frac{y^5}{5!}+\frac{y^7}{7!}+\cdots \tag{2.119}$$

Dividing the cosh y expansion by the sinh y expansion, inverting, repeating the division, and continuing this process, we obtain an infinite continued-fraction expansion of the coth y function:

$$\coth y=\frac{1}{y}+\cfrac{1}{\cfrac{3}{y}+\cfrac{1}{\cfrac{5}{y}+\cfrac{1}{\cfrac{7}{y}+\ddots}}} \tag{2.120}$$

The approximation is obtained by truncating this series with the $(2n-1)/y$ term. This results in a quotient of polynomials where the numerator is identified with cosh y and the denominator with sinh y. The sum of the numerator and denominator polynomials is therefore the approximation to e^y in (2.116). For example, for $n=4$, the truncated continued fraction becomes

$$\coth y=\frac{1}{y}+\cfrac{1}{\cfrac{3}{y}+\cfrac{1}{\cfrac{5}{y}+\cfrac{1}{\cfrac{7}{y}}}} \tag{2.121}$$

or

$$\coth y=\frac{y^4+45y^2+105}{10y^3+105y} \tag{2.122}$$

Adding the numerator and denominator yields an approximation function for $H(y)$ shown in (2.116):

$$H(y)=\frac{105}{y^4+10y^3+45y^2+105y+105} \tag{2.123}$$

where the number 105 is introduced in the numerator to ensure that $H(0)=1$. Note that the numerator of (2.122) is not equal to the series expansion (2.118) truncated with the $y^4/4!$ term. Likewise, the denominator of (2.122) is different from that of (2.119) truncating with the $y^3/3!$ term.

In general, after truncating the infinite continued-fraction expansion (2.120) with the $(2n-1)/y$ term, let the general numerator polynomial be M and the denominator be N so that $\coth y$ is approximated by

$$\coth y = \frac{\cosh y}{\sinh y} \approx \frac{M(y)}{N(y)} \tag{2.124}$$

Let us identify M with $\cosh y$ and N with $\sinh y$ so that (2.116) becomes

$$H(y) = \frac{c_0}{M(y) + N(y)} \tag{2.125}$$

where c_0 is introduced so that $H(0) = 1$. By truncating the expansion (2.120) at various values of n, we generate the denominator polynomial

$$B_n(y) = M(y) + N(y) = c_n y^n + c_{n-1} y^{n-1} + \cdots + c_1 y + c_0 \tag{2.126}$$

as follows:

$$\begin{aligned}
B_0(y) &= 1 \\
B_1(y) &= y + 1 \\
B_2(y) &= y^2 + 3y + 3 \\
B_3(y) &= y^3 + 6y^2 + 15y + 15 \\
B_4(y) &= y^4 + 10y^3 + 45y^2 + 105y + 105 \\
B_5(y) &= y^5 + 15y^4 + 105y^3 + 420y^2 + 945y + 945
\end{aligned} \tag{2.127}$$

Polynomials of higher order may be found from the recursion formula

$$B_n = (2n-1)B_{n-1} + y^2 B_{n-2} \tag{2.128}$$

In fact, the coefficients of $B_n(y)$ can be found directly by formula

$$c_k = \frac{(2n-k)!}{2^{n-k}k!(n-k)!}, \qquad k = 0, 1, 2, \ldots, n \tag{2.129}$$

The polynomial $B_n(y)$ is called the *Bessel polynomial of order n*. We shall now check to see how these polynomials approximate the desired phase and so the time delay characteristic. For each choice of n, the transfer function with a Bessel–Thomson response assumes the general form

$$H(y) = B_n(0)/B_n(y) \tag{2.130}$$

the delay of which can be calculated as a function of ωT. For $n = 3$, the transfer function becomes

$$H(y) = \frac{15}{y^3 + 6y^2 + 15y + 15} \tag{2.131}$$

For this function the phase is

$$\operatorname{Arg} H(j\omega T) = -\tan^{-1}\frac{15\omega T - \omega^3 T^3}{15 - 6\omega^2 T^2} \tag{2.132}$$

Differentiating with respect to ωT gives the normalized delay function

$$D(\omega T) = \frac{6(\omega T)^4 + 45(\omega T)^2 + 225}{(\omega T)^6 + 6(\omega T)^4 + 45(\omega T)^2 + 225} \tag{2.133}$$

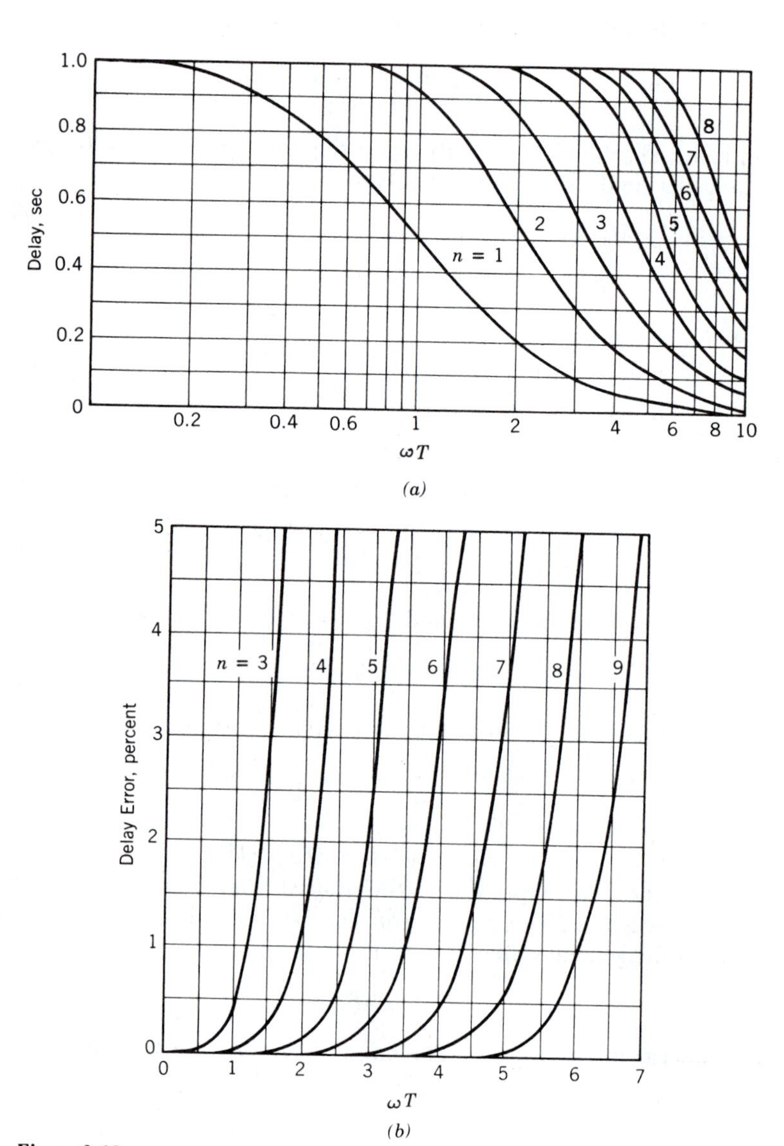

Figure 2.15

A plot of $D(\omega T)$ as a function of the normalized frequency ωT is shown in Figure 2.15(a) with $n=3$. This figure also contains plots for other values of n. Observe that for small ωT, $D(\omega T) \approx 1$ for all n. The range of flatness depends on the values of n: The larger the values of n the larger band of frequencies for flat delay response. Because the first n derivatives of the delay are zero at $\omega=0$, the Bessel–Thomson response is also known as the *maximally-flat delay response.* When specifications are given in terms of delay time T and maximum deviation at a given frequency, the minimum

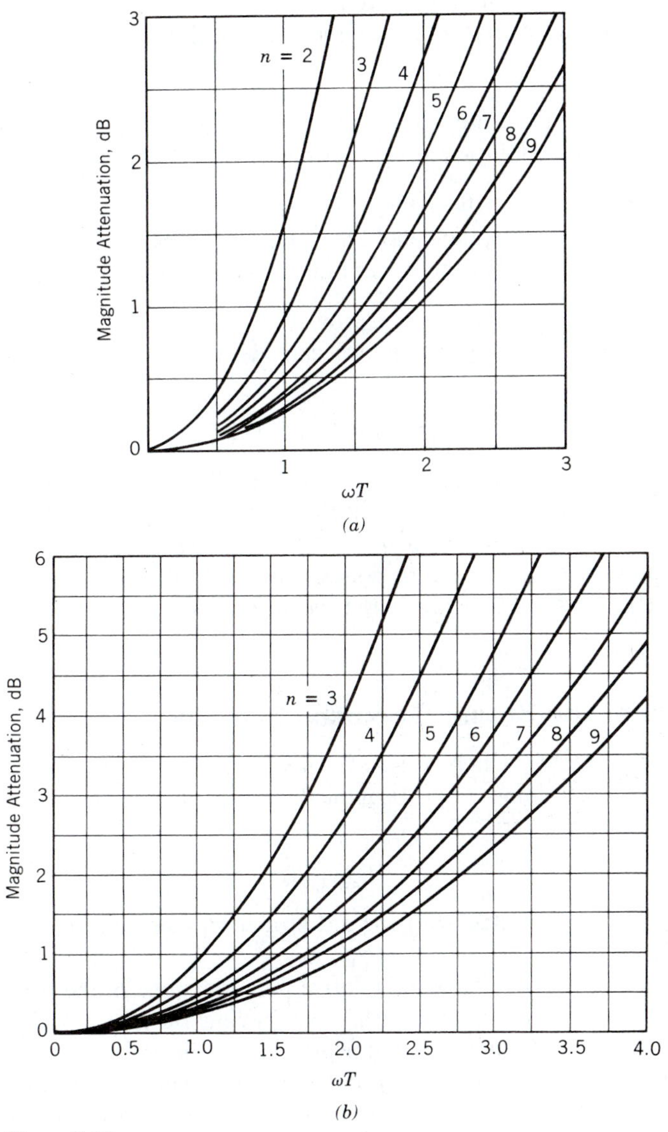

Figure 2.16

order n of the Bessel polynomial required to satisfy the specifications can be determined from Figure 2.15(*b*). This n together with the value of T is sufficient to specify the desired transfer function $H(y)$, from which a network can be synthesized.

So far we have concentrated on the requirement of flatness of the time delay. In some situations, the flatness of the magnitude function $|H(\omega T)|$ is specified in terms of maximum deviation at a given frequency. For this we can plot a family of curves of $|H(\omega T)|$ as functions of ωT. Figure 2.16 gives the magnitude error in decibels as a function of the normalized frequency ωT for various values of n. When specifications are given in terms of delay and magnitude deviations, it is possible to choose a value of n satisfying the more stringent requirement. We illustrate this by the following example.

Example 2.8 Determine the minimum order of the Bessel–Thomson response that realizes the following specifications:

(i) At zero frequency the delay is 10 ms.
(ii) For $\omega \leqslant 250$ rad/s, the delay error should be less than 1 percent.
(iii) For $\omega \leqslant 250$ rad/s, the maximum deviation in magnitude cannot exceed 2 dB.

From the above specifications, we can identify $T = 10$ ms. The normalized frequency at 250 rad/s is

$$\omega T = 10^{-2} \times 250 = 2.5 \tag{2.134}$$

From Figure 2.15 we see that a fifth-order Bessel–Thomson response will meet the requirement. However, Figure 2.16 indicates that at $\omega T = 2.5$ a fifth-order response would have much more than the allowed 2-dB variation in magnitude. Instead, Figure 2.16 shows that an eighth-order response will satisfy the magnitude requirement. Thus, we must choose $n = 8$ to satisfy the specifications.

2.4.2 Poles of the Bessel–Thomson Function

Unlike the Butterworth, Chebyshev, or inverse Chebyshev response, there is no simple formula to determine the poles of the Bessel–Thomson response, which are the roots of the equation

$$B_n(y) = 0 \tag{2.135}$$

These roots can be computed with the aid of a computer and are tabulated up to $n = 8$ in Table 2.7.

It is interesting to compare the relative pole locations of the transfer function having Bessel–Thomson response with those having Butterworth and Chebyshev responses. As shown in Figure 2.17, the relative positions of the poles for the Bessel–Thomson response have more nearly the same real part in comparison with the Butterworth and Chebyshev responses. As the poles move away from the $j\omega$-axis, the maximally-flat magnitude response changes to maximally-flat delay characteristic;

Table 2.7 Pole Locations for the Normalized Bessel–Thomson Response

n	
1	-1.0000000
2	$-1.5000000 \pm j0.8660254$
3	$-2.3221854;\ -1.8389073 \pm j1.7543810$
4	$-2.8962106 \pm j0.8672341;\ -2.1037894 \pm j2.6574180$
5	$-3.6467386;\ -3.3519564 \pm j1.7426614;\ -2.3246743 \pm j3.5710229$
6	$-4.2483594 \pm j0.8675097;\ -3.7357084 \pm j2.6262723;\ -2.5159322 \pm j4.4926730$
7	$-4.9717869;\ -4.7582905 \pm j1.7392861;\ -4.0701329 \pm j3.5171740;$ $-2.6856769 \pm j5.4206941$
8	$-5.5878860 \pm j0.8676144;\ -2.8389840 \pm j6.3539113;$ $-4.3682892 \pm j4.4144425;\ -5.2048408 \pm j2.6161751$

as they move toward the $j\omega$-axis, the magnitude becomes more like the equal-ripple response.

In addition to the four types of approximation discussed in the foregoing, another interesting response whose pole and zero distribution is shown in Figure 2.18(b) is called the *elliptic* or *Cauer-parameter response.* This response gives equal-ripple characteristic in both the passband and stopband as depicted in Figure 2.18(a). The pole and zero locations are found through the use of the Jacobian elliptic function. Detailed discussions of the elliptic functions and their use in the design of elliptic filters can be found in Chen.†

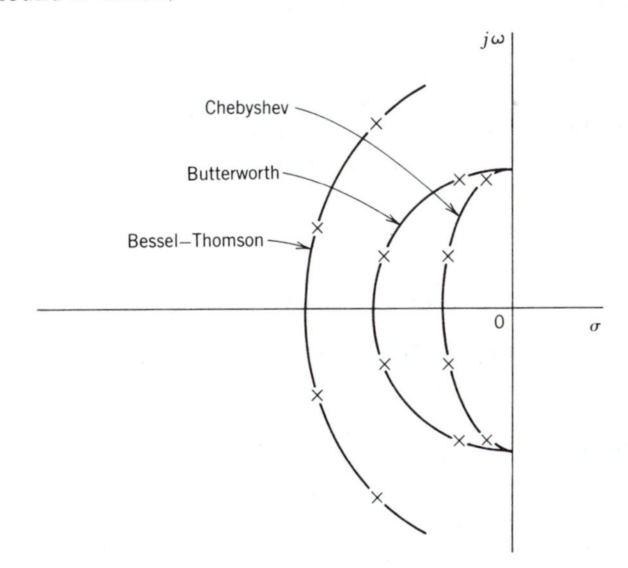

Figure 2.17

†W. K. Chen 1976, *Theory and Design of Broadband Matching Networks*, pp. 152–196. Cambridge, England: Pergamon Press.

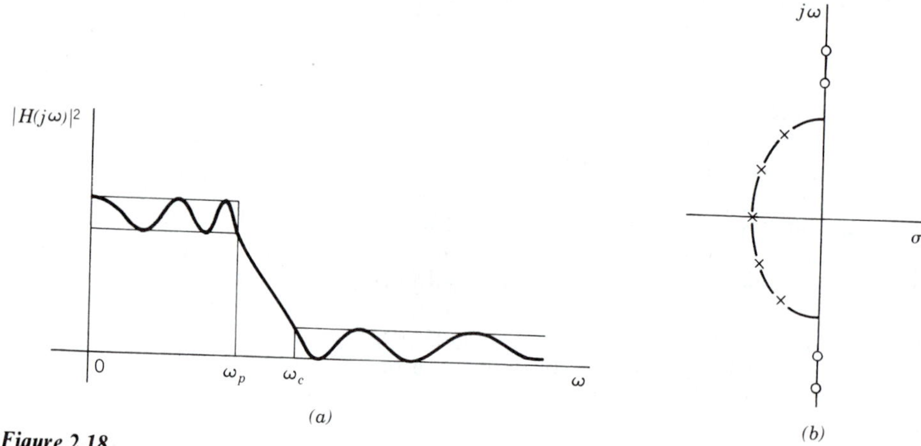

Figure 2.18

2.5 FREQUENCY TRANSFORMATIONS

So far we have dealt exclusively with approximating the ideal low-pass brick-wall type of response. In this section, we show that the results obtained in the foregoing for the low-pass approximation can readily be adapted to other cases such as high-pass, band-pass, band-elimination, etc., by means of transformations of the frequency variable. In this way, it permits the design to be made in terms of the low-pass characteristic, and then to be transformed to attain other forms of response.

The basic idea is that if we could find a frequency transformation that would map the desired passband and stopband onto the corresponding passband and stopband of the low-pass characteristic, we could then solve the equivalent low-pass problem. Applying the inverse transformation would lead to the solution of the original problem.

To discuss the transformation, let

$$s' = \sigma' + j\omega' \tag{2.136}$$

be the complex frequency for the low-pass function and let

$$s = \sigma + j\omega \tag{2.137}$$

be the new complex frequency variable. Then a frequency transformation is a function

$$s' = f(s) \tag{2.138}$$

which maps one or several frequency ranges of interest to the frequency range of the passband of the low-pass characteristic. The remainder of the frequencies is to be mapped to the stopband of the low-pass response. Thus, in moving the passband and stopband, we obtain the different types of filter characteristics. After realizing the network of the equivalent low-pass problem, the desired network is obtained from this low-pass network by replacing each inductance L by a one-port whose impedance is $Lf(s)$, and each capacitance C by a one-port whose admittance is $Cf(s)$. Since the impedance of a resistor is not a function of frequency, the resistances are not altered by the transformation. In this way, the low-pass filter and its derived network have the same value at the corresponding frequencies defined by (2.138).

We now proceed to discuss three important transformations that result in *high-pass*, *band-pass*, and *band-elimination characteristics*. To facilitate our discussion, it is convenient to normalize the low-pass characteristic to its cutoff frequency. With this, s' denotes the normalized low-pass frequency variable.

2.5.1 **Transformation to High-Pass**

Consider the transformation

$$s' = \omega_0/s \tag{2.139}$$

which maps the interval from $j\omega_0$ to $+\infty$ in the s-plane to the interval $-j1$ to 0 in the s'-plane and from $-j\omega_0$ to $-\infty$ in the s-plane to $j1$ to 0 in the s'-plane, and vice versa, as indicated in Figure 2.19. This transformation will convert a low-pass characteristic such as shown in Figure 2.20(a) to the frequency response shown in (b). Figure 2.20(b) is an approximation to the ideal brick-wall type of *high-pass* frequency response of Figure 2.21.

Suppose that a high-pass frequency characteristic is prescribed. This characteristic can be transformed into a corresponding low-pass response by means of (2.139). We now solve the equivalent low-pass problem and obtain its network realization. To obtain the desired high-pass filter, we replace each branch of the low-pass filter by a branch whose impedance at a point in the high-pass interval is the same as the impedance of the replaced branch at the corresponding point in the low-pass interval. Thus, if L and C represent the inductance and capacitance in the low-pass filter, we

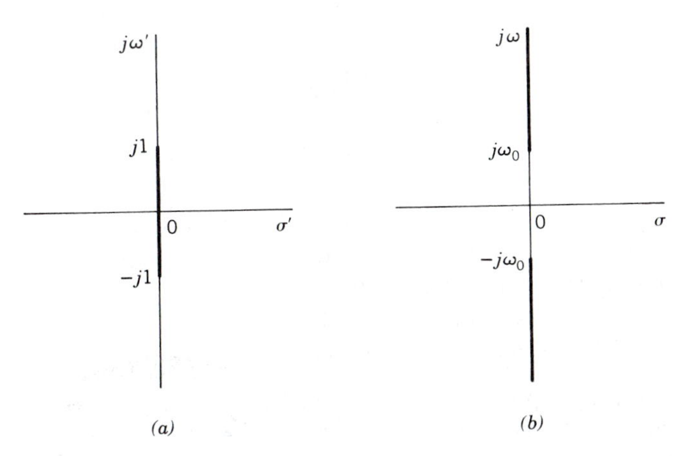

Figure 2.19

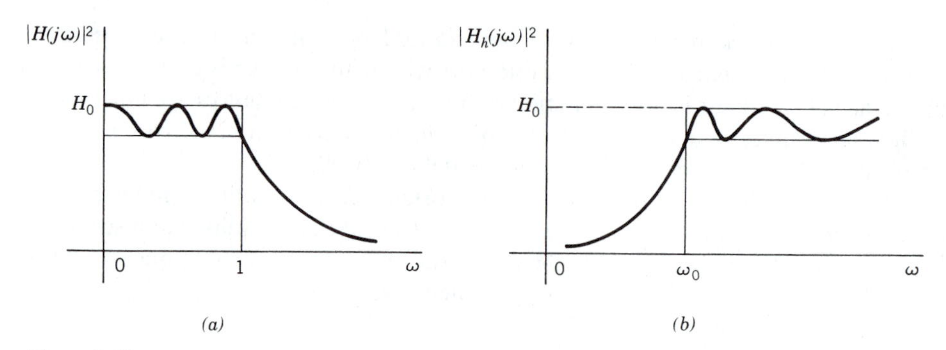

Figure 2.20

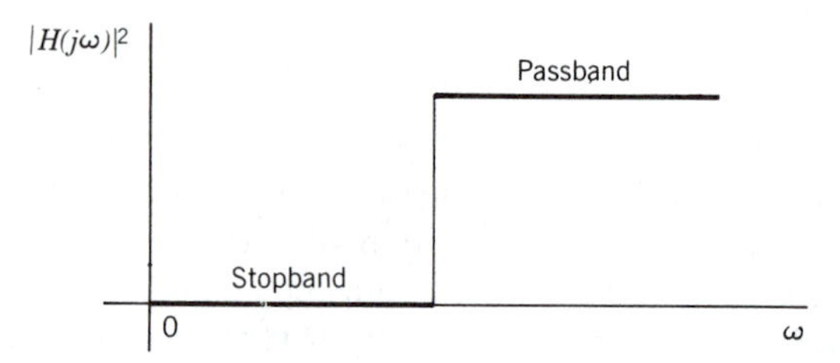

Figure 2.21

require that

$$Ls' = L\frac{\omega_0}{s} = \frac{1}{C_h s} \tag{2.140a}$$

$$\frac{1}{Cs'} = \frac{s}{C\omega_0} = L_h s \tag{2.140b}$$

where

$$C_h = \frac{1}{L\omega_0} \tag{2.141a}$$

$$L_h = \frac{1}{C\omega_0} \tag{2.141b}$$

In other words, to obtain a high-pass filter from its low-pass prototype, we simply replace each inductance by a capacitance and each capacitance by an inductance with the element values as given in (2.141). This is illustrated in Figure 2.22. Note that the radian cutoff frequency of the high-pass filter is ω_0, while its low-pass prototype is 1. Also, in making the replacement resistances are not affected.

Example 2.9 Suppose that we wish to design a high-pass filter whose radian cutoff frequency is 10^5 rad/s. The filter is to be operated between a resistive generator of internal resistance 400 Ω and a 100-Ω load and is required to have an equal-ripple

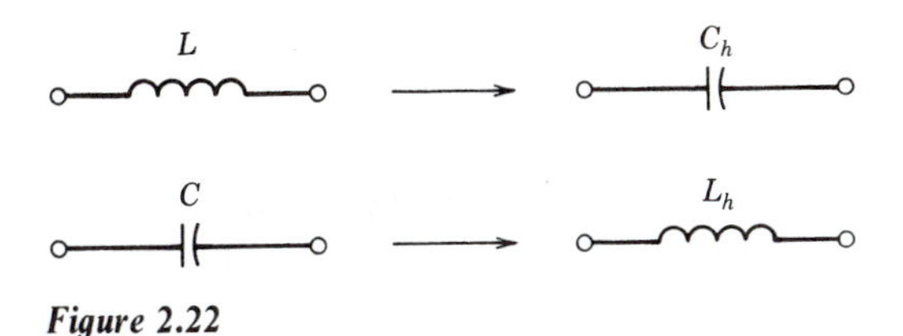

Figure 2.22

characteristic in the passband. The peak-to-peak ripple in the passband must not exceed 1 dB and at $\omega = 2 \times 10^4$ rad/s or less the response must be at least 60 dB down from its peak value in the passband.

To proceed with the design, the first step is to translate the high-pass specifications into the equivalent low-pass requirements. To this end, let $\omega_0 = 10^5$ rad/s. From (2.139) or from the plots of Figure 2.20, it is clear that, in terms of the low-pass specifications, the peak-to-peak ripple in the passband must not exceed 1 dB and at $\omega' = 5$ and beyond the response must be at least 60 dB down from its peak value in the passband. To determine the ripple factor, we use (2.41) and obtain

$$\varepsilon = \sqrt{10^{0.1} - 1} = 0.50885 \tag{2.142}$$

The minimum order of the Chebyshev response required is from (2.45)

$$n = \frac{\cosh^{-1}[(10^6 - 1)^{1/2}/0.50885]}{\cosh^{-1} 5} = 3.61 \tag{2.143}$$

which is rounded up to $n = 4$. Note that the low-pass response is normalized to its cutoff frequency. This is equivalent to setting $\omega_c = 1$ in (2.45). For $n = 4$ and $\varepsilon = 0.50885$, The required low-pass transfer function may be determined by formula (2.62) or directly from Table 2.5:

$$H(s') = \frac{H_0^{1/2}}{s'^4 + 0.95281 s'^3 + 1.45392 s'^2 + 0.74262 s' + 0.27563} \tag{2.144}$$

A ladder realizing this transfer voltage ratio $H(s')$ is shown in Figure 2.23. Using (2.141) yields the element values of the desired high-pass filter:

$$L_{h1} = 10^{-5}/C_1 = 0.88 \text{ mH} \tag{2.145a}$$

$$C_{h2} = 10^{-5}/L_2 = 46 \text{ nF} \tag{2.145b}$$

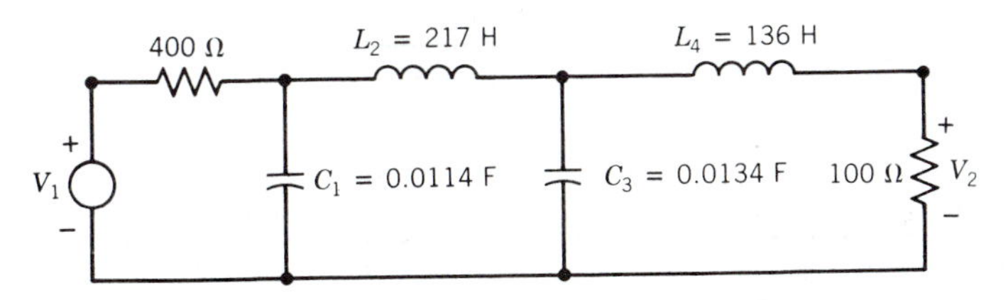

Figure 2.23

$$L_{h3} = 10^{-5}/C_3 = 0.75 \text{ mH} \tag{2.145c}$$

$$C_{h4} = 10^{-5}/L_4 = 74 \text{ nF} \tag{2.145d}$$

The high-pass filter together with its terminations is presented in Figure 2.24. The constant $H_0^{1/2}$ is found from Figure 2.23 or 2.24 to be 0.05513.

2.5.2 Transformation to Band-Pass

Transformation from low-pass to band-pass can be handled in a similar manner. We seek a function that will transform the pass- and stopbands of the ideal low-pass brick-wall response of Figure 2.1 to the *band-pass* brick-wall response of Figure 2.25. Such a function will transform the pass- and stopbands of the prototype low-pass response to those that will approximate the ideal band-pass characteristic of Figure 2.25.

Consider the transformation

$$s' = \frac{\omega_0}{B}\left[\frac{s}{\omega_0} + \frac{\omega_0}{s}\right] \tag{2.146}$$

where

$$\omega_0^2 = \omega_1 \omega_2 \tag{2.147}$$

$$B = \omega_2 - \omega_1 \tag{2.148}$$

This transformation maps the intervals $j\omega_1$ to $j\omega_2$ and $-j\omega_1$ to $-j\omega_2$ in the s-plane to the interval $-j1$ to $j1$ in the s'-plane, as indicated in Figure 2.26. To see this we

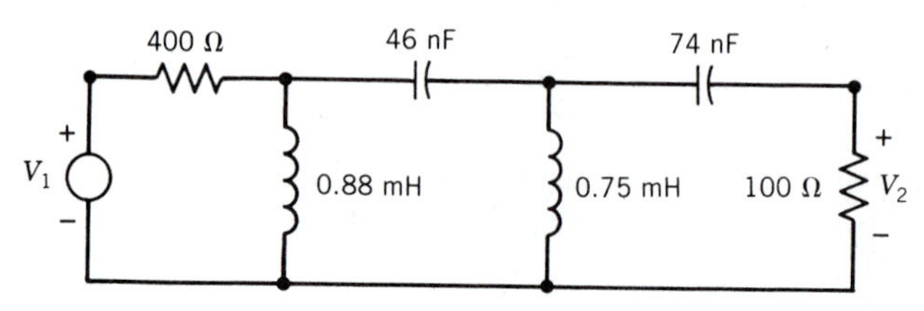

Figure 2.24

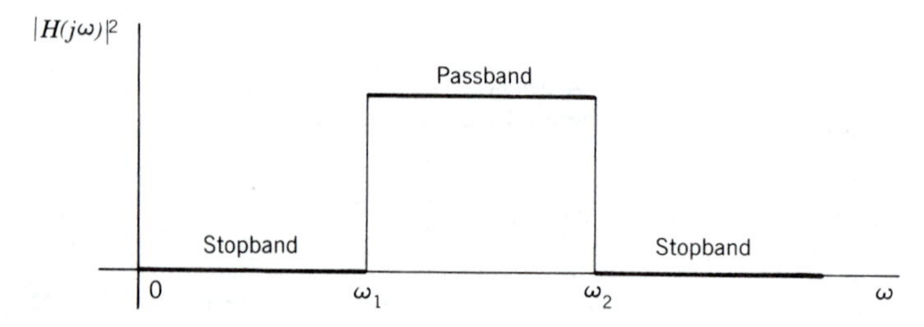

Figure 2.25

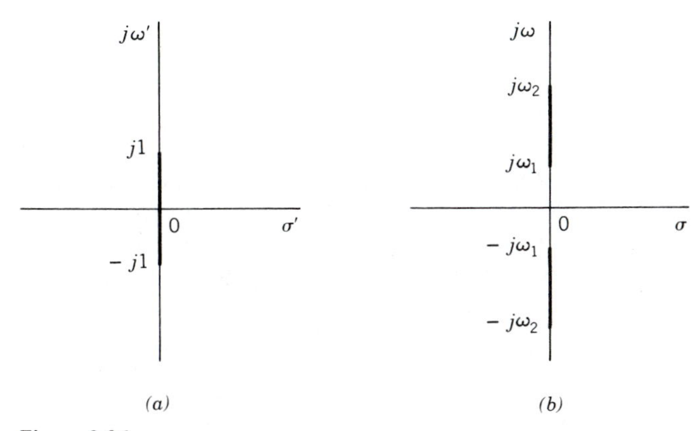

Figure 2.26

substitute $s' = \pm j1$ and $s = j\omega$ in (2.146) and obtain

$$\omega^2 \pm B\omega - \omega_0^2 = 0 \tag{2.149}$$

the solutions of which are given by $\pm\omega_1$ and $\pm\omega_2$, where

$$\omega_1, \omega_2 = \mp\tfrac{1}{2}B + \sqrt{\tfrac{1}{4}B^2 + \omega_0^2} \tag{2.150}$$

From these it is straightforward to confirm (2.147) and (2.148). Thus, B is the *bandwidth* of the band-pass filter and ω_0 is the geometric mean of the cutoff frequencies ω_1 and ω_2 and is referred to as the *mid-band frequency.* The reason for this name is that if frequency is plotted on a logarithmic scale, ω_0 falls midway between ω_1 and ω_2. In a similar way, we recognize that this transformation yields a response characteristic that is geometrically symmetric with respect to the mid-band frequency. Note that for each point in the low-pass interval in the s'-plane, there correspond two points in the s-plane, one in the positive passband and one in the negative passband.

Using the band-pass transformation (2.146), we can translate the band-pass specifications to those of a low-pass. After realizing the low-pass filter, the required band-pass filter is obtained by replacing each branch in the low-pass realization by a one-port, the impedance of which at a point in the band-pass interval is the same as the impedance of the replaced branch at the corresponding point in the low-pass interval. This requires that

$$Ls' = \frac{Ls}{B} + \frac{L\omega_0^2}{Bs} = L_{b1}s + \frac{1}{C_{b1}s} \tag{2.151}$$

$$\frac{1}{Cs'} = \frac{1}{\dfrac{Cs}{B} + \dfrac{C\omega_0^2}{Bs}} = \frac{1}{C_{b2}s + \dfrac{1}{L_{b2}s}} \tag{2.152}$$

where

$$L_{b1} = L/B \tag{2.153a}$$

$$C_{b1} = B/L\omega_0^2 \tag{2.153b}$$

$$L_{b2} = B/C\omega_0^2 \tag{2.153c}$$

$$C_{b2} = C/B \tag{2.153d}$$

Therefore, to obtain a band-pass filter from its low-pass prototype, each inductance L in the low-pass realization is replaced by a series combination of an inductor with inductance L_{b1} and a capacitor with capacitance C_{b1} and each capacitance C in the low-pass realization is replaced by a parallel combination of an inductor with inductance L_{b2} and a capacitor with capacitance C_{b2} as shown in Figure 2.27.

Example 2.10 It is desired to design a band-pass filter with a passband that extends from $\omega = 10^5$ rad/s to $\omega = 4 \times 10^5$ rad/s. The filter is required to have an equal-ripple characteristic in the passband with peak-to-peak ripple not to exceed 1 dB, and is to be operated between a resistive generator of internal resistance 400 Ω and a 100-Ω load. At $\omega = 15.263 \times 10^5$ rad/s the response must be at least 60 dB down from its peak value in the passband.

We first translate the band-pass specifications to the equivalent low-pass requirements. For this we compute the bandwidth and the mid-band frequency as

$$B = \omega_2 - \omega_1 = 4 \times 10^5 - 10^5 = 3 \times 10^5 \text{ rad/s} \tag{2.154a}$$

$$\omega_0 = \sqrt{\omega_1 \omega_2} = \sqrt{4 \times 10^5 \times 10^5} = 2 \times 10^5 \text{ rad/s} \tag{2.154b}$$

At $\omega = 15.263 \times 10^5$ rad/s, the corresponding frequency in the low-pass characteristic is found from (2.146) to be

$$\omega' = \frac{2 \times 10^5}{3 \times 10^5}\left[\frac{15.263 \times 10^5}{2 \times 10^5} - \frac{2 \times 10^5}{15.263 \times 10^5}\right] = 5 \tag{2.155}$$

Hence, the equivalent low-pass specifications are that the passband ripple must not exceed 1 dB and at five times the normalized frequency, which is one, and beyond, the response must be at least 60 dB down from its peak value. This is precisely the same low-pass problem solved in Example 2.9. A ladder realization is shown in Figure 2.23. To obtain the desired band-pass filter, we use (2.153), giving

$$L_{b1} = B/C_1\omega_0^2 = 0.66 \text{ mH}, \qquad C_{b1} = C_1/B = 38 \text{ nF} \tag{2.156a}$$

$$L_{b2} = L_2/B = 0.72 \text{ mH}, \qquad C_{b2} = B/L_2\omega_0^2 = 35 \text{ nF} \tag{2.156b}$$

$$L_{b3} = B/C_3\omega_0^2 = 0.56 \text{ mH}, \qquad C_{b3} = C_3/B = 45 \text{ nF} \tag{2.156c}$$

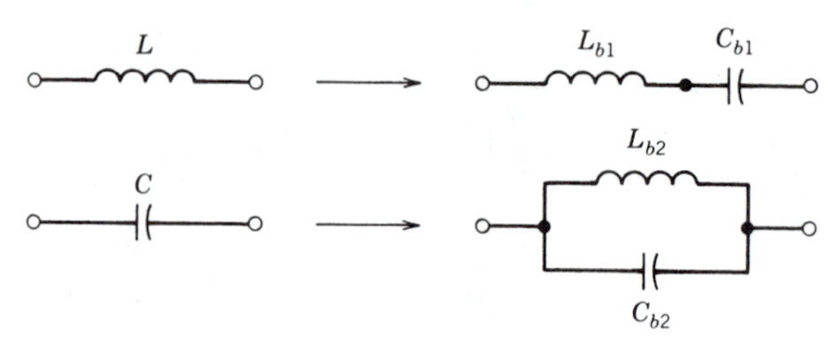

Figure 2.27

$$L_{b4} = L_4/B = 0.45 \text{ mH}, \qquad C_{b4} = B/L_4\omega_0^2 = 55 \text{ nF} \tag{2.156d}$$

The band-pass filter together with its terminations is given in Figure 2.28.

2.5.3 Transformation to Band-Elimination

The ideal *band-elimination* brick-wall type of response is shown in Figure 2.29. We seek a transformation that will convert the low-pass response of Figure 2.1 to the band-elimination of Figure 2.29. From our previous experience, the required transformation is seen to be

$$s' = \frac{1}{\dfrac{\omega_0}{B}\left[\dfrac{s}{\omega_0} + \dfrac{\omega_0}{s}\right]} \tag{2.157}$$

which maps the desired intervals in the s-plane to the interval $-j1$ to $j1$ in the s'-plane as depicted in Figure 2.30. The result is rather obvious because the transformation relating the low-pass to the high-pass will transform a band-pass to a band-elimination. Proceeding as in the transformation to band-pass, we can show that the points $\pm j\omega_1$ and $\pm j\omega_2$ in the s-plane map to the points $\pm j1$ in the s'-plane with B denoting the rejection bandwidth. The mid-band frequency ω_0 becomes the middle frequency in the rejection band on a logarithmic scale.

To obtain a band-elimination filter from its low-pass prototype, it is straightforward to show that we need only to replace each inductance L in the low-pass realization by a parallel combination of an inductor with inductance L_{e1} and a capacitor with capacitance C_{e1} and each capacitance C by a series combination of an

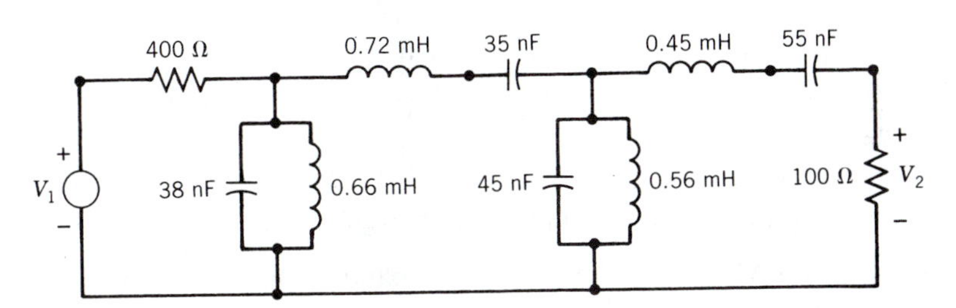

Figure 2.28

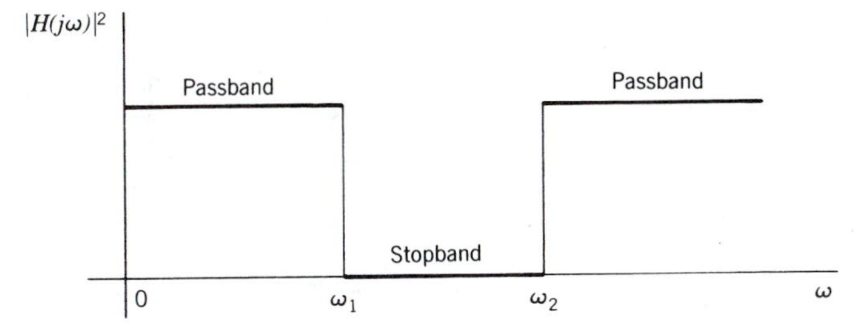

Figure 2.29

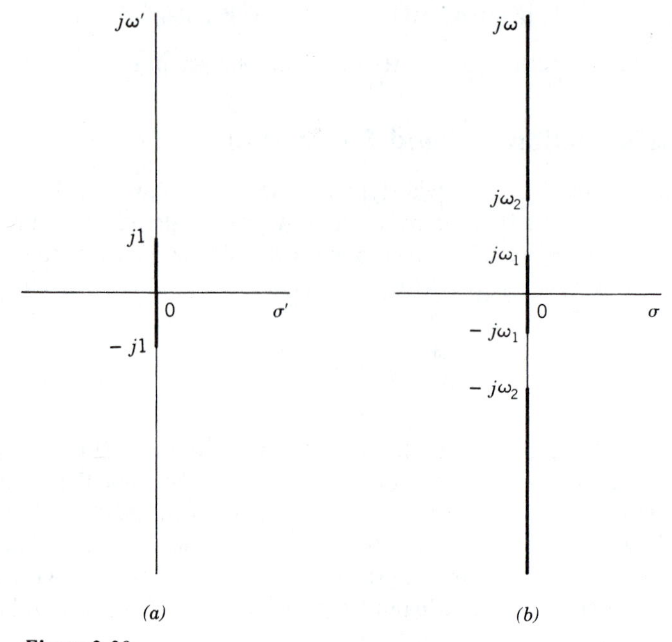

Figure 2.30

inductor with inductance L_{e2} and a capacitor with capacitance C_{e2}, whose values are given by

$$L_{e1}=LB/\omega_0^2 \tag{2.158a}$$

$$C_{e1}=1/LB \tag{2.158b}$$

$$L_{e2}=1/CB \tag{2.158c}$$

$$C_{e2}=CB/\omega_0^2 \tag{2.158d}$$

This replacement process is also illustrated in Figure 2.31.

Example 2.11 It is desired to design a band-elimination filter with a rejection band that extends from $\omega=10^5$ rad/s to $\omega=4\times10^5$ rad/s. The filter is to be operated between a resistive generator of internal resistance 400 Ω and a 100-Ω load, and is

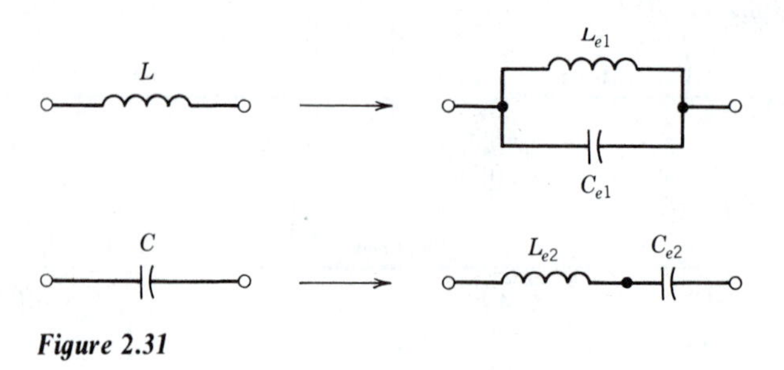

Figure 2.31

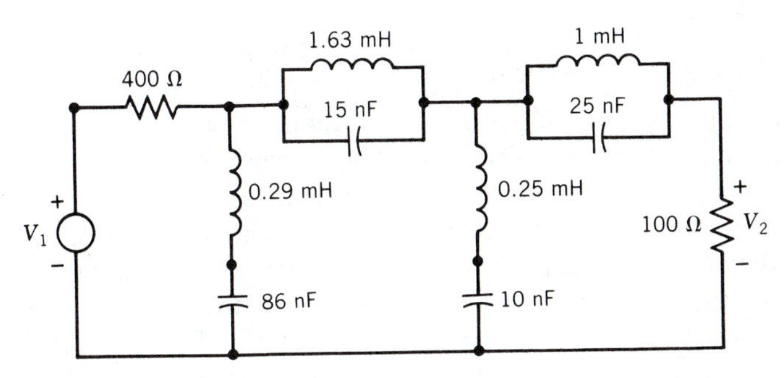

Figure 2.32

required to have an equal-ripple characteristic in the passband with peak-to-peak ripple not to exceed 1 dB. At $\omega = 1.722 \times 10^5$ rad/s the response must be at least 60 dB down from its peak value in the passband.

From specifications we compute the rejection bandwidth B and mid-band frequency ω_0:

$$B = \omega_2 - \omega_1 = 4 \times 10^5 - 10^5 = 3 \times 10^5 \text{ rad/s} \tag{2.159a}$$

$$\omega_0 = \sqrt{\omega_1 \omega_2} = \sqrt{4 \times 10^5 \times 10^5} = 2 \times 10^5 \text{ rad/s} \tag{2.159b}$$

At $\omega = \pm 1.722 \times 10^5$ rad/s, the corresponding frequency in the low-pass characteristic is found from (2.157) to be

$$\omega' = \frac{1}{\dfrac{2 \times 10^5}{3 \times 10^5}\left[\dfrac{\mp 1.722 \times 10^5}{2 \times 10^5} \pm \dfrac{2 \times 10^5}{1.722 \times 10^5}\right]} = \pm 5 \tag{2.160}$$

Thus, the equivalent low-pass specifications are that the passband ripple must not exceed 1 dB and at five times the normalized frequency, which is one, and beyond, the response must be at least 60 dB down from its peak value. A low-pass filter that satisfies these specifications is shown in Figure 2.23. To obtain the desired band-elimination filter, we use (2.158), obtaining

$$L_{e1} = 1/C_1 B = 0.29 \text{ mH}, \qquad C_{e1} = C_1 B/\omega_0^2 = 86 \text{ nF} \tag{2.161a}$$

$$L_{e2} = L_2 B/\omega_0^2 = 1.63 \text{ mH}, \qquad C_{e2} = 1/L_2 B = 15 \text{ nF} \tag{2.161b}$$

$$L_{e3} = 1/C_3 B = 0.25 \text{ mH}, \qquad C_{e3} = C_3 B/\omega_0^2 = 10 \text{ nF} \tag{2.161c}$$

$$L_{e4} = L_4 B/\omega_0^2 = 1 \text{ mH}, \qquad C_{e4} = 1/L_4 B = 25 \text{ nF} \tag{2.161d}$$

The band-elimination filter together with its terminations is shown in Figure 2.32.

2.6 SUMMARY AND SUGGESTED READINGS

In this chapter, we considered four popular rational function approximation schemes for the ideal low-pass brick-wall type of response. They are the *Butterworth* (*maximally-flat*) *response*, the *Chebyshev* (*equiripple*) *response*, the *inverse Chebyshev response*, and the *Bessel–Thomson* (*maximally-flat delay*) *response*. The Butterworth approximation yields a maximally-flat response near the origin and infinity and is monotonic throughout. The Chebyshev response gives an equal-ripple characteristic in the passband and is maximally-flat near infinity in the stopband. The inverse Chebyshev, on the other hand, results in a maximally-flat response near the origin in the passband and an equal-ripple characteristic in the stopband.

Unlike the other three functions that approximate the magnitude of a transfer function, the Bessel–Thomson response gives a maximally-flat time delay characteristic. This characteristic is determined when the specifications are given in terms of delay time and maximum deviation in magnitude and/or in time delay at a given frequency.

As to the distribution of the poles and zeros of these approximation functions, it was shown that the poles obtained from a Butterworth response lie on a circle while those obtained from a Chebyshev response lie on an ellipse. The poles for the Bessel–Thomson response have more nearly the same real part in comparison with the Butterworth and Chebyshev responses. As the poles move away from the $j\omega$-axis, the maximally-flat magnitude changes to maximally-flat delay; as they move toward the $j\omega$-axis, the magnitude becomes more like the equal-ripple response. All the zeros of the Butterworth, Chebyshev, and the Bessel–Thomson responses are at the infinity, being termed *all-pole functions*, while those of the inverse Chebyshev response lie on the $j\omega$-axis, infinity included. In the case of Butterworth response, apart from a constant one parameter—the degree of the Butterworth polynomial—is sufficient to specify its characteristic. For a Chebyshev or inverse Chebyshev response, two parameters are required, one being the degree of the Chebyshev polynomial and the other being determined by the amplitude of the passband or stopband ripple.

Finally, we mentioned that confining attention to the low-pass characteristic is not so restrictive as it appears at first glance. We demonstrated this by considering *frequency transformations* that permit a low-pass response to be converted to a high-pass, band-pass, or band-elimination characteristic.

A discussion of the approximation problem in greater detail is given by Daniels [3]. An excellent introduction to various aspects of approximation can be found in Van Valkenburg [7]. For a detailed treatment of the elliptic (Cauer-parameter) response, see Chen [2].

REFERENCES

1. A. Budak 1974, *Passive and Active Network Analysis and Synthesis*, Chapter 17. Boston: Houghton Mifflin.
2. W. K. Chen 1976, *Theory and Design of Broadband Matching Networks*, Chapter 3. Cambridge, England: Pergamon Press.
3. R. W. Daniels 1974, *Approximation Methods for Electronic Filter Design*, Chapters 2–4 and 6. New York: McGraw–Hill.

4. H. Y-F. Lam 1979, *Analog and Digital Filters: Design and Realization*, Chapter 8. Englewood Cliffs, NJ: Prentice–Hall.

5. A. S. Sedra and P. O. Brackett 1978, *Filter Theory and Design: Active and Passive*, Chapters 2–6. Champaign, IL: Matrix Publishers.

6. G. C. Temes and J. W. LaPatra 1977, *Introduction to Circuit Synthesis and Design*, Chapter 12. New York: McGraw–Hill.

7. M. E. Van Valkenburg 1982, *Analog Filter Design*, Chapters 6–8 and 10–13. New York: Holt, Rinehart & Winston.

8. L. Weinberg 1962, *Network Analysis and Synthesis*. New York: McGraw–Hill. Reissued by R. E. Krieger Publishing Co., Melbourne, FL, 1975, Chapter 11.

PROBLEMS

2.1 Show that the order of the Butterworth response can be determined by

$$n=\frac{\log[(10^{\alpha_{\min}/10}-1)/(10^{\alpha_{\max}/10}-1)]}{2\log(\omega_s/\omega_p)} \qquad (2.162)$$

where the parameters $\alpha_{\min}$, $\alpha_{\max}$, ω_s, and ω_p are defined in Figure 2.3.

2.2 Locate the poles of a transfer function having a maximally-flat response for $n=7$ and $n=8$. Compare the results with those in Table 2.1.

2.3 It is required to find a transfer function for a low-pass filter to satisfy the following specifications:

- **(i)** The end of the passband is 10 kHz.
- **(ii)** The peak-to-peak ripple in the passband must not exceed 1 dB.
- **(iii)** At 60 kHz the response must be down at least 50 dB from its peak value in the passband.

2.4 Find the transfer function of a high-pass filter satisfying the following specifications:

- **(i)** The passband extends from $\omega=10^4$ rad/s to $\omega=\infty$.
- **(ii)** The peak-to-peak ripple in the passband must not exceed 2 dB.
- **(iii)** For $\omega \leqslant 2000$ rad/s the response must be down at least 50 dB from its peak value in the passband.

2.5 Show that the quantities ε, n, and a of the Chebyshev response are related by the equation

$$na=\ln[1/\varepsilon+(1+\{1/\varepsilon^2\})^{1/2}] \qquad (2.163)$$

2.6 It is desired to design a band-elimination filter satisfying the following specifications:

- **(i)** The stopband extends from 10 kHz to 100 kHz.
- **(ii)** The magnitude characteristic is at least 30 dB down from its peak value in the passband at 20 kHz.
- **(iii)** The peak-to-peak ripple in the passband is not to exceed 1 dB.

Determine the corresponding low-pass specifications and the desired low-pass filter transfer function.

2.7 An equiripple band-pass filter is required satisfying the following specifications:

- **(i)** The passband extends from $\omega=1000$ rad/s to $\omega=4000$ rad/s.
- **(ii)** The peak-to-peak ripple in the passband is not to exceed $\frac{1}{2}$ dB.
- **(iii)** The magnitude characteristic is to be at least 30 dB down at $\omega=12\times10^3$ rad/s from its peak value in the passband.

Determine the corresponding low-pass specifications and the desired low-pass filter transfer function.

2.8 Find the minimum order of a Butterworth filter to meet the following requirements:

- **(i)** The passband extends from $\omega=0$ to $\omega=1000$ rad/s with maximum attenuation $\alpha_{max}=0.1$ dB.
- **(ii)** The stopband attenuation is at least 60 dB for $\omega\geqq2500$ rad/s.

2.9 Find the minimum order of a Bessel–Thomson filter to meet the following specifications:

- **(i)** The time delay at dc is $T=2$ ms.
- **(ii)** The delay error must be less than 1% for $\omega\leqslant2500$ rad/s.
- **(iii)** The magnitude error should not exceed 5 dB for $\omega\leqslant2000$ rad/s.

2.10 An inverse Chebyshev band-elimination filter is required to satisfy the following requirements:

- **(i)** The rejection band extends from $\omega=1000$ rad/s to $\omega=5000$ rad/s.
- **(ii)** The filter must give at least 60 dB of attenuation at $\omega=1800$ rad/s from its peak value in the passband.
- **(iii)** The passband attenuation must not exceed 1 dB.

Determine the corresponding low-pass specifications and the desired low-pass filter transfer function. What is the transfer function of the band-elimination filter?

2.11 Repeat Problem 2.10 for a Chebshev band-elimination filter with (iii) being replaced by the peak-to-peak ripple in the passband not to exceed 1 dB.

2.12 Determine the minimum order of a Butterworth band-elimination filter satisfying the specifications of Problem 2.10 with passband attenuation not to exceed 3 dB.

2.13 Determine the transfer function of a Bessel–Thomson filter having at most 10% deviation in delay at the frequency $\omega=3000$ rad/s and at most a 1-dB deviation of loss at $\omega=1500$ rad/s. The filter is to provide 100 μs of delay.

2.14 A Bessel–Thomson response is to be flat within 4% of the dc delay of 1 ms for $\omega\leqslant2000$ rad/s. Determine the desired transfer function. What is the attenuation at $\omega=2000$ rad/s?

2.15 A sixth-order low-pass Butterworth filter characteristic has the values $\omega_p=1500$ rad/s and $\alpha_{max}=0.2$ dB as defined in Figure 2.3. Find the attenuation at $\omega=3000$ rad/s.

2.16 Determine the transfer function of a Butterworth filter meeting the following specifications (see Figure 2.3): $\alpha_{max}=0.1$ dB, $\alpha_{min}=30$ dB, $\omega_p=1000$ rad/s, and $\omega_s=2500$ rad/s.

2.17 A low-pass filter is required to satisfy the following specifications (see Figure 2.3): $\alpha_{max}=0.25$ dB, $\alpha_{min}=20$ dB, $\omega_p=1000$ rad/s, and $\omega_s=1500$ rad/s. Determine the transfer function of the filter that is achieved by a Butterworth response; by a Chebyshev response; and by an inverse Chebyshev response.

2.18 Find the transfer function of a high-pass filter with a Chebyshev response. The filter must have at least 50 dB of attenuation for $\omega \leqslant 1000$ rad/s and no more than 0.25 dB of attenuation for $\omega \geqq 4500$ rad/s.

2.19 Show that the Chebyshev polynomial of degree $2n$ and the square of the Chebyshev polynomial of degree n are related by

$$2C_n^2(\omega) = C_{2n}(\omega) + 1 \tag{2.164}$$

2.20 The magnitude characteristic

$$|H(j\omega)|^2 = \frac{H_0}{1 + \varepsilon^2 \omega^{2m} C_{n-m}^2(\omega)} \tag{2.165}$$

is called the *transitional Butterworth–Chebyshev response*. For $0 < m < n$, show that this response possesses both Butterworth-like and Chebyshev-like characteristics. Determine the dB attenuation produced in the stopband by this response.

2.21 A band-elimination filter is required to satisfy the specifications shown in Figure 2.33. Determine the minimum order of a filter having (a) a Butterworth response, (b) a Chebyshev response, and (c) an inverse Chebyshev response. Calculate the transfer functions of the corresponding low-pass responses.

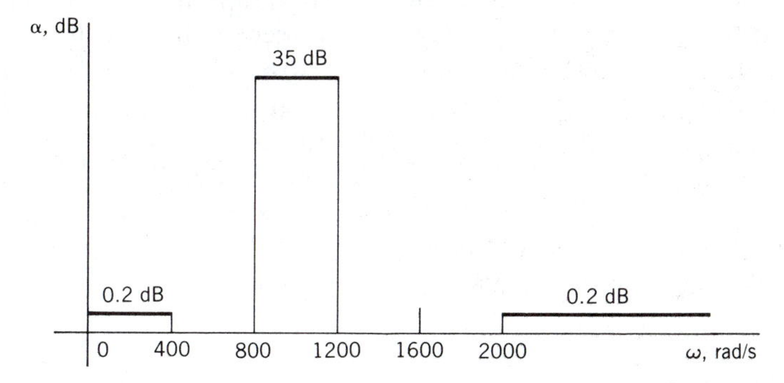

Figure 2.33

2.22 A band-pass filter is required to satisfy the specifications shown in Figure 2.34. Determine the minimum order of a filter having (a) a Butterworth response, (b) a Chebyshev response, and (c) an inverse Chebyshev response.

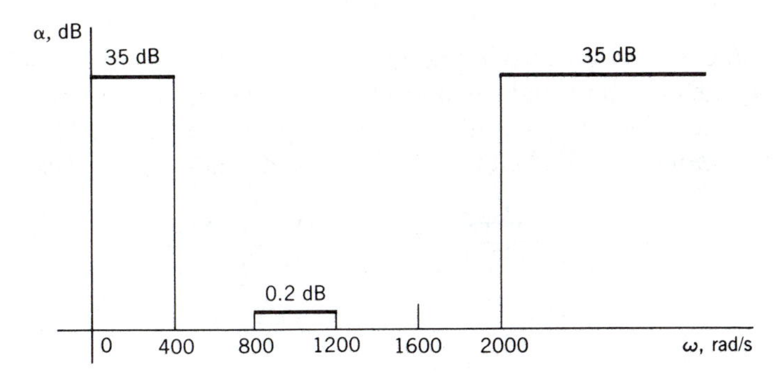

Figure 2.34

chapter 3

PASSIVE FILTER SYNTHESIS

In the preceding chapter, we showed how to find an approximation function that meets all the specifications. In this chapter we present techniques for synthesis of transfer functions using passive elements. We shall study in some detail a popular structure known as the ladder network. A salient feature of this structure is its very low sensitivity to element variations. As a result, this topology is also used in the realization of low-sensitivity active filters to be discussed in Chapter 8. In fact, some of these realizations are just active *RC* equivalents of the passive *LC* ladder filters. Because the synthesis of transfer functions requires a knowledge of one-port synthesis, we begin this chapter by considering one-port *LC* synthesis.

3.1 SYNTHESIS OF *LC* ONE-PORT NETWORKS

Consider an impedance function that can be written in the form

$$Z(s)=\frac{m_1+n_1}{m_2+n_2} \tag{3.1}$$

where m_1 and n_1 denote the even part and the odd part of the numerator polynomial, and m_2 and n_2 the even part and the odd part of the denominator polynomial, respectively. The choice of the symbols m and n is appropriate in that M is even with respect to its midpoint and N is odd with respect to its midpoint. For example, if

$$Z(s)=\frac{2s^3+s^2+4s+1}{s^3+3s^2+3s+1}=\frac{(s^2+1)+(2s^3+4s)}{(3s^2+1)+(s^3+3s)} \tag{3.2}$$

then

$$m_1=s^2+1, \qquad n_1=2s^3+4s \tag{3.3a}$$

$$m_2=3s^2+1, \qquad n_2=s^3+3s \tag{3.3b}$$

If we multiply $Z(s)$ by $(m_2-n_2)/(m_2-n_2)$, there results

$$Z(s)=\frac{(m_1m_2-n_1n_2)+(m_2n_1-m_1n_2)}{m_2^2-n_2^2} \tag{3.4}$$

Now since the product of two even functions or two odd functions is itself an even function, while the product of an even and an odd function is odd, we see that

$$\text{even part of } Z(s) \triangleq \text{Ev } Z(s)=\frac{m_1m_2-n_1n_2}{m_2^2-n_2^2} \tag{3.5a}$$

$$\text{odd part of } Z(s) \triangleq \text{Od } Z(s)=\frac{m_2n_1-m_1n_2}{m_2^2-n_2^2} \tag{3.5b}$$

The substitution of $s=j\omega$ into Ev $Z(s)$ gives the real part of $Z(j\omega)$:

$$\text{Ev } Z(s)\Big|_{s=j\omega}=\text{Re } Z(j\omega) \tag{3.6}$$

If $Z(s)$ is the driving-point impedance of a lossless network, then

$$\text{Re } Z(j\omega)=0 \qquad \text{for all } \omega \tag{3.7}$$

To make Re $Z(j\omega)=0$, there are three nontrivial ways: (i) $m_1=0$ and $n_2=0$, (ii) $m_2=0$ and $n_1=0$, and (iii) $m_1m_2-n_1n_2=0$. The first possibility leads $Z(s)$ into the form n_1/m_2, the second to m_1/n_2. For the third possibility, we require $m_1m_2=n_1n_2$ or

$$(m_1+n_1)m_2=(m_2+n_2)n_1 \tag{3.8}$$

This is equivalent to

$$Z(s)=\frac{m_1+n_1}{m_2+n_2}=\frac{n_1}{m_2} \tag{3.9}$$

The conclusion is that the driving-point impedance of a lossless network is always the quotient of even to odd or odd to even polynomials. Since m_2 is an even polynomial in s, the complex zeros of $m_2(s)$ have *quadrantal symmetry* as shown in Figure 3.1, i.e., if s_0 is a complex zero of $m_2(s)$, so are $-s_0$, $\bar{s}_0$, and $-\bar{s}_0$, where the bar denotes complex conjugate. Thus, the zeros of m_2 appear in pairs, one always being the negative of the other. As a result, if $Z(s)$ assumes the form (3.9), all of its poles must be simple and purely imaginary from stability considerations. Similarly, by considering the admittance function and the fact that n_1 is equal to s times an even polynomial, the zeros of $Z(s)$ are also simple and purely imaginary. The same conclusions hold for the other two situations. Thus, $Z(s)$ can be written in the form

$$Z(s)=H\frac{(s^2+\omega_{z1}^2)(s^2+\omega_{z2}^2)(s^2+\omega_{z3}^2)\ldots}{s(s^2+\omega_{p1}^2)(s^2+\omega_{p2}^2)\ldots} \tag{3.10}$$

where $\omega_{z1} \geqq 0$. The partial-fraction expansion of this equation is

$$Z(s)=Hs+\frac{K_0}{s}+\frac{K_{j\omega_{p1}}}{s-j\omega_{p1}}+\frac{K_{-j\omega_{p1}}}{s+j\omega_{p1}}+\frac{K_{j\omega_{p2}}}{s-j\omega_{p2}}+\cdots \tag{3.11}$$

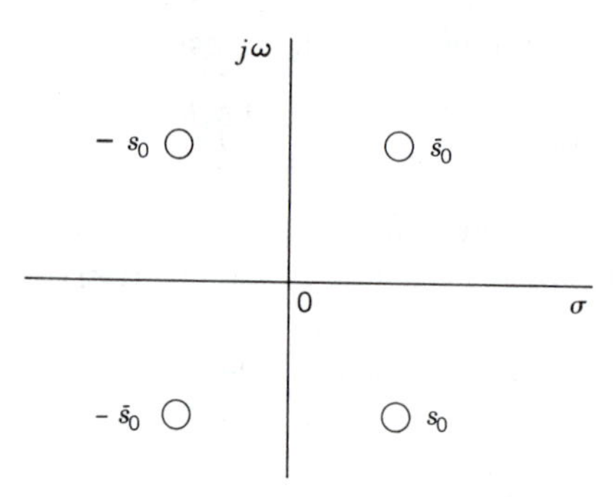

Figure 3.1

Since $Z(s)$ is positive real, all residues evaluated at the $j\omega$-axis poles are real and positive, giving $K_{j\omega_{pi}} = K_{-j\omega_{pi}} = K_i$. Let $\omega_{pi} = \omega_i$. Equation (3.11) becomes

$$Z(s) = Hs + \frac{K_0}{s} + \sum_{i=1}^{n} \frac{2K_i s}{s^2 + \omega_i^2} \tag{3.12}$$

Substituting $s = j\omega$ in (3.12) and writing $Z(j\omega) = \text{Re } Z(j\omega) + j \text{ Im } Z(j\omega)$ yields an odd function known as the *reactance function* $X(\omega)$:†

$$X(\omega) = \text{Im } Z(j\omega) = H\omega - \frac{K_0}{\omega} + \sum_{i=1}^{n} \frac{2K_i\omega}{-\omega^2 + \omega_i^2} \tag{3.13}$$

Taking the derivatives on both sides of (3.13) gives

$$\frac{dX}{d\omega} = H + \frac{K_0}{\omega^2} + \sum_{i=1}^{n} \frac{2K_i(\omega^2 + \omega_i^2)}{(-\omega^2 + \omega_i^2)^2} \tag{3.14}$$

Observe that every factor in this equation is positive for all positive and negative values of ω. We conclude that

$$\frac{dX}{d\omega} > 0 \qquad \text{for } -\infty < \omega < \infty \tag{3.15}$$

In words, it states that the slope of the reactance function versus frequency curve is always positive, as shown in Figure 3.2. As a result, the poles and zeros of $Z(s)$ alternate along the $j\omega$-axis. This is known as the *separation property* for reactance functions.‡ Because of this, the pole and zero frequencies of (3.10) are related by

$$0 \leqslant \omega_{z1} < \omega_{p1} < \omega_{z2} < \omega_{p2} < \cdots \tag{3.16}$$

In the above analysis, we deal exclusively with the impedance functions. The same conclusion can be reached for the admittance functions.

†A *reactance function* is frequently defined as a positive-real function that maps the imaginary axis into the imaginary axis. Using this definition, $Z(j\omega) = j \text{ Im } Z(j\omega) = jX(\omega)$ is the reactance function instead of $X(\omega)$.

‡This result is due to Ronald M. Foster (1896–) and Otto J. Zobel (1887–).

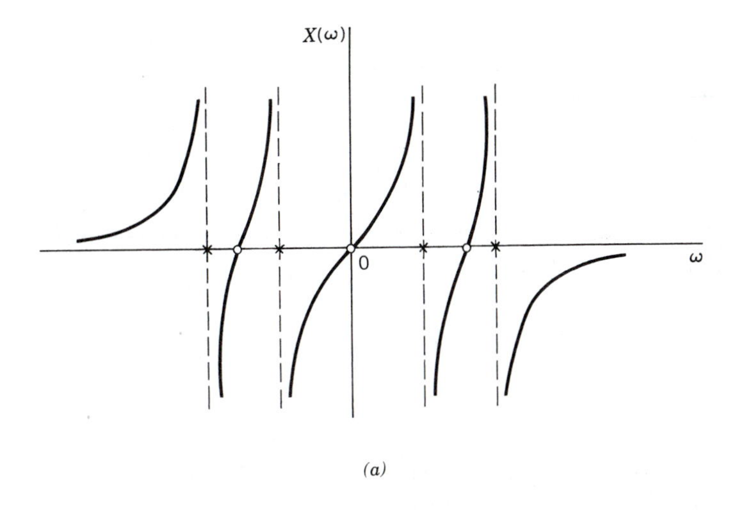

(a)

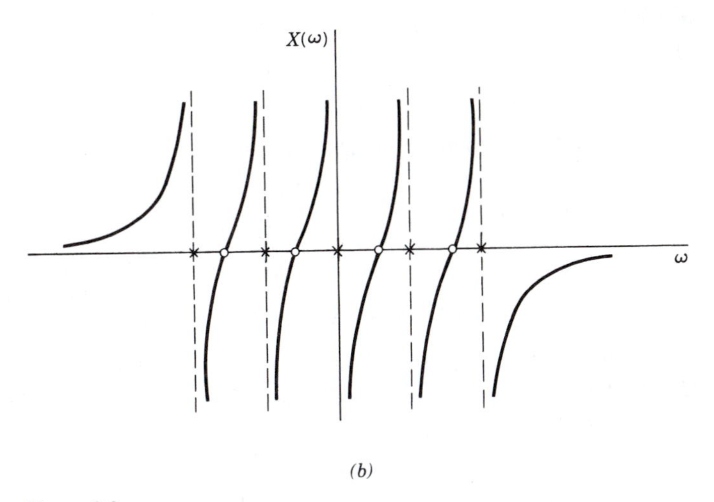

(b)

Figure 3.2

3.1.1 The Foster Canonical Forms

We now consider the realization of $Z(s)$. If each term on the right-hand side of (3.12) is identified as the input impedance of a one-port LC network, the series connection of these one-ports would yield the desired realization. Proceeding in this fashion, the first term is seen to be the impedance of an inductor of inductance H. The second term corresponds to a capacitor of capacitance $1/K_0$. Each of the remaining terms can be realized as a parallel combination of an inductor of inductance $2K_i/\omega_i^2$ and a capacitor of capacitance $1/2K_i$. The resulting realization is shown in Figure 3.3, and is known as the *first Foster canonical form*. Likewise, if we consider the admittance

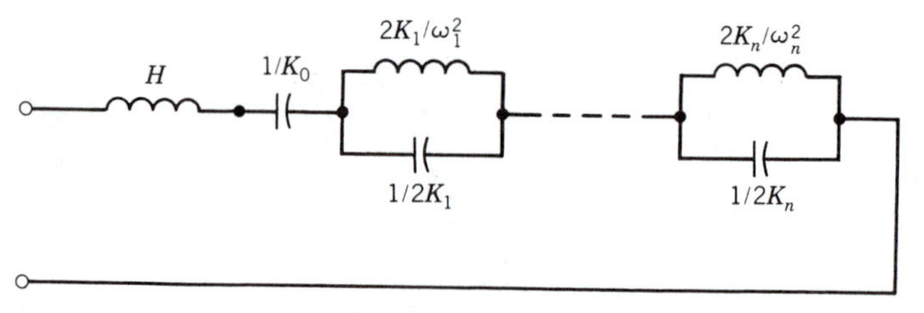

Figure 3.3

function $Y(s)$ and expand it in partial fractions, we obtain

$$Y(s)=Hs+\frac{K_0}{s}+\sum_{i=1}^{n}\frac{2K_i s}{s^2+\omega_i^2} \tag{3.17}$$

which can be realized by the network configuration of Figure 3.4 known as the *second Foster canonical form*. We illustrate this by the following example.

Example 3.1 Realize the following impedance function:

$$Z(s)=2\frac{s(s^2+4)(s^2+16)}{(s^2+1)(s^2+9)} \tag{3.18}$$

This function is first expanded in partial fractions, as follows:

$$Z(s)=2s+\frac{45/8}{s+j}+\frac{45/8}{s-j}+\frac{35/8}{s+j3}+\frac{35/8}{s-j3}$$

$$=2s+\frac{45s}{4(s^2+1)}+\frac{35s}{4(s^2+9)} \tag{3.19}$$

The network realization of (3.19) is presented in Figure 3.5(a), and is the first Foster canonical form. For the second Foster canonical form, we expand $Y(s)=1/Z(s)$ in partial fractions and obtain

$$Y(s)=\frac{9/128}{s}+\frac{5/64}{s+j2}+\frac{5/64}{s-j2}+\frac{35/256}{s+j4}+\frac{35/256}{s-j4}$$

$$=\frac{9}{128s}+\frac{5s}{32(s^2+4)}+\frac{35s}{128(s^2+16)} \tag{3.20}$$

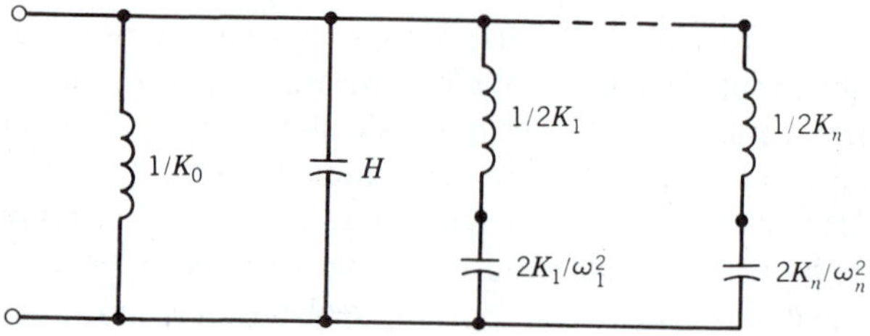

Figure 3.4

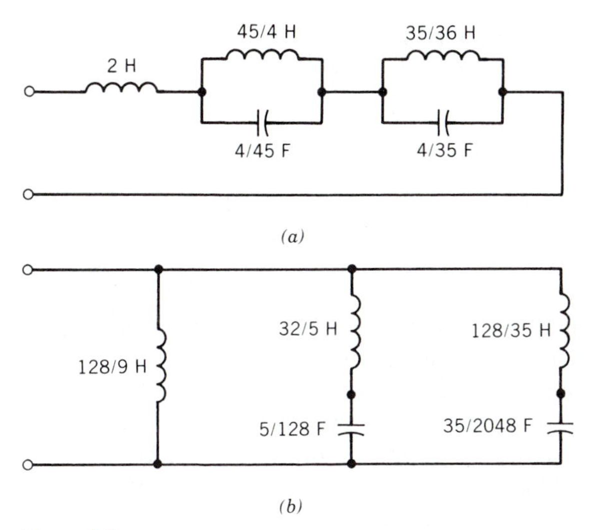

Figure 3.5

This equation can be realized by the network of Figure 3.5(*b*).

The preceding discussion shows that a reactance function can be realized by an *LC* one-port, which may assume either one of the two Foster canonical forms. The term *canonical form* refers to a network containing the minimum number of elements to meet given specifications. A reactance function as expressed in (3.10) requires $k+1$ pieces of information to make the function unique, where k is the number of zeros and poles in the interval $0<\omega<\infty$. Using the first Foster form of Figure 3.3 for a model, we recognize that the number of elements in the network equal the total number p of poles on the entire $j\omega$-axis, and that the number of poles equal the number of zeros. Excluding the pole and zero at the origin and infinity, there are $2p-2$ finite nonzero poles and zeros on the $j\omega$-axis. Since poles and zeros occur in complex conjugate pairs, we have $2k=2p-2$ or $p=k+1$. Thus, the two Foster forms are canonical.

We summarize the above results by stating the following theorem.

Theorem 3.1 A real rational function is the input immittance (impedance or admittance) function of an *LC* one-port network if and only if all of its zeros and poles are simple, lie on the $j\omega$-axis, and alternate with each other.

3.1.2 The Cauer Canonical Forms

In addition to the two Foster canonical forms, there is another synthesis procedure that gives rise to networks known as the *Cauer canonical forms*.† The process of obtaining Cauer forms is as follows: Let

$$Z(s)=m(s)/n(s) \tag{3.21}$$

†Named after the German engineer Wilhelm Adolf Eduard Cauer (1900–1945).

where m is assumed to be of higher degree than n. Otherwise, we consider $Y(s)=1/Z(s)$ instead of $Z(s)$. We next expand $Z(s)$ in a continued fraction

$$Z(s)=\frac{m(s)}{n(s)}=L_1s+\cfrac{1}{C_2s+\cfrac{1}{L_3s+\cfrac{1}{C_4s+\cfrac{1}{\ddots}}}} \tag{3.22}$$

which can be realized as the driving-point impedance of the ladder network of Figure 3.6. This LC ladder realization is known as the *first Cauer canonical form*. The physical process associated with (3.22) is equivalent to the successive removal of poles at infinity, as will be demonstrated in the following example.

Suppose now that we rearrange the numerator and denominator polynomials m and n in ascending order of s, and expand the resulting function in a continued fraction. Such an expansion gives

$$Z(s)=\frac{m(s)}{n(s)}=\frac{a_0+a_2s^2+\cdots+a_{k-2}s^{k-2}+a_ks^k}{b_1s+b_3s^3+\cdots+b_{k-1}s^{k-1}} \tag{3.23}$$

$$=\frac{1}{C_1s}+\cfrac{1}{\frac{1}{L_2s}+\cfrac{1}{\frac{1}{C_3s}+\cfrac{1}{\frac{1}{L_4s}+\cfrac{1}{\ddots}}}}$$

The expansion can be realized by the general ladder of Figure 3.7. The resulting LC ladder is called the *second Cauer canonical form*. Physically, this ladder is attained by the successive removal of poles at the origin. Both Cauer forms are canonical, because they contain the same number of elements as in the Foster forms. We illustrate the above procedures by the following example.

Example 3.2 Consider the reactance function

$$Z(s)=\frac{2s^5+40s^3+128s}{s^4+10s^2+9} \tag{3.24}$$

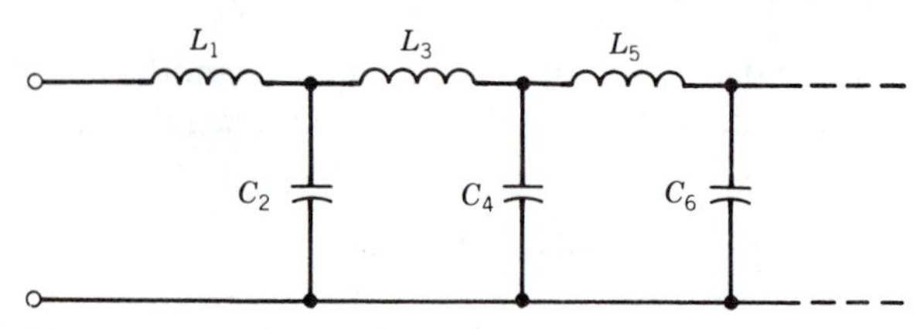

Figure 3.6

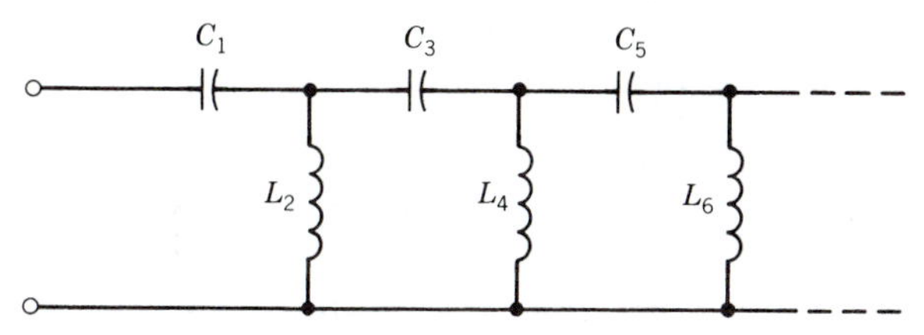

Figure 3.7

which can be expanded in a continued fraction

$$Z(s)=2s+\cfrac{1}{s/20+\cfrac{1}{40s/9+\cfrac{1}{9s/140+\cfrac{1}{70s/9}}}} \tag{3.25}$$

A ladder realization is shown in Figure 3.8 and is known as the first Cauer canonical form. To obtain the second Cauer canonical form, we rearrange the numerator and denominator polynomials of (3.24) in ascending order of s and then expand the resulting function in a continued fraction, obtaining

$$Z(s)=\frac{128s+40s^3+2s^5}{9+10s^2+s^4}$$

$$=\cfrac{1}{1/14.2s+\cfrac{1}{1/0.056s+\cfrac{1}{1/3.44s+\cfrac{1}{1/0.011s+\cfrac{1}{1/7.25s}}}}} \tag{3.26}$$

The desired LC ladder is presented in Figure 3.9. The two Cauer forms together with the two Foster forms obtained in Example 3.1 are completely equivalent in that if they were enclosed within black boxes, one cannot be distinguished from another. Often one realization is preferred over another because of other considerations such as element size, network configuration, compensation for parasitic effects, sensitivity to variations of element values, etc.

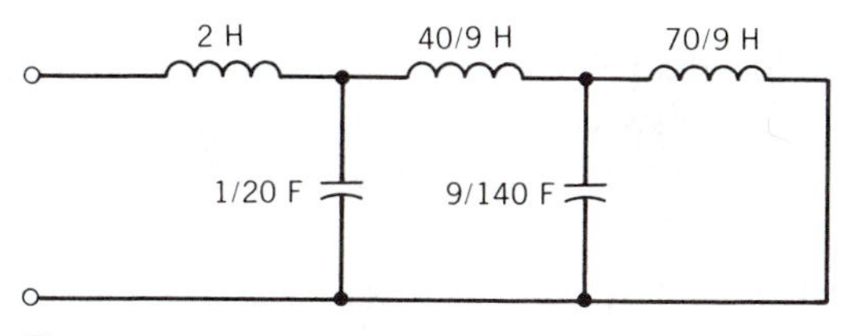

Figure 3.8

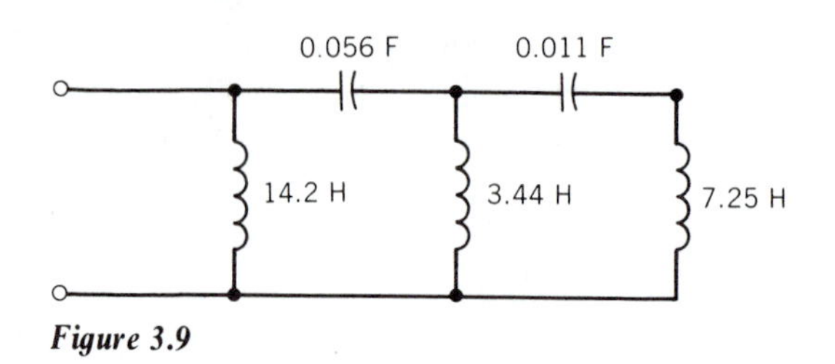

Figure 3.9

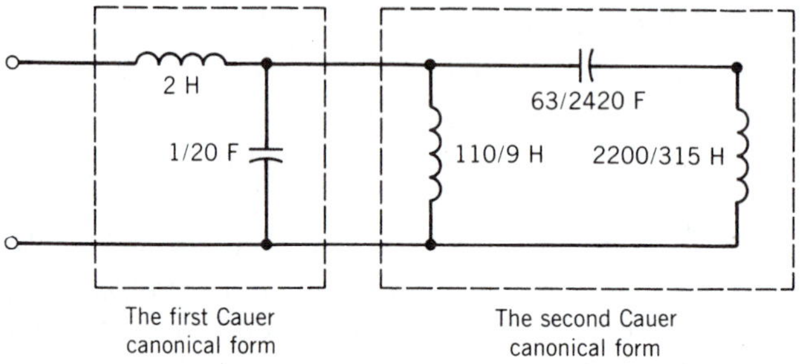

2 H
1/20 F
63/2420 F
110/9 H
2200/315 H
The first Cauer canonical form
The second Cauer canonical form

Figure 3.10

In addition to the four canonical forms, a mixed form is also possible. It results from a continued-fraction expansion alternating between two Cauer forms and/or from a partial-fraction expansion alternating between two Foster forms or any mix of these. For example, the reactance function (3.24) can be expanded in a continued fraction as

$$Z(s)=2s+\cfrac{1}{s/20+9/110s+\cfrac{1}{2420/63s+\cfrac{1}{315/2200s}}} \tag{3.27}$$

This expansion can be realized by the ladder of Figure 3.10. The first two elements correspond to the first Cauer form, and the remaining three to the second Cauer form.

3.2 SYNTHESIS OF *RC* ONE-PORT NETWORKS

In this section, we exploit the properties of impedance functions of one-ports composed only of resistors and capacitors from the known properties of *LC* networks. To this end we consider the *RC* network of Figure 3.11, the loop equations of which are given by

$$\begin{bmatrix} R_1+R_2+1/sC_1+1/sC_2 & -R_2-1/sC_2 \\ -R_2-1/sC_2 & R_2+R_3+1/sC_2+1/sC_3 \end{bmatrix}\begin{bmatrix} I_{m1} \\ I_{m2} \end{bmatrix}=\begin{bmatrix} V_s \\ 0 \end{bmatrix} \tag{3.28}$$

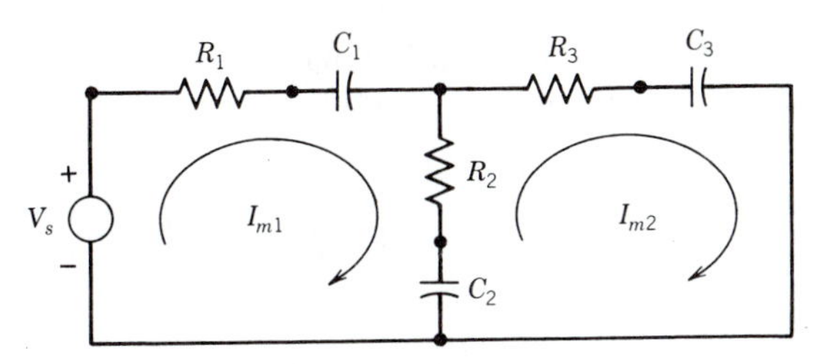

Figure 3.11

The input impedance facing the voltage source V_s is found to be

$$Z_{RC}(s)=\frac{\Delta(s)}{\Delta_{11}(s)} \tag{3.29}$$

where Δ is the determinant of the coefficient matrix of (3.28) and Δ_{11} is the cofactor of its (1,1)-element.

We next construct an *LC* network from the *RC* network of Figure 3.11 by replacing each resistor of resistance R_i by an inductor of inductance $L_i=R_i$. The resulting network is shown in Figure 3.12, the loop equations of which are given by

$$\begin{bmatrix} sL_1+sL_2+1/sC_1+1/sC_2 & -sL_2-1/sC_2 \\ -sL_2-1/sC_2 & sL_2+sL_3+1/sC_2+1/sC_3 \end{bmatrix}\begin{bmatrix} I'_{m1} \\ I'_{m2} \end{bmatrix}=\begin{bmatrix} V_s \\ 0 \end{bmatrix} \tag{3.30}$$

The input impedance facing the voltage source in the *LC* network of Figure 3.12 is determined by the equation

$$Z_{LC}(s)=\frac{\Delta'(s)}{\Delta'_{11}(s)} \tag{3.31}$$

where Δ' is the determinant of the coefficient matrix of (3.30) and Δ'_{11} is the cofactor of its (1,1)-element. Comparing the coefficient matrix $\mathbf{M}_{LC}(s)$ of (3.30) with that $\mathbf{M}_{RC}(s)$ of (3.28), we obtain the relation

$$\mathbf{M}_{RC}(s)=\frac{1}{p}\,\mathbf{M}_{LC}(p)\bigg|_{p^2=s} \tag{3.32}$$

The process outlined above clearly can be extended to general *RC* networks. In the

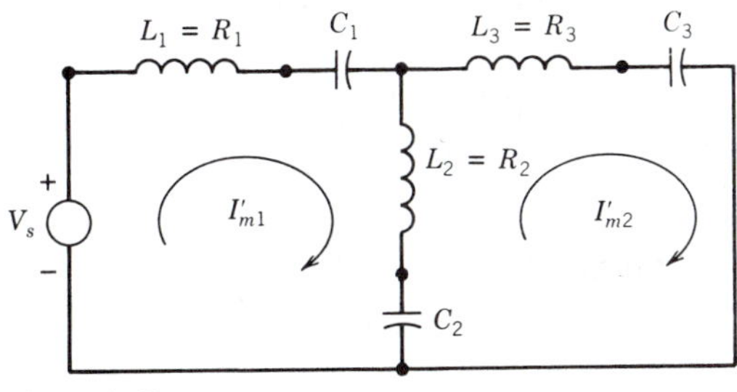

Figure 3.12

general case, the order of the matrices will equal the number r of linearly independent loop equations. This results in

$$Z_{RC}(s)=\frac{\Delta(s)}{\Delta_{11}(s)}=\left.\frac{(1/p^r)\Delta'(p)}{(1/p^{r-1})\Delta'_{11}(p)}\right|_{p^2=s}=\left.\frac{1}{p}\cdot\frac{\Delta'(p)}{\Delta'_{11}(p)}\right|_{p^2=s} \tag{3.33a}$$

or from (3.31)

$$Z_{RC}(s)=\left[\frac{1}{p}Z_{LC}(p)\right]_{p^2=s} \tag{3.33b}$$

This relation allows us to deduce the properties of RC networks from those of LC networks. It states that we form $Z_{LC}(p)$ and then divide this expression by p. When p^2 is replaced by s, the result is the impedance of the RC network. Let us carry out this operation for the most general form of $Z_{LC}(s)$ of (3.10) having the partial-fraction expansion (3.12). We obtain

$$Z_{RC}(s)=H\frac{(s+\sigma_{z1})(s+\sigma_{z2})(s+\sigma_{z3})\ldots}{s(s+\sigma_{p1})(s+\sigma_{p2})\ldots} \tag{3.34a}$$

$$=H+\frac{K_0}{s}+\sum_{i=1}^{n}\frac{\hat{K}_i}{s+\sigma_i} \tag{3.34b}$$

where $\sigma_{zj}=\omega_{zj}^2$, $\sigma_{pi}=\omega_{pi}^2$, $\sigma_i=\omega_i^2$, $\hat{K}_i=2K_i$, and from (3.16)

$$0\leqq\sigma_{z1}<\sigma_{p1}<\sigma_{z2}<\sigma_{p2}<\cdots \tag{3.35}$$

Thus, the zeros and poles of an RC impedance alternate along the negative real axis.† This property is sufficient to characterize the RC impedance functions. We state this result as

Theorem 3.2 A real rational function is the driving-point impedance of an RC one-port network if and only if all the poles and zeros are simple, lie on the negative real axis, and alternate with each other, the first *critical frequency* (pole or zero) being a pole.

To find the slope of $Z_{RC}(\sigma)$, we use (3.34b) and obtain

$$\frac{dZ_{RC}(\sigma)}{d\sigma}=-\frac{K_0}{\sigma^2}-\sum_{i=1}^{n}\frac{\hat{K}_i}{(\sigma+\sigma_i)^2} \tag{3.36}$$

which is negative for all values of σ, both positive and negative, since K_0 and $\hat{K}_i$ are positive. This shows

$$\frac{dZ_{RC}(\sigma)}{d\sigma}<0 \tag{3.37}$$

A plot of $Z_{RC}(\sigma)$ as a function of σ is shown in Figure 3.13. From this plot we see that since there are no poles and zeros along the positive real axis, then

$$Z_{RC}(\infty)\leqq Z_{RC}(0) \tag{3.38}$$

†It also includes the origin of the s-plane.

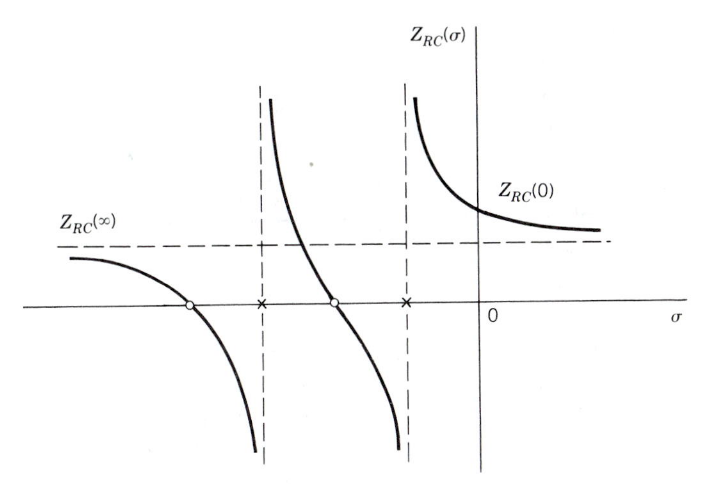

Figure 3.13

We next turn our attention to the synthesis of RC networks. Suppose that we are given $Z_{RC}(s)$ as in (3.34). By analogy to the LC case, this impedance can be realized by the network of Figure 3.14. This network structure is called the *first Foster canonical form* for RC impedance. To obtain the *second Foster canonical form*, we expand $Y_{RC}(s)/s$, where $Y_{RC}(s)=1/Z_{RC}(s)$, in a partial fraction and then multiply the resulting equation by s. The reason is that a direct partial-fraction expansion of $Y_{RC}(s)$ will result in negative residues. Proceeding as proposed, we obtain

$$Y_{RC}(s)=K_0+K_\infty s+\sum_{i=1}^{n}\frac{K_i s}{s+\sigma_i} \tag{3.39}$$

Equation (3.39) tells us that three kinds of one-ports are connected in parallel. The first term represents a resistor of resistance $1/K_0$. The second term is the admittance of a capacitor of capacitance K_∞. Each of the remaining terms represents the admittance of a series-connected RC one-port. The final realization is given in Figure 3.15. We illustrate the above result by the following example.

Example 3.3 Consider the impedance function

$$Z(s)=4\frac{(s+2)(s+5)}{(s+1)(s+4)} \tag{3.40}$$

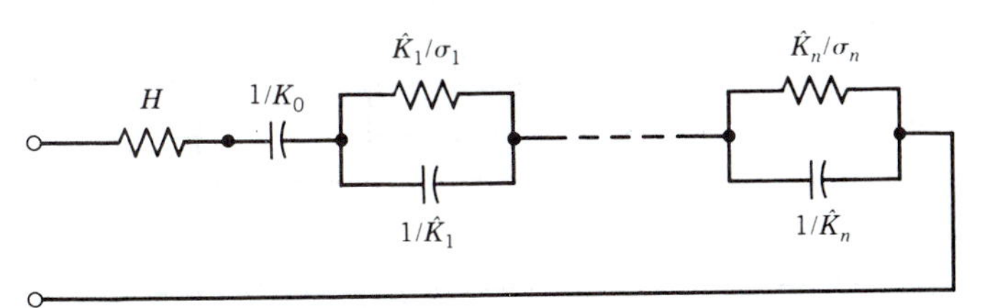

Figure 3.14

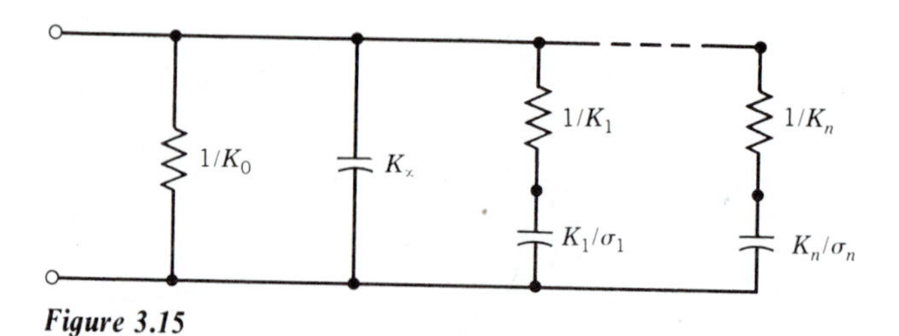

Figure 3.15

the poles and zeros of which alternate along the negative real axis. According to Theorem 3.2, it represents the input impedance of an RC one-port. To obtain the first Foster canonical form, we expand $Z(s)$ in a partial fraction

$$Z(s) = 4 + \frac{16/3}{s+1} + \frac{8/3}{s+4} \tag{3.41}$$

The network realization is shown in Figure 3.16. For the second Foster canonical form, we consider the admittance $Y(s) = 1/Z(s)$. If we expand $Y(s)$ directly by partial fractions, we obtain

$$Y(s) = \frac{(s+1)(s+4)}{4(s+2)(s+5)} = \frac{1}{4} + \frac{-1/6}{s+2} + \frac{-1/3}{s+5} \tag{3.42}$$

showing that the residues of $Y(s)$ at its poles are negative. The proper function to expand is $Y(s)/s$, yielding

$$\frac{Y(s)}{s} = \frac{(s+1)(s+4)}{4s(s+2)(s+5)} = \frac{1/10}{s} + \frac{1/12}{s+2} + \frac{1/15}{s+5} \tag{3.43}$$

Multiplying both sides by s gives

$$Y(s) = \frac{1}{10} + \frac{s}{12s+24} + \frac{s}{15s+75} \tag{3.44}$$

The realization of (3.44) is presented in Figure 3.17.

In addition to the two Foster forms, ladder realization for the RC impedance is also possible. The process is similar to the LC case by performing a continued-

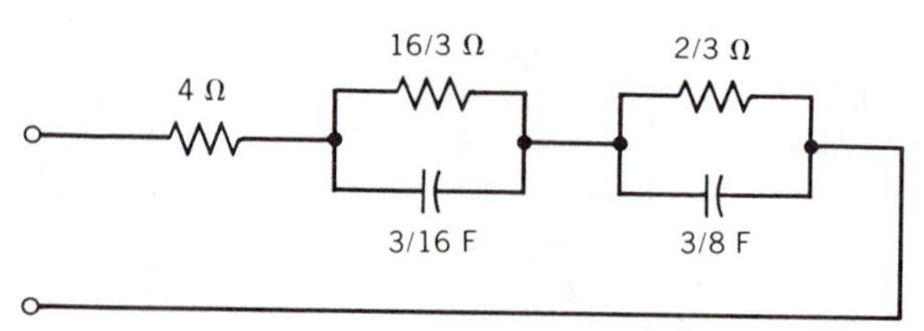

Figure 3.16

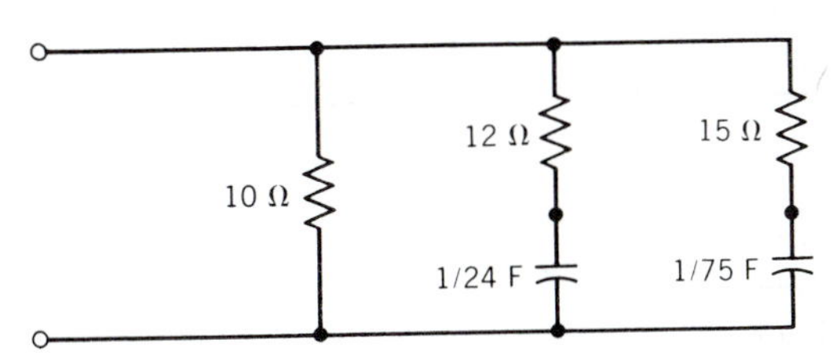

Figure 3.17

fraction expansion of $Z_{RC}(s)$, yielding

$$Z_{RC}(s)=R_1+\cfrac{1}{C_2s+\cfrac{1}{R_3+\cfrac{1}{C_4s+\cfrac{1}{\ddots}}}} \tag{3.45}$$

which can be realized by the ladder network of Figure 3.18. This ladder is called the *first Cauer canonical form* for *RC* impedance. If we rearrange terms of $Z_{RC}(s)$ so that the numerator and denominator polynomials appear in ascending order of s, the resulting continued-fraction expansion takes the general form

$$Z_{RC}(s)=\frac{1}{C_1s}+\cfrac{1}{\cfrac{1}{R_2}+\cfrac{1}{\cfrac{1}{C_3s}+\cfrac{1}{\cfrac{1}{R_4}+\cfrac{1}{\cfrac{1}{C_5s}+\cfrac{1}{\ddots}}}}} \tag{3.46}$$

This yields the *second Cauer canonical form* as shown in Figure 3.19.

Example 3.4 Consider the impedance function

$$Z(s)=\frac{4(s^2+7s+10)}{s^2+5s+4} \tag{3.47}$$

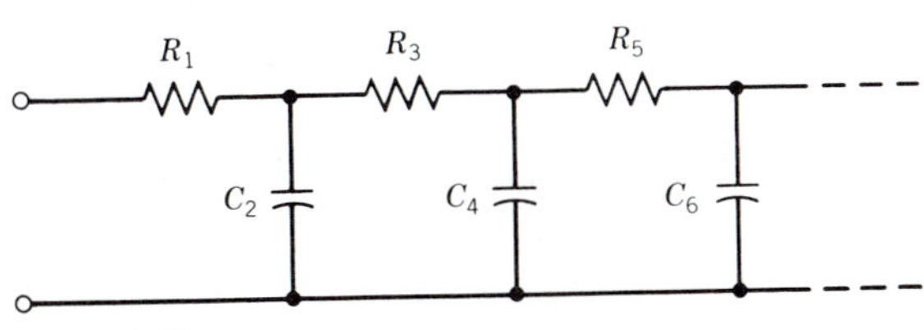

Figure 3.18

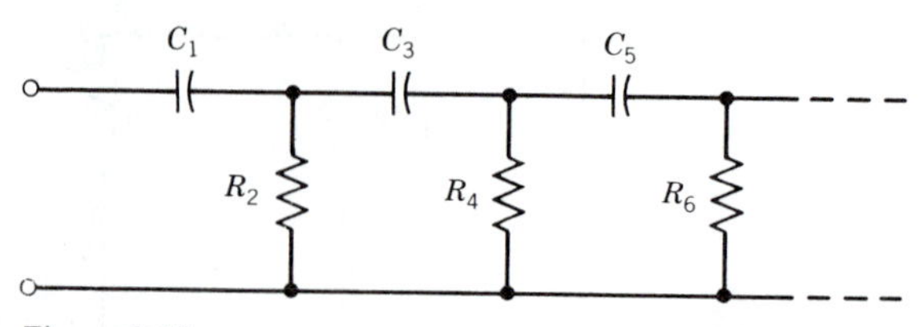

Figure 3.19

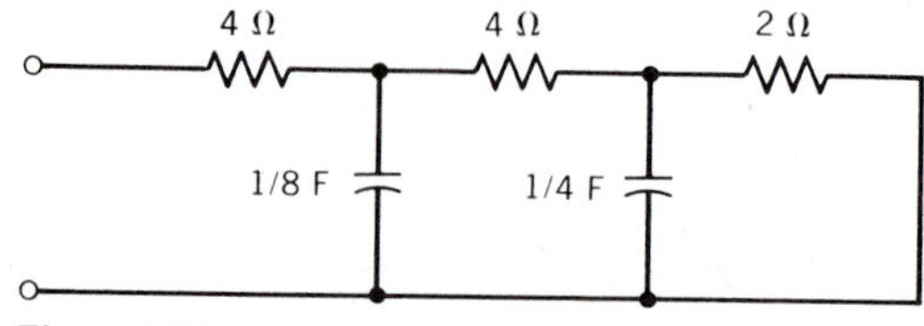

Figure 3.20

which can be expanded in continued-fraction form as

$$Z(s)=4+\cfrac{1}{s/8+\cfrac{1}{4+\cfrac{1}{s/4+\cfrac{1}{2}}}} \tag{3.48}$$

The first Cauer canonical form is shown in Figure 3.20. To obtain the second Cauer canonical form, we first rearrange the numerator and denominator polynomials of (3.47) in ascending order of s, and then expand the resulting function in a continued fraction, as follows:

$$\begin{aligned} Z(s)&=\frac{40+28s+4s^2}{4+5s+s^2} \\ &=\cfrac{1}{1/10+\cfrac{1}{200/11s+\cfrac{1}{121/940+\cfrac{1}{2209/11s+\cfrac{1}{1/47}}}}} \end{aligned} \tag{3.49}$$

The *RC* ladder is shown in Figure 3.21.

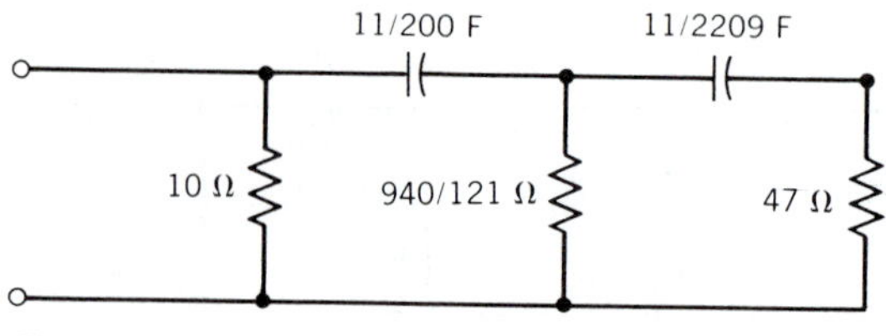

Figure 3.21

3.3 TWO-PORT SYNTHESIS BY LADDER DEVELOPMENT

In two-port synthesis, specifications are often given in terms of the transfer functions such as the transfer voltage ratio, transfer current ratio, transfer admittance, or transfer impedance as defined in Chapter 1. However, realization is usually accomplished in terms of the y- or z-parameters. Figure 3.22 shows a network driven by a voltage source with output terminating in Z_2. Let the two-port network N be characterized by its y-parameters y_{ij} or z-parameters z_{ij}. Then the transfer voltage ratio G_{12} can be expressed in terms of y_{ij} and Z_2 by the equation†

$$G_{12}(s)=\frac{V_2}{V_1}=\frac{-y_{21}}{y_{22}+Y_2} \tag{3.50}$$

where $Y_2=1/Z_2$. In the case where the output terminals are open-circuited, (3.50) reduces to

$$G_{12}(s)=\frac{V_2}{V_1}=\frac{-y_{21}}{y_{22}}=\frac{z_{21}}{z_{11}} \tag{3.51}$$

In the preceding two sections, we demonstrated the realization of an impedance function as the input impedance of an LC or RC ladder network. In the present section, we require the simultaneous realizations of two functions $-y_{21}$ and y_{22}, or z_{21} and z_{22} or z_{11}, so that the resulting network will yield the desired open-circuit transfer voltage ratio G_{12}.

In the network of Figure 3.22, there is *transmission* through the network when, for finite input, there results an output. The network is said to have zero transmission when, for finite input, zero output occurs. The frequencies at which a two-port network yields zero output for a finite input are referred to as the *zeros of transmission*. Zeros of transmission play a major role in ladder development, to be discussed in this section. In a two-port network, there are many ways of producing zeros of transmission. One possible way for preventing the input signal from reaching the output is by shorting together all transmission paths or by opening all transmission paths by means of a series or parallel resonance. Another possibility is that signals transmitted by different paths cancel at the output.

Our immediate question is how do the zeros of transmission relate to the network functions? From (3.50) and (3.51) we see that zero output, $V_2=0$, implies a zero for each of these functions. Therefore, zeros of transmission are zeros of $-y_{21}$ or

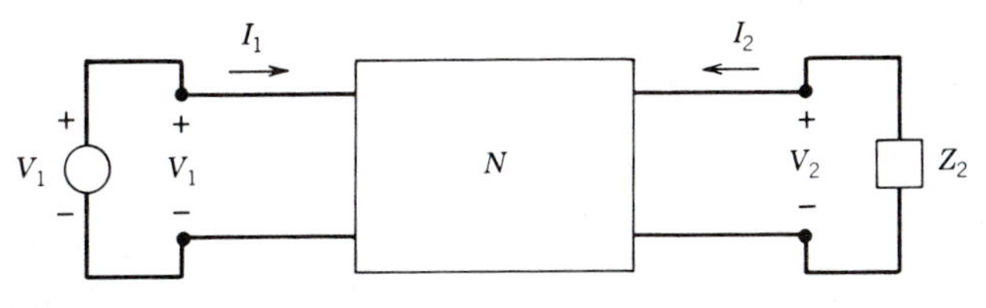

Figure 3.22

†The transfer current ratio α_{12} can be expressed by the equation $\alpha_{12}=-I_2/I_1=z_{21}/(z_{22}+Z_2)$.

z_{21} provided y_{21} and y_{22} or z_{21} and z_{11} have the same poles. For the ladder network, the transmission can be interrupted only by a short in a shunt arm or an open circuit in a series arm. The short circuit of a shunt arm corresponds to the pole frequencies of its admittances, whereas the open circuit of a series arm corresponds to the pole frequencies of its impedances. This means that the zeros of transmission of a ladder network can be identified directly with the zeros of the impedances of the shunt arms and the poles of the impedances of the series arms. For the LC ladder, all zeros of transmission are on the $j\omega$-axis, and for the RC ladder they lie on the negative real axis of the s-plane.

3.3.1 The *LC* Ladder

For LC ladders, the requirements for $-y_{21}$ and y_{22} or z_{21} and z_{22} are that the driving-point functions y_{22} and z_{22} be positive real with poles and zeros interlaced on the $j\omega$-axis. The transfer function $-y_{21}$ or z_{21}, assumed to have the same poles as y_{22} or z_{22}, must have all of its zeros on the $j\omega$-axis. However, these zeros need not be interlaced with the poles and they may not be simple. Our strategy in realization is that of carrying out the driving-point synthesis of y_{22} or z_{22}, using Cauer ladder development method, in such a way that the zeros of transmission are realized at the same time. The procedure consists of two steps: a zero-shifting step and a zero-producing step. These will be discussed below.

ZERO SHIFTING BY PARTIAL REMOVAL. Consider an impedance $Z(s)$ of (3.10), the partial-fraction expansion of which is shown in (3.12). The first term Hs on the right-hand side of (3.12) is due to the contribution of the pole at the infinity. If this term Hs is subtracted from $Z(s)$, the resulting function $Z(s) - Hs$ is devoid of the pole at the infinity. We say that the pole at infinity has been removed. Instead of complete removal of the pole at infinity, suppose that we subtract a fraction of the term Hs from $Z(s)$ by introducing a constant k_p such that

$$Z_1(s) = Z(s) - k_p Hs, \qquad k_p < 1 \tag{3.52}$$

We say that the pole at infinity has been *partially removed* or *weakened*. The function $Z_1(s)$ that results from the partial removal of the pole at infinity still possesses a pole at infinity. Since all the zeros of $Z_1(s)$ are again located on the $j\omega$-axis of the s-plane, these zeros are found by substituting $s = j\omega$ in (3.52),

$$X_1(\omega) = X(\omega) - k_p H\omega \tag{3.53}$$

where $Z_1(j\omega) = jX_1(\omega)$. The zeros of X_1 are the values of ω satisfying the equation

$$X(\omega) = k_p H\omega \tag{3.54}$$

Solutions to this equation are found graphically from the intersections of the curves $X(\omega)$ and $k_p H\omega$ as shown in Figure 3.23. Observe that all of the zeros in the resulting function are shifted toward the pole being weakened, a direct consequence of the positive slope property of reactance functions. The amount of shift of the zeros depends on the value of k_p and the proximity of a zero to the pole being weakened.

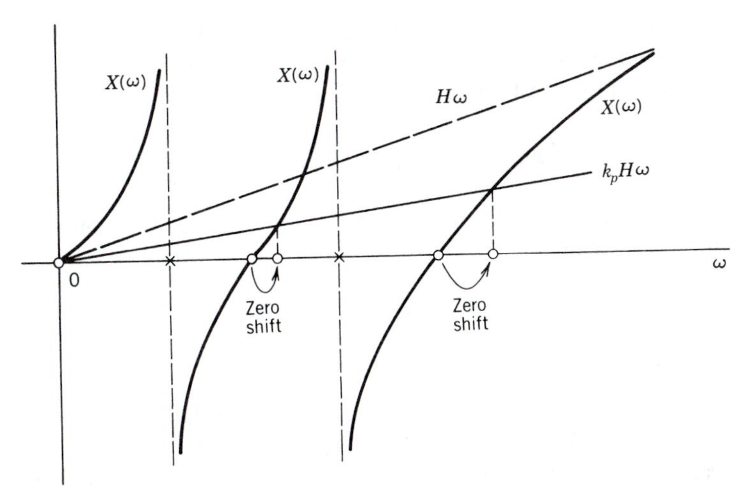

Figure 3.23

We next consider the term K_0/s in (3.12), which is due to the pole at the origin. The partial removal of this pole is equivalent to the operation

$$Z_2(s)=Z(s)-k_pK_0/s, \qquad k_p<1 \tag{3.55}$$

As before, the zeros of $Z_2(s)$ are defined by the intersections of the curves $X(\omega)$ and $-k_pK_0/\omega$ with ω,

$$X(\omega)=-k_pK_0/\omega \tag{3.56}$$

as illustrated in Figure 3.24. Observe again that the zeros are shifted toward the pole being weakened, which is at the origin.

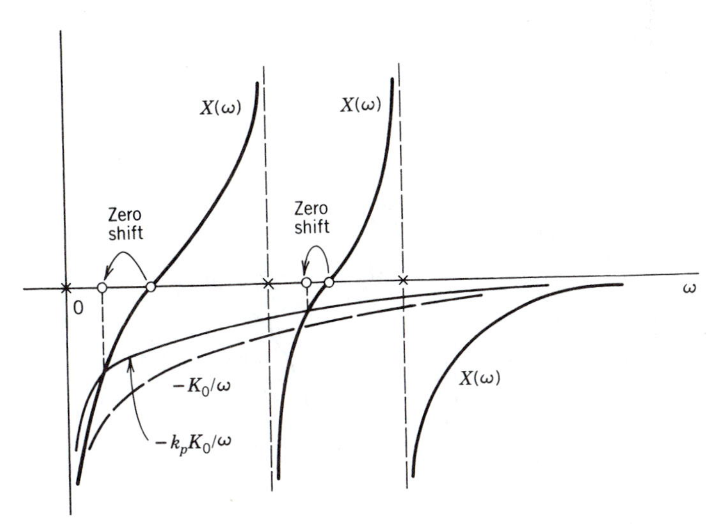

Figure 3.24

Finally, for the finite nonzero poles, the corresponding factors take the general form $2K_i s/(s^2+\omega_i^2)$ as previously given in (3.12). The partial removal of this pair of complex conjugate poles results in the new function

$$Z_3(s)=Z(s)-k_p\frac{2K_i s}{s^2+\omega_i^2}, \qquad k_p<1 \tag{3.57}$$

The zeros of this function are defined by the intersections of the plots of $X(\omega)$ and $-k_p 2K_i\omega/(\omega^2-\omega_i^2)$ with ω,

$$X(\omega)=-k_p\frac{2K_i\omega}{\omega^2-\omega_i^2} \tag{3.58}$$

as indicated in Figure 3.25.

As a result of the above discussion, we conclude that the partial removal of a pole shifts the zero toward that pole. The amount of shift depends on the value of k_p and the proximity of a zero to that pole, but in no case can a zero be shifted beyond an adjacent pole.

ZERO PRODUCING BY COMPLETE POLE REMOVAL. After a zero of transmission has been shifted to a desired location by the partial removal of an appropriate pole, the realization of this zero of transmission is accomplished by the complete removal of the pole of the reciprocal function corresponding to the shifted zero. For the LC case, a series combination of L and C produces a zero at its resonant frequency $\omega=1/\sqrt{LC}$, and this network is used in the shunt arm in the ladder to produce the desired zero of transmission. Likewise, the parallel connection of L and C yields an infinite impedance at its resonant frequency, and is used in the series arm of the ladder. These networks are shown in Figure 3.26 in appropriate positions. We illustrate the above results by the following examples.

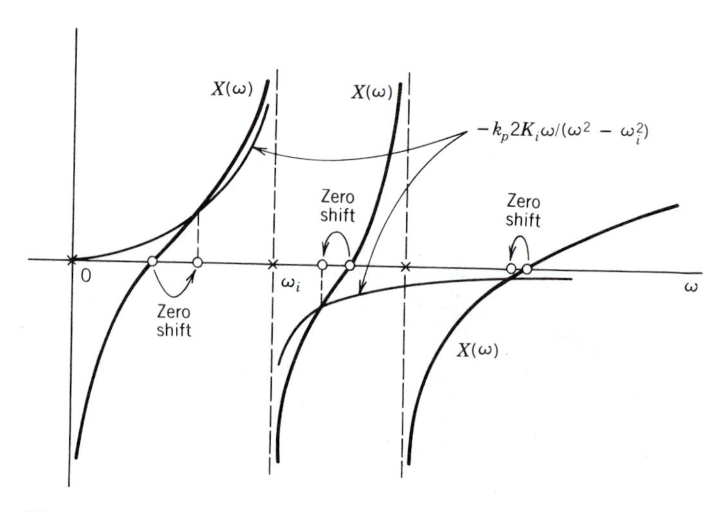

Figure 3.25

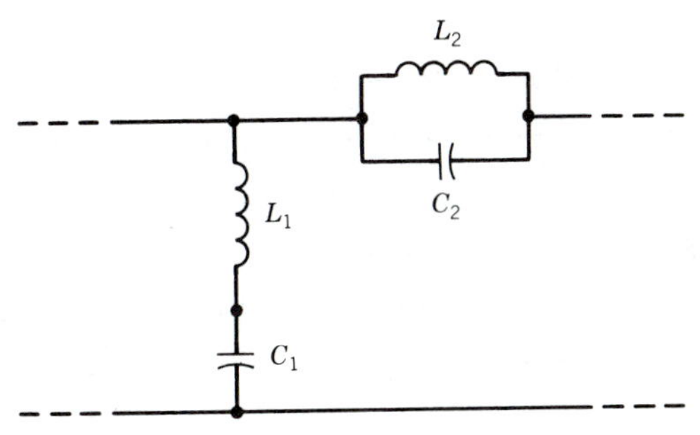

Figure 3.26

Example 3.5 The transfer voltage-ratio function $G_{12}(s)$ of a lossless two-port network when terminated in a 1-Ω resistor is given by

$$G_{12}(s)=K\frac{s^2+4}{s^3+6s^2+11s+6} \tag{3.59}$$

We wish to realize an LC two-port network to meet this specification within a multiplicative constant K.

Our first step is to identify the required y_{22} and $-y_{21}$. For this we divide the numerator and denominator of (3.59) by the odd part of the denominator polynomial and use (3.50) with $Y_2=1$. The result is

$$G_{12}(s)=\frac{K\dfrac{s^2+4}{s^3+11s}}{\dfrac{6(s^2+1)}{s^3+11s}+1}=\frac{-y_{21}}{y_{22}+1} \tag{3.60}$$

giving

$$-y_{21}(s)=K\frac{s^2+4}{s(s^2+11)} \tag{3.61}$$

$$y_{22}(s)=\frac{6(s^2+1)}{s(s^2+11)} \tag{3.62}$$

Both functions have the same poles at $s=0$ and $s=\pm j\sqrt{11}$. The zeros of transmission are zeros of $-y_{21}$ and are located at $s=\pm j2$ and $s=\infty$. These poles and zeros are shown in Figure 3.27. The zero of transmission at the infinity can be realized by the complete removal of the pole of the reciprocal function $z_1=1/y_{22}$ at the infinity. This is equivalent to the operation

$$z_2(s)=z_1(s)-K_\infty s \tag{3.63}$$

where $K_\infty=1/6$ is the residue at the pole $s=\infty$ of $1/y_{22}$, giving

$$z_2(s)=\frac{5s}{3(s^2+1)} \tag{3.64}$$

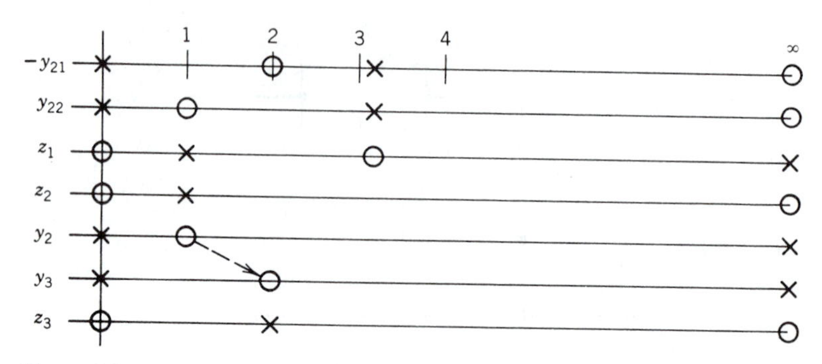

Figure 3.27

The poles and zeros of z_2 are again shown in Figure 3.27.

To realize the zero of transmission at $s=j2$, we consider the function $y_2=1/z_2$, which has a zero at $s=j1$. Clearly, this zero at $s=j1$ can be shifted to $s=j2$ by the partial removal of the pole at infinity or

$$y_3(s)=y_2(s)-k_p 3s/5 \tag{3.65}$$

where 3/5 is the residue of y_2 at the pole $s=\infty$. To determine the value of k_p, we set $y_3(s)=0$ at $s=j2$ or

$$y_2(j2)=jk_p 3\times 2/5 \tag{3.66}$$

giving $k_p=3/4$, which is less than 1 as required. The new function becomes

$$y_3(s)=\frac{3s^2+3}{5s}-\frac{9s}{20}=\frac{3(s^2+4)}{20s} \tag{3.67}$$

corresponding to the removal of a shunt capacitor of capacitance 9/20 F, as shown in Figure 3.28. The factor (s^2+4) in the numerator was anticipated because our objective was to produce a zero in the driving-point function y_3 at $s=j2$. To realize this zero, we consider the reciprocal function $z_3=1/y_3$ by complete removal of its pole at $s=j2$. This yields a parallel connection of L and C as shown in Figure 3.28. Turning around the network of Figure 3.28 and terminating the output with a 1-Ω resistor

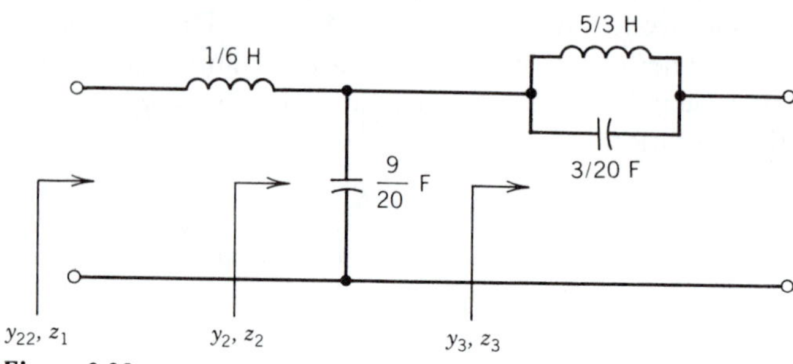

Figure 3.28

gives the complete network presented in Figure 3.29. This network will meet the given specification of (3.59) with $K=3/2$.

Example 3.6 Synthesize an LC ladder to satisfy the following open-circuit impedance parameters:

$$z_{21}(s)=K\frac{(s^2+4)(s^2+9)}{s(s^2+2)(s^2+5)} \tag{3.68}$$

$$z_{22}(s)=\frac{(s^2+1)(s^2+3)}{s(s^2+2)(s^2+5)} \tag{3.69}$$

The poles and zeros of z_{21} and z_{22} are shown in Figure 3.30. As before, the zeros of z_{21} are the zeros of transmission. From this figure we see that the zero of the reciprocal function $y_1=1/z_{22}$ at $s=j\sqrt{5}$ can be shifted to $s=j3$ by the partial removal of the pole at $s=\infty$,

$$y_2(s)=y_1(s)-k_p s \tag{3.70}$$

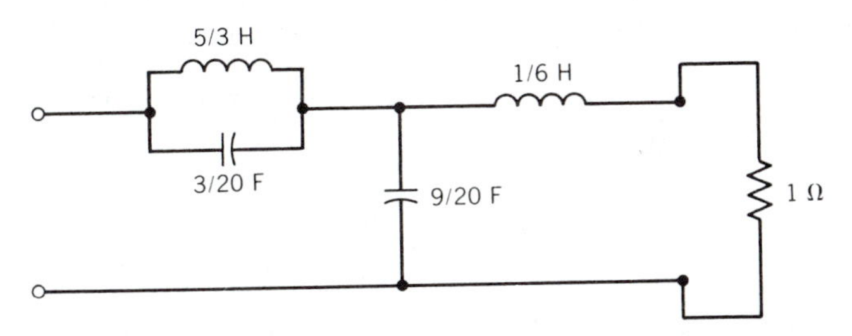

Figure 3.29

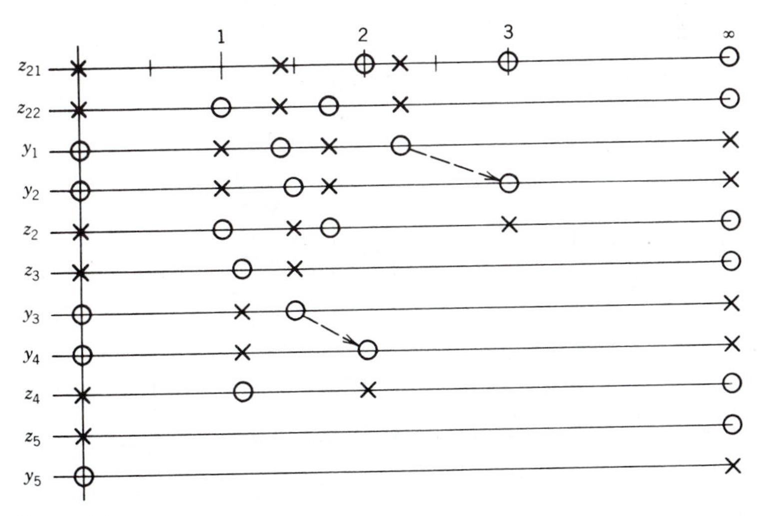

Figure 3.30

where the residue of y_1 at $s = \infty$ is 1. To ascertain k_p we set $y_2(j3)=0$ in (3.70) and obtain

$$k_p=7/12 \tag{3.71}$$

yielding

$$y_2(s)=\frac{5s(s^2+9)(s^2+11/5)}{12(s^2+1)(s^2+3)} \tag{3.72}$$

which has a zero at $s=j3$, as expected. To realize this zero, we consider the reciprocal function $z_2=1/y_2$ and remove the pair of poles at $s=\pm j3$. The resulting function is

$$z_3(s)=z_2(s)-\frac{32s/17}{s^2+9}=\frac{2.6s^2+4}{5s(s^2+11/5)} \tag{3.73}$$

corresponding to the removal of a parallel combination of an inductor of value 32/153 H and a capacitor of value 17/32 F as indicated in Figure 3.31.

We next realize the zero of transmission at $s=j2$. For this we consider the reciprocal function $y_3=1/z_3$. From Figure 3.30 we see that the zero at $s=j\sqrt{11/5}$ can be shifted to $s=j2$ by weakening the pole at the infinity or

$$y_4(j2)=y_3(j2)-k_p\,j10/2.6=0 \tag{3.74}$$

giving $k_p=0.732$. The new function becomes

$$y_4(s)=0.515\,\frac{s(s^2+4)}{s^2+1.54} \tag{3.75}$$

containing a pair of zeros at $s=\pm j2$. To realize this pair of zeros, we consider the reciprocal function $z_4=1/y_4$ by the complete removal of its poles at $s=\pm j2$,

$$z_5(s)=z_4(s)-\frac{1.2s}{s^2+4}=\frac{0.746}{s} \tag{3.76}$$

corresponding to a parallel connection of L and C as shown in Figure 3.31.

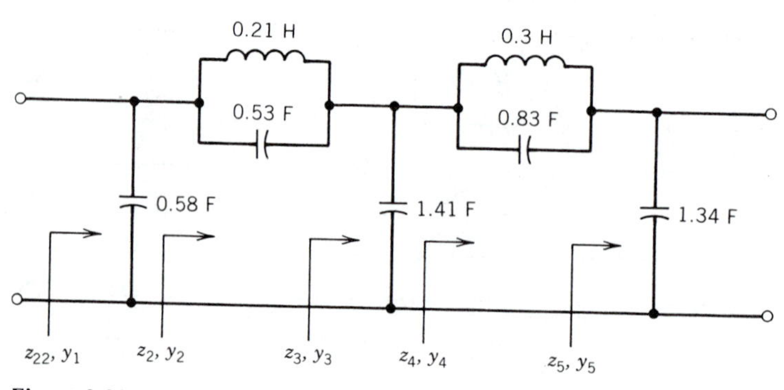

Figure 3.31

Finally, to realize the zero of transmission at the infinity, we consider $y_5 = 1/z_5 = 1.34s$ corresponding to a shunt capacitor of value 1.34 F as shown in Figure 3.31. The constant K for this realization is found directly from Figure 3.31 to be $K = 0.0833$.

In the synthesis of the two-port network of Figure 3.31, the zeros of transmission are realized in the following order: $\pm j3$, $\pm j2$, and ∞. If we realize this sequence by first realizing the zero of transmission at $s = \infty$, then at $s = \pm j3$, and finally at $s = \pm j2$, in accordance with the pole-zero configuration of Figure 3.32, the resulting network is shown in Figure 3.33 with $K = 1/12 = 0.0833$.

On the other hand, if the zeros of transmission are realized in the order of $\pm j2$, $\pm j3$, and ∞, in accordance with the pole-zero pattern of Figure 3.34, the corresponding two-port network is shown in Figure 3.35 with $K = 0.0833$. So, all these give virtually identical K's, a result that is not typical.

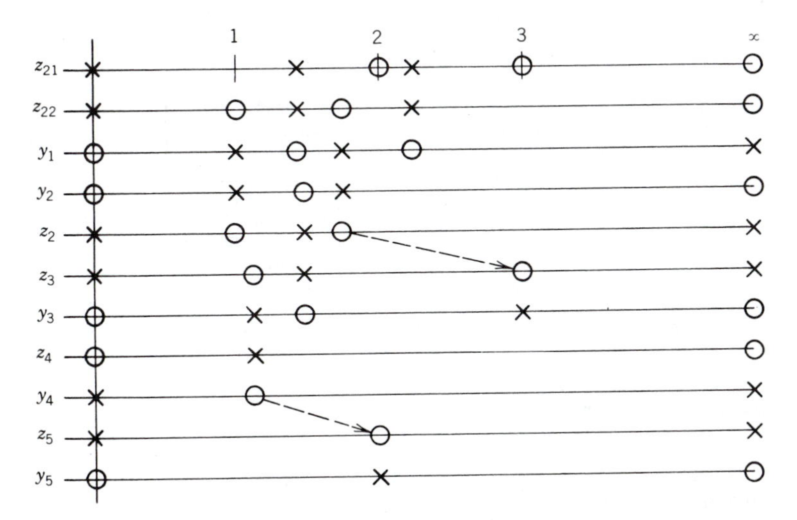

Figure 3.32

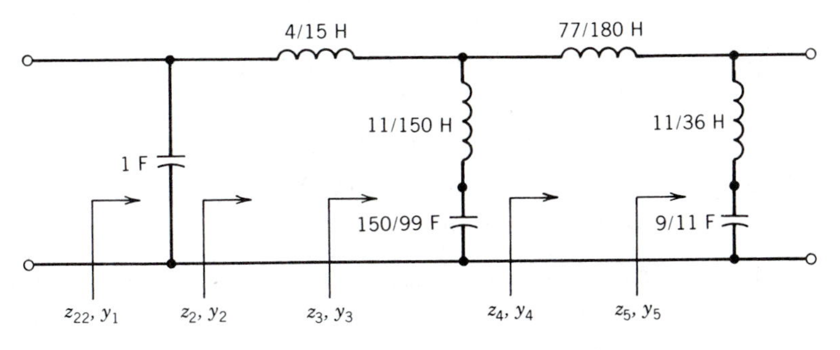

Figure 3.33

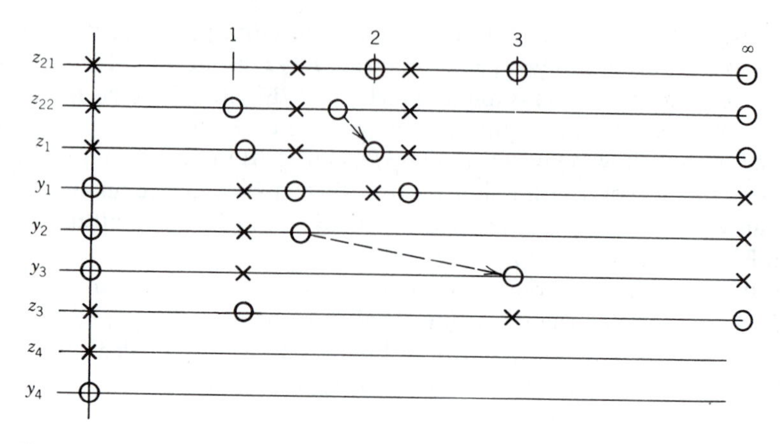

Figure 3.34

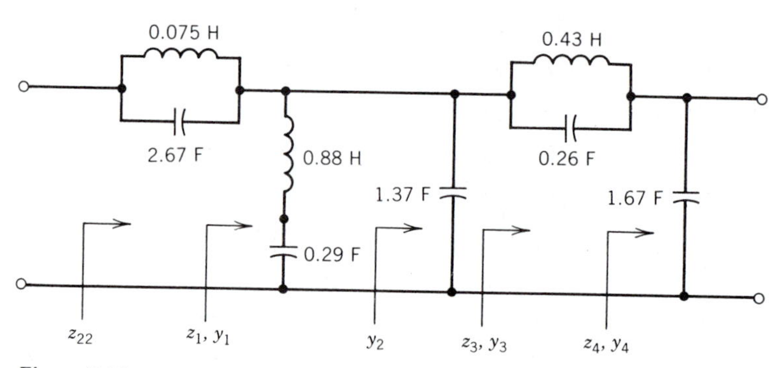

Figure 3.35

3.3.2 The *RC* Ladder

In this section we consider the realization of the *RC* ladder with prescribed $-y_{21}$ and y_{22} or z_{21} and z_{22}. The *RC* ladders are regarded to be of more interest than the *RL* ladders because of practical considerations where size and weight preclude the use of inductors.

As in the *LC* case, the zero shifting for the *RC* driving-point functions is accomplished by the following three operations:

(i) The partial removal of a constant $Z(\infty)$ from $Z(s)$.

(ii) The partial removal of a constant $Y(0)$ from $Y(s)$.

(iii) The partial removal of a pole from $Z(s)$ or $Y(s)$.

The first operation permits a series resistance to be removed so that the resulting impedance is still positive real:

$$Z_1(s) = Z(s) - k_p Z(\infty), \qquad k_p \leqslant 1 \tag{3.77}$$

The zeros of $Z_1(s)$, known to be real and nonpositive, are defined by the intersections of the plots of $Z(\sigma)$ and the horizontal line $k_p Z(\infty)$ with σ,

$$Z(\sigma) = k_p Z(\infty) \tag{3.78}$$

as shown in Figure 3.36. Observe that all zeros are shifted toward $s = -\infty$ by the partial removal of $Z(\infty)$, a direct consequence of the negative slope property of the RC impedances.

The second operation permits a shunt resistance to be removed so that the resulting function

$$Y_1(s) = Y(s) - k_p Y(0), \qquad k_p \leqslant 1 \tag{3.79}$$

remains positive real. The zeros of $Y_1(s)$, again real and nonpositive, are defined by the equation

$$Y(\sigma) = k_p Y(0) \tag{3.80}$$

which is plotted in Figure 3.37. Observe again that the zeros of $Y_1(s)$ are shifted toward $s = 0$ in comparison with the zeros of $Y(s)$.

The partial removal of a pole from $Z(s)$ is equivalent to the operation

$$Z_2(s) = Z(s) - k_p \frac{K_i}{s + \sigma_i}, \qquad k_p < 1 \tag{3.81}$$

The zeros of $Z_2(s)$ are defined by the equation

$$Z(\sigma) = k_p \frac{K_i}{\sigma + \sigma_i} \tag{3.82}$$

as shown in Figure 3.38. These zeros are again shifted toward the pole at $s = -\sigma_i$, which is partially weakened.

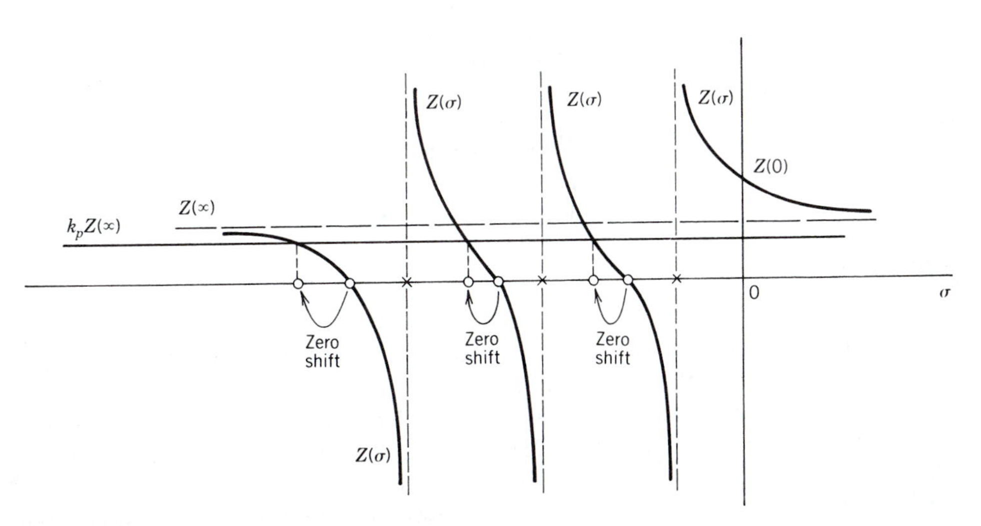

Figure 3.36

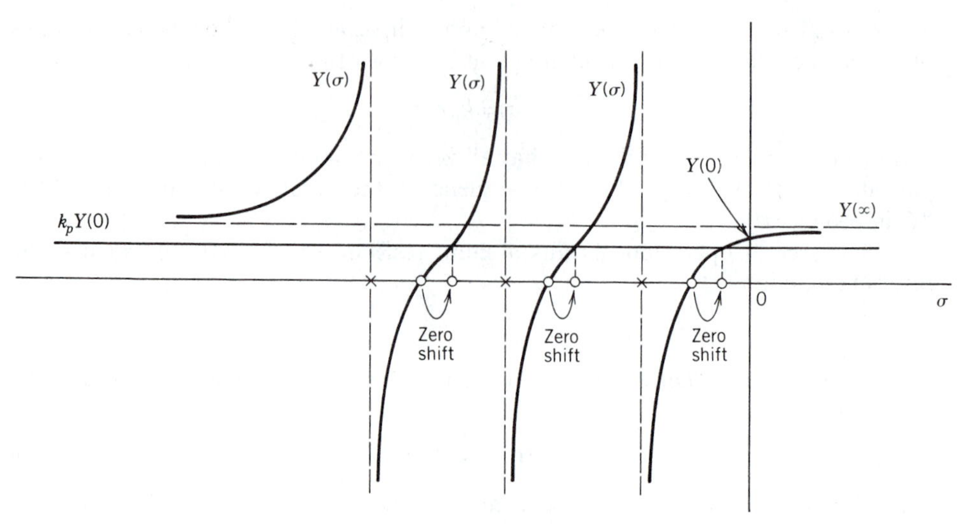

Figure 3.37

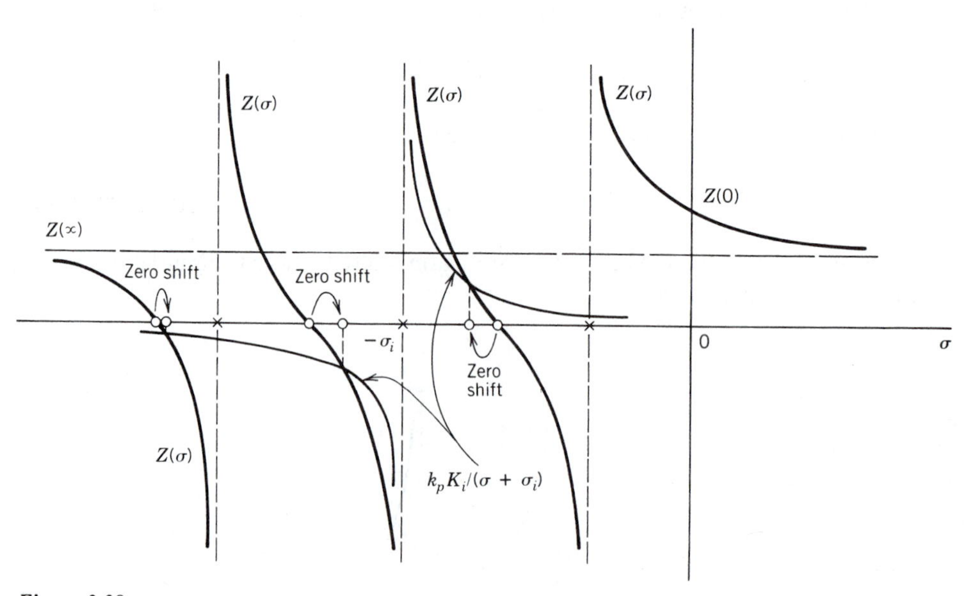

Figure 3.38

Our conclusion is that the partial removal of a constant or a pole results in the shifting of the zeros of the remaining function toward the quantity being weakened, be it $Z(\infty)$, $Y(0)$, or a pole. The amount of shift depends on the value of k_p and the proximity of a zero to that quantity. Once a zero is shifted to an appropriate location, the realization of this zero is accomplished by the complete removal of the pole of the reciprocal function corresponding to the shifted zero. We illustrate this procedure by the following example.

Example 3.7 Consider the following specification functions:

$$z_{21}(s)=K\frac{(s+1)(s+4)}{(s+2)(s+5)} \tag{3.83}$$

$$z_{22}(s)=\frac{(s+3)(s+6)}{(s+2)(s+5)} \tag{3.84}$$

The zeros of transmission are located at $s=-1$ and $s=-4$. Suppose that we wish to realize these zeros in the order of -1 and -4. For this we consider the reciprocal function $y_1=1/z_{22}$ whose poles and zeros are shown in Figure 3.39. Clearly, the zero at $s=-2$ of y_1 can be shifted to $s=-1$ by partial removal of the constant $y_1(0)=5/9$,

$$y_2(s)=y_1(s)-k_p y_1(0) \tag{3.85}$$

To ascertain the value k_p, we set $y_2(-1)=0$, giving $k_p=18/25$ and

$$y_2(s)=\frac{(s+1)(3s+14)}{5(s+3)(s+6)} \tag{3.86}$$

The function y_2 has a zero at $s=-1$, as expected. To realize this zero, we consider the reciprocal function $z_2=1/y_2$ by the complete removal of the pole at $s=-1$. This results in a new function

$$z_3(s)=z_2(s)-\frac{50/11}{s+1}=\frac{55s+290}{33s+154} \tag{3.87}$$

The removed term corresponds to a parallel connection of a resistor of value 50/11 Ω and a capacitor of capacitance 11/50 F, as indicated in Figure 3.40.

For the zero of transmission at $s=-4$, we consider the reciprocal function $y_3=1/z_3$. From Figure 3.39 we see that the zero of y_3 at $s=-14/3$ can be shifted to -4 by partial removal of the constant $y_3(0)=77/145$:

$$y_4(s)=y_3(s)-k_p 77/145 \tag{3.88}$$

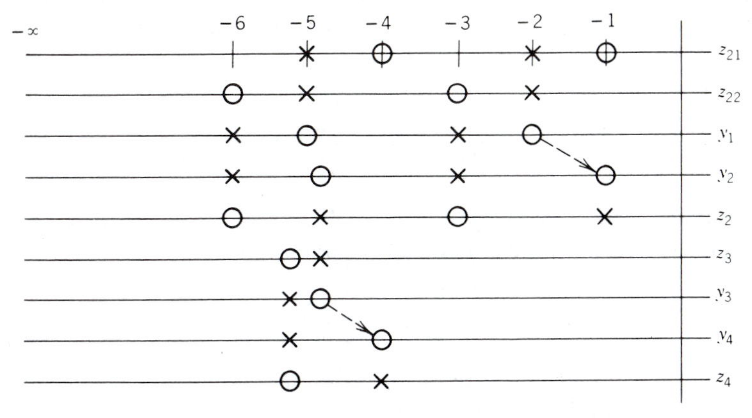

Figure 3.39

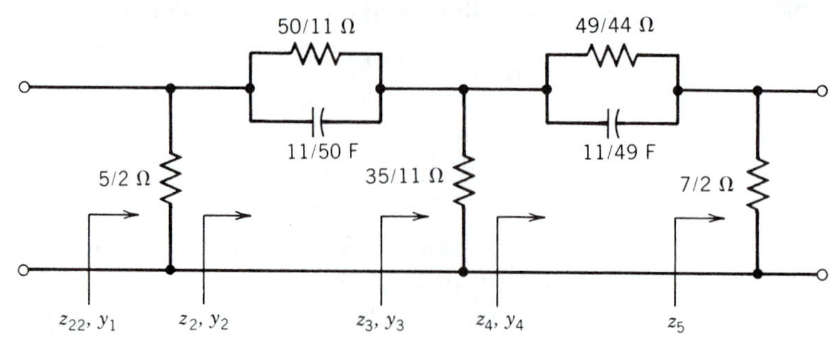

Figure 3.40

To ascertain k_p we set $y_4(-4)=0$ and obtain $k_p=29/49$ and

$$y_4(s)=\frac{110(s+4)}{35(11s+58)} \tag{3.89}$$

showing a zero at $s=-4$ as anticipated. This zero is realized by the complete removal of the pole of $z_4=1/y_4$ at $s=-4$, yielding

$$z_5(s)=z_4(s)-\frac{49/11}{s+4}=7/2\,\Omega \tag{3.90}$$

The complete realization of the network is shown in Figure 3.40 with $K=1$.

As an alternative, we wish to realize the zeros of transmission in the order of -4 and -1. From Figure 3.39 we might attempt to shift the zero of y_1 at $s=-5$ to -4 by partial removal of $y_1(0)=5/9$. Proceeding along this line, the desired value of k_p is determined by the equation

$$y_1(-4)-k_p y_1(0)=0 \tag{3.91}$$

yielding $k_p=1.8$, which is greater than 1. This means that the zero of y_1 at $s=-5$ cannot be shifted to -4 by the partial or complete removal of the constant $y_1(0)$ without destroying the positive realness of the remaining function. To avoid this difficulty, let us first make the complete removal of the constant $y_1(0)$. The remaining function becomes

$$y_2'(s)=y_1(s)-y_1(0)=\frac{2s(2s+9)}{9(s+3)(s+6)} \tag{3.92}$$

whose poles and zeros are shown in Figure 3.41. To shift the zero of y_2' at $s=-4.5$ to -4, we now weaken the pole of y_2' at $s=-3$ and obtain a new function

$$y_3'(s)=y_2'(s)-k_p\frac{2s/9}{s+3} \tag{3.93}$$

Note that the term $2s/[9(s+3)]$ is obtained by first expanding y_2'/s in a partial fraction and then multiplying both sides by s, as discussed in Section 3.2. Setting $y_3'(-4)=0$

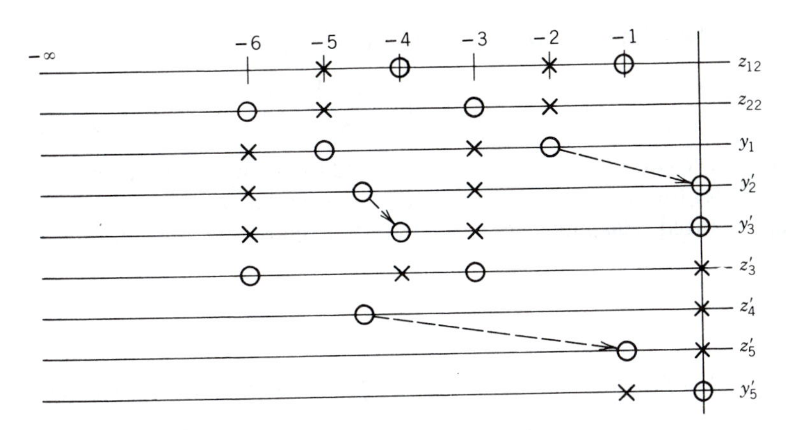

Figure 3.41

in (3.93) gives $k_p = \frac{1}{2}$ and

$$y_3'(s) = \frac{s(s+4)}{3(s+3)(s+6)} \tag{3.94}$$

This equation has a desired zero at $s = -4$, which can be realized by the complete removal of the pole of the reciprocal function $z_3' = 1/y_3'$ at $s = -4$:

$$z_4'(s) = z_3'(s) - \frac{3/2}{s+4} = \frac{3(2s+9)}{2s} \tag{3.95}$$

The removed pole corresponds to a parallel connection of a resistor of value 3/8 Ω and a capacitor of capacitance 2/3 F, as indicated in Figure 3.42.

For the zero of transmission at $s = -1$, we partially remove the pole at $s = 0$ from z_4'. This results in a new function

$$z_5'(s) = z_4'(s) - k_p 27/2s \tag{3.96}$$

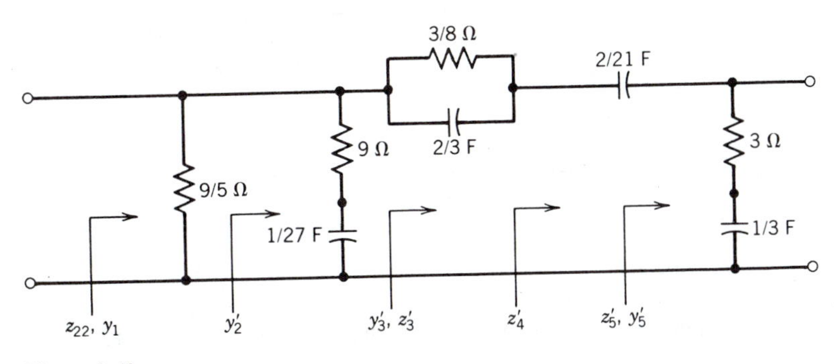

Figure 3.42

Setting $z_5'(-1)=0$ in (3.96) yields $k_p=7/9$ and

$$z_5'(s)=\frac{3(s+1)}{s} \tag{3.97}$$

which has a zero at $s=-1$. This zero is realized by considering the function $y_5'=1/z_5'$, which denotes the series connection of a resistor and a capacitor as shown in Figure 3.42. The realized constant K is found directly from the network of Figure 3.42 to be $K=1$.

3.4 TWO-PORT SYNTHESIS WITH PARALLEL LADDERS

In the synthesis of LC ladders, all zeros of transmission are required to be located on the $j\omega$-axis. For RC ladders, they are restricted to the negative real axis of the s-plane. Now some applications require that the zeros of transmission be complex, not necessarily restricted to the two axes of the s-plane. Such complex zeros of transmission are needed for certain phase-correction applications, and cannot be realized by a single LC or RC ladder network, because there is only a single transmission path between input and output terminals. The use of parallel ladders, on the other hand, provides the multiple paths of transmission that are capable of producing complex zeros of transmission. This structure was first suggested by Guillemin.†

Consider the parallel connection of the ladder networks N_α and N_β of Figure 3.43. The y-parameters y_{ij} of the composite two-port N can be expressed in terms of those $y'_{ij\alpha}$ and $y'_{ij\beta}$ of the individual ones N_α and N_β by the equation‡

$$y_{ij}=y'_{ij\alpha}+y'_{ij\beta}, \qquad i, j=1, 2 \tag{3.98}$$

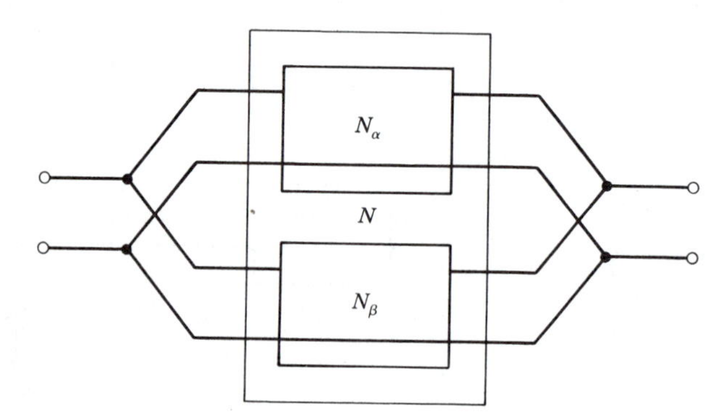

Figure 3.43

†E. A. Guillemin, "Synthesis of *RC* networks," *J. Math. Phys.*, Vol. 28, 1949, pp. 22–42.

‡The primes on the y-parameters are for algebraic convenience in a later step of the solution process.

Thus, to realize $-y_{21}$ and y_{22} we may divide them into pairs $-y'_{21\alpha}$, $y'_{22\alpha}$ and $-y'_{21\beta}$, $y'_{22\beta}$, and realize each pair separately as an LC or RC ladder. Connecting these individual ladders in parallel will yield the desired realization for $-y_{21}$ and y_{22}. However, there is a problem that we will have to deal with before the procedure is considered to be successful. Recall that in Cauer ladder development discussed in the foregoing for the LC or RC case, $-y_{21}$ is realized only within the multiplicative constant k. Thus, the transfer admittances actually realized by the component ladders will be $-k_\alpha y'_{21\alpha}$ and $-k_\beta y'_{21\beta}$. The sum of these may not be the desired $-ky_{21}$ unless $k=k_\alpha=k_\beta$. To circumvent this difficulty, let us adjust the admittance level of the α-ladder N_α by a factor of b_α and the β-ladder N_β by b_β. Then the functions of the resulting realizations become

$$-y'_{21\alpha}=-b_\alpha k_\alpha y_{21\alpha}, \qquad y'_{22\alpha}=b_\alpha y_{22\alpha} \tag{3.99a}$$

$$-y'_{21\beta}=-b_\beta k_\beta y_{21\beta}, \qquad y'_{22\beta}=b_\beta y_{22\beta} \tag{3.99b}$$

where $y_{ij}=y_{ij\alpha}+y_{ij\beta}$, $i, j=1, 2$. Substituting these into (3.98) gives

$$y_{21}=b_\alpha k_\alpha y_{21\alpha}+b_\beta k_\beta y_{21\beta} \tag{3.100}$$

$$y_{22}=b_\alpha y_{22\alpha}+b_\beta y_{22\beta} \tag{3.101}$$

There are many ways to satisfy these equations. One simple way is to let $y'_{22\alpha}$ and $y'_{22\beta}$ have the same poles and zeros as y_{22} but different scale factors such that $y'_{22\alpha}=b_\alpha y_{22}$ and $y'_{22\beta}=b_\beta y_{22}$, obtaining from (3.101)

$$b_\alpha+b_\beta=1 \tag{3.102}$$

and for (3.100) we set

$$b_\alpha k_\alpha=b_\beta k_\beta=k \tag{3.103}$$

For m ladders in parallel, the conditions become

$$b_1+b_2+\cdots+b_m=1 \tag{3.104}$$

$$b_1k_1=b_2k_2=\cdots=b_mk_m=k \tag{3.105}$$

where b_i is the admittance scale factor for the ith ladder, and k_i is the realized multiplicative constant of the ith ladder. Equations (3.104) and (3.105) constitute m simultaneous equations in m unknowns b_i, once realized values of the k's have been determined. When these equations are solved for b_i. the admittance level of each ladder can be scaled accordingly. Finally, we connect the m ladders in parallel to realize the $-y_{21}$ specifications within the multiplicative constant k of (3.105), and the y_{22} specifications exactly.

We next consider the decomposition of $-y_{21}$ into the sum of m terms. Let us write $y_{22}=p(s)/q(s)$ and

$$-y_{21}(s)=\frac{r(s)}{q(s)}=\frac{a_0+a_1s+a_2s^2+\cdots+a_{n-1}s^{n-1}+a_ns^n}{q(s)} \tag{3.106}$$

This equation can be decomposed into the sum in at least two simple ways:

$$-y_{21}(s)=\frac{a_0}{q(s)}+\frac{a_1 s}{q(s)}+\frac{a_2 s^2}{q(s)}+\cdots+\frac{a_{n-1}s^{n-1}}{q(s)}+\frac{a_n s^n}{q(s)} \tag{3.107}$$

where $m=n$, or

$$-y_{21}(s)=\frac{a_0+a_1 s}{q(s)}+\frac{a_2 s^2+a_3 s^3}{q(s)}+\cdots+\frac{a_{n-1}s^{n-1}+a_n s^n}{q(s)} \tag{3.108}$$

Observe that in each of these two forms, the zeros of transmission are all at the origin, at infinity, or on the negative real axis of the s-plane, so long as all of the coefficients are real and positive. In such a case, each component of $-y_{21}$ can be realized as an RC ladder network. Furthermore, if the zeros of transmission are at the origin or at infinity, the corresponding RC ladder can be obtained by a simple continued-fraction expansion. This is because the first and second Cauer canonical forms are identified with all zeros of transmission at infinity and at origin, respectively. We illustrate the above procedure by the following examples.

Example 3.8 Consider the following specification functions:

$$-y_{21}(s)=\frac{s^2+1}{(s+4)(s+8)} \tag{3.109}$$

$$y_{22}(s)=\frac{(s+2)(s+6)}{(s+4)(s+8)} \tag{3.110}$$

The zeros of transmission are located at $s=\pm j1$. Clearly, these specifications cannot be realized by a single RC or LC ladder. We shall realize this pair by the method of parallel ladders. First, we arbitrarily select the transfer admittances

$$-y_{21\alpha}(s)=\frac{s^2}{(s+4)(s+8)}, \qquad -y_{21\beta}(s)=\frac{1}{(s+4)(s+8)} \tag{3.111}$$

to satisfy the specifications (3.109). For $y_{22\alpha}$ and $y_{22\beta}$ we set

$$y_{22\alpha}(s)=y_{22\beta}(s)=y_{22}(s)=\frac{(s+2)(s+6)}{(s+4)(s+8)} \tag{3.112}$$

We now proceed to synthesize the required ladders called the α-ladder and the β-ladder by the technique developed in the foregoing.

THE α-LADDER. The specifications for the α-ladder are $-y_{21\alpha}$ and $y_{22\alpha}$. Since the zeros of transmission are all at the origin, the second Cauer canonical form is required. For this we expand $y_{22\alpha}$ in a continued fraction as

$$y_{22\alpha}(s)=\frac{12+8s+s^2}{32+12s+s^2}=3/8+\cfrac{1}{64/7s+\cfrac{1}{49/88+\cfrac{1}{1936/21s+\cfrac{1}{3/44}}}} \tag{3.113}$$

The desired α-ladder is shown in Figure 3.44(a). The realized value of the constant k_α is found to be $k_\alpha = 3/44$.

THE β-LADDER. The specifications for the β-ladder are $-y_{21\beta}$ and $y_{22\beta}$. Since the zeros of transmission are all at infinity, we require the first Cauer canonical form. For this we expand $y_{22\beta}$ in a continued fraction as

$$y_{22\beta} = \frac{s^2+8s+12}{s^2+12s+32} = \cfrac{1}{1+\cfrac{1}{s/4+\cfrac{1}{4/3+\cfrac{1}{3s/4+\cfrac{1}{1/3}}}}} \tag{3.114}$$

yielding the β-ladder of Figure 3.44(b) with $k_\beta = 12$.

Our next task is to adjust the admittance level of the individual ladders so that when they are connected in parallel, the y_{22} specifications are realized exactly, and the $-y_{21}$ specifications to within a multiplicative constant k. Thus, from (3.104) and (3.105) we require

$$b_\alpha + b_\beta = 1 \tag{3.115a}$$

$$b_\alpha k_\alpha = b_\beta k_\beta = k \tag{3.115b}$$

where $k_\alpha = 3/44$ and $k_\beta = 12$ or

$$b_\alpha + b_\beta = 1 \tag{3.116a}$$

$$b_\alpha - 176 b_\beta = 0 \tag{3.116b}$$

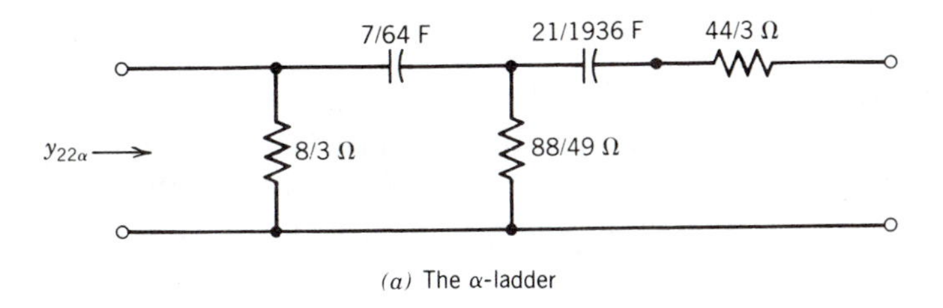

(a) The α-ladder

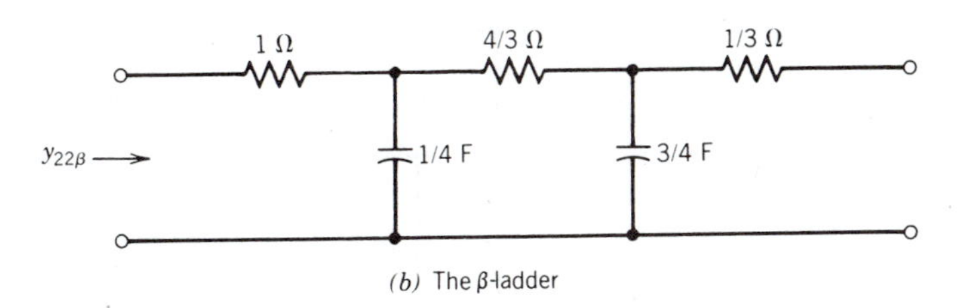

(b) The β-ladder

Figure 3.44

giving

$$b_\alpha = 176/177, \qquad b_\beta = 1/177 \tag{3.117a}$$

and

$$k = b_\alpha k_\alpha = b_\beta k_\beta = 12/177 \tag{3.117b}$$

We now adjust the admittance level of the α-ladder by a factor of 176/177, and the β-ladder by 1/177. The final realization is obtained by the parallel connection of the resulting α-ladder and the β-ladder as shown in Figure 3.45. Our realization of Figure 3.45 gives y_{22} specifications exactly and $-y_{21}$ to within a multiplicative constant:

$$-y_{21}(s) = \frac{12}{177} \cdot \frac{s^2+1}{(s+4)(s+8)} \tag{3.118}$$

Observe that in the network of Figure 3.45 there is a wide spread of element values ranging from 1.81 to 236 Ω. This may limit the practical utility of the realization. One possible solution is to make a different decomposition of $-y_{21}$ and to realize the resulting zeros of transmission in different orders. This will result in a number of realizations from which one can select what appears to be the best solution.

Refer to Figure 3.46. The y- and z-parameters of the reciprocal two-port networks $\hat{N}$ and N are related by the equations

$$y_{11} = \hat{y}_{11} + Y_A \tag{3.119a}$$

$$y_{22} = \hat{y}_{22} + Y_B \tag{3.119b}$$

$$y_{12} = y_{21} = \hat{y}_{12} = \hat{y}_{21} \tag{3.119c}$$

$$z_{11} = \hat{z}_{11} + Z_C \tag{3.120a}$$

$$z_{22} = \hat{z}_{22} + Z_D \tag{3.120b}$$

$$z_{12} = z_{21} = \hat{z}_{12} = \hat{z}_{21} \tag{3.120c}$$

For the network of Figure 3.46(a), a one-port admittance Y_A connected across the input terminals appears only in y_{11}, and one Y_B connected across the output terminals appears only in y_{22}. This indicates that a pole of y_{11} not present in $-y_{21}$ can be realized by a partial-fraction expansion as a one-port in the position of that of Y_A,

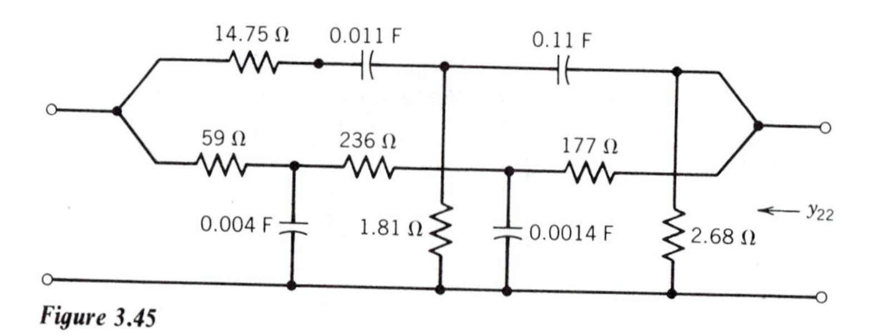

Figure 3.45

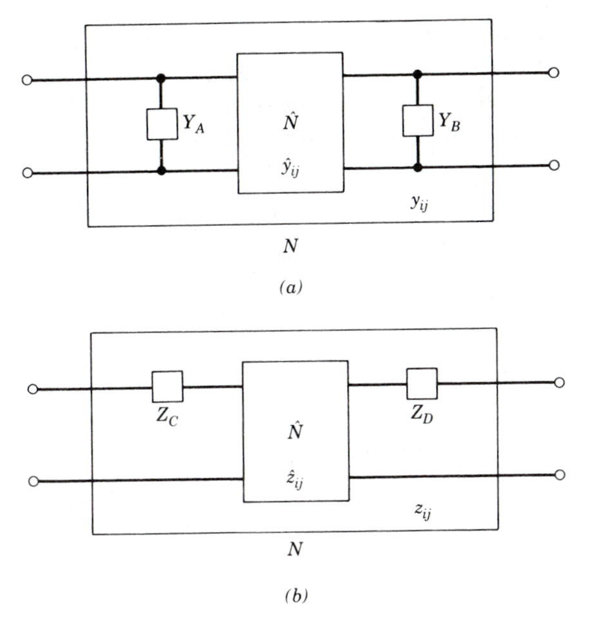

Figure 3.46

and a pole of y_{22} not present in $-y_{21}$ is realized in the position of Y_B. Poles that are present in y_{11} and/or y_{22} but not in $-y_{21}$ are distinguished by the name *private poles*. Likewise, poles in z_{11} and/or z_{22} but not in z_{12} are also called the *private poles* and can be realized by the one-port impedances in the positions of Z_C and Z_D as shown in Figure 3.46(*b*).

Example 3.9 It is required to design a lossless two-port network terminated in a 1-Ω resistor to meet the specifications for the transfer voltage ratio

$$G_{12}(s)=\frac{s^4+3s^2+5}{s^4+s^3+4s^2+2s+3} \tag{3.121}$$

To identify the desired $-y_{21}$ and y_{22}, we divide numerator and denominator of (3.121) by the odd part of the denominator polynomial and use (3.50):

$$G_{12}(s)=\frac{\dfrac{s^4+3s^2+5}{s^3+2s}}{\dfrac{s^4+4s^2+3}{s^3+2s}+1}=\frac{-y_{21}(s)}{y_{22}(s)+1} \tag{3.122}$$

From this we can make the following identification:

$$-y_{21}(s)=\frac{s^4+3s^2+5}{s(s^2+2)} \tag{3.123}$$

$$y_{22}(s)=\frac{(s^2+1)(s^2+3)}{s(s^2+2)} \tag{3.124}$$

Observe that the zeros of transmission of G_{12}, being the zeros of $-y_{21}$, are complex and occur in quad. These zeros cannot be realized by a single *LC* ladder. We shall apply the method of parallel ladder networks.

To proceed we arbitrarily decompose the transfer admittance $-y_{21}$ into the form

$$-y_{21}(s)=\frac{(s^2+1)(s^2+2)+3}{s(s^2+2)}=\frac{s^2+1}{s}+\frac{3}{s(s^2+2)} \tag{3.125}$$

from which we can make the following selection:

$$-y_{21\alpha}(s)=\frac{s^2+1}{s}, \qquad -y_{21\beta}(s)=\frac{3}{s(s^2+2)} \tag{3.126}$$

The required driving-point admittance is

$$y_{22\alpha}(s)=y_{22\beta}(s)=y_{22}(s)=\frac{(s^2+1)(s^2+3)}{s(s^2+2)} \tag{3.127}$$

The two desired ladders are next synthesized.

THE α-LADDER. $y_{22\alpha}$ has a pair of poles at $s=\pm j\sqrt{2}$, which are not contained in $-y_{21\alpha}$. These poles are the private poles of $y_{22\alpha}$. Private poles should be removed before the zero-shifting and zero-producing processes begin. The function that results after the complete removal of the poles at $s=\pm j\sqrt{2}$ is

$$y_1(s)=y_{22\alpha}(s)-\frac{\frac{1}{2}s}{s^2+2}=\frac{2s^2+3}{2s} \tag{3.128}$$

From Figure 3.47 we see that the zero of transmission at $s=j1$ can be produced by the partial removal of the pole of y_1 at $s=0$:

$$y_2(s)=y_1(s)-k_p 3/2s \tag{3.129}$$

Setting $y_2(j1)=0$ gives $k_p=1/3$ and

$$y_2(s)=\frac{s^2+1}{s} \tag{3.130}$$

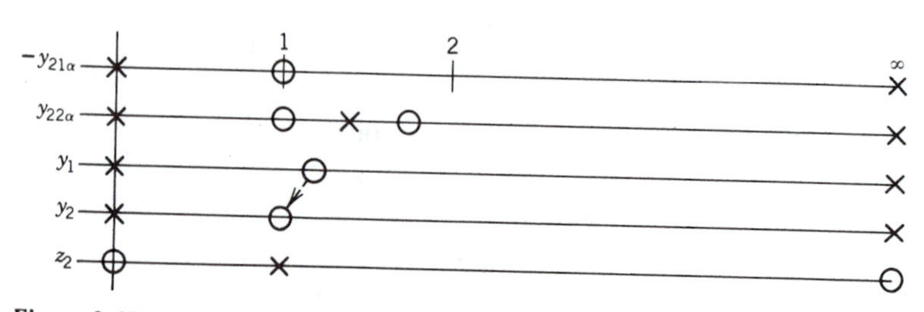

Figure 3.47

This zero is realized by the complete removal of the pole of $z_2 = 1/y_2$ at $s = j1$. The desired α-ladder is shown in Figure 3.48(a) with the value of the constant found to be $k_\alpha = 1$.

THE β-LADDER. The required β-ladder has a third-order zero of transmission at infinity, which can be realized by the first Cauer canonical form. However, before we do this the private pole of $y_{22\beta}$ at infinity, not belonging to $-y_{21\beta}$, must be removed first. The remaining function is then expanded in a continued fraction. The result is

$$y_3(s) = y_{22\beta}(s) - s = \frac{2s^2+3}{s(s^2+2)}$$

$$= \cfrac{1}{s/2 + \cfrac{1}{4s + \cfrac{1}{s/6}}} \tag{3.131}$$

The LC network is shown in Figure 3.48(b) with constant k_β found to be $k_\beta = 1$.

We next adjust the admittance level of the ladders so that when they are connected in parallel, the desired specifications are realized. From (3.104) and (3.105) we require that

$$b_\alpha + b_\beta = 1 \tag{3.132}$$

$$k_\alpha b_\alpha = k_\beta b_\beta = k \tag{3.133}$$

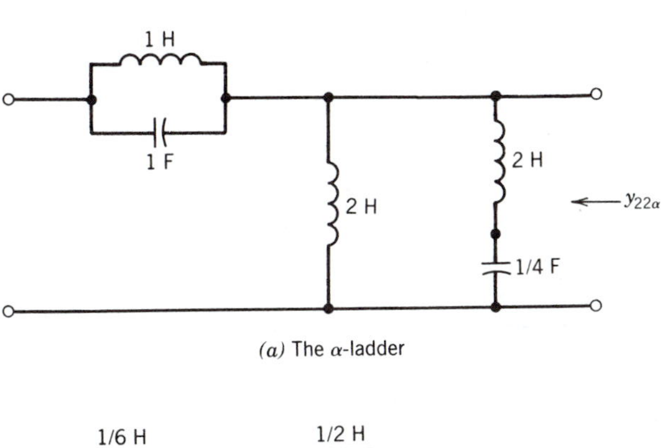

(a) The α-ladder

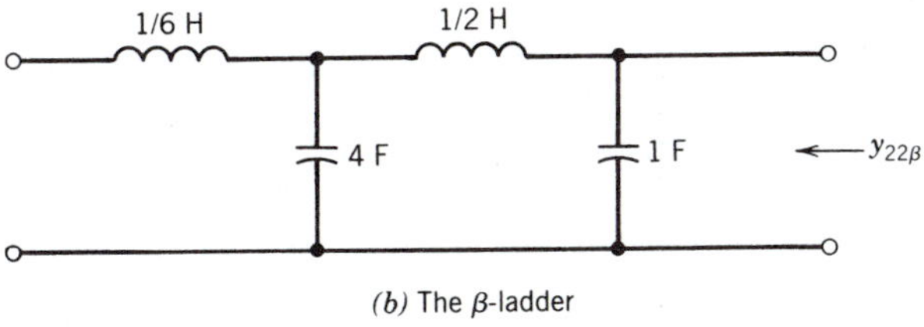

(b) The β-ladder

Figure 3.48

with $k_\alpha = k_\beta = 1$, yielding

$$b_\alpha = b_\beta = \tfrac{1}{2} \tag{3.134}$$

and $k = \frac{1}{2}$. The final realization is shown in Figure 3.49 with $k = \frac{1}{2}$.

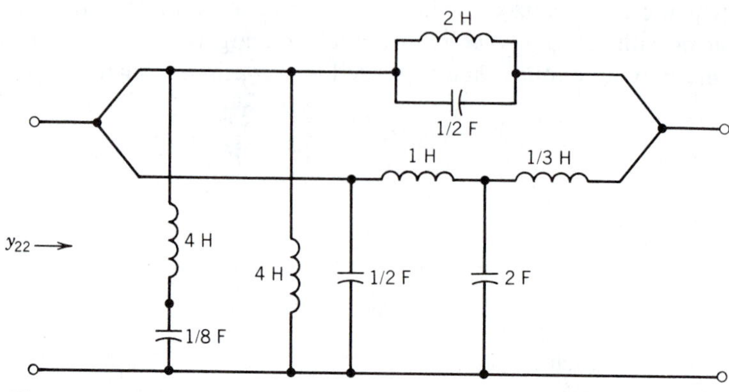

Figure 3.49

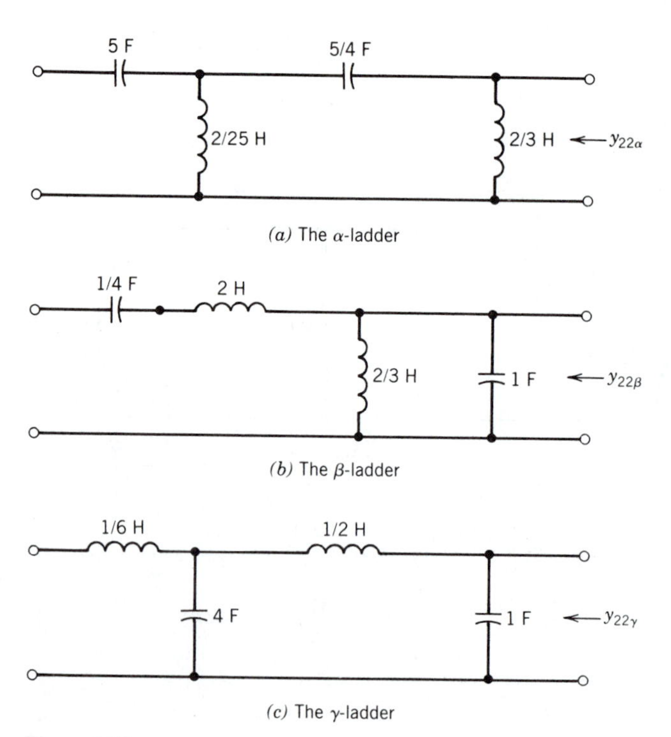

(a) The α-ladder

(b) The β-ladder

(c) The γ-ladder

Figure 3.50

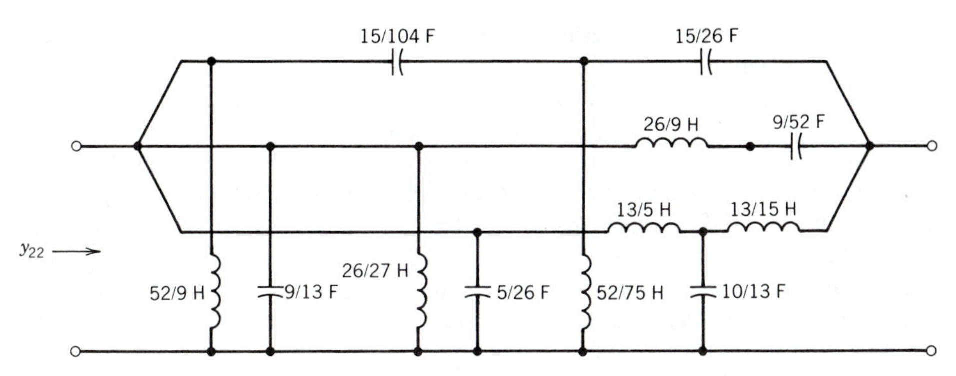

Figure 3.51

On the other hand, if $-y_{21}$ is decomposed in accordance with (3.107),

$$-y_{21\alpha}(s)=\frac{s^4}{s(s^2+2)}, \qquad -y_{21\beta}(s)=\frac{3s^2}{s(s^2+2)}, \qquad -y_{21\gamma}(s)=\frac{5}{s(s^2+1)} \tag{3.135}$$

the resulting *LC* ladders are shown in Figure 3.50 with

$$k_\alpha=1, \qquad k_\beta=1/6, \qquad k_\gamma=3/5 \tag{3.136}$$

To adjust admittance level of the individual components, we require that

$$b_\alpha+b_\beta+b_\gamma=1 \tag{3.137}$$

$$b_\alpha=b_\beta/6=3b_\gamma/5=k \tag{3.138}$$

yielding

$$b_\alpha=3/26, \qquad b_\beta=9/13, \qquad b_\gamma=5/26, \qquad k=3/26 \tag{3.139}$$

The final realization is shown in Figure 3.51. We remark that the component ladders are either the first or the second Cauer canonical form or a mix of these two structures, and can easily be obtained by performing continued-fraction expansions.

3.5 SUMMARY AND SUGGESTED READINGS

We began this chapter by considering one-port *LC* synthesis. We showed that a real rational function represents the input immittance of a lossless one-port network if and only if all of its poles and zeros are simple, lie on the $j\omega$-axis, and alternate with each other. This is a direct consequence of the positive slope property of the reactance function. Such a function can be realized by at least four canonical forms: two Foster and two Cauer forms. From the known properties of the *LC* one-ports, we next derived the properties for the *RC* networks. We showed that a function is the input impedance of an *RC* one-port network if and only if all of its poles and zeros are simple, lie on the negative real axis, and alternate with each other, the first critical frequency being a pole. As in the *LC* case, at least four canonical forms are possible for the realization. These results are needed for two-port synthesis by ladder or parallel ladder development.

In two-port synthesis, specifications are often given in terms of the transfer functions, from which the y- or z-parameters are identified. We presented a procedure for the synthesis of LC or RC ladder networks that realizes the specification functions $-y_{21}$ and y_{22} or z_{21} and z_{22} simultaneously. The process consists of two steps: the zero-shifting step and the zero-producing step. The ladder network thus generated realizes the y_{22} or z_{22} specifications exactly, and the $-y_{21}$ or z_{21} specifications within a multiplicative constant. These structures can only realize the zeros of transmission on the $j\omega$-axis or the negative real axis of the s-plane, because there is only a single transmission path between the input and the output terminals.

For the complex zeros of transmission, we introduced the method of parallel ladders, where multiple paths of transmission are possible between the input and the output terminals. In fact, we can use the parallel RC ladders to realize the zeros of transmission on the $j\omega$-axis.

A salient feature of the ladder structure is its very low sensitivity to element variations. A detailed discussion on this problem will be taken up in Section 8.1 of Chapter 8. As a result, this topology is also used in the realization of low-sensitivity active filters to be discussed in Chapter 8, where some of the realizations are just active RC equivalents of the passive LC ladder filters. The RC ladders are attractive because of the practical considerations where size and weight preclude the use of inductors. Nevertheless, inductors can be simulated with the RC networks and the operational amplifiers to be considered in Chapter 8.

For a lucid treatment of LC and RC ladder networks, see Van Valkenburg [6]. For a more extended discussion of the Cauer ladder development, see Cauer's original work [2].

REFERENCES

1. A. Budak 1974, *Passive and Active Network Analysis and Synthesis*, Chapters 3–5. Boston: Houghton Mifflin.
2. W. Cauer 1958, *Synthesis of Linear Communication Networks*, Chapters 5 and 6. New York: McGraw–Hill.
3. H. Y-F. Lam 1979, *Analog and Digital Filters: Design and Realization*, Chapters 5–7. Englewood Cliffs, NJ: Prentice–Hall.
4. A. S. Sedra and P. O. Brackett 1978, *Filter Theory and Design: Active and Passive*, Chapter 7. Champaign, IL: Matrix Publishers.
5. G. C. Temes and J. W. LaPatra 1977, *Introduction to Circuit Synthesis and Design*, Chapters 3 and 5. New York: McGraw–Hill.
6. M. E. Van Valkenburg 1960, *Modern Network Synthesis*, Chapters 5, 6, 10, and 11. New York: John Wiley.
7. L. Weinberg 1962, *Network Analysis and Synthesis*. New York: McGraw–Hill. Reissued by R. E. Krieger Publishing Co., Melbourne, FL, 1975, Chapters 7 and 9.

PROBLEMS

3.1 Find the two Foster and two Cauer canonical forms for the impedance

$$Z(s)=\frac{s(s^2+4)(s^2+36)}{(s^2+1)(s^2+25)(s^2+81)} \tag{3.140}$$

3.2 Given the impedance function

$$Z(s)=\frac{s^4+17s^2+16}{s^5+34s^3+225s} \tag{3.141}$$

which is to be realized by a network having the structure as shown in Figure 3.52, ascertain the element values of the network.

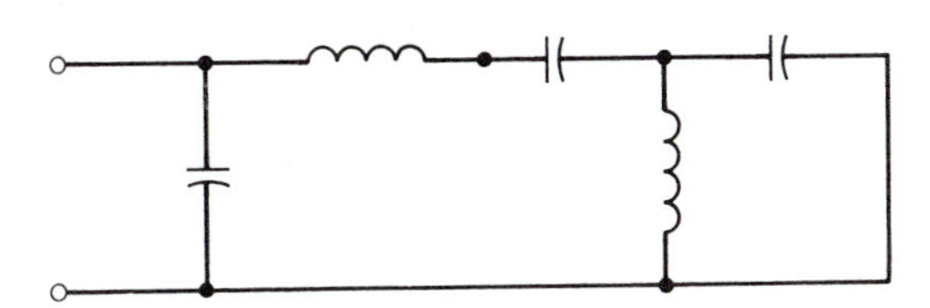

Figure 3.52

3.3 For each function given below, find the range of α within which the function represents the input immittance of an LC one-port network.

(a)

$$F(s)=\frac{s^3+\alpha s}{s^4+6s^2+2.25} \tag{3.142a}$$

(b)

$$F(s)=\frac{s^3+4s}{s^4+\alpha s+3} \tag{3.142b}$$

3.4 Find the two Foster and two Cauer canonical forms for the impedance

$$Z(s)=\frac{s^2+12s+35}{s^2+10s+24} \tag{3.143}$$

3.5 An impedance function has simple poles at $s=0$ and $s=\pm j4$ and simple zeros at $s=\pm j2$ and $s=\pm j5$. Also, at $s=1$, $Z(s)=260/17$. Find the second Cauer canonical form for this impedance.

3.6 Find the element values for a ladder network having the same driving-point impedance as the network shown in Figure 3.53 but with only three elements.

3.7 Synthesize the input impedance function

$$Z(s)=\frac{6s^4+42s^2+48}{s^5+18s^3+48s} \tag{3.144}$$

in the form as shown in Figure 3.54, and ascertain the element values of the network.

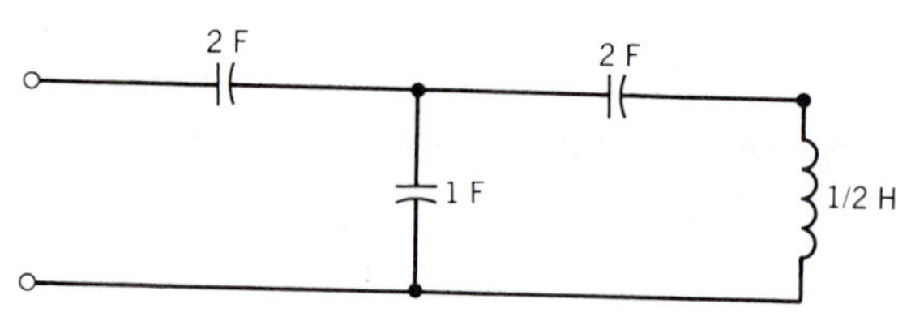

Figure 3.53

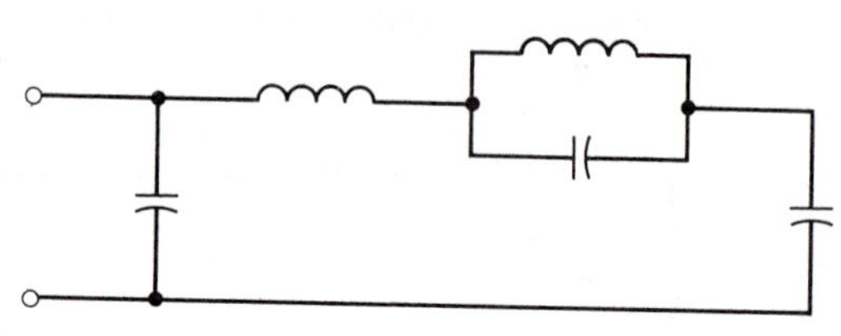

Figure 3.54

3.8 It is desired to realize a one-port impedance that is known to have poles at $s=-1$ and $s=-5$ and zeros at $s=-3$ and $s=-7$. It is also known that the function has value $16/3$ at $s=1$. Synthesize the two Cauer canonical forms for this impedance.

3.9 The network shown in Figure 3.55 is required to have poles at $s=-2$ and $s=-4$ and zeros at $s=-3$ and $s=-5$. The first element is a resistor of resistance $R=2\,\Omega$. Find the other element values of this network.

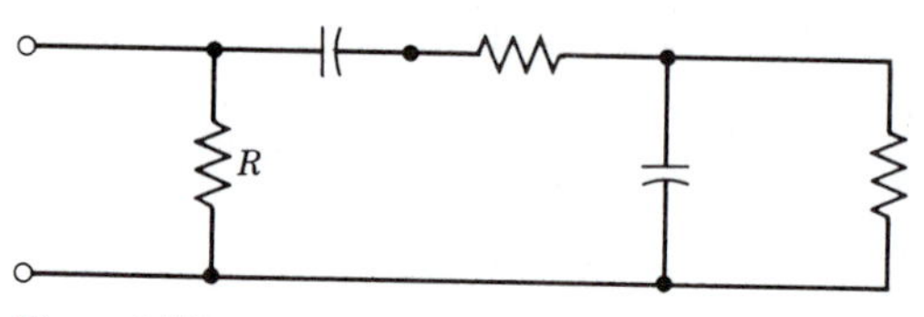

Figure 3.55

3.10 Specifications for an electronic amplifier require that the device terminate in a network having the impedance

$$Z(s)=\frac{s^2+6s+8}{s^2+4s+3} \tag{3.145}$$

Realize this impedance.

3.11 It is required to design a lossless two-port network terminated in a 1-Ω resistor to meet the specifications for the transfer voltage-ratio function (Figure 3.22)

$$G_{12}(s)=\frac{V_2}{V_1}=K\,\frac{s^2+1}{s^3+s^2+9s+4} \tag{3.146}$$

Synthesize a ladder network with these specifications and determine the value of the constant K of the realization.

3.12 Design an RC ladder network satisfying the following specifications for the transfer impedance when the two-port network is terminated in a 1-Ω resistor (Figure 3.22):

$$Z_{12}(s)=\frac{V_2}{I_1}=K\frac{(s+1)^2}{2s^2+20s+44} \tag{3.147}$$

[*Hint*: $Z_{12}=z_{21}/(z_{22}Y_2+1)$ and $2s^2+20s+44=(s+4)(s+8)+(s+2)(s+6)$.]

3.13 Repeat Problem 3.12 for the transfer impedance function

$$Z_{12}(s)=\frac{s(s+1)}{(s+2)(s+5)+(s+3)(s+8)} \tag{3.148}$$

3.14 It is desired to design a lossless two-port network terminated in a 100-Ω resistor to meet the specifications for the transfer voltage-ratio function (Figure 3.22)

$$G_{12}(s)=\frac{V_2}{V_1}=K\frac{s^2+4}{s^3+4s^2+9s+4} \tag{3.149}$$

Realize an LC ladder network to meet the specifications and determine the constant K of the realization.

3.15 Find an RC two-port network to meet the following specifications:

$$-y_{21}(s)=\frac{s^2+1}{(s+5)(s+9)} \tag{3.150a}$$

$$y_{22}(s)=\frac{(s+3)(s+7)}{(s+5)(s+9)} \tag{3.150b}$$

3.16 Find the range of the values of α within which the function

$$F(s)=\frac{s^2+4s+\alpha}{s^2+3s+2} \tag{3.151}$$

represents the input impedance of an RC one-port network.

3.17 Find an RC network to meet the specifications given by the open-circuit transfer voltage ratio (Figure 3.22)

$$G_{12}(s)=\frac{V_2}{V_1}=-\frac{y_{21}}{y_{22}}=K\frac{s^2-s+1}{s^2+8s+15} \tag{3.152}$$

to within a multiplicative constant K. Determine the constant K of the realization. [*Hint*: Multiply numerator and denominator by a common factor to remove the minus sign in the numerator.]

3.18 The transfer impedance of an LC two-port network terminated in a 1-Ω resistor is given by (Figure 3.22)

$$Z_{12}(s)=\frac{V_2}{I_1}=K\frac{s^2+1}{s^3+2s^2+2s+1} \tag{3.153}$$

Find an LC ladder network to realize $Z_{12}(s)$ and determine the constant K of the realization. [*Hint*: $Z_{12}=z_{21}/(z_{22}Y_2+1)$.]

3.19 Repeat Problem 3.17 for the open-circuit transfer voltage ratio

$$G_{12}(s)=\frac{z_{21}}{z_{11}}=K\frac{s^2-2s+10}{s^2+8s+15} \tag{3.154}$$

chapter 4
DESIGN OF RESISTIVELY TERMINATED NETWORKS

In the design of communication systems, a basic problem is to design a coupling network between a given source and a given load so that the transfer of power from the source to the load is maximized over a given frequency band of interest. A problem of this type invariably involves the design of a coupling network that transforms a given load impedance into another specified one. This, of course, is not always possible. Then what are the necessary and sufficient conditions for the coupling network to exist? In this chapter, we first show that any nonnegative resistance can be transformed by a lossless reciprocal two-port network to any prescribed rational positive-real immittance. In other words, any rational positive-real immittance can be realized as the driving-point immittance of a lossless reciprocal two-port network terminated in a nonnegative resistance, which may be zero. We next apply this result to the synthesis of lossless terminated networks. We then discuss conditions under which a lossless two-port exists that, when it is operated between a resistive generator and a resistive load, yields a preassigned transducer power-gain characteristic. Finally, explicit formulas for the design of double-terminated *LC* ladder networks will be presented.

4.1 THE PROBLEM OF COMPATIBLE IMPEDANCES

We begin our discussion by introducing the concepts of compatible impedances.

Definition 4.1: ***Compatible impedances.*** Two positive-real rational impedances $Z_1(s)$ and $Z_2(s)$ are said to be *compatible* if $Z_1(s)$ can be realized as the driving-point impedance of a linear lossless two-port network terminated in $Z_2(s)$.

As implied by the definition, the lossless two-port may be either reciprocal or nonreciprocal. The concept can be extended to also include the situation where lossy two-port networks are allowed, but we do not find it necessary. We recognize that the choice of a lossy two-port would not only lessen the transducer power gain but also severely hamper our ability to manipulate. To illustrate this concept, consider the two positive-real impedances:

$$Z_1(s)=\frac{14s^2+9s+4}{14s^3+9s^2+6s+1} \tag{4.1}$$

$$Z_2(s)=\frac{2s+4}{2s+1} \tag{4.2}$$

These two impedances are compatible since there exists a lossless two-port N as shown in Figure 4.1 whose input impedance is Z_1 when the output is terminated in Z_2. In the following, we shall present Darlington's theory† by showing that if the terminating impedance Z_2 is a pure resistance, the impedances Z_1 and $Z_2=R_2$ are always compatible. Procedures for realizing the desired two-port network will be outlined.

Refer to the network of Figure 4.2, where the load resistance is normalized to 1. If z_{ij} are the z-parameters of the reciprocal two-port network N, then the driving-point impedance Z can be expressed in terms of z_{ij} by

$$Z(s)=z_{11}-\frac{z_{12}^2}{z_{22}+1} \tag{4.3}$$

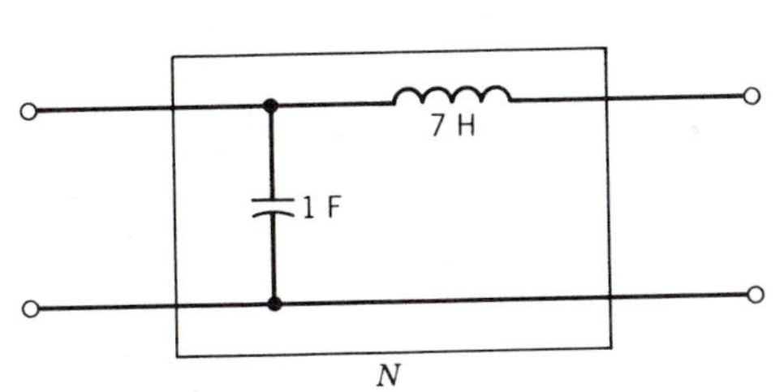

Figure 4.1

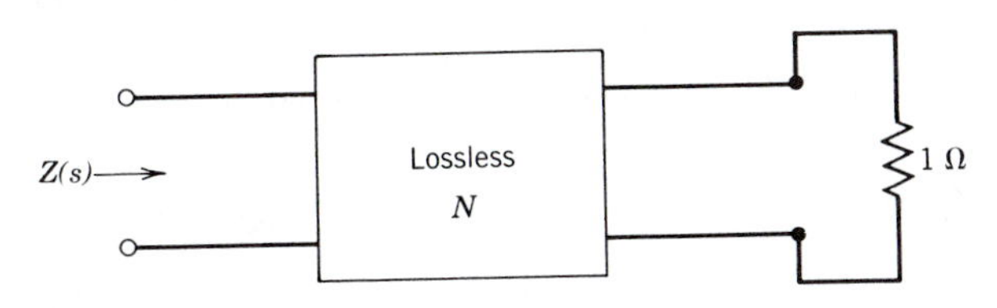

Figure 4.2

†S. Darlington, "Synthesis of reactance 4-poles which produce prescribed insertion loss characteristics," *J. Math. Phys.*, vol. 18, Sept. 1939, pp. 257–353.

with $z_{12}=z_{21}$. Equation (4.3) can be simplified by means of the well-known relation

$$y_{22}=\frac{z_{11}}{z_{11}z_{22}-z_{12}^2} \tag{4.4}$$

where y_{ij} are the y-parameters of N. This gives

$$Z(s)=z_{11}\frac{1/y_{22}+1}{z_{22}+1} \tag{4.5}$$

Writing Z in terms of the even and odd parts of its numerator and denominator polynomials as in (3.1)

$$Z(s)=\frac{m_1+n_1}{m_2+n_2} \tag{4.6}$$

we see that if we factor m_1 out of the numerator and n_2 out of the denominator, we have

$$Z(s)=\frac{m_1}{n_2}\cdot\frac{n_1/m_1+1}{m_2/n_2+1} \qquad \text{(Case A)} \tag{4.7}$$

Alternatively, factoring out n_1 from the numerator and m_2 from the denominator yields

$$Z(s)=\frac{n_1}{m_2}\cdot\frac{m_1/n_1+1}{n_2/m_2+1} \qquad \text{(Case B)} \tag{4.8}$$

By comparing (4.7) and (4.8) with (4.5), we can make the following identifications:

Case A	Case B	
$z_{11}=\dfrac{m_1}{n_2}$,	$z_{11}=\dfrac{n_1}{m_2}$	(4.9a)
$z_{22}=\dfrac{m_2}{n_2}$,	$z_{22}=\dfrac{n_2}{m_2}$	(4.9b)
$y_{22}=\dfrac{m_1}{n_1}$,	$y_{22}=\dfrac{n_1}{m_1}$	(4.9c)

We now proceed to show that these immittances are always compatible in representing a lossless reciprocal two-port network. To facilitate our discussion, we consider the following situations, each being presented in a separate section.

4.1.1 z_{11}, z_{22}, and y_{22} Being Reactance Functions

For the parameters (4.9) to be realizable by a lossless reciprocal two-port network, we must show that z_{11}, z_{22}, and y_{22} are reactance functions, being the input immittances of some lossless one-port networks. To this end, consider the function

$$Z_1(s)=\frac{m_1+n_2}{m_2+n_1} \tag{4.10}$$

where the odd parts of the two polynomials in Z have been interchanged. Our objective is to show that Z_1 is positive real. Appealing to Theorem 1.4 indicates that Z_1 is positive real if and only if

(i) $Z_1(s)$ is real when s is real,

(ii) $m_1+n_2+m_2+n_1$ is strictly Hurwitz,

(iii) $m_1m_2-n_1n_2|_{s=j\omega} \geqq 0$ for all ω.

Clearly, these conditions are all satisfied since Z, being positive real, possesses all three properties. Thus, Z_1 is a positive-real function. It follows that the polynomials m_1+n_1, m_2+n_2, m_1+n_2, and m_2+n_1 are all Hurwitz. Since the ratio of the even part to the odd part or the odd part to the even part of a Hurwitz polynomial yields a continued-fraction expansion with real and positive coefficients, it can be identified as the input immittance of a lossless (LC) one-port network. This shows that the functions

$$(m_1/n_1)^{\pm 1}, \qquad (m_2/n_2)^{\pm 1}, \qquad (m_1/n_2)^{\pm 1}, \qquad (m_2/n_1)^{\pm 1} \tag{4.11}$$

are all reactance functions.

4.1.2 Rationalization of z_{12}

For z_{12} to be physically realizable, it must be a rational function. Thus, we must demonstrate that a consistent rational z_{12} can always be found from the sets given in (4.9). For this we substitute (4.9) in (4.4) and solve for z_{12}, giving

$$z_{12}=\frac{\sqrt{m_1m_2-n_1n_2}}{n_2} \qquad \text{(Case A)} \tag{4.12a}$$

$$z_{12}=\frac{\sqrt{n_1n_2-m_1m_2}}{m_2} \qquad \text{(Case } B\text{)} \tag{4.12b}$$

Since z_{12} must be an odd rational function, the two cases are distinguished by the even or odd character of the numerators of (4.12). If $\sqrt{m_1m_2-n_1n_2}$ is even, then Case A applies, whereas if $\sqrt{n_1n_2-m_1m_2}$ is odd, Case B applies. In other words, if the even polynomial $m_1m_2-n_1n_2$ contains a factor s^2, then Case B applies; if it does not, Case A applies.

For z_{12} to be rational, it is necessary that $m_1m_2-n_1n_2$ for Case A and its negative for Case B be a full square, meaning that its s^2 zeros must be of even multiplicity. But even for the positive-real impedance Z, this is not always true. Fortunately, it is possible to remedy this situation by multiplying the polynomial $m_1m_2-n_1n_2$ or its negative by an *auxiliary even polynomial*

$$m_0^2-n_0^2=(m_0+n_0)(m_0-n_0) \tag{4.13}$$

which is formed by all the first-order factors that occur in $m_1m_2-n_1n_2$ with odd multiplicity. That this is always possible follows from the fact that the zeros of $m_1m_2-n_1n_2$ have quadrantal symmetry, being symmetric with respect to both axes as shown in Figure 3.1. For our purposes, let m_0+n_0 be the *auxiliary Hurwitz polynomial* formed by the left-half of the s-plane (LHS) zeros of (4.13), so that $m_0 - n_0$ has all the zeros at the mirror-image points in the right-half of the s-plane (RHS). It remains to be shown that this manipulation does not affect the given Z; it merely alters the two-port parameters z_{ij}.

To see this we consider the function

$$Z(s) = \frac{m_1 + n_1}{m_2 + n_2} \cdot \frac{m_0 + n_0}{m_0 + n_0}$$

$$= \frac{(m_0 m_1 + n_0 n_1) + (m_0 n_1 + n_0 m_1)}{(m_0 m_2 + n_0 n_2) + (n_0 m_2 + m_0 n_2)} \triangleq \frac{m'_1 + n'_1}{m'_2 + n'_2} \tag{4.14}$$

From the new even and odd parts, we see that

$$m'_1 m'_2 - n'_1 n'_2 = (m_1 m_2 - n_1 n_2)(m_0^2 - n_0^2) \tag{4.15}$$

showing that multiplying $m_1 m_2 - n_1 n_2$ by $m_0^2 - n_0^2$ to make it a full square does not affect the given impedance Z; it merely alters the associated two-port parameters. We illustrate this by the following examples.

Example 4.1 Consider the input impedance of the network of Figure 4.3:

$$Z(s) = \frac{s+2}{2s+1} = \frac{m_1 + n_1}{m_2 + n_2} \tag{4.16}$$

We determine the associated two-port parameters z_{ij}. For this we first compute the quantity

$$m_1 m_2 - n_1 n_2 = 2 - 2s^2 = 2(1 - s^2) \tag{4.17}$$

which has simple zeros at $s = 1$ and $s = -1$. The auxiliary Hurwitz polynomial is found to be

$$m_0 + n_0 = s + 1 \tag{4.18}$$

Now we multiply the numerator and denominator polynomials of Z by $s + 1$, obtaining the new even and odd parts of the polynomials as

$$m'_1 = s^2 + 2, \qquad n'_1 = 3s \tag{4.19a}$$

$$m'_2 = 2s^2 + 1, \qquad n'_2 = 3s \tag{4.19b}$$

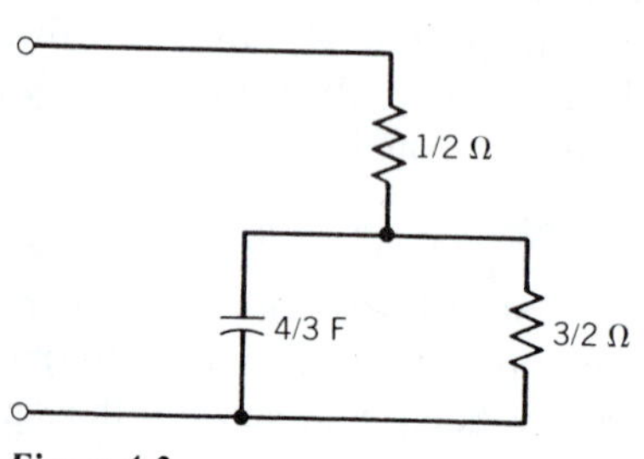

Figure 4.3

Since $m_1'm_2'-n_1'n_2'$ does not contain a factor s^2, Case A applies, and from (4.9) and (4.12a) with m_1' replacing m_1, etc., we obtain

$$z_{11}=\frac{m_1'}{n_2'}=\frac{s^2+2}{3s} \tag{4.20a}$$

$$z_{22}=\frac{m_2'}{n_2'}=\frac{2s^2+1}{3s} \tag{4.20b}$$

$$z_{12}=\frac{\sqrt{m_1'm_2'-n_1'n_2'}}{n_2'}=\frac{\sqrt{2}(1-s^2)}{3s} \tag{4.20c}$$

This represents a set of compatible z-parameters for a lossless reciprocal two-port network.

Example 4.2 Determine the associated two-port parameters z_{ij} for the positive-real impedance

$$Z(s)=\frac{6s^2+5s+6}{2s^2+4s+4}=\frac{m_1+n_1}{m_2+n_2} \tag{4.21}$$

As before, we first compute the quantity

$$m_1m_2-n_1n_2=(6s^2+6)(2s^2+4)-20s^2=4(3s^4+4s^2+6) \tag{4.22}$$

which is not a full square, and so we must choose an auxiliary even polynomial

$$m_0^2-n_0^2=3s^4+4s^2+6 \tag{4.23}$$

Write the auxiliary Hurwitz polynomial as

$$m_0+n_0=\sqrt{3}s^2+\alpha s+\sqrt{6} \tag{4.24}$$

Then we have

$$m_0^2-n_0^2=3s^4+(6\sqrt{2}-\alpha^2)s+6=3s^4+4s^2+6 \tag{4.25}$$

from which we set

$$6\sqrt{2}-\alpha^2=4 \tag{4.26}$$

and obtain $\alpha=2.11785$. The desired auxiliary Hurwitz polynomial becomes

$$m_0+n_0=1.73205s^2+2.11785s+2.44949 \tag{4.27}$$

Multiplying the numerator and denominator of (4.21) by m_0+n_0 results in the new even and odd parts, as follows:

$$m_1'=10.39230s^4+35.67849s^2+14.69694 \tag{4.28a}$$

$$n_1'=21.36735s^3+24.95455s \tag{4.28b}$$

$$m_2'=3.46410s^4+20.29858s^2+9.79796 \tag{4.28c}$$

$$n_2'=11.16390s^3+18.26936s \tag{4.28d}$$

Since $m_1'm_2'-n_1'n_2'$ is devoid of the factor s^2, Case A applies. The associated two-port parameters are found to be

$$z_{11}=\frac{m_1'}{n_2'}=\frac{10.39230s^4+35.67849s^2+14.69694}{11.16390s^3+18.26936s} \tag{4.29a}$$

$$z_{22}=\frac{m_2'}{n_2'}=\frac{3.46410s^4+20.29858s^2+9.79796}{11.16390s^3+18.26936s} \tag{4.29b}$$

$$z_{12}=\frac{\sqrt{m_1'm_2'-n_1'n_2'}}{n_2'}=\frac{6s^4+8s^2+12}{11.16390s^3+18.26936s} \tag{4.29c}$$

This represents a set of compatible z-parameters for a lossless reciprocal two-port network.

4.1.3 The Residue Condition

To establish the residue condition, we expand the z-parameters z_{ij} in partial fractions and the results can be written as

$$z_{ij}=k_{ij}^{(\infty)}s+\frac{k_{ij}^{(0)}}{s}+\sum_{r=1}^{q}\frac{2k_{ij}^{(r)}s}{s^2+\omega_r^2}, \qquad i, j=1, 2 \tag{4.30}$$

q being the number of finite nonzero poles of z_{ij}, where the superscript identifies the residue associated with a given pole. To simplify our notation, let $k_{ij}=k_{ij}^{(\infty)}$, $k_{ij}^{(0)}$, or $k_{ij}^{(r)}$. Since z_{11} and z_{22} are reactance functions, the residues at the $j\omega$-axis poles must be real and positive,

$$k_{11}\geqq 0 \tag{4.31a}$$

$$k_{22}\geqq 0 \tag{4.31b}$$

In addition, we show that the residues of z_{12} are related to those of z_{11} and z_{22} by the equation

$$k_{11}k_{22}\geqq k_{12}^2 \tag{4.32}$$

This equation is known as the *residue condition*. To see that condition (4.32) holds, we apply the well-known formula

$$k=\left.\frac{m}{\dfrac{dn}{ds}}\right|_{s=s_0} \tag{4.33}$$

for the evaluation of the residue at a simple finite pole s_0 of m/n. Using this in conjunction with (4.9) and (4.12a) for Case A, we compute

$$k_{12}^2-k_{11}k_{22}=\left[\frac{m_1m_2-n_1n_2}{\left(\dfrac{dn_2}{ds}\right)^2}-\frac{m_1}{\dfrac{dn_2}{ds}}\cdot\frac{m_2}{\dfrac{dn_2}{ds}}\right]_{s=s_0}=\frac{-n_1(s_0)n_2(s_0)}{\left(\left.\dfrac{dn_2}{ds}\right|_{s=s_0}\right)^2}=0 \tag{4.34}$$

since s_0, being a pole of z_{ij}, is a zero of n_2, i.e., $n_2(s_0)=0$; and dn_2/ds is not zero at $s=s_0$. Thus, equation (4.32) is satisfied with the equality sign. In a similar manner, the same conclusion can be reached for Case B. A pole possessing this property is termed a *compact pole*. In other words all the finite nonzero poles of the z-parameters obtained by Darlington's procedure are compact.

As to the pole at infinity, it must be investigated independently, because formula (4.33) for the evaluation of the residue at a simple pole does not apply in this case. For Case A assume that both z_{11} and z_{22} have a pole at infinity. Then m_1 and m_2 are of the same degree, being one degree higher than n_2. The residue of z_{11} at infinity is the ratio of the coefficients of the highest power of s in m_1 and n_2, and that of z_{22} is the corresponding ratio in m_2 and n_2. They are denoted by k_{11} and k_{22}. If n_1 is of the same degree as n_2, being one degree less than both m_1 and m_2, m_1m_2 will be two degrees higher than n_1n_2. Thus, from (4.12a) the residue of z_{12} at infinity becomes $\sqrt{k_{11}k_{22}}$, showing that the pole at infinity will be compact. If n_1 is of two degrees higher than n_2, it will be one degree higher than both m_1 and m_2. Under this condition, Z will have a pole at infinity and the terms m_1m_2 and n_1n_2 will have the same degree. The residue condition (4.32) becomes

$$k_{11}k_{22}-k_{12}^2=k_{11}k_{22}-(k_{11}k_{22}-L_{12})=L_{12}>0 \tag{4.35}$$

where L_{12} denotes the ratio of the coefficients of the highest power of s in n_1 and n_2. We conclude that the pole at infinity is compact if and only if Z has no pole at infinity. A similar conclusion can be reached for Case B.

Before we turn our attention to the realization of the impedance parameters z_{ij}, we note that from inspection of the expressions (4.9) for Case B, it would appear that $s=0$ will be a zero of these functions, not a pole. However, in Case B the even polynomial $n_1n_2-m_1m_2$ contains a factor s^2, meaning that the constant term in either m_1 or m_2 and hence the numerator or denominator of Z is missing. Suppose that it is missing in m_2. Then, instead of having a zero at $s=0$, the impedance parameters z_{ij} have a pole there. Appealing to (4.33), the residue condition (4.32) can be expressed as

$$k_{11}k_{22}-k_{12}^2=\left.\frac{m_1m_2}{\left(\dfrac{dm_2}{ds}\right)^2}\right|_{s=0}=\frac{m_1(0)}{4b_2}>0 \tag{4.36a}$$

where

$$m_2=b_2s^2+b_4s^4+\cdots+b_{2u}s^{2u}, \qquad b_2\neq 0 \tag{4.36b}$$

To summarize the above discussion, we can state that all the poles of the z-parameters z_{ij} obtained by Darlington's procedure are compact except possibly the pole at infinity and the pole at the origin. The pole at infinity is noncompact if and only if Z has a pole there, and the pole at the origin is noncompact if and only if Z has a pole there, the latter being associated with Case B.

Finally, we note that if Z has a pair of finite nonzero poles on the $j\omega$-axis, say at $s=\pm j\omega_0$, then $s^2+\omega_0^2$ will be a factor of both m_2 and n_2. From (4.9) we see that there will be a cancellation in z_{22} but not in z_{11} for the factor $s^2+\omega_0^2$. Hence the poles at $s=\pm j\omega_0$ become a pair of private poles of z_{11}, corresponding to a reactance one-port connected in series at the input port. It is easy to verify that the residue condition at these poles is trivially satisfied with the equality sign. Hence they are all compact, as expected.

Example 4.3 Consider the positive-real impedance

$$Z(s)=\frac{s^3+2s^2+3s+6}{3s^2+2s+6} \tag{4.37}$$

As before we first compute the quantity

$$m_1m_2-n_1n_2=4(s^2+3)^2 \tag{4.38}$$

which is already in a full square. Thus, no auxiliary Hurwitz polynomial is required. Since (4.38) does not contain a factor s^2, Case A applies and from (4.9) and (4.12a) we obtain

$$z_{11}=\frac{m_1}{n_2}=\frac{2s^2+6}{2s}=s+\frac{3}{s} \tag{4.39a}$$

$$z_{22}=\frac{m_2}{n_2}=\frac{3s^2+6}{2s}=\frac{3s}{2}+\frac{3}{s} \tag{4.39b}$$

$$z_{12}=\frac{\sqrt{m_1m_2-n_1n_2}}{n_2}=\frac{2(s^2+3)}{2s}=s+\frac{3}{s} \tag{4.39c}$$

It is easy to verify that the pole at the origin is compact, while the pole at infinity is not.

Example 4.4 Consider the *RC* one-port impedance function

$$Z(s)=\frac{s^2+4s+3}{s^2+2s} \tag{4.40}$$

We first form the quantity

$$m_1m_2-n_1n_2=s^2(s^2-5) \tag{4.41}$$

Since it is not in a full square, we choose the auxiliary Hurwitz polynomial

$$m_0+n_0=s+\sqrt{5} \tag{4.42}$$

The modified impedance becomes

$$Z(s)=\frac{[(4+\sqrt{5})s^2+3\sqrt{5}]+[s^3+(3+4\sqrt{5})s]}{(2+\sqrt{5})s^2+(s^3+2\sqrt{5}s)} \tag{4.43}$$

Since $m_1'm_2'-n_1'n_2'$ contains a factor s^2, Case B applies and the associated z-parameters are found to be

$$z_{11}=\frac{n_1'}{m_2'}=\frac{s^3+(3+4\sqrt{5})s}{(2+\sqrt{5})s^2}=\frac{s}{2+\sqrt{5}}+\frac{14-5\sqrt{5}}{s} \tag{4.44a}$$

$$z_{22}=\frac{n_2'}{m_2'}=\frac{s^3+2\sqrt{5}s}{(2+\sqrt{5})s^2}=\frac{s}{2+\sqrt{5}}+\frac{10-4\sqrt{5}}{s} \tag{4.44b}$$

$$z_{12}=\frac{\sqrt{n_1'n_2'-m_1'm_2'}}{m_2'}=\frac{s(s^2-5)}{(2+\sqrt{5})s^2}=\frac{s}{2+\sqrt{5}}+\frac{10-5\sqrt{5}}{s} \tag{4.44c}$$

It is easy to verify that the pole at infinity is compact, while the pole at the origin is not since

$$k_{11}k_{22}-k_{12}^2=15-6\sqrt{5}=1.584 \tag{4.45}$$

4.2 TWO-PORT REALIZATION

We now demonstrate that the z parameters obtained by Darlington's procedure can be realized by a lossless two-port network. For this we consider the partial-fraction expansions of z_{ij} as in (4.30). To discuss all three types of poles of (4.30) simultaneously, let $f(s)$ represent any one of the following three functions:

$$f(s)=1/s, \qquad s, \qquad \text{or} \qquad 2s/(s^2+\omega_r^2) \tag{4.46}$$

Then each term on the right-hand side of (4.30) corresponds to a set of three parameters:

$$k_{11}f(s), \qquad k_{22}f(s), \qquad k_{12}f(s) \tag{4.47}$$

If each set of like terms in the expansion (4.30) can be realized as a lossless two-port network, the desired two-port network would be realized by the series connection of the individual realizations when they are properly isolated by ideal transformers. For the procedure to succeed, we must realize (4.47).

Consider the two-port network of Figure 4.4, the z-parameters of which are found to be

$$\tilde{z}_{11}=z, \qquad \tilde{z}_{22}=n^2z, \qquad \tilde{z}_{12}=nz \tag{4.48}$$

If they realize the set (4.47), we must have

$$z=\tilde{z}_{11}=k_{11}f(s) \tag{4.49a}$$

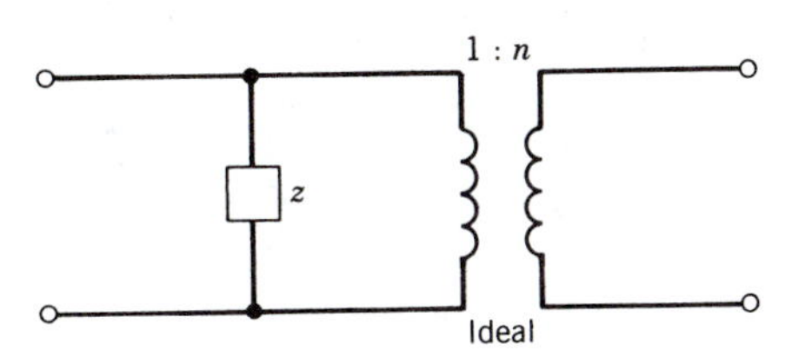

Figure 4.4

$$n^2 z = \tilde{z}_{22} = k_{22} f(s) \tag{4.49b}$$

$$nz = \tilde{z}_{12} = k_{12} f(s) \tag{4.49c}$$

Eliminating n and z in (4.49) gives

$$k_{11}k_{22} - k_{12}^2 = 0 \tag{4.50}$$

requiring that all the poles of the z-parameters be compact, a fact that was demonstrated in the preceding section. The impedance z is clearly realizable as a capacitor, inductor, or the parallel combination of a capacitor and an inductor. Connecting the realizations in series at the input and output ports, as depicted in Figure 4.5, results in a lossless reciprocal two-port network that realizes the z-parameters z_{ij} obtained by Darlington's procedure. To complete the Darlington realization of the given Z, all that is necessary is to terminate the lossless two-port network in a 1-Ω resistor.

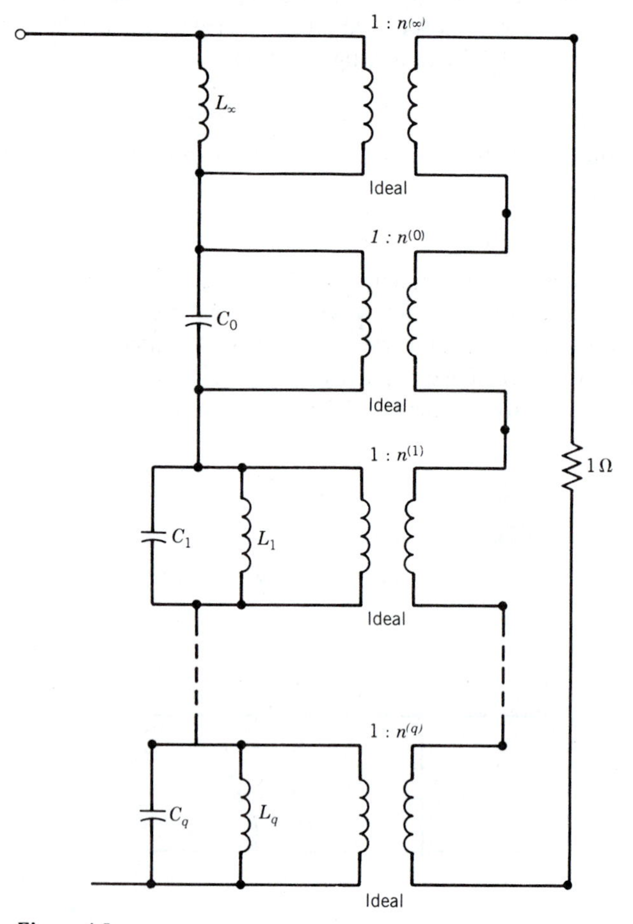

Figure 4.5

The value of n for each component two-port is computed from (4.49) by the equation

$$n=\frac{k_{22}}{k_{12}}=\frac{k_{12}}{k_{11}} \tag{4.51}$$

The realization is not very attractive because of the presence of a large number of ideal transformers. However, with the use of equivalent networks as shown in Figure 4.6, the ideal transformers of Figure 4.5 together with their shunt inductances can be replaced by physical transformers with unity coupling. To remove the ideal transformer associated with the pole at the origin, we consider the equation (4.3). If we multiply the numerator and denominator of the second term on the right-hand side of (4.3) by a factor α, it does not affect the input impedance Z. This means that we can multiply z_{22} by α and z_{12} by $\sqrt{\alpha}$, and then raise the terminating resistance from one to α ohms without affecting the given impedance Z. Thus, setting $n=1$ in (4.51) with αk_{22} replacing k_{22} and $\sqrt{\alpha}k_{12}$ replacing k_{12} yields the desired value

$$\alpha=k_{12}^2/k_{22}^2 \tag{4.52}$$

We illustrate the above procedure by the following examples.

Example 4.5 Suppose that we wish to convert the one-port impedance of Figure 4.3 to be the input impedance of a lossless two-port network terminated in a resistor. For this we first compute the z-parameters. These were obtained in Example 4.1 and are given by (4.20). Their partial-fraction expansions are found to be

$$z_{11}=\frac{s^2+2}{3s}=\frac{s}{3}+\frac{2}{3s} \tag{4.53a}$$

$$z_{22}=\frac{2s^2+1}{3s}=\frac{2s}{3}+\frac{1}{3s} \tag{4.53b}$$

$$z_{12}=\frac{\sqrt{2}(1-s^2)}{3s}=-\frac{\sqrt{2}s}{3}+\frac{\sqrt{2}}{3s} \tag{4.53c}$$

confirming that the two poles are compact. From (4.49) and (4.51) the required turns

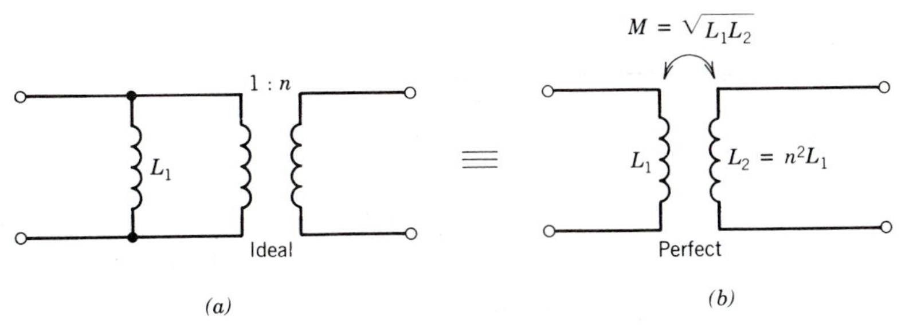

Figure 4.6

ratio n and impedance z are found to be

$$z^{(\infty)} = s/3, \qquad n^{(\infty)} = -\sqrt{2} \tag{4.54a}$$

$$z^{(0)} = 2/3s, \qquad n^{(0)} = 1/\sqrt{2} \tag{4.54b}$$

The complete network realization is presented in Figure 4.7. To remove the ideal transformer associated with the pole at the origin, we multiply z_{22} by a factor

$$\alpha = k_{12}^2/k_{22}^2 = 2 \tag{4.55}$$

and z_{12} by $\sqrt{2}$ in (4.53), yielding a network realization shown in Figure 4.8. With the aid of Figure 4.6, the other ideal transformer in Figure 4.8 can be replaced by a perfect transformer as illustrated in Figure 4.9. Comparing this network with that of Figure 4.3 shows that the former involves a perfectly coupled transformer while the latter has none. This is precisely the price that we paid in replacing the two resistors by one at the output port.

Example 4.6 Realize the impedance Z of Example 4.2 by Darlington's procedure. From (4.29) the partial-fraction expansions of the z-parameters z_{ij} are

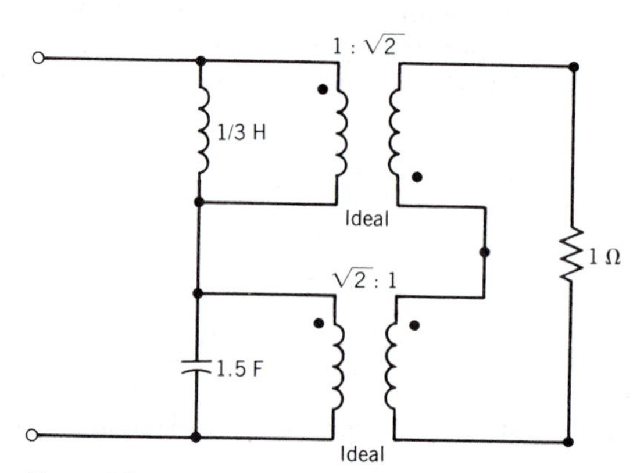

Figure 4.7

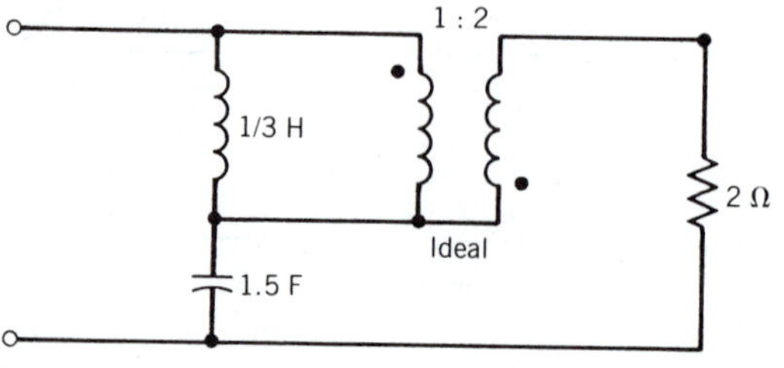

Figure 4.8

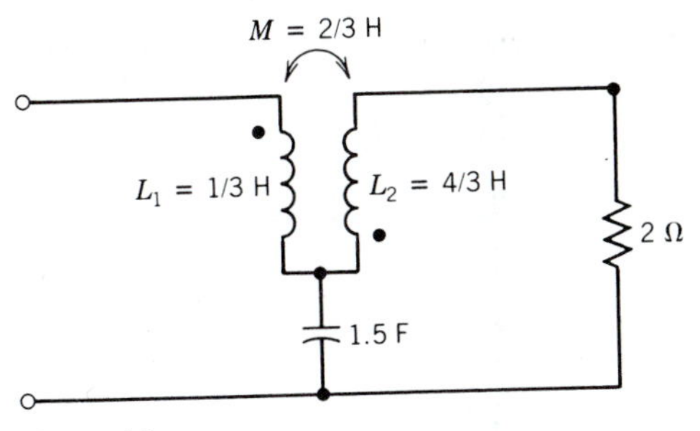

Figure 4.9

obtained as follows:

$$z_{11}=0.93088s+\frac{0.80446}{s}+\frac{0.86806s}{s^2+1.63649} \tag{4.56a}$$

$$z_{22}=0.31030s+\frac{0.53631}{s}+\frac{0.77416s}{s^2+1.63649} \tag{4.56b}$$

$$z_{12}=0.53745s+\frac{0.65684}{s}-\frac{0.81977s}{s^2+1.63649} \tag{4.56c}$$

confirming that all the poles of z_{ij} are compact. From (4.49) and (4.51), the transformer turns ratio n and the impedance z are obtained as

$$z^{(\infty)}=0.93088s, \qquad n^{(\infty)}=0.57736 \tag{4.57a}$$

$$z^{(0)}=\frac{0.80446}{s}, \qquad n^{(0)}=0.81650 \tag{4.57b}$$

$$z^{(1)}=\frac{0.86806s}{s^2+1.63649}, \qquad n^{(1)}=-0.94436 \tag{4.57c}$$

The complete network realization is shown in Figure 4.10. To remove the ideal transformer associated with the pole at the origin, we multiply z_{22} by a factor

$$\alpha=\frac{(0.65684)^2}{(0.53631)^2}=1.5 \tag{4.58}$$

and z_{12} by 1.22474. After these multiplications and with the aid of the equivalent networks of Figure 4.6, we obtain the network of Figure 4.11.

The above examples demonstrated how the compact poles are realized. For the noncompact poles, the realizations are essentially the same. First, we divide the residues of z_{11} and z_{22} at a noncompact pole into two parts:

$$k_{11}=k'_{11}+k''_{11} \tag{4.59a}$$

$$k_{22}=k'_{22}+k''_{22} \tag{4.59b}$$

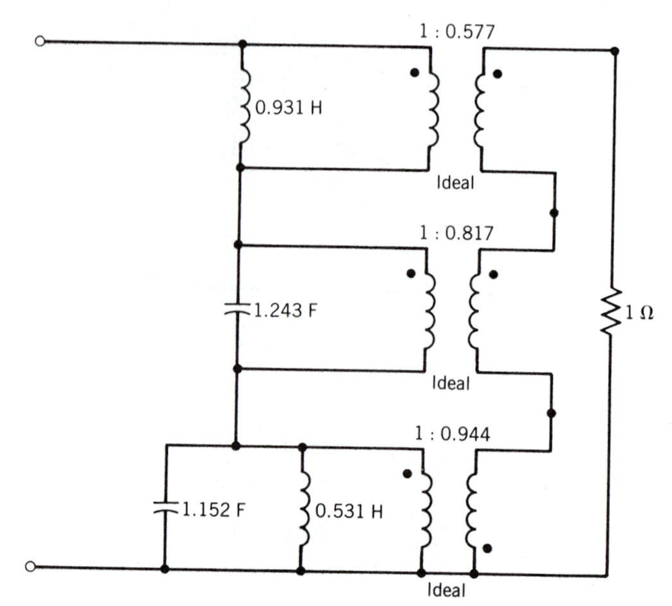

Figure 4.10

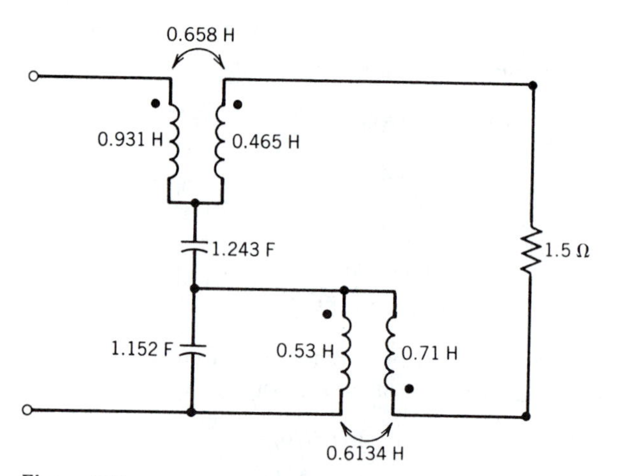

Figure 4.11

so that using the first terms on the right-hand side of (4.59) in the residue condition satisfies it by the equality sign, i.e.,

$$k'_{11}k'_{22}-k'^2_{12}=0 \tag{4.60}$$

The terms corresponding to k'_{11} and k'_{22} in the partial-fraction expansions of z_{ij} can be realized as a compact pole, while those corresponding to k''_{11} and k''_{22} can be treated as the private poles of z_{11} and z_{22}, which lead to reactive networks connected in series at the input and the output terminals of the component two-port networks, or equivalently, at the input and the output terminals of the overall two-port network.

Example 4.7 Realize the RC impedance function Z of Example 4.4. Let $k_{11}^{(0)}=k'_{11}+k''_{11}$ and $k_{22}^{(0)}=k'_{22}$ so that (4.60) is satisfied. This yields

$$k'_{11}=12.5-5\sqrt{5}, \qquad k''_{11}=1.5 \tag{4.61}$$

From (4.44) the partial-fraction expansions of z_{ij} can be written as

$$z_{11}=\frac{s}{2+\sqrt{5}}+\frac{12.5-5\sqrt{5}}{s}+\frac{3}{2s} \tag{4.62a}$$

$$z_{22}=\frac{s}{2+\sqrt{5}}+\frac{10-4\sqrt{5}}{s} \tag{4.62b}$$

$$z_{12}=\frac{s}{2+\sqrt{5}}+\frac{10-5\sqrt{5}}{s} \tag{4.62c}$$

The complete realization is shown in Figure 4.12, in which the series capacitor at the input port corresponds to the last term in (4.62a).

In the foregoing, we have demonstrated that a given impedance Z can be realized by the series configuration of Figure 4.5. By working with the reciprocal of $Z=1/Y$ and using the dual expression

$$Y(s)=y_{11}-\frac{y_{12}^2}{y_{22}+1} \tag{4.63}$$

where the terminating resistance is again 1 Ω (Figure 4.2), an alternative set of realizations in the form of parallel connection results. Since the two approaches are very similar, there is hardly any point in going over the same ground once again, and so we shall blandly state the results, leaving the details as obvious.

From (4.63) we first identify the admittance parameters y_{ij}, using an auxiliary Hurwitz polynomial if necessary. We then expand the admittances y_{ij} in partial fractions as those shown in (4.30) with y_{ij} replacing z_{ij}. Since the corresponding

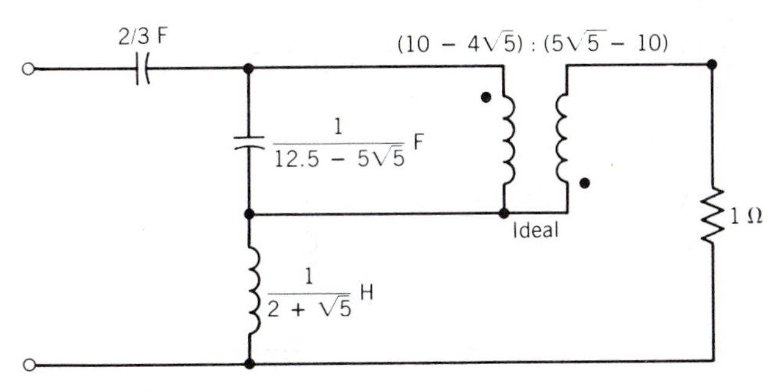

Figure 4.12

y-parameters add for two-ports connected in parallel when properly isolated by ideal transformers, the desired two-port network is shown in Figure 4.13.

Example 4.8 Realize the admittance function

$$Y(s)=\frac{4s^2+3s+18}{3s^2+2s+6} \tag{4.64}$$

by the parallel configuration of Figure 4.13. As before, we first compute

$$m_1m_2-n_1n_2=12(s^2+3)^2 \tag{4.65}$$

which is already in a full square. Thus, no auxiliary Hurwitz polynomial is required. Since (4.65) does not contain a factor s^2, Case A applies. From (4.9) and (4.12a) with y_{ij} replacing z_{ij}, we obtain

$$y_{11}=\frac{m_1}{n_2}=\frac{4s^2+18}{2s}=2s+\frac{9}{s} \tag{4.66a}$$

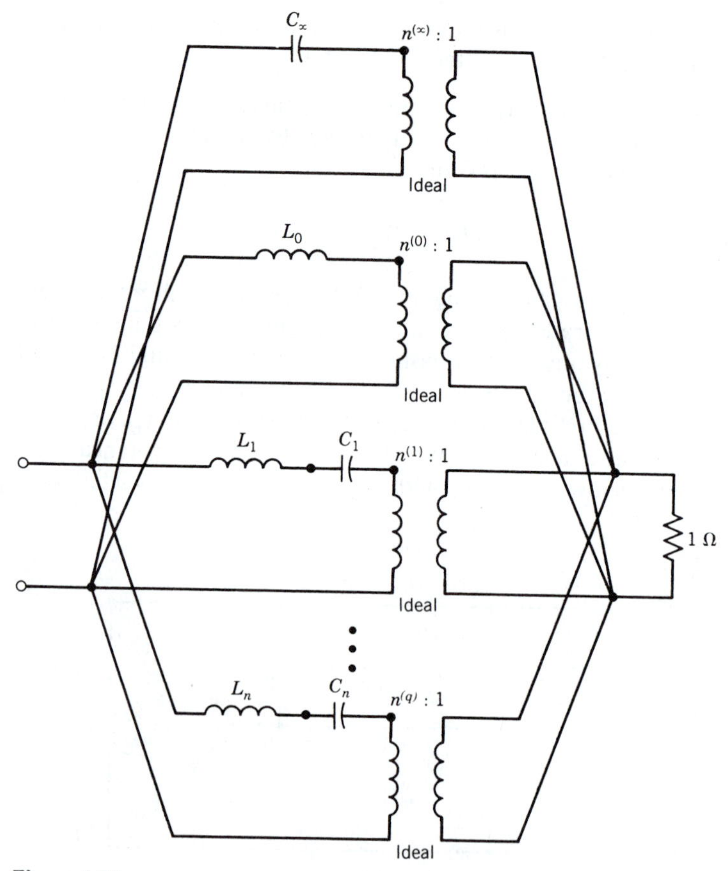

Figure 4.13

$$y_{22}=\frac{m_2}{n_2}=\frac{3s^2+6}{2s}=\frac{3s}{2}+\frac{3}{s} \tag{4.66b}$$

$$y_{12}=\frac{\sqrt{m_1m_2-n_1n_2}}{n_2}=\frac{2\sqrt{3}(s^2+3)}{2s}=\sqrt{3}s+\frac{3\sqrt{3}}{s} \tag{4.66c}$$

showing that all the poles are compact. From (4.49) and (4.51) with y replacing z, we obtain

$$y^{(\infty)}=2s, \qquad n^{(\infty)}=\sqrt{3}/2 \tag{4.67a}$$

$$y^{(0)}=9/s, \qquad n^{(0)}=1/\sqrt{3} \tag{4.67b}$$

The complete network realization is presented in Figure 4.14. To remove the ideal transformer associated with the pole at infinity, we multiply y_{22} by a factor $\alpha=k_{12}^2/k_{22}^2=4/3$ and y_{12} by $2/\sqrt{3}$ in (4.66). This results in a network realization as shown in Figure 4.15, in which the terminating resistance has been lowered from 1 Ω to 3/4 Ω.

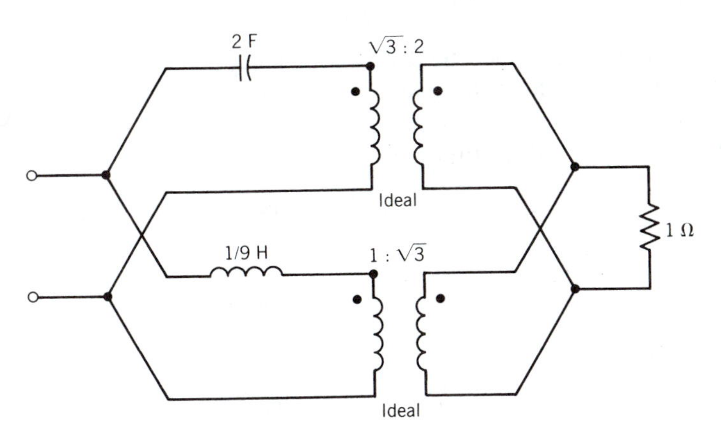

Figure 4.14

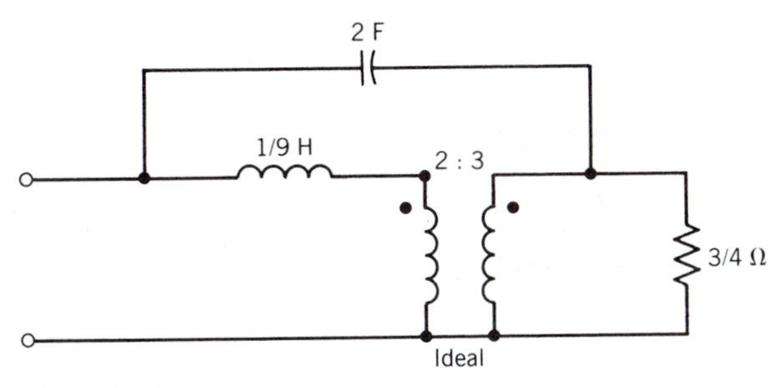

Figure 4.15

4.3 SYNTHESIS OF RESISTIVELY TERMINATED TWO-PORT NETWORKS

In the foregoing, we have demonstrated that any rational positive-real function can be realized as the input impedance of a lossless two-port terminated in a resistor. In the present section, we shall apply this result to the synthesis of a resistively terminated lossless two-port network having desired specifications.

The various specification functions for a terminated two-port network of Figure 4.16 are the transfer impedance $Z_{12}=V_2/I_1$, the transfer admittance $Y_{12}=I_2/V_1$, the transfer voltage ratio $G_{12}=V_2/V_1$, or the transfer current ratio $\alpha_{12}=I_2/I_1$. Since the impedance level of a network can be adjusted by normalization or impedance scaling, for simplicity and without loss of generality we may assume that the terminating resistance is always 1 Ω, as depicted in Figure 4.16. Under this condition, we have

$$G_{12}(s)=\frac{-y_{21}}{y_{22}+1}=-Y_{12}(s) \tag{4.68}$$

$$Z_{12}(s)=\frac{z_{21}}{z_{22}+1}=-\alpha_{12} \tag{4.69}$$

Thus, it suffices to consider only the realization of Z_{12} and Y_{12}.

Refer to Figure 4.16. Since the two-port is lossless, the average input power to the network must equal the average power delivered to the 1-Ω resistor for sinusoidal driving functions. This leads to the expression

$$|I_1(j\omega)|^2 \text{ Re } Z_{11}(j\omega)=|V_2(j\omega)|^2 \tag{4.70}$$

giving

$$|Z_{12}(j\omega)|^2=\text{Re } Z_{11}(j\omega) \tag{4.71}$$

or

$$|V_1(j\omega)|^2 \text{ Re } Y_{11}(j\omega)=|I_2(j\omega)|^2 \tag{4.72}$$

giving

$$|Y_{12}(j\omega)|^2=\text{Re } Y_{11}(j\omega) \tag{4.73}$$

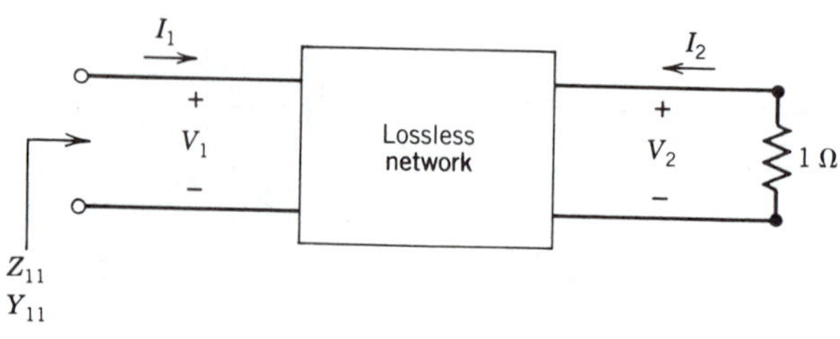

Figure 4.16

Equations (4.71) and (4.73) form the foundation of a synthesis procedure to be discussed here. Putting them in words, they state that the magnitude squared of the transfer immittance of a lossless two-port terminated in a 1-Ω resistor is numerically equal to the real part of the driving-point immittance at all frequencies. Thus, from a prescribed $|Z_{12}(j\omega)|$ or $|Y_{12}(j\omega)|$ if we can find the input immittance Z_{11} or Y_{11}, we can use Darlington's procedure to obtain the desired realization. For the method to succeed, we must be able to determine the immittance Z_{11} or Y_{11} from its real part. We shall focus our attention on the impedance, leaving the other possibility as obvious.

We now discuss the possibility of determining an impedance or admittance function from a specified rational function of ω that is to be the real part of the function. A question that naturally arises at this point is whether or not an immittance function can be uniquely determined from its $j\omega$-axis real part. We can quickly think of several one-port networks whose input immittances have the same real part, so the answer to this question is negative. For example, the impedances $Z_1=R+sL_1$ and $Z_2=R+sL_2$ have the same $j\omega$-axis real part but they are different if $L_1 \neq L_2$. In fact, there are an infinite number of impedances that are different from Z_1 and that have the same real part as Z_1. However, we can show that an immittance function can be determined from its $j\omega$-axis real part to only within a reactance function. We now describe two procedures for the solution of this problem.

4.3.1 **Bode Method**

Consider the impedance Z, whose real part can be expressed as

$$\text{Re } Z(j\omega)=\tfrac{1}{2}[Z(j\omega)+Z(-j\omega)] \tag{4.74}$$

Suppose that we replace ω by $-js$ on the right-hand side of this equation. We obtain the even part of Z, which is related to Z by the equation

$$\text{Ev } Z(s)=\tfrac{1}{2}[Z(s)+Z(-s)] \tag{4.75}$$

When the real part of $Z(j\omega)$ is specified, the even part of $Z(s)$ is determined. The question now is how to find $Z(s)$ from its even part. A solution to this was proposed by Bode† as follows: We first expand Ev $Z(s)$ by partial fractions, and then combine all the terms contributed by its LHS poles. If there is a constant term in the expansion, we add half of this to the combined term. The result is $\tfrac{1}{2}Z(s)$. Note that we do not really need to compute the entire partial-fraction expansion of Ev $Z(s)$, only those terms corresponding to its LHS poles. Bode's procedure follows directly from the facts that the poles of Ev $Z(s)$ are those of $Z(s)$ and $Z(-s)$, that those of $Z(s)$ are in the LHS and those of $Z(-s)$ are in the RHS, and that if $Z(s)$ is nonzero at infinity, $Z(-s)$ will have the same value there. Furthermore, as can be seen from (4.75), if $Z(s)$ has a pair of $j\omega$-axis poles, the partial-fraction expansion at these poles will cancel out

†H. W. Bode 1945, *Network Analysis and Feedback Amplifier Design*, Chapter 14. Princeton, NJ: Van Nostrand. Hendrik Wade Bode (1905–1982) was a professor at Harvard University and a member of the technical staff of Bell Laboratories.

with the corresponding terms in $Z(-s)$, and consequently these poles will not appear in Ev $Z(s)$. Thus, Bode's procedure will recover $Z(s)$ from Ev $Z(s)$ to within a reactance function corresponding to the poles on the $j\omega$-axis, a fact that was pointed out earlier. We illustrate this by the following example.

Example 4.9 Realize the magnitude squared of a transfer impedance

$$|Z_{12}(j\omega)|^2 = \frac{1+\omega^2}{1+\omega^4} \tag{4.76}$$

by a lossless two-port network terminated in a 1-Ω resistor.

From (4.71), (4.76) also represents the real part of the input impedance of a two-port terminated in a 1-Ω resistor. Replacing ω by $-js$ in (4.76) yields

$$\text{Ev } Z_{11}(s) = \frac{1-s^2}{1+s^4} \tag{4.77}$$

which can be expanded in partial fractions. The result is

$$\text{Ev } Z_{11}(s) = \frac{\sqrt{2}/4}{s+(1+j1)/\sqrt{2}} + \frac{\sqrt{2}/4}{s+(1-j1)/\sqrt{2}} + \frac{-\sqrt{2}/4}{s-(1-j1)/\sqrt{2}} + \frac{-\sqrt{2}/4}{s-(1+j1)/\sqrt{2}} \tag{4.78}$$

The sum of the first two terms on the right-hand side of (4.78) is $\frac{1}{2}Z_{11}(s)$ or

$$Z_{11}(s) = \frac{\sqrt{2}s+1}{s^2+\sqrt{2}s+1} \tag{4.79}$$

As remarked earlier, to determine $Z_{11}(s)$ there is no need to compute the last two terms of (4.78) corresponding to the RHS poles of Ev $Z_{11}(s)$.

Now we can use Darlington's procedure to realize the desired two-port. The auxiliary Hurwitz polynomial is chosen to be $m_0+n_0=s+1$, and the resulting z-parameters are obtained as

$$z_{11} = \frac{\sqrt{2}s^2+1}{s(s^2+1+\sqrt{2})} = \frac{1}{(1+\sqrt{2})s} + \frac{s}{s^2+1+\sqrt{2}} \tag{4.80a}$$

$$z_{22} = \frac{(1+\sqrt{2})s^2+1}{s(s^2+1+\sqrt{2})} = \frac{1}{(1+\sqrt{2})s} + \frac{2s}{s^2+1+\sqrt{2}} \tag{4.80b}$$

$$z_{12} = \frac{s^2-1}{s(s^2+1+\sqrt{2})} = \frac{-1}{(1+\sqrt{2})s} + \frac{\sqrt{2}s}{s^2+1+\sqrt{2}} \tag{4.80c}$$

The two-port network together with its termination is shown in Figure 4.17.

4.3.2 Brune and Gewertz Method

Bode's technique for determining $Z(s)$ from its even part is conceptually simple; it amounts to making the partial-fraction expansion of Ev $Z(s)$. However, it is not always

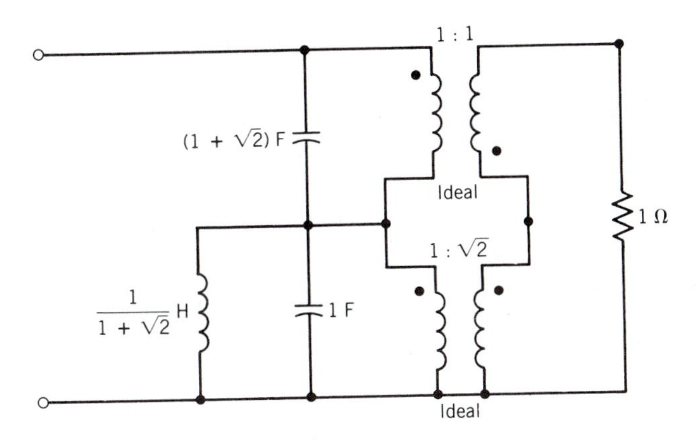

Figure 4.17

a simple matter to make such an expansion when Ev $Z(s)$ possesses the high-order poles. As an alternative, the procedure developed by Brune and Gewertz† circumvents this difficulty.

As in (4.6) write

$$Z(s)=\frac{m_1+n_1}{m_2+n_2}=\frac{a_0+a_1s+\cdots+a_ms^m}{b_0+b_1s+\cdots+b_ns^n} \tag{4.81}$$

the even part of which can be expressed as

$$\text{Ev } Z(s)=\frac{m_1m_2-n_1n_2}{m_2^2-n_2^2}=\frac{A_0+A_1s^2+\cdots+A_ms^{2m}}{B_0+B_1s^2+\cdots+B_ns^{2n}} \tag{4.82}$$

Our problem is to determine $Z(s)$ from (4.82). Recall that the poles of Ev $Z(s)$ are those of $Z(s)$ and $Z(-s)$. Those belonging to $Z(s)$ lie in the LHS and those belonging to $Z(-s)$ in the RHS. Thus, the denominator of $Z(s)$ can be determined by assigning all the LHS factors of those of Ev $Z(s)$ to $Z(s)$. At this point, the denominator polynomial of (4.81) is known. To determine the numerator, we form the expression $m_1m_2-n_1n_2$ and set it equal to the numerator of (4.82), obtaining

$$(a_0+a_2s^2+\cdots)(b_0+b_2s^2+\cdots)-(a_1s+a_3s^3+\cdots)(b_1s+b_3s^3+\cdots)$$
$$=A_0+A_1s^2+\cdots+a_ms^{2m} \tag{4.83}$$

Equating the coefficients of the same power of s yields

$$A_0=a_0b_0 \tag{4.84a}$$

$$A_1=a_0b_2+b_0a_2-a_1b_1 \tag{4.84b}$$

†O. Brune, "Synthesis of a finite two-terminal network whose driving-point impedance is a prescribed function of frequency," *J. Math. Phys.*, vol. 10, Aug. 1931, pp. 191–236. C. M. Gewertz, "Synthesis of a finite, four-terminal network from its prescribed driving point functions and transfer functions," *J. Math. Phys.*, vol. 12, Jan. 1933, pp. 1–257.

$$A_2 = a_0 b_4 + a_2 b_2 + a_4 b_0 - a_1 b_3 - a_3 b_1 \tag{4.84c}$$

$$\cdots\cdots\cdots\cdots\cdots\cdots\cdots\cdots\cdots\cdots$$

$$A_k = \sum_{j=-k}^{k} (-1)^{k+j} a_{k+j} b_{k-j} \tag{4.84d}$$

To find the numerator of $Z(s)$, we solve the unknown a's from this system of linear equations.

Example 4.10 Consider the same problem as discussed in Example 4.9. Clearly we can write

$$Z_{11}(s) = \frac{a_0 + a_1 s}{1 + \sqrt{2}s + s^2} \tag{4.85}$$

since the given real part is zero at infinity, so must be $Z_{11}(s)$. Using (4.84) in conjunction with (4.77) and (4.85) gives

$$1 = a_0 \tag{4.86a}$$

$$-1 = a_0 - \sqrt{2}a_1 \tag{4.86b}$$

Solving these two equations for a_0 and a_1, we obtain $a_0 = 1$ and $a_1 = \sqrt{2}$, confirming (4.79).

Example 4.11 Design a lossless two-port network realizing a given magnitude squared of a transfer admittance when terminated in a 1-Ω resistor. The magnitude squared is specified by the function

$$|Y_{12}(j\omega)|^2 = \frac{\omega^2}{1 + \omega^4} \tag{4.87}$$

which according to (4.73) is also the real part of the input admittance Y_{11} when the output of the two-port is terminated in a 1-Ω resistor. Thus, by replacing ω by $-js$ in (4.87), we obtain

$$\text{Ev } Y_{11}(s) = \frac{-s^2}{1 + s^4} = \frac{-s^2}{(s^2 + \sqrt{2}s + 1)(s^2 - \sqrt{2}s + 1)} \tag{4.88}$$

Since the given magnitude squared is zero at infinity, the admittance $Y_{11}(s)$ takes the

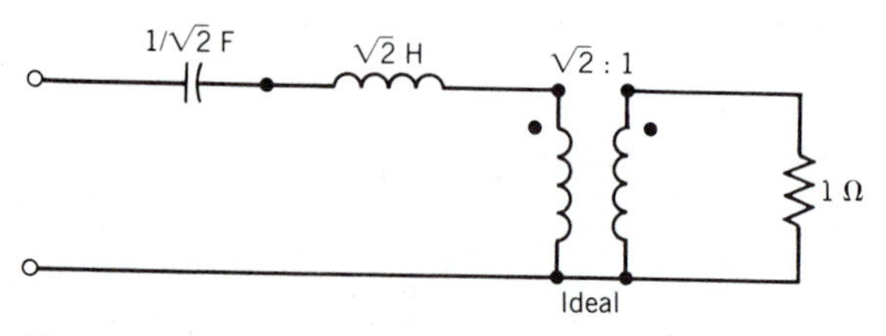

Figure 4.18

form

$$Y_{11}(s)=\frac{a_0+a_1 s}{1+\sqrt{2}s+s^2} \tag{4.89}$$

Using (4.84) in conjunction with (4.88) and (4.89) yields

$$0=a_0 \tag{4.90a}$$

$$-1=a_0-\sqrt{2}a_1 \tag{4.90b}$$

from which we obtain $a_1=1/\sqrt{2}$. Thus, the desired input admittance is

$$Y_{11}(s)=\frac{s}{\sqrt{2}s^2+2s+\sqrt{2}} \tag{4.91}$$

Using Darlington's procedure, the desired network is shown in Figure 4.18.

4.4 BROADBAND MATCHING NETWORKS: PRELIMINARY CONSIDERATIONS

So far we have shown that any nonnegative resistance can be transformed by a lossless reciprocal two-port network to any prescribed rational positive-real immittance. A more general problem in the design of communication systems is to synthesize a coupling network that will transform a given frequency-dependent load impedance into another specified one. We refer to this operation as *impedance matching* or *equalization*, and the resulting coupling network as *matching network* or *equalizer*. We recognize that the choice of a lossy equalizer would not only lessen the transfer of power from the source to the load but also severely hamper our ability to manipulate. Hence, we shall deal exclusively with the design of lossless equalizers.

Refer to the network configurations of Figure 4.19 where the source is represented either by its Thévenin equivalent or by its Norton equivalent. The load impedance z_2 is assumed to be strictly passive over a frequency band of interest, the reason being that the matching problem cannot be meaningfully defined if the load is purely reactive. Our objective is to design an "optimum" lossless two-port network (equalizer) N to match out the load impedance z_2 to the source and to achieve a preassigned transducer power-gain characteristic G over the entire sinusoidal frequency spectrum, where the interpretation of "optimum" becomes evident as the study is developed.

For our purposes, we shall only consider the special situation where the source impedance z_1 is resistive. The general matching problem is beyond the scope of this book. Thus, let $z_1=R_1$. Under this situation, the maximum power transfer from the source to the load over a frequency band of interest is achieved when the impedance presented to the generator is equal to the source resistance R_1. With the exception that the load impedance is purely resistive, it will be shown in Chapter 9 that it is not always possible to match an arbitrary passive load to a resistive generator over a

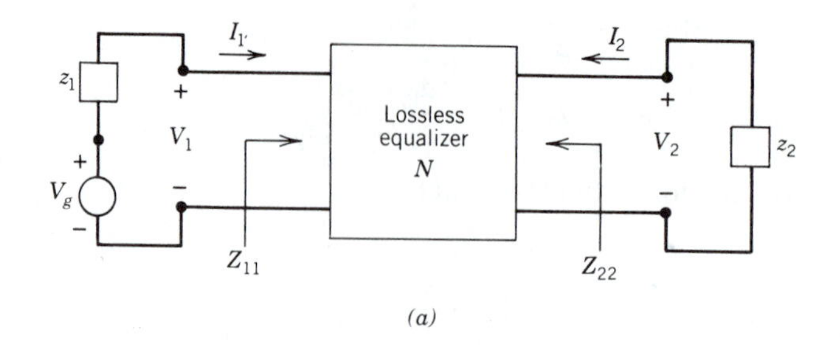

(a)

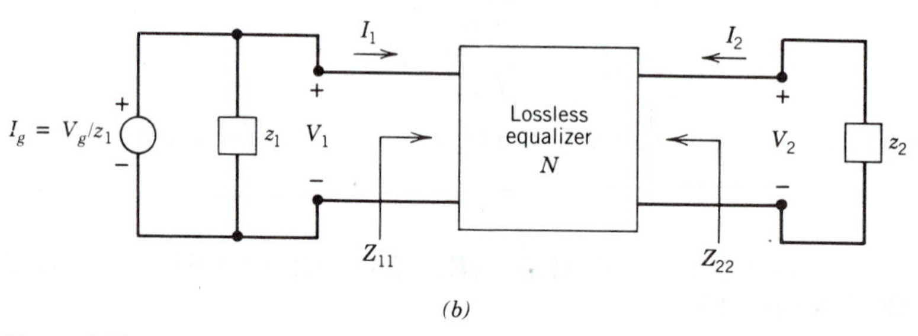

(b)

Figure 4.19

desired frequency band with a prescribed gain characteristic, and that the limitations originate from the physical realizability of the matching network that, in turn, is dictated by the load impedance. Thus, any matching problem must include the maximum tolerance on the match as well as the minimum bandwidth within which the match is to be obtained.

In the networks of Figure 4.19, let Z_{11} and Z_{22} be the impedances looking into the input and output ports when the output and input ports are terminated in z_2 and z_1, respectively. The input and output *reflection coefficients* are defined by the equations

$$\rho_1(s)=\frac{Z_{11}(s)-z_1(-s)}{Z_{11}(s)+z_1(s)} \tag{4.92}$$

$$\rho_2(s)=\frac{Z_{22}(s)-z_2(-s)}{Z_{22}(s)+z_2(s)} \tag{4.93}$$

respectively. We next show that the transducer power gain G, which is defined as the ratio of average power delivered to the load to the maximum available average power at the source, is precisely

$$G(\omega^2)=|S_{21}(j\omega)|^2\triangleq 1-|\rho_1(j\omega)|^2 \tag{4.94}$$

where S_{21} is known as the *transmission coefficient*. To verify this, let us compute the quantity $1-\rho_1(s)\rho_1(-s)$ from (4.92), obtaining

$$1-\rho_1(s)\rho_1(-s)=\frac{4r_1(s)R_{11}(s)}{[Z_{11}(s)+z_1(s)][Z_{11}(-s)+z_1(-s)]} \tag{4.95}$$

where

$$r_1(s) = \text{Ev } z_1(s) = \tfrac{1}{2}[z_1(s) + z_1(-s)] \tag{4.96}$$

$$R_{11}(s) = \text{Ev } Z_{11}(s) = \tfrac{1}{2}[Z_{11}(s) + Z_{11}(-s)] \tag{4.97}$$

Referring to Figure 4.19(a), we see that the impedance of the entire network from the source is

$$\frac{V_g(s)}{I_1(s)} = Z_{11}(s) + z_1(s) \tag{4.98}$$

If we let $s = j\omega$, then $r_1(j\omega) = \text{Re } z_1(j\omega)$, $R_{11}(j\omega) = \text{Re } Z_{11}(j\omega)$, and (4.95) reduces to

$$1 - |\rho_1(j\omega)|^2 = \frac{4r_1(j\omega)R_{11}(j\omega)}{|Z_{11}(j\omega) + z_1(j\omega)|^2} = \frac{|I_1(j\omega)|^2 R_{11}(j\omega)}{|V_g(j\omega)|^2/4r_1(j\omega)} \tag{4.99}$$

In the equation, we carry the argument $j\omega$ to emphasize the fact that all the quantities are to be evaluated at the sinusoidal frequency ω.

The power input in the sinusoidal steady state to the network of Figure 4.19(a) is

$$P_{\text{in}} = |I_1(j\omega)|^2 R_{11}(j\omega) \tag{4.100}$$

while the power output to the terminating load z_2 is

$$P_{\text{out}} = |I_2(j\omega)|^2 r_2(j\omega) \tag{4.101}$$

where

$$r_2(s) = \text{Ev } z_2(s) = \tfrac{1}{2}[z_2(s) + z_2(-s)] \tag{4.102}$$

Since the network N is lossless, the power input and power output must be equal, or

$$|I_1(j\omega)|^2 R_{11}(j\omega) = |I_2(j\omega)|^2 r_2(j\omega) \tag{4.103}$$

The maximum power that the source combination is capable of delivering to the network occurs when the input is conjugately matched so that $Z_{11}(j\omega) = \bar{z}_1(j\omega)$. Under this condition, the network is equivalent to one with two impedances $z_1(j\omega)$ and $\bar{z}_1(j\omega)$ in series with the voltage source. The average power consumed by $\bar{z}_1(j\omega)$ is

$$P_{\text{ava}} = \frac{|V_g(j\omega)|^2}{4r_1(j\omega)} \tag{4.104}$$

Combining this with (4.103) and (4.99), we arrive at the important relationship

$$1 - |\rho_1(j\omega)|^2 = \frac{|I_2(j\omega)|^2 r_2(j\omega)}{|V_g(j\omega)|^2/4r_1(j\omega)} = \frac{P_{\text{out}}}{P_{\text{ava}}} = G(\omega^2) \tag{4.105}$$

Using (4.94) the last equation tells us that

$$|\rho_1(j\omega)|^2 + |S_{21}(j\omega)|^2 = 1 \tag{4.106}$$

and that

$$|\rho_1(j\omega)|^2 = 1 - \frac{\text{power to load}}{\text{power available}} = \frac{\text{power available} - \text{power to load}}{\text{power available}}$$

$$= \frac{\text{“reflected” power}}{\text{power available}} \tag{4.107}$$

$$|S_{21}(j\omega)|^2 = \frac{\text{power to load}}{\text{power available}} \tag{4.108}$$

Thus, the magnitude squared of the reflection coefficient $|\rho_1(j\omega)|^2$ denotes the fraction of the maximum available average power that is reflected back to the source, and the magnitude squared of the transmission coefficient $|S_{21}(j\omega)|^2$ represents the fraction of the maximum available average power that is transmitted to the load from the source. Indeed, the names reflection coefficient and transmission coefficient are suggested by these interpretations. Observe from (4.105) that all the terms are even functions of ω. As a result, the transducer power gain G is a function of ω^2. To emphasize this we write G as $G(\omega^2)$. Also, from (4.105) if $|\rho_1|=0$ then $P_{\text{out}}=P_{\text{ava}}$ and there is no reflection. But when $|\rho_1|=1$, there is complete reflection and $P_{\text{out}}=0$.

We shall next show the relationships of the transmission coefficient and several other possible specification functions. By (4.105),

$$|S_{21}(j\omega)|^2 = 4r_1(j\omega)r_2(j\omega)\left|\frac{I_2(j\omega)}{V_g(j\omega)}\right|^2 \tag{4.109}$$

Since $V_g = z_1 I_g$, we have

$$|S_{21}(j\omega)|^2 = \frac{4r_1(j\omega)r_2(j\omega)}{|z_1(j\omega)|^2}\cdot\left|\frac{I_2(j\omega)}{I_g(j\omega)}\right|^2 \tag{4.110}$$

In terms of voltage ratio and transfer impedance, we use $V_2 = -I_2 z_2$ in (4.109) and (4.110),

$$|S_{21}(j\omega)|^2 = \frac{4r_1(j\omega)r_2(\omega)}{|z_2(j\omega)|^2}\cdot\left|\frac{V_2(j\omega)}{V_g(j\omega)}\right|^2 \tag{4.111}$$

$$|S_{21}(j\omega)|^2 = \frac{4r_1(j\omega)r_2(j\omega)}{|z_1(j\omega)z_2(j\omega)|^2}\cdot\left|\frac{V_2(j\omega)}{I_g(j\omega)}\right|^2 \tag{4.112}$$

Similar relations and interpretations can be derived for the output reflection coefficient ρ_2 and the transmission coefficient magnitude squared:

$$|S_{12}(j\omega)|^2 = 1 - |\rho_2(j\omega)|^2 \tag{4.113}$$

In fact, for the lossless reciprocal two-port network N we have

$$|S_{21}(j\omega)|^2 = |S_{12}(j\omega)|^2 = 1 - |\rho_1(j\omega)|^2 = 1 - |\rho_2(j\omega)|^2 \tag{4.114}$$

With this background, we may now outline the steps to be followed in the design of equalizers.

(i) From any of the specification functions $|I_2/V_g|^2$, $|I_2/I_g|^2$, $|V_2/V_g|^2$, or $|V_2/I_g|^2$, together with the given impedances $z_1 = R_1$ and z_2, determine the squared magnitude of the transmission coefficient $|S_{21}(j\omega)|^2$. Then determine the squared magnitude of the reflection coefficient $|\rho_2(j\omega)|^2$.

(ii) From $|\rho_2(j\omega)|^2$, determine $\rho_2(s)$.

(iii) From $\rho_2(s)$ and the value of z_2, determine $Z_{22}(s)$.

(iv) From $Z_{22}(s)$, synthesize a lossless two-port network terminated in a resistor, using Darlington's procedure if necessary.

Before we proceed to consider the general load, we discuss the special situation where z_2 is purely resistive. The special case is of sufficient importance in practical applications and merits a separate consideration.

4.5 DOUBLE-TERMINATED NETWORKS

The problem considered in this section is to derive formulas pertinent to the design of a lossless two-port network operating between a resistive generator with internal resistance R_1 and a resistive load with resistance R_2 and having a preassigned transducer power-gain characteristic, as depicted in Figure 4.20. The general problem of matching an arbitrary passive load to a resistive generator having a prescribed transducer power-gain characteristic will be considered in Chapter 9.

4.5.1 Butterworth Networks

We show that an LC ladder can always be obtained that when inserted between the resistive generator and the load R_2 will yield an nth-order Butterworth transducer power-gain characteristic

$$|S_{21}(j\omega)|^2 = G(\omega^2) = \frac{K_n}{1 + (\omega/\omega_c)^{2n}} \tag{4.115}$$

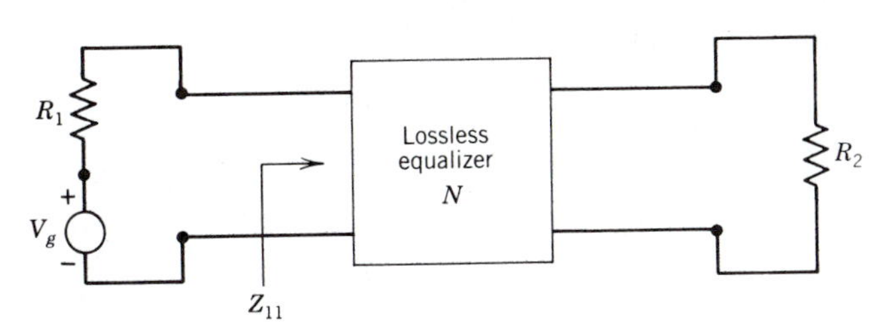

Figure 4.20

as in (2.1). Since for a passive network $G(\omega^2)$ is bounded between 0 and 1, the dc gain K_n is restricted by

$$0 \leqslant K_n \leqslant 1 \tag{4.116}$$

From (4.114) the squared magnitude of the input reflection coefficient is

$$|\rho_1(j\omega)|^2 = 1 - |S_{21}(j\omega)|^2 = \frac{1 - K_n + (\omega/\omega_c)^{2n}}{1 + (\omega/\omega_c)^{2n}} \tag{4.117}$$

or

$$\rho_1(j\omega)\rho_1(-j\omega) = \alpha^{2n} \frac{1 + (\omega/\alpha\omega_c)^{2n}}{1 + (\omega/\omega_c)^{2n}} \tag{4.118}$$

where

$$\alpha = (1 - K_n)^{1/2n} \tag{4.119}$$

Substituting ω by $-js$ in (4.118) gives

$$\rho_1(s)\rho_1(-s) = \alpha^{2n} \frac{1 + (-1)^n x^{2n}}{1 + (-1)^n y^{2n}} \tag{4.120}$$

where

$$y = s/\omega_c \tag{4.121a}$$

$$x = y/\alpha \tag{4.121b}$$

Now with the zeros and poles of (4.120) specified as in (2.13), the input reflection coefficient $\rho_1(s)$ can be identified. Since $\rho_1(s)$ is devoid of poles in the closed RHS,† we must assign all of the LHS poles of (4.120) to $\rho_1(s)$, resulting in a unique decomposition of the denominator polynomial of (4.120). However, the zeros of $\rho_1(s)$ may lie in the RHS, so that in general a number of different numerators are possible for $\rho_1(s)$. For reasons to be given in later sections, we choose only the LHS zeros for $\rho_1(s)$. Define a *minimum-phase reflection coefficient* as one that is devoid of zeros in the open RHS. Then a minimum-phase solution of (4.120) is found from (2.14) and (2.15) to be

$$\rho_1(s) = \pm \alpha^n \frac{q(x)}{q(y)} \tag{4.122}$$

From (4.92) the impedance Z_{11} looking into the input port of the network of Figure 4.20 with the output port terminating in R_2 is obtained as

$$Z_{11}(s) = R_1 \frac{1 + \rho_1(s)}{1 - \rho_1(s)} \tag{4.123}$$

Substituting (4.122) in (4.123) yields

$$Z_{11}(s) = R_1 \frac{q(y) \pm \alpha^n q(x)}{q(y) \mp \alpha^n q(x)} \tag{4.124}$$

†The right-half of the s-plane, including the $j\omega$-axis. Open RHS means that the $j\omega$-axis is excluded.

For some design applications, it may be desirable to specify both R_1 and R_2 in advance. In this case, the dc gain K_n cannot be chosen independently. In fact, substituting $s=0$ in (4.124) gives the desired relationship for an LC two-port network with $K_n \neq 0$:

$$\frac{R_2}{R_1} = \left(\frac{1+\alpha^n}{1-\alpha^n}\right)^{\pm 1} \tag{4.125}$$

The $\pm$ signs are determined, respectively, according to $R_2 \geqq R_1$ and $R_2 \leqq R_1$. Thus, if any two of the three quantities R_1, R_2, and K_n are specified, the third one is fixed.

We now show that Z_{11} can be realized as a resistively terminated LC ladder network. The ladder networks are attractive from an engineering viewpoint in that they are unbalanced and contain no coupling coils. Also, explicit formulas for their element values are available, which reduce the design problem to simple arithmetic. Depending upon the choice of the plus and minus signs in (4.122), two cases are distinguished.

Case A: $\rho_1(0) \geqq 0$. With the choice of the plus sign in (4.122), the input impedance becomes

$$Z_{11}(s) = R_1 \frac{q(y)+\alpha^n q(y/\alpha)}{q(y)-\alpha^n q(y/\alpha)} = R_1 \frac{\sum_{m=0}^{n} a_m(1+\alpha^{n-m})y^m}{\sum_{m=0}^{n} a_m(1-\alpha^{n-m})y^m} \tag{4.126}$$

where the coefficient a's are defined in (2.15). This impedance can be expanded in a continued fraction about infinity, as in the first Cauer canonical form, and results in a lossless ladder network terminated in a resistor. Hence, we can write

$$Z_{11}(s) = L_1 s + \cfrac{1}{C_2 s + \cfrac{1}{L_3 s + \cfrac{1}{\ddots + \cfrac{1}{W}}}} \tag{4.127}$$

W being a constant representing either a resistance or a conductance. Depending on whether n is odd or even, the LC ladder network has the configurations as shown in Figure 4.21. The first element L_1 can easily be determined from (4.126) in conjunction with (2.18) after a simple division. The result is given by

$$L_1 = \frac{2R_1}{(1-\alpha)a_{n-1}\omega_c} = \frac{2R_1 \sin \pi/2n}{(1-\alpha)\omega_c} \tag{4.128}$$

where $a_{n-1} = 1/(\sin \pi/2n)$. Moreover, it can be shown that the values of other elements can be computed by the recurrence formulas

$$L_{2m-1}C_{2m} = \omega_c^{-2} \frac{4 \sin \gamma_{4m-3} \sin \gamma_{4m-1}}{1-2\alpha \cos \gamma_{4m-2}+\alpha^2} \tag{4.129a}$$

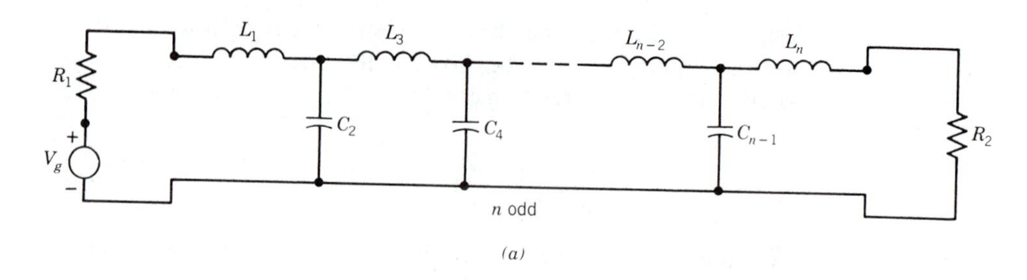

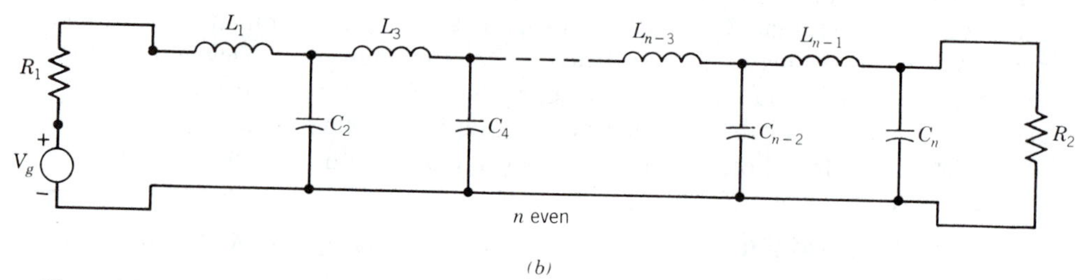

Figure 4.21

$$L_{2m+1}C_{2m}=\omega_c^{-2}\frac{4\sin\gamma_{4m-1}\sin\gamma_{4m+1}}{1-2\alpha\cos\gamma_{4m}+\alpha^2} \tag{4.129b}$$

for $m=1, 2, \ldots, [\frac{1}{2}n]$, where

$$\gamma_m=m\pi/2n \tag{4.130}$$

and $[\frac{1}{2}n]$ denotes the largest integer not greater than $\frac{1}{2}n$. This symbol will be used throughout the remainder of the chapter. The final element of the ladder is L_n as shown in Figure 4.21(*a*) if n is odd, or C_n of Figure 4.21(*b*) if n is even. In addition, the values of the final elements are related to R_2 by

$$L_n=\frac{2R_2\sin\gamma_1}{(1+\alpha)\omega_c} \tag{4.131a}$$

for n odd, and

$$C_n=\frac{2\sin\gamma_1}{R_2(1+\alpha)\omega_c} \tag{4.131b}$$

for n even. The above formulas can be derived deductively by carrying out the calculations in detail for the cases of low order and then guessing the final result. A formal complete proof was first given by Bossé.† Hence we can calculate the element values starting from either the first or the last element.

†G. Bossé, "Siebketten ohne Dämpfungsschwankungen im Durchlassbereich (Potenzketten)," *Frequenz*, vol. 5, 1951, pp. 279–284.

In particular, for $R_1 = R_2$ we have $\alpha = 0$ and the above formulas are amazingly simple, and (4.129) reduces to

$$L_{2m-1} = \frac{2R_1}{\omega_c} \sin \gamma_{4m-3} \tag{4.132a}$$

$$C_{2m} = \frac{2}{R_1 \omega_c} \sin \gamma_{4m-1} \tag{4.132b}$$

We illustrate the use of above formulas by the following example.

Example 4.12 Given

$$R_1 = 100\ \Omega, \qquad R_2 = 200\ \Omega \tag{4.133a}$$

$$\omega_c = 10^4 \text{ rad/s}, \qquad n = 5 \tag{4.133b}$$

obtain a Butterworth LC ladder network to meet these specifications.

Since $R_2 > R_1$, we choose the plus sign in (4.125) and obtain $\alpha = 0.80274$. Thus, from (4.128) we compute

$$L_1 = \frac{2 \times 100 \times \sin 18^\circ}{(1 - 0.8027) \times 10^4} = 31.331 \text{ mH} \tag{4.134}$$

Using (4.129a) with $m = 1$ in conjunction with (4.134) yields

$$C_2 = \frac{10^{-6} \times 4 \times \sin 18^\circ \sin 54^\circ}{3.1331 \times (1 - 1.6055 \times \cos 36^0 + 0.6444)} = 0.9237\ \mu\text{F} \tag{4.135a}$$

Now using (4.129b) with $m = 1$ and (4.135a), we obtain

$$L_3 = \frac{10^{-2} \times 4 \times \sin 54^\circ \sin 90^\circ}{0.9237 \times (1 - 1.6055 \times \cos 72^\circ + 0.6444)} = 30.510 \text{ mH} \tag{4.135b}$$

Repeat the above process for $m = 2$ to give

$$C_4 = \frac{10^{-6} \times 4 \times \sin 90^\circ \sin 126^\circ}{3.051 \times (1 - 1.6055 \times \cos 108^\circ + 0.6444)} = 0.4955\ \mu\text{F} \tag{4.135c}$$

$$L_5 = \frac{10^{-2} \times 4 \times \sin 126^\circ \sin 162^\circ}{0.4955 \times (1 - 1.6055 \times \cos 144^\circ + 0.6444)} = 6.857 \text{ mH} \tag{4.135d}$$

Alternatively, L_5 can be computed directly from (4.131a):

$$L_5 = \frac{2 \times 200 \times \sin 18^\circ}{(1 + 0.8027) \times 10^4} = 6.857 \text{ mH} \tag{4.136}$$

The ladder network together with its termination is presented in Figure 4.22. This network possesses the fifth-order Butterworth transducer power-gain response with a dc gain

$$K_5 = 1 - \alpha^{10} = 8/9 \tag{4.137}$$

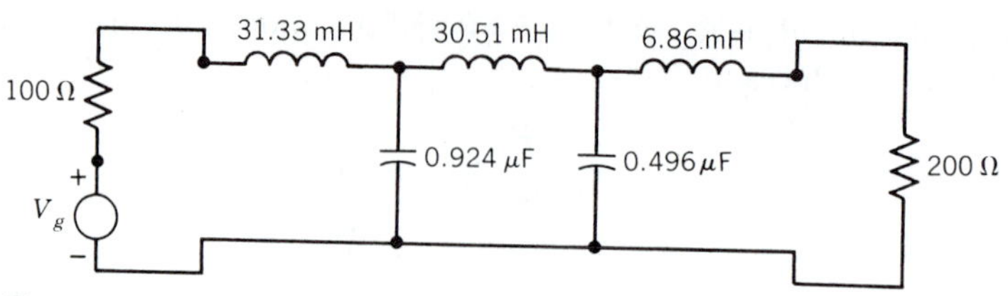

Figure 4.22

Now suppose that we change R_2 from 200 Ω to 100 Ω, everything else being the same. Then from (4.132) the element values for the desired *LC* ladder network are found to be

$$L_1 = 200 \times 10^{-4} \times \sin 18° = 6.1804 \text{ mH} \tag{4.138a}$$

$$C_2 = 2 \times 10^{-6} \times \sin 54° = 1.618\ \mu\text{F} \tag{4.138b}$$

$$L_3 = 200 \times 10^{-4} \times \sin 90° = 20 \text{ mH} \tag{4.138c}$$

$$C_4 = 2 \times 10^{-6} \times \sin 126° = 1.618\ \mu\text{F} \tag{4.138d}$$

$$L_5 = 200 \times 10^{-4} \times \sin 162° = 6.1804 \text{ mH} \tag{4.138e}$$

The corresponding *LC* ladder network and its terminations are shown in Figure 4.23 with dc gain $K_5 = 1$.

Case B: $\rho_1(0) < 0$. With the choice of the minus in (4.122), the input impedance becomes

$$Z_{11}(s) = R_1 \frac{\sum_{m=0}^{n} a_m(1 - \alpha^{n-m})y^m}{\sum_{m=0}^{n} a_m(1 + \alpha^{n-m})y^m} \tag{4.139}$$

which, aside from the constant R_1, is the reciprocal of that given in (4.126). Hence it

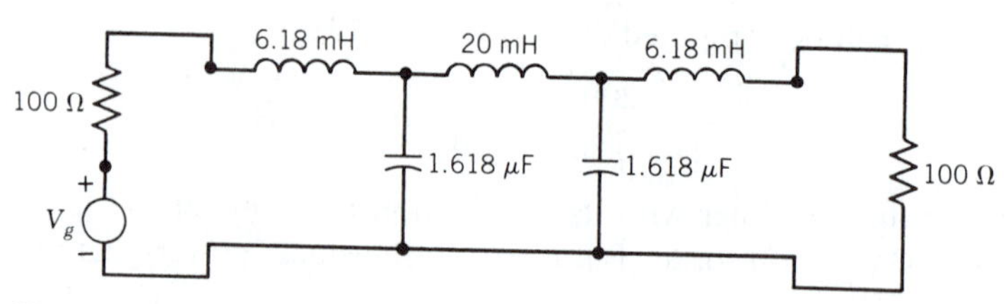

Figure 4.23

can be expanded as

$$\frac{1}{Z_{11}(s)}=C_1s+\cfrac{1}{L_2s+\cfrac{1}{C_3s+\cfrac{1}{\ddots+\cfrac{1}{W}}}} \tag{4.140}$$

which can be realized by the ladder networks of Figure 4.24, depending on whether n is even or odd, where W is the terminating resistance or conductance. Formulas for the element values are the same as those given in (4.128)–(4.132) except that the roles of C's and L's are interchanged and R_1 and R_2 are replaced by their reciprocals:

$$C_1=\frac{2\sin\gamma_1}{R_1(1-\alpha)\omega_c} \tag{4.141a}$$

$$C_{2m-1}L_{2m}=\frac{4\sin\gamma_{4m-3}\sin\gamma_{4m-1}}{(1-2\alpha\cos\gamma_{4m-2}+\alpha^2)\omega_c^2} \tag{4.141b}$$

$$C_{2m+1}L_{2m}=\frac{4\sin\gamma_{4m-1}\sin\gamma_{4m+1}}{(1-2\alpha\cos\gamma_{4m}+\alpha^2)\omega_c^2} \tag{4.141c}$$

for $m=1, 2, \ldots, [\frac{1}{2}n]$, and

$$C_n=\frac{2\sin\gamma_1}{R_2(1+\alpha)\omega_c} \tag{4.142a}$$

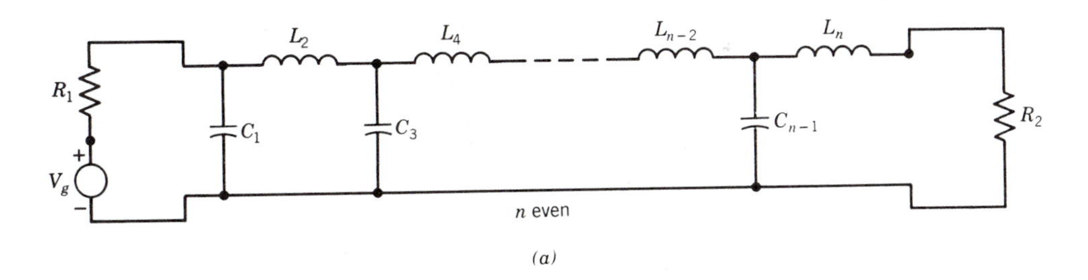

(a)

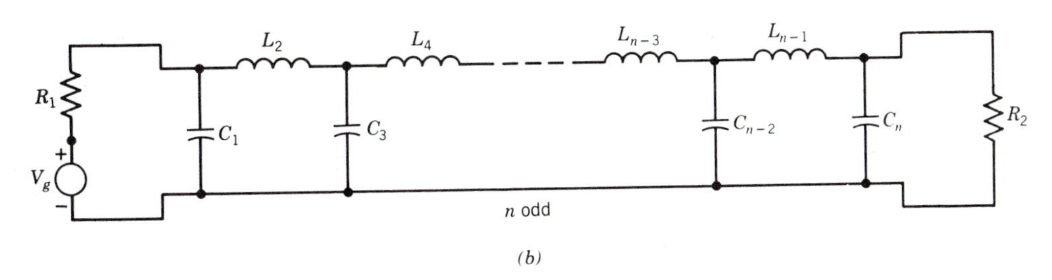

(b)

Figure 4.24

for n odd, and

$$L_n = \frac{2R_2 \sin \gamma_1}{(1+\alpha)\omega_c} \tag{4.142b}$$

for n even.

In Example 4.12, if we choose the minus sign in (4.122) instead of the plus sign, we obtain two ladder networks as shown in Figure 4.25, the normalized input impedances Z_{11}/R_1 of which are reciprocals of those obtained for the corresponding ladders of Figures 4.22 and 4.23. Note that in selecting the minus sign in (4.122), we must choose the minus sign in (4.125) for the same values of R_1 and α. This means that for $R_1 = 100\,\Omega$ and $\alpha = 0.80274$, the terminating resistance R_2 is determined by (4.125) to be $50\,\Omega$ and cannot be specified independently.

4.5.2 Chebyshev Networks

Now consider the synthesis of a two-port network possessing a preassigned Chebyshev transducer power-gain response,

$$|S_{21}(j\omega)|^2 = G(\omega^2) = \frac{K_n}{1+\varepsilon^2 C_n^2(\omega/\omega_c)} \tag{4.143}$$

as in (2.23), with K_n bounded between 0 and 1. Following (4.117), the squared magnitude of the input reflection coefficient can be written as

$$\rho_1(j\omega)\rho_1(-j\omega) = 1 - |S_{21}(\omega)|^2 = \frac{1 - K_n + \varepsilon^2 C_n^2(\omega/\omega_c)}{1+\varepsilon^2 C_n^2(\omega/\omega_c)} \tag{4.144}$$

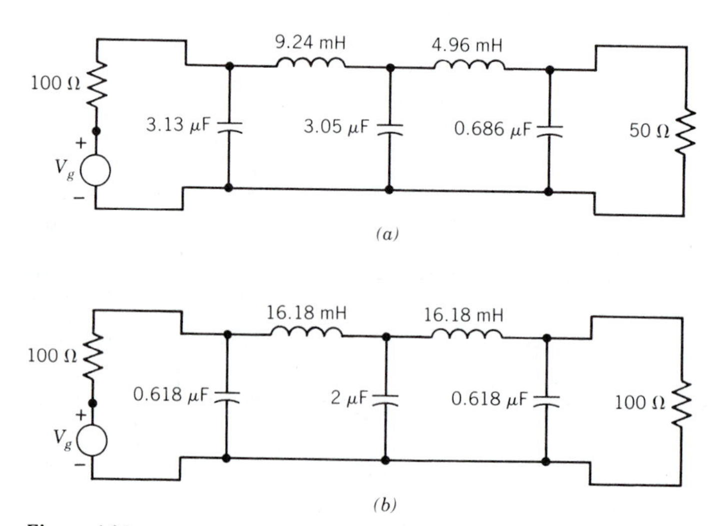

Figure 4.25

giving

$$\rho_1(s)\rho_1(-s)=(1-K_n)\,\frac{1+\hat{\varepsilon}^2C_n^2(-jy)}{1+\varepsilon^2C_n^2(-jy)} \tag{4.145}$$

where $y=s/\omega_c$, as in (4.121a), and

$$\hat{\varepsilon}=\varepsilon(1-K_n)^{-1/2} \tag{4.146}$$

As in the Butterworth response, the denominator of (4.145) can be uniquely decomposed as in (2.60). However, the zeros of ρ_1 may lie in the RHS and there are no known reasons why they cannot, the only restriction being that the zeros of a complex-conjugate pair must be assigned together. For our purposes, we choose only the LHS zeros. With this restriction, we obtain a minimum-phase decomposition of (4.145):

$$\rho_1(s)=\pm\frac{\hat{p}(y)}{p(y)} \tag{4.147}$$

where $p(y)$ is defined in (2.61) and

$$\hat{p}(y)=\hat{b}_0+\hat{b}_1y+\cdots+\hat{b}_{n-1}y^{n-1}+\hat{b}_ny^n=\sum_{m=0}^{n}\hat{b}_my^m \tag{4.148}$$

with $\hat{b}_n=1$ is the Hurwitz polynomial formed by the LHS roots of the equation $1+\hat{\varepsilon}^2C_n^2(-jy)=0$. Thus from (4.92) with $z_1=R_1$, the input impedance Z_{11} of N with the output port terminating in R_2 is found to be

$$Z_{11}(s)=R_1\,\frac{p(y)\pm\hat{p}(y)}{p(y)\mp\hat{p}(y)} \tag{4.149}$$

By setting $y=0$ in (4.149) and using (4.146) and the known formula for the coefficient

$$b_0=2^{1-n}\sinh na,\qquad n\text{ odd} \tag{4.150a}$$

$$=2^{1-n}\cosh na,\qquad n\text{ even} \tag{4.150b}$$

where $na=\sinh^{-1}1/\varepsilon$ as in (2.58c), we obtain a relationship among the quantities R_1, R_2, and K_n:

$$\frac{R_2}{R_1}=\left[\frac{b_0+\hat{b}_0}{b_0-\hat{b}_0}\right]^{\pm 1} \tag{4.151a}$$

$$=\left[\frac{1+\sqrt{1-K_n}}{1-\sqrt{1-K_n}}\right]^{\pm 1},\qquad n\text{ odd} \tag{4.151b}$$

$$=\left[\frac{\sqrt{1+\varepsilon^2}+\sqrt{1+\varepsilon^2-K_n}}{\sqrt{1+\varepsilon^2}-\sqrt{1+\varepsilon^2-K_n}}\right]^{\pm 1},\qquad n\text{ even} \tag{4.151c}$$

The plus and minus signs in the equations are determined, respectively, according to $R_2 \geqq R_1$ and $R_2<R_1$. Thus, if n is odd and the dc gain is specified, the ratio of the terminating resistances is fixed by (4.151b). On the other hand, if n is even and the

peak-to-peak ripple in the passband and K_n or the dc gain is specified, the ratio of the resistances is given by (4.151c).

Like the Butterworth response, the input impedance Z_{11} of (4.149) can be realized by an LC ladder terminating in a resistor. Depending on the choice of the signs in (4.147), two cases are distinguished.

Case A: $\rho_1(0) \geqq 0$. With the choice of the plus sign in (4.147), the input impedance becomes

$$Z_{11}(s)=R_1\frac{\sum_{m=0}^{n}(b_m+\hat{b}_m)y^m}{\sum_{m=0}^{n}(b_m-\hat{b}_m)y^m} \tag{4.152}$$

which can be expanded in a continued fraction as in (4.127). Depending on whether n is odd or even, the corresponding LC ladder network has the configurations as shown in Figure 4.21. The first element L_1 is determined from (4.152) with the aid of the formula $b_{n-1}=(\sinh a)/(\sin\gamma_1)$, and is given by

$$L_1=\frac{2R_1}{(b_{n-1}-\hat{b}_{n-1})\omega_c}=\frac{2R_1\sin\gamma_1}{\omega_c(\sinh a-\sinh\hat{a})} \tag{4.153}$$

where $\gamma_1=\pi/2n$ as in (4.130) and

$$\hat{a}=\frac{1}{n}\sinh^{-1}\left(\frac{\sqrt{1-K_n}}{\varepsilon}\right) \tag{4.154}$$

Moreover, the values of other elements can be computed by the recurrence formulas

$$L_{2m-1}C_{2m}=\frac{4\sin\gamma_{4m-3}\sin\gamma_{4m-1}}{\omega_c^2 f_{2m-1}(\sinh a,\sinh\hat{a})} \tag{4.155a}$$

$$L_{2m+1}C_{2m}=\frac{4\sin\gamma_{4m-1}\sin\gamma_{4m+1}}{\omega_c^2 f_{2m}(\sinh a,\sinh\hat{a})} \tag{4.155b}$$

for $m=1, 2, \ldots, [\frac{1}{2}n]$, terminating in L_n as shown in Figure 4.21(a) if n is odd or in C_n as in Figure 4.21(b) if n is even, where

$$f_m(\sinh a,\sinh\hat{a})=\sinh^2 a+\sinh^2\hat{a}+\sin^2\gamma_{2m}-2\sinh a\sinh\hat{a}\cos\gamma_{2m} \tag{4.156}$$

and $\gamma_m=m\pi/2n$ as in (4.130). In addition, the values of the last elements are related to R_2 by the equations

$$L_n=\frac{2R_2\sin\gamma_1}{\omega_c(\sinh a+\sinh\hat{a})},\qquad n\text{ odd} \tag{4.157a}$$

$$C_n=\frac{2\sin\gamma_1}{\omega_c R_2(\sinh a+\sinh\hat{a})},\qquad n\text{ even} \tag{4.157b}$$

Again the above formulas can be derived deductively by carrying out the calculations in detail for the cases of low order and then guessing the final result. A formal elegant proof was first given by Takahasi.† Hence we can calculate the element values starting from either the first or the last element. We illustrate the use of these formulas by the following example.

Example 4.13 Given

$$R_1 = 100\,\Omega, \qquad R_2 = 200\,\Omega, \qquad \omega_c = 10^4 \text{ rad/s}, \qquad n = 4 \tag{4.158}$$

find a Chebyshev network to meet these specifications with peak-to-peak ripple in the passband not exceeding 1 dB.

Since R_1 and R_2 are both specified, the minimum passband gain $G_{\min}$ is fixed by (4.151c) in accordance with (2.37). For $R_2 > R_1$ we must choose the plus sign in (4.151c), obtaining

$$\frac{200}{100} = \frac{1 + \sqrt{1 - G_{\min}}}{1 - \sqrt{1 - G_{\min}}} \tag{4.159}$$

where $G_{\min} = K_4/(1 + \varepsilon^2)$. Solving for $G_{\min}$ yields

$$G_{\min} = \frac{K_4}{1 + \varepsilon^2} = \frac{8}{9} \tag{4.160}$$

According to (2.41), the 1-dB ripple in the passband corresponds to a ripple factor $\varepsilon = 0.509$. If we use this value of ε in (4.160), we obtain $K_4 = 1.119$, which is too large for the network to be physically realizable. Thus, let $K_4 = 1$, the maximum permissible value. This according to (4.160) gives $\varepsilon = 0.354$ or 0.51-dB ripple, well within the 1-dB specification. With this selection of parameters, we next compute the following quantities:

$$\hat{a} = 0 \tag{4.161a}$$

$$a = \frac{1}{4} \sinh^{-1} \frac{1}{0.354} = 0.441 \tag{4.161b}$$

$$f_m(\sinh 0.441, 0) = 0.207 + \sin^2 \gamma_{2m} \tag{4.161c}$$

From (4.153) the value of the first inductance is given by

$$L_1 = \frac{200 \sin 22.5^\circ}{10^4 \sinh 0.441} = 16.82 \text{ mH} \tag{4.162a}$$

Using (5.155a) with $m = 1$ in conjunction with (4.162a), we obtain

$$C_2 = \frac{4 \sin 22.5^\circ \sin 67.5^\circ}{1.682 \times 10^6 \times (0.207 + \sin^2 45^\circ)} = 1.1892\ \mu\text{F} \tag{4.162b}$$

Now use (4.155b) with $m = 1$ and (4.162b) to give

$$L_3 = \frac{4 \sin 67.5^\circ \sin 112.5^\circ}{1.1892 \times 10^2 \times (0.207 + \sin^2 90^\circ)} = 23.786 \text{ mH} \tag{4.162c}$$

†H. Takahasi, "On the ladder-type filter network with Tchebysheff response," *J. Inst. Elec. Commun. Engrs. Japan*, vol. 34, 1951, pp. 65–74.

Repeating the above process for $m=2$ in (4.155a), we obtain

$$C_4=\frac{4\sin 112.5^\circ \sin 157.5^\circ}{2.3786\times 10^6\times(0.207+\sin^2 135^\circ)}=0.841\ \mu\text{F} \tag{4.162d}$$

The last capacitance can also be computed directly from (4.157b) as

$$C_4=\frac{2\sin 22.5^\circ}{200\times 10^4\times \sinh 0.441}=0.841\ \mu\text{F} \tag{4.162e}$$

coinciding with (4.162d). The LC ladder network together with its terminations is presented in Figure 4.26. The network possesses the fourth-order Chebyshev transducer power-gain characteristic.

For illustrative purposes, we also compute the element values by expanding the input impedanze Z_{11} as a continued fraction as in (4.127). To this end, we first determined the poles of the Chebyshev response by means of formula (2.62) with $n=4$ and $a=0.441$, giving

$$y_1,\ y_4=-0.1741\pm j1.015 \tag{4.163a}$$

$$y_2,\ y_3=-0.4204\pm j0.4204 \tag{4.163b}$$

The associated Hurwitz polynomial is found to be

$$p(y)=y^4+1.189y^3+1.7068y^2+1.0147y+0.375 \tag{4.164}$$

For $K_4=1$, $\hat{p}(y)$ is simply the Hurwitz polynomial formed by the LHS roots of $C_4^2(-jy)=0$, and from (2.31) we obtain

$$\hat{p}(y)=y^4+y^2+0.125 \tag{4.165}$$

Substituting these in (4.149) yields the input impedance

$$Z_{11}(s)=100\,\frac{2y^4+1.189y^3+2.7068y^2+1.0147y+0.5}{1.189y^3+0.7068y^2+1.0147y+0.25} \tag{4.166}$$

This impedance is next expanded in a continued fraction

$$Z_{11}(s)=16.82\times 10^{-3}s+\cfrac{1}{1.189\times 10^{-6}s+\cfrac{1}{23.786\times 10^{-3}s+\cfrac{1}{0.841\times 10^{-6}s+\cfrac{1}{200}}}} \tag{4.167}$$

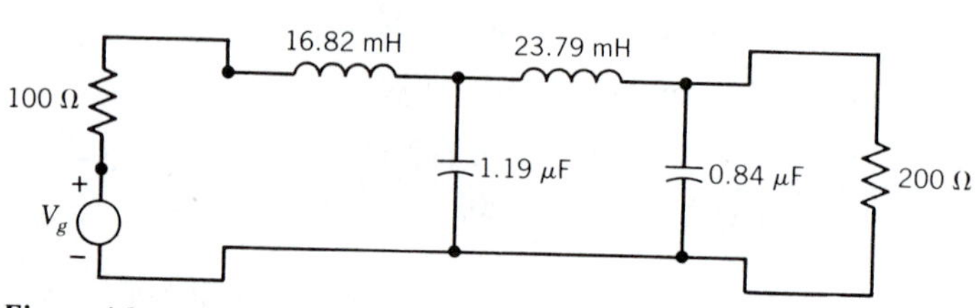

Figure 4.26

which can be identified as an LC ladder terminating in a 200-Ω resistor, as required, the element values of which are given in (4.162) and Figure 4.26.

Case B: $\rho_1(0) < 0$. With the choice of the minus sign in (4.147), the input impedance, aside from the constant R_1, becomes the reciprocal of that given in (4.152):

$$Z_{11}(s) = R_1 \frac{\sum_{m=0}^{n} (b_m - \hat{b}_m) y^m}{\sum_{m=0}^{n} (b_m + \hat{b}_m) y^m} \tag{4.168}$$

the continued-fraction expansion of which is shown in (4.140). Depending on whether n is even or odd, the LC ladder network has the configurations of Figure 4.24. Formulas for the element values are the same as those given in (4.153)–(4.157) except that the roles of C's and L's are interchanged and R_1 and R_2 are replaced by their reciprocals:

$$C_1 = \frac{2 \sin \gamma_1}{\omega_c R_1(\sinh a - \sinh \hat{a})} \tag{4.169a}$$

$$C_{2m-1} L_{2m} = \frac{4 \sin \gamma_{4m-3} \sin \gamma_{4m-1}}{\omega_c^2 f_{2m-1}(\sinh a, \sinh \hat{a})} \tag{4.169b}$$

$$C_{2m+1} L_{2m} = \frac{4 \sin \gamma_{4m-1} \sin \gamma_{4m+1}}{\omega_c^2 f_{2m}(\sinh a, \sinh \hat{a})} \tag{4.169c}$$

for $m = 1, 2, \ldots, [\frac{1}{2}n]$, and

$$C_n = \frac{2 \sin \gamma_1}{\omega_c R_2(\sinh a + \sinh \hat{a})}, \qquad n \text{ odd} \tag{4.169d}$$

$$L_n = \frac{2R_2 \sin \gamma_1}{\omega_c(\sinh a + \sinh \hat{a})}, \qquad n \text{ even} \tag{4.169e}$$

where γ_m and $f_m(\sinh a, \sinh \hat{a})$ are defined in (4.130) and (4.156).

As an illustration of the use of the formulas, let us choose the minus sign in (4.147) for the problem considered in Example 4.13. The resulting LC ladder network together with its terminations is shown in Figure 4.27, the normalized input impedance Z_{11}/R_1 of which is the reciprocal of that obtained for the corresponding ladder of Figure 4.26. We recognize that in selecting the minus sign in (4.147), we must choose the minus sign in (4.151c). Thus, for a specified minimum passband gain and a fixed generator resistance R_1, the terminating resistance R_2 is determined by (4.151c). In the present situation, we have $R_1 = 100\,\Omega$, $K_4 = 1$, and $\varepsilon = 0.354$ (0.51-dB ripple in the passband). Substituting these in (4.151c) yields a terminating resistance $R_2 = 50\,\Omega$, as indicated in Figure 4.27.

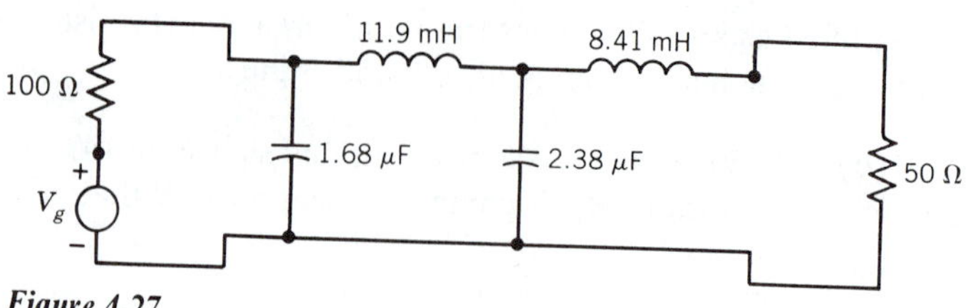

Figure 4.27

4.6 SUMMARY AND SUGGESTED READINGS

We began this chapter by introducing the concept of compatible impedances. Two impedances are said to be *compatible* if one can be realized as the input impedance of a lossless two-port network terminated in the other. We then presented *Darlington's theory*, which states that any positive-real impedance is compatible with any nonnegative resistance, which may be zero. This is justified by showing that for the given impedance we can identify a set of immittance parameters that are always compatible in representing a lossless reciprocal two-port network.

The application of Darlington's theory to the synthesis of resistively terminated lossless two-port networks was indicated. We first demonstrated that the magnitude squared of a transfer function is numerically equal to the real part of the input immittance at all frequencies. To apply Darlington's procedure, we must find the immittance function from its real part. For this we introduced two procedures. One is based on the partial-fraction expansion of the even part of the immittance, and the other on solving a system of linear equations.

We then presented formulas pertinent to the design of a lossless two-port network operating between a resistive generator and a resistive load and having a preassigned transducer power-gain characteristic. In particular, we demonstrated that an *LC* ladder two-port can always be realized that when inserted between the given resistive generator and load will yield an nth-order Butterworth or Chebyshev transducer power-gain characteristic. Also, explicit formulas for the element values of this *LC* ladder are given, which reduce the design problem to simple arithmetic.

The general problem of matching between a frequency-dependent load and a resistive generator having a preassigned transducer power-gain characteristic will be taken up in Chapter 9.

For additional study on topics of this chapter, the reader is referred to Guillemin [3] and Van Valkenburg [6]. For explicit formulas on double-terminated *LC* ladder networks, see Chen [1] and Weinberg [7]. Extensive tables are available for the design of networks of the type discussed in this chapter. For an introduction to the use of these tables, see Weinberg [7], Saal and Ulbrich [4], and Green [2].

REFERENCES

1. W. K. Chen 1976, *Theory and Design of Broadband Matching Networks*, Chapter 3. New York: Pergamon Press.

2. E. Green 1954, *Amplitude-Frequency Characteristics of Ladder Networks.* Chelmsford, England: Marconi's Wireless Telegraph Co.
3. E. A. Guillemin 1957, *Synthesis of Passive Networks,* Chapter 11. New York: John Wiley.
4. R. Saal and E. Ulbrich, "On the design of filters by synthesis," *IRE Trans. Circuit Theory,* vol. CT-5, 1958, pp. 284–328.
5. G. C. Temes and J. W. LaPatra 1977, *Introduction to Circuit Synthesis and Design,* Chapter 6. New York: McGraw–Hill.
6. M. E. Van Valkenburg 1960, *Modern Network Synthesis,* Chapters 14 and 15. New York: John Wiley.
7. L. Weinberg 1962, *Network Analysis and Synthesis.* New York: McGraw–Hill. Reissued by R. E. Krieger Publishing Co., Melbourne, FL, 1975, Chapters 12 and 13.
8. A. I. Zverev 1967, *Handbook of Filter Synthesis.* New York: John Wiley.

PROBLEMS

4.1 Show that the zeros of the even part of a positive-real function are always of even multiplicity on the $j\omega$-axis.

4.2 Justify that all the finite nonzero poles of the impedance parameters obtained by Darlington's procedure for Case B are compact.

4.3 Show that the two two-port networks of Figure 4.6 are equivalent.

4.4 Assume that the even part of a positive-real impedance Z is a nonnegative constant k. Show that this impedance must be of the form

$$Z(s) = Z_{LC}(s) + k \tag{4.170}$$

where Z_{LC} is a reactance function.

4.5 Show that for Case B in Darlington theory the pole at infinity is compact if and only if the impedance Z is devoid of pole at infinity.

4.6 Realize the following impedance using Darlington's procedure:

$$Z(s) = \frac{s^2 + s + 2}{s^2 + s + 1} \tag{4.171}$$

4.7 Realize the following impedance with 1-Ω termination:

$$Z(s) = \frac{2s + 1}{s^2 + s + 1} \tag{4.172}$$

4.8 Move the ideal transformer of the network of Figure 4.12 to the load, so that the input impedance of the network remains unaltered.

4.9 In the network of Figure 4.17, move one of the ideal transformers to the load and replace the other by an equivalent perfect transformer so that the input impedance remains unaltered.

4.10 Realize a lossless two-port network terminated in a resistor to meet the specification for the transfer current-ratio function

$$|\alpha_{12}(j\omega)|^2 = \frac{1 + \omega^2}{1 + \omega^2 + \omega^4} \tag{4.173}$$

4.11 Given the RC impedance

$$Z(s)=\frac{(s+2)(s+4)}{(s+1)(s+3)} \tag{4.174}$$

synthesize an LC two-port terminating in a resistor. Remove as many ideal transformers as possible.

4.12 Repeat Problem 4.11 if (4.174) represents an admittance function.

4.13 Synthesize a lossless two-port network terminated in a resistor to meet the specification for the transfer voltage-ratio function

$$|G_{12}(j\omega)|^2=\frac{1}{1+\omega^8} \tag{4.175}$$

4.14 Given the positive-real impedance

$$Z(s)=\frac{8s^2+9s+10}{2s^2+4s+4} \tag{4.176}$$

realize a lossless two-port network terminated in a resistor. Remove as many ideal transformers as possible.

4.15 Using Darlington's procedure, synthesize the impedance

$$Z(s)=\frac{2s^2+s+1}{s^2+s+2} \tag{4.177}$$

4.16 Repeat Problem 4.15 if (4.177) represents an admittance function.

4.17 Using Darlington's procedure, synthesize the impedance

$$Z(s)=\frac{2s^2+2s+5}{s^2+s+4} \tag{4.178}$$

4.18 Realize a lossless two-port network, which when terminated in a 1-Ω resistor meets the specification for the transfer impedance function

$$|Z_{12}(j\omega)|^2=\frac{\omega^2}{(1+\omega^2)^2} \tag{4.179}$$

4.19 Suppose that we wish to design a low-pass filter that has a maximally-flat transducer power-gain characteristic. The filter is to be operated between a resistive generator of internal resistance 100 Ω and a 200-Ω load and such that it gives at least 60-dB attenuation in gain at the frequency four times the radian cutoff frequency $\omega_c=10^4$ rad/s and beyond.

4.20 Repeat Problem 4.19 if generator resistance is 200 Ω and load resistance is 100 Ω, everything else being the same.

4.21 Repeat Problem 4.19 if generator and load resistances are both 200 Ω.

4.22 Suppose that we wish to design a low-pass filter that has an equiripple transducer power-gain characteristic. The filter is to be operated between a resistive generator of internal resistance 100 Ω and a 200-Ω load and such that it gives at least 60-dB attenuation in gain at the frequency four times the radian cutoff frequency $\omega_c = 10^4$ rad/s and beyond. The peak-to-peak ripple in the passband must not exceed 2 dB.

4.23 Repeat Problem 4.22 if the generator resistance is 200 Ω and the load resistance is 100 Ω, everything else being the same.

4.24 Repeat Problem 4.22 if generator and load resistances are both 200 Ω.

4.25 Repeat Problem 4.22 if passband ripple cannot exceed 0.5 dB.

chapter 5

ACTIVE FILTER SYNTHESIS: FUNDAMENTALS

In the preceding chapters we discussed the synthesis of transfer functions using passive components. In this and Chapters 7 and 8 we will consider the general subject of active filter synthesis. More specifically, we shall present techniques for the synthesis of transfer functions through the use of both active and passive elements, the latter being restricted exclusively to resistors and capacitors. Such filters are known as *active RC* or *inductorless* filters. They are attractive in that they usually weight less and require less space than their passive counterparts, and that they can be fabricated in microminiature form using integrated-circuit technology. As a result, they can be mass-produced inexpensively. Because of the smaller size and because all the processing steps can be automated, active *RC* filters have reduced parasitic effects and increased circuit reliability and performance. On the other hand, since it is not yet possible to integrate an inductor, passive circuits using inductors can only be produced using discrete components, which is usually far more expensive. Among the drawbacks of active *RC* filters is the finite bandwidth of the active devices, which places a limit on the high-frequency performance. Most of the active *RC* filters are used up to approximately 30 kHz. This is quite adequate for use in voice and data communication system. We emphasize that the upper frequency bound of 30 kHz reflects the state of the present-day technology and it is reasonable to expect that with advances in integrated-circuit technology, a higher bound is possible. In fact, active *RC* filters can provide high reliability in instrumentation that is subjected to environmental changes and allow savings in size and weight even at frequencies approaching 1 MHz.

On the other hand, passive filters do not have such a limitation, and they can be employed up to approximately 500 MHz, the limitation being the parasitics associated with the passive elements. Another important criterion for comparing filters is sensitivity due to variation in the element values. It will be shown that the sensitivity of passive filters is much less than that for active filters. Finally, active filters need power supplies while passive filters do not. Nevertheless, the economic and performance advantages of active *RC* filters far outweigh the disadvantages. The trend is to use active *RC* filters in most voice and data communication systems. The passive filters are used as the basis for active simulation. In fact, some of the active realizations are just active *RC* equivalents of the corresponding passive filters. We will demonstrate that doubly terminated *LC* ladder filters have very low sensitivities, and that if they are simulated by active *RC* or digital circuits, the low-sensitivity properties are preserved.

In general, there are two methods to realize transfer functions using active *RC* circuits. The first technique is known as the *cascade* method. The transfer function to be realized is first factored into a product of first-order and/or second-order terms. Each term is then individually realized by an active *RC* circuit. A cascade connection of these individual circuits realizes the overall transfer function. The second general method is the *direct* method, where a single circuit is used to realize the entire transfer function, as was done for the passive filter design discussed in the preceding chapters. A discussion of the direct method will be given in Section 5.5 and also in Chapter 8.

5.1 THE CASCADE APPROACH

Consider a general transfer function that can be factored into the form

$$H(s)=K\frac{(s-z_1)(s-z_2)\dots(s-z_n)}{(s-p_1)(s-p_2)\dots(s-p_m)} \tag{5.1a}$$

$$=\prod_{i=1}^{q} K_i\frac{\alpha_i s^2+c_i s+d_i}{\beta_i s^2+a_i s+b_i} \tag{5.1b}$$

Note that the typical term in the product (5.1b) represents either a real root or a pair of complex-conjugate roots. For example, in the case of a real zero and a real pole, we set $\alpha_i=\beta_i=0$ and $c_i=a_i=1$. The corresponding term becomes

$$H_i(s)=K_i\frac{s+d_i}{s+b_i} \tag{5.2}$$

For a complex-conjugate zero-pole pair, we set $\alpha_i=\beta_i=1$ and obtain

$$H_i(s)=K_i\frac{s^2+c_i s+d_i}{s^2+a_i s+b_i} \tag{5.3}$$

Our objective is to study techniques for connecting the first-order and/or second-order networks in cascade so that the overall transfer function is related to the transfer functions of the individual networks by the chain rule

$$H(s)=H_1(s)H_2(s)\dots H_q(s)=\prod_{i=1}^{q} H_i(s) \tag{5.4}$$

where H_i takes either the linear form (5.2) or the biquadratic form (5.3).

Refer to Figure 5.1. The transfer voltage-ratio functions or simply the transfer

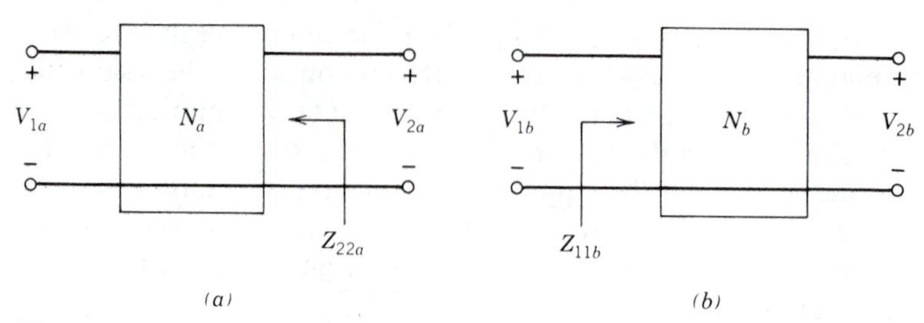

Figure 5.1

functions of the networks N_a and N_b are defined by the equation

$$H_i(s)=V_{2i}(s)/V_{1i}(s), \qquad i=a \text{ or } b \tag{5.5}$$

Now we consider the effect of connecting these two networks in cascade as shown in Figure 5.2. If the output impedance Z_{22a} of N_a is negligibly small compared with the input impedance Z_{11b} of N_b, then the output voltage of N_a will not be adversely affected or loaded down when N_b is connected to it. Under this situation, the output voltage of N_a in the composite network of Figure 5.2 remains approximately the same:

$$V_x(s)=V_{2a}(s)=V_{1b}(s) \tag{5.6}$$

The transfer voltage-ratio function of the overall network becomes

$$H(s)=\frac{V_{2b}}{V_{1a}}=\frac{V_{2a}}{V_{1a}}\cdot\frac{V_{2b}}{V_{1b}}=H_1(s)H_2(s) \tag{5.7}$$

The ideal situation is depicted in Figure 5.3, where a VCVS (*v*oltage-*c*ontrolled *v*oltage *s*ource) is introduced as a buffer. If this VCVS is considered to be part of the second network, then the output impedance of N_a, being finite, is negligibly small

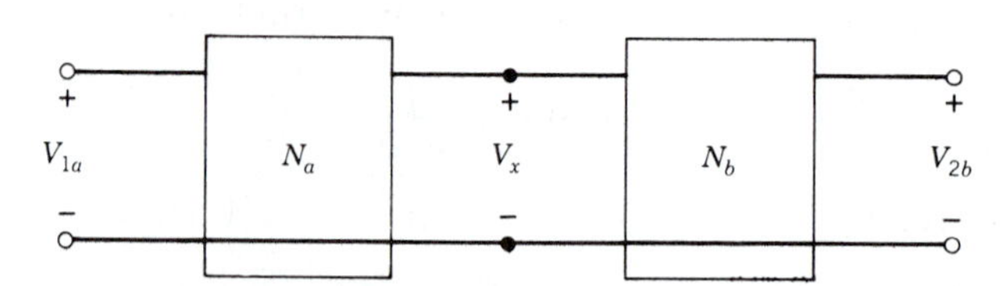

Figure 5.2

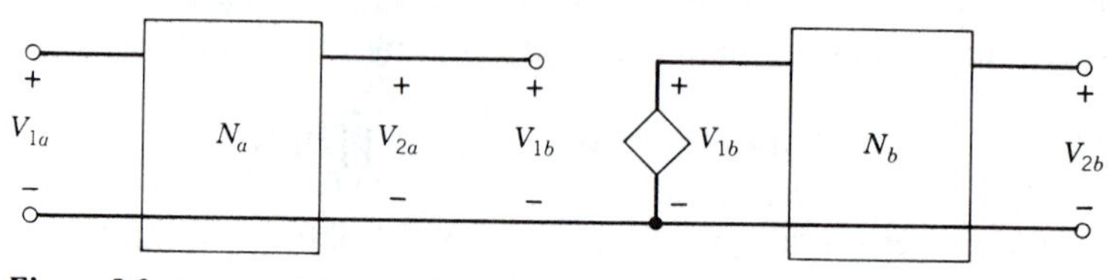

Figure 5.3

compared with the input impedance of the second network, which is infinite. Extending this argument to the cascade of q sections as shown in Figure 5.4, having transfer functions $H_1, H_2, \ldots, H_q$, we obtain

$$H(s) = H_1(s)H_2(s)\ldots H_q(s) \tag{5.8}$$

Thus, the transfer function of a cascade of networks is the product of the individual transfer functions provided that the input impedance of each successive section is very large compared with the output impedance of the preceding section. This requirement is readily fulfilled in most op-amp circuit realizations, as will be demonstrated shortly.

Observe that in deriving (5.8), the zeros and poles of the transfer function (5.1) were appropriately pairwise grouped together to form the section transfer functions $H_i(s)$. This pairwise grouping, often known as *pole-zero pairing*, is an important phase in filter design using cascade approach. The reason is that both the pole-zero pairing and the subsequent order of sections in the overall structure have important implications.† It often affects the filter's sensitivity and ultimately determines the dynamic range of the filter. Two contradicting rules have been proposed for pole-zero pairing to reduce sensitivity in the overall structure:

1 Pair the high-Q poles p_i with those zeros z_i farthest away such that $|p_i - z_i|$ is maximum.

2 Pair the high-Q poles p_i with the closest zeros such that $|p_i - z_i|$ is minimum.

Much of the reasoning for proposing rule 1 is that the more sensitive sections are those that possess the highest pole Q's. There is no general rule for pole-zero pairings in cascade synthesis to minimize the filter's overall sensitivity.

An equally important problem in filter realization is its dynamic range. By *dynamic range* we mean that range of input signal levels, usually expressed in dB, that will not produce a distorted output. A filter that permits input signals varying in level from 1 mV to 1 V without distortion is said to have a dynamic range of 60 dB. Active filters typically have a dynamic range between 70 and 100 dB. The pole-zero pairings, the choice of section gain constants, and the subsequent order of sections usually have a significant effect on the overall dynamic range.

As an illustration of transfer-function decomposition, consider the *RC* ladder

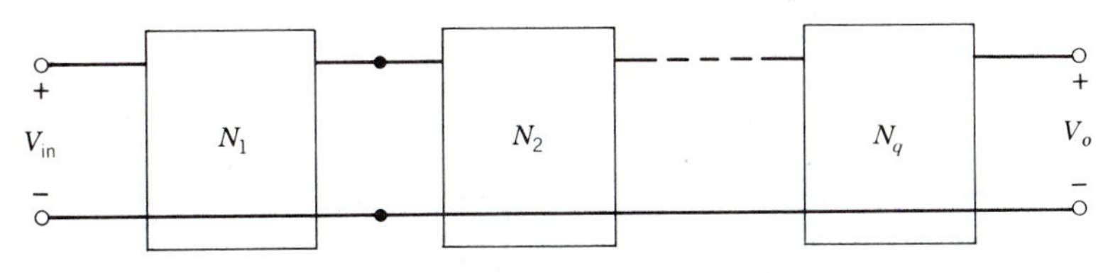

Figure 5.4

†See, for example, M. S. Ghausi and K. R. Laker 1981, *Modern Filter Design: Active RC and Switched Capacitor*, Chapter 5. Englewood Cliffs, NJ: Prentice-Hall.

network of Figure 5.5. The transfer function of the network is found to be

$$H(s)=\frac{V_o(s)}{V_{in}(s)}=\frac{1/R_1R_3C_2C_4}{s^2+(1/R_3C_4+1/R_3C_2+1/R_1C_2)s+1/R_1R_3C_2C_4} \tag{5.9}$$

Now suppose that we break the ladder network in half and insert a VCVS as shown in Figure 5.6. This VCVS serves as a buffer between the two sections in the sense that the current $I_{2a}=0$. Under this condition, the transfer function of the first section is obtained as

$$H_1(s)=\frac{V_{2a}(s)}{V_{in}(s)}=\frac{1/R_1C_2}{s+1/R_1C_2} \tag{5.10}$$

The transfer function of the second section is

$$H_2(s)=\frac{V_o(s)}{V_{2a}(s)}=\frac{1/R_3C_4}{s+1/R_3C_4} \tag{5.11}$$

The transfer function of the overall network is obtained by forming the product of the individual transfer functions:

$$H(s)=\frac{V_o(s)}{V_{in}(s)}=H_1(s)H_2(s)=\frac{1/R_1R_3C_2C_4}{s^2+(1/R_1C_2+1/R_3C_4)s+1/R_1R_3C_2C_4} \tag{5.12}$$

Observe that this equation differs from that of (5.9) for the network without a VCVS. However, by choosing the appropriate parameter values in the individual networks, the overall transfer function can be made the same. The network with the VCVS of Figure 5.6 can be realized by the op-amp network of Figure 5.7, in which the VCVS is simulated by the voltage follower. The voltage follower has infinite input impedance and zero output impedance, and behaves as a VCVS, as previously discussed in Figure 1.23.

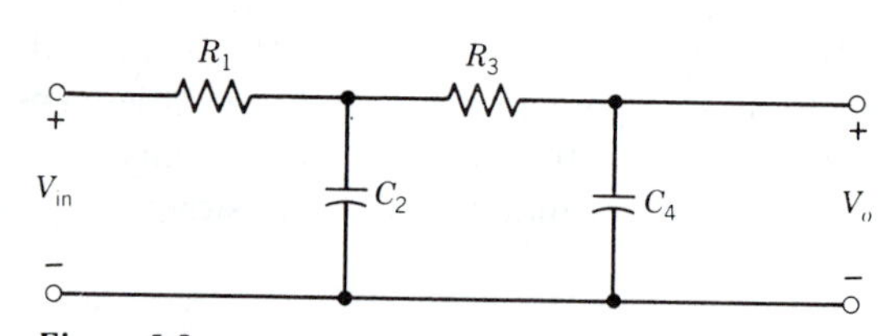

Figure 5.5

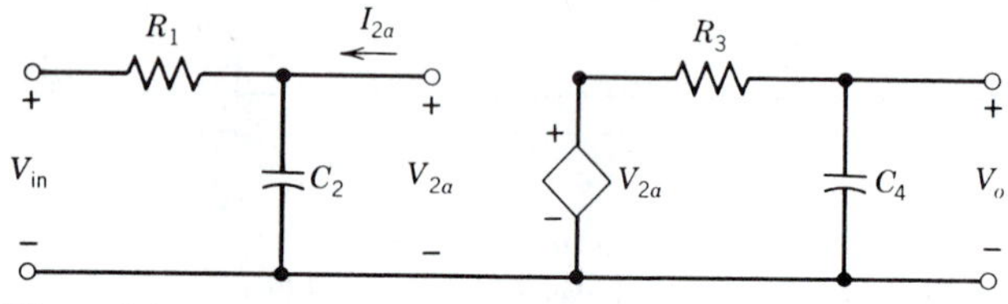

Figure 5.6

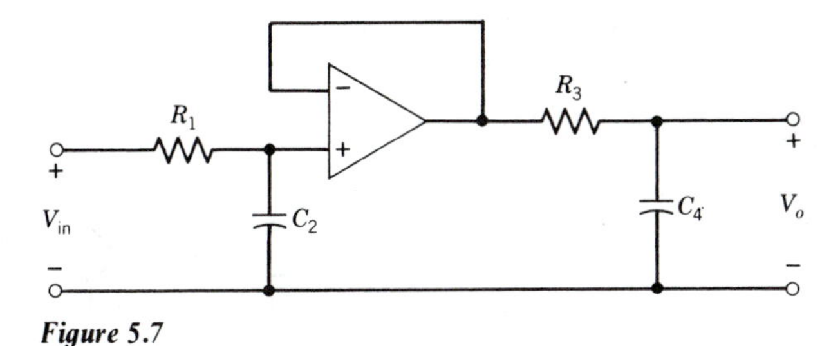

Figure 5.7

Our conclusion is that a general transfer function of (5.1) can be realized by a cascade of component networks, each of which realizes either the *first-order factor* of (5.2) or the *biquadratic function* of (5.3), provided that they are isolated from each other in the sense that each successive network does not load the previous one. This condition is readily satisfied in most op-amp network realizations, but is not for passive networks. Consequently, the realization of a higher-order transfer function is reduced to the much simpler realization of a general second-order function. The network block that realizes a biquadratic function is commonly referred to as a *biquad*. We recognize that the synthesis procedure needed for determining the element values of a biquad is relatively simple, and the implementation of additional constraints such as the use of standard element values and the minimization of sensitivity is usually easy to attain. Furthermore, since individual biquads are totally isolated, each active *RC* network may be individually tuned without affecting other networks. This feature is extremely useful in adjusting network performance at the time of manufacture, and is, of course, far easier than trying to tune a network in which all the elements interact, as in the case of passive synthesis or the direct method of realization to be presented in Section 5.5 and also in Chapter 8.

The biquad is therefore the basic building block in active filter synthesis. Its network configurations and realizations are of fundamental importance and we will devote a portion of this chapter and all of Chapter 7 to this subject after introducing filter sensitivity in Chapter 6.

5.2 THE FIRST-ORDER NETWORKS

In this section we describe two simple op-amp networks that realize a first-order factor of the form

$$H(s)=K\frac{s+a}{s+b} \tag{5.13}$$

where a and b are real. The first configuration is the inverting amplifier structure shown in Figure 5.8(a), the equivalent network of which is shown in (b). As previously shown in (1.9), the transfer voltage-ratio function is

$$H(s)=\frac{V_o(s)}{V_{in}(s)}=-\frac{Z_2(s)}{Z_1(s)} \tag{5.14}$$

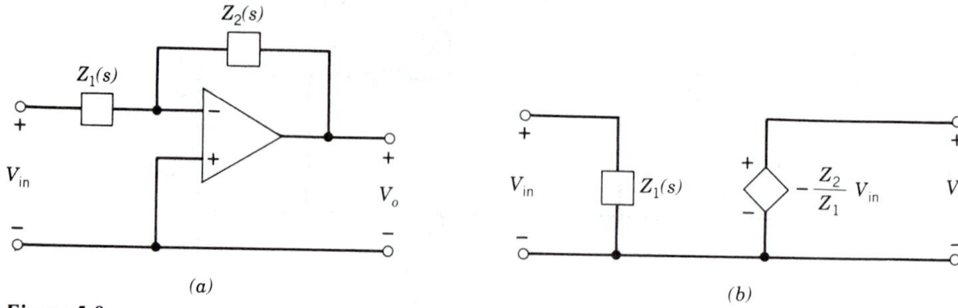

Figure 5.8

Apart from the minus sign in (5.14), our objective is to identify the impedances Z_1 and Z_2 from the right-hand side of (5.13) so that they represent the input impedances of the RC one-port networks. According to Theorem 3.2, an RC impedance requires that all of its zeros and poles be simple, lie on the negative real axis, and alternate with each other, the first critical frequency being a pole. For negative K and nonnegative a and b, we can make the following identifications:

$$Z_1(s)=\frac{1}{s+a}, \qquad Z_2(s)=\frac{-K}{s+b} \tag{5.15}$$

The realization is shown in Figure 5.9. This solution applies whether a is larger or smaller than b.

Another possible assignment is achieved by making use of division of both the numerator and the denominator of (5.13) by s, giving

$$H(s)=-\frac{K_1(s+a)/s}{K_2(s+b)/s}=-\frac{Z_2(s)}{Z_1(s)} \tag{5.16}$$

where $|K|=K_1/K_2$. This results in the following assignments:

$$Z_1(s)=K_2\frac{s+b}{s}, \qquad Z_2(s)=K_1\frac{s+a}{s} \tag{5.17}$$

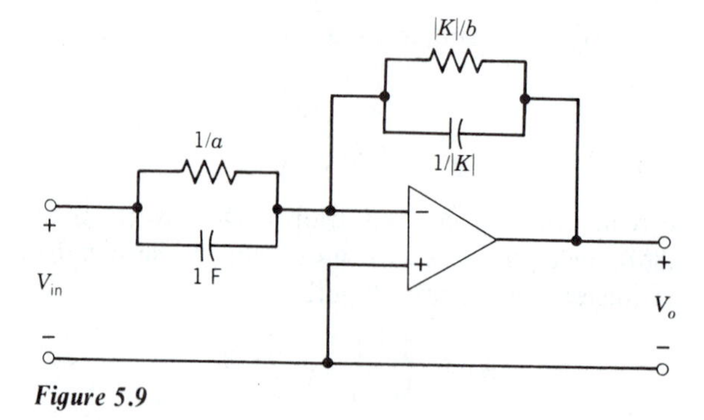

Figure 5.9

The realization is shown in Figure 5.10. This solution is again valid for either relationship between a and b. For positive K, a unity-gain sign inversion section of Figure 5.11 is required. This section is obtained from Figure 5.8 by selecting $Z_1 = Z_2 = R$.

Example 5.1 Realize the normalized transfer voltage-ratio function

$$H(s) = -2\frac{s+1}{s+3} \tag{5.18}$$

From the specifications, we can make the following identifications:

$$K = -2, \qquad a = 1, \qquad b = 3 \tag{5.19}$$

Using the design of Figure 5.9, the element values are shown in Figure 5.12(a). Suppose

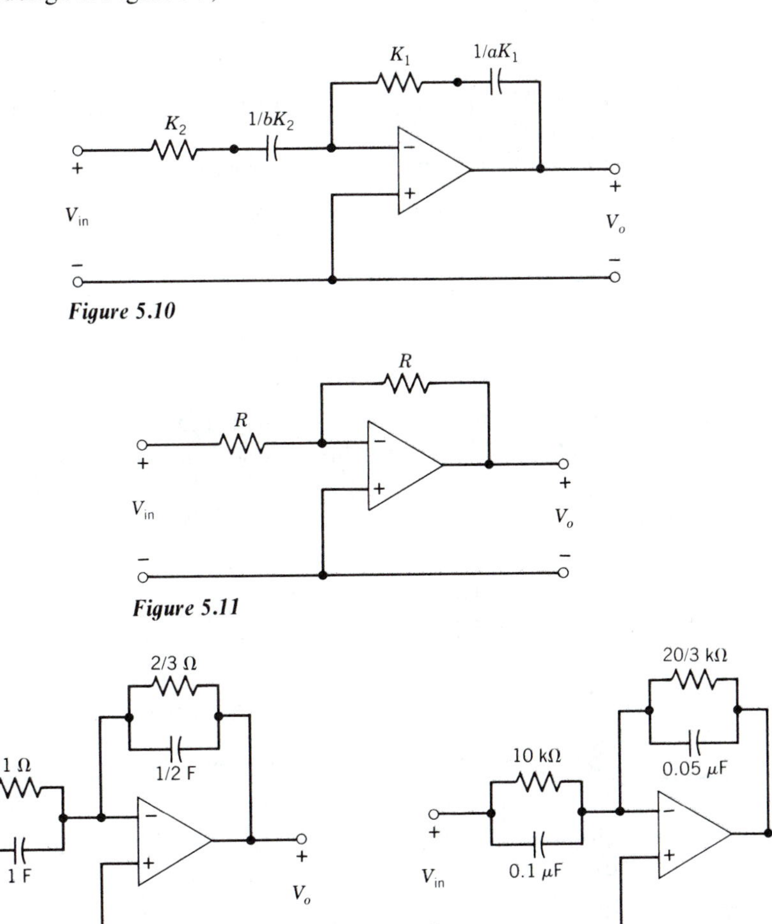

Figure 5.10

Figure 5.11

Figure 5.12

that we make a frequency-scaling by a factor of 1000 and magnitude-scaling by a factor of 10^4, the resulting element values are as indicated in Figure 5.12(*b*). Note that these scalings have no effect on the op amp.

In addition to the inverting amplifier configuration of Figure 5.8, the first-order factor of (5.13) can also be realized by the noninverting op-amp structure of Figure 5.13(*a*). From (1.18) the transfer voltage-ratio function is

$$H(s)=\frac{V_o(s)}{V_{in}(s)}=1+\frac{Z_2(s)}{Z_1(s)} \tag{5.20}$$

This network behaves like a VCVS as shown in Figure 5.13(*b*) with infinite input impedance. Equating (5.13) and (5.20) yields

$$\frac{Z_2(s)}{Z_1(s)}=K\frac{s+a}{s+b}-1=\frac{(K-1)s+(Ka-b)}{s+b} \tag{5.21}$$

For Z_1 and Z_2 to be the RC input impedances, we see that there are constraints on K and on the relationship for a and b.

Suppose that we let $K=1$ so that (5.21) can be written as

$$\frac{Z_2(s)}{Z_1(s)}=\frac{a-b}{s+b}=\frac{(a-b)/s}{(s+b)/s} \tag{5.22}$$

identifying

$$Z_1(s)=\frac{s+b}{s}, \qquad Z_2(s)=\frac{a-b}{s} \tag{5.23}$$

The circuit realization is shown in Figure 5.14 provided that $a>b\geqq 0$.

On the other hand, if we select the value of K so that in (5.21)

$$Ka-b=0 \tag{5.24}$$

or $K=b/a$,

$$\frac{Z_2(s)}{Z_1(s)}=\frac{(b-a)s/a}{s+b}=\frac{(b-a)/a}{1+b/s} \tag{5.25}$$

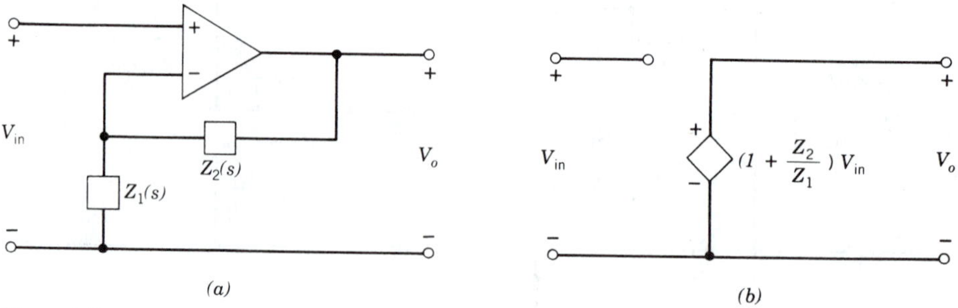

Figure 5.13

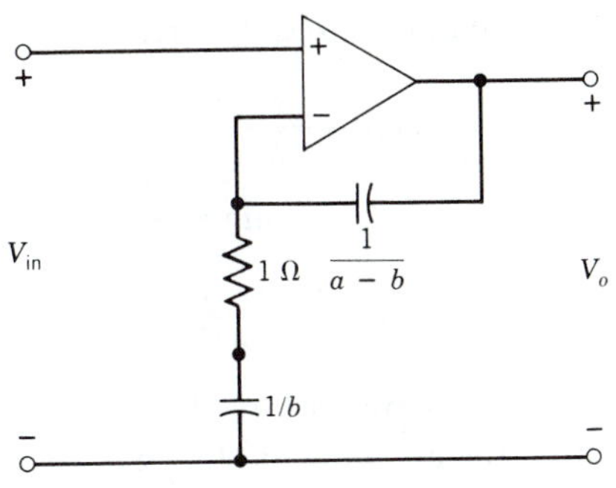

Figure 5.14

From this we can make the following assignments:

$$Z_1(s) = 1 + b/s, \qquad Z_2(s) = b/a - 1 \tag{5.26}$$

which can be realized by the network of Figure 5.15 provided that $b \geqq a > 0$. Another possible identification in (5.25) leads to

$$Z_1(s) = \frac{a}{(b-a)s}, \qquad Z_2(s) = \frac{1}{s+b} \tag{5.27}$$

giving the network realization of Figure 5.16, provided that $b > a > 0$.

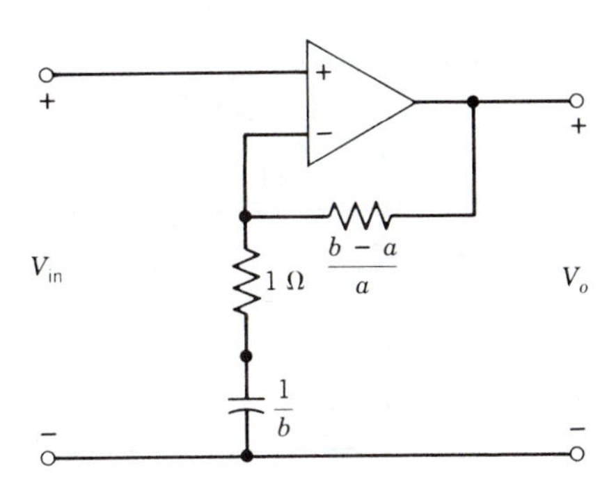

Figure 5.15

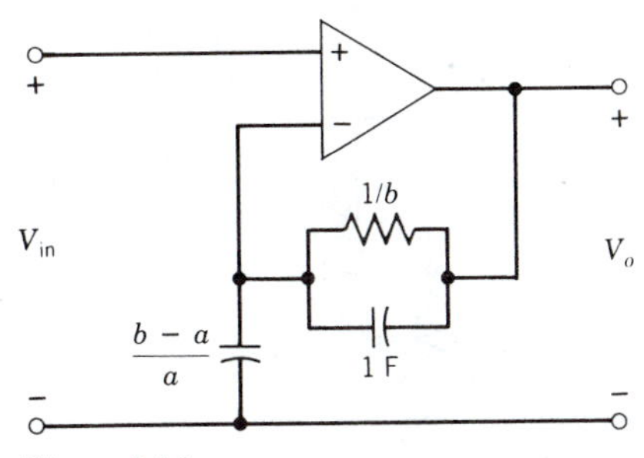

Figure 5.16

Example 5.2 Realize the normalized transfer function

$$H(s)=4\frac{s+1}{s+3} \tag{5.28}$$

From (5.28) we identify $a=1$ and $b=3$. Choosing $K=b/a=3$, (5.28) can be rewritten as

$$H(s)=\frac{4}{3}\,3\,\frac{s+1}{s+3} \tag{5.29}$$

Apart from the constant 4/3, the remainder of the function can be realized by the network of Figure 5.15 with element values as shown in Figure 5.17.

For the constant 4/3, we again consider the noninverting amplifier of Figure 5.13 by setting

$$1+\frac{Z_2(s)}{Z_1(s)}=\frac{4}{3}=1+\frac{1}{3} \tag{5.30}$$

obtaining the network realization of Figure 5.18. Suppose that we frequency-scale the networks by a factor of 10^3 and magnitude-scale them by 10^4, the resulting element values and the final realization are shown in Figure 5.19.

Before we turn our attention to the realization of the biquad circuits, we consider the realizations of several simple functions that are frequently used in the

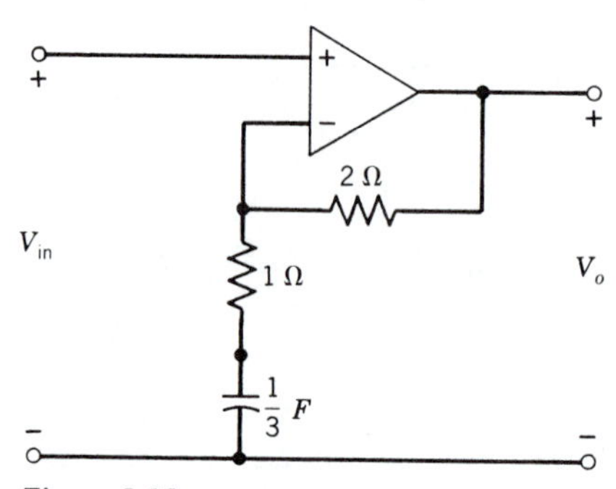

Figure 5.17

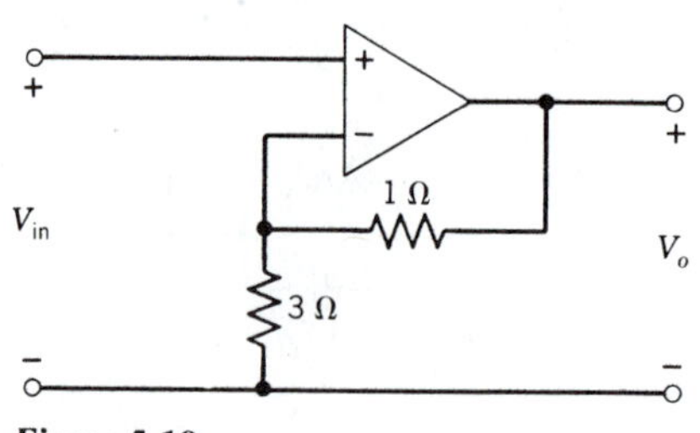

Figure 5.18

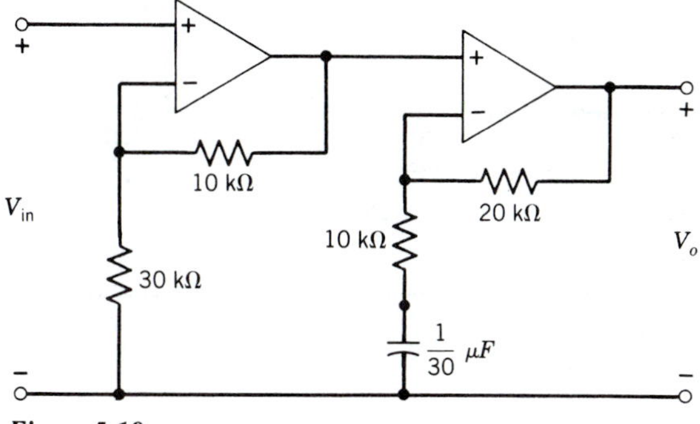

Figure 5.19

cascade synthesis. First, we realize the transfer voltage-ratio function

$$H(s) = K\frac{1}{s+b} \tag{5.31}$$

by the inverting amplifier structure of Figure 5.8(*a*). For negative K and nonnegative b, we let

$$H(s) = -\frac{Z_2(s)}{Z_1(s)} = K\frac{1}{s+b} \tag{5.32}$$

and make the following identifications:

$$Z_1(s) = 1, \qquad Z_2(s) = \frac{-K}{s+b} \tag{5.33}$$

The realization of the network is shown in Figure 5.20(*a*). For positive K, a unity-gain sign inversion section of Figure 5.11 is required.

We next realize the transfer voltage-ratio function

$$H(s) = K\frac{s}{s+b} \tag{5.34}$$

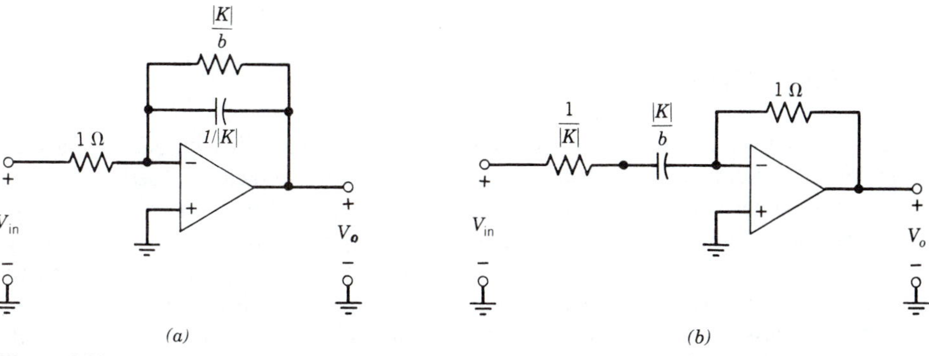

Figure 5.20

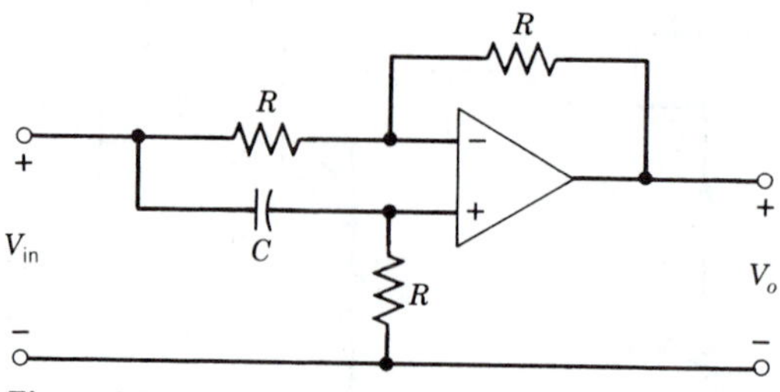

Figure 5.21

again by the inverting amplifier of Figure 5.8(a). To this end, we let

$$H(s) = -\frac{Z_2(s)}{Z_1(s)} = K\frac{s}{s+b} \tag{5.35}$$

from which we can make the following assignments:

$$Z_1 = \frac{s+b}{-Ks}, \qquad Z_2(s) = 1 \tag{5.36}$$

If K is negative, the network realization is shown in Figure 5.20(b). For positive K, an additional unity-gain sign inversion section of Figure 5.11 is required.

Finally, we mention that the network of Figure 5.21 can realize the first-order all-pass function

$$H(s) = \frac{s-a}{s+a}, \qquad a > 0 \tag{5.37}$$

To see this we compute the transfer voltage-ratio function for the network of Figure 5.21 and obtain

$$H(s) = \frac{V_o(s)}{V_{in}(s)} = \frac{s - 1/RC}{s + 1/RC} \tag{5.38}$$

which identifies

$$a = 1/RC \tag{5.39}$$

5.3 THE BIQUAD NETWORK

In this section, we introduce the commonly used single-amplifier biquad topologies. These networks consist of an RC network and an op amp, and can be used to realize a complex pole-zero pair. Many networks can accomplish this, and the majority of them can be classified as negative feedback topology or positive feedback topology, depending on which input terminal of the op amp the RC network is connected to. The fundamental properties of these topologies together with illustrative networks are to be discussed in this section. The analysis and design of various biquad circuits will be studied in Chapter 7 after the introduction of sensitivity functions.

5.3.1 **Negative Feedback Topology**

Consider the *RC* amplifier configuration of Figure 5.22, where the *RC* network provides a feedback path to the negative input terminal of the op amp. Such a structure is capable of realizing a biquadratic function and is termed the *negative feedback topology*. The transfer function of this general structure can be expressed in terms of the *feedforward* and *feedback transfer functions* of the *RC* network defined by the equations

$$\text{Feedforward transfer function} = H_{ff}(s) = \left.\frac{V_3(s)}{V_1(s)}\right|_{V_2(s)=0} \tag{5.40}$$

$$\text{Feedback transfer function} = H_{fb}(s) = \left.\frac{V_3(s)}{V_2(s)}\right|_{V_1(s)=0} \tag{5.41}$$

From Figure 5.22, the terminal voltages are related by the equations

$$V^-(s) = V_3(s) = H_{ff}(s)V_{in}(s) + H_{fb}(s)V_o(s) \tag{5.42a}$$

$$V_o(s) = A[V^+(s) - V^-(s)] = -AV^-(s) \tag{5.42b}$$

Eliminating V^- in (5.42) gives the transfer voltage-ratio function of the network of Figure 5.22 as

$$H(s) = \frac{V_o(s)}{V_{in}(s)} = -\frac{H_{ff}(s)}{H_{fb}(s) + 1/A} \tag{5.43}$$

For the ideal op amp, $A = \infty$ and the transfer function reduces to

$$H(s) = -H_{ff}(s)/H_{fb}(s) \tag{5.44}$$

Writing H_{ff} and H_{fb} explicitly as ratios of two polynomials with common denominator D,

$$H_{ff}(s) = N_{ff}(s)/D(s), \qquad H_{fb}(s) = N_{fb}(s)/D(s) \tag{5.45}$$

we obtain

$$H(s) = -N_{ff}(s)/N_{fb}(s) \tag{5.46}$$

Equation (5.46) states that the zeros of the feedforward network determine the zeros of the transfer function, while the zeros of the feedback network determine its

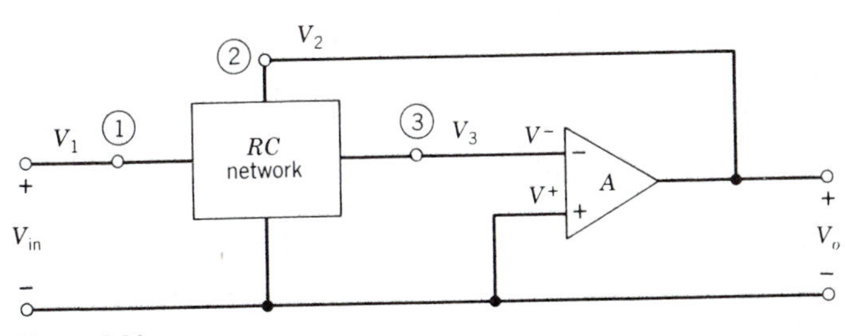

Figure 5.22

poles. The poles of the RC network do not contribute in any way to the transfer function, assuming that the op amp is ideal. For a stable network, the poles of (5.46) are restricted to the open LHS. Figure 5.23 shows some of the RC networks used in the design of biquads with negative feedback topology.

Example 5.3 Consider an RC amplifier filter with a specific form of passive network shown in Figure 5.24. The terminals are labeled as those in Figure 5.22. To compute the feedforward transfer function, we use the network of Figure 5.25, which is obtained by shorting terminal 2 to ground, and obtain

$$H_{ff}(s)=\frac{\tilde{V}_3(s)}{\tilde{V}_1(s)}=\frac{y_1 y_2}{(y_2+y_4)(y_3+y_5)+y_4(y_1+y_2)+y_1 y_2} \tag{5.47}$$

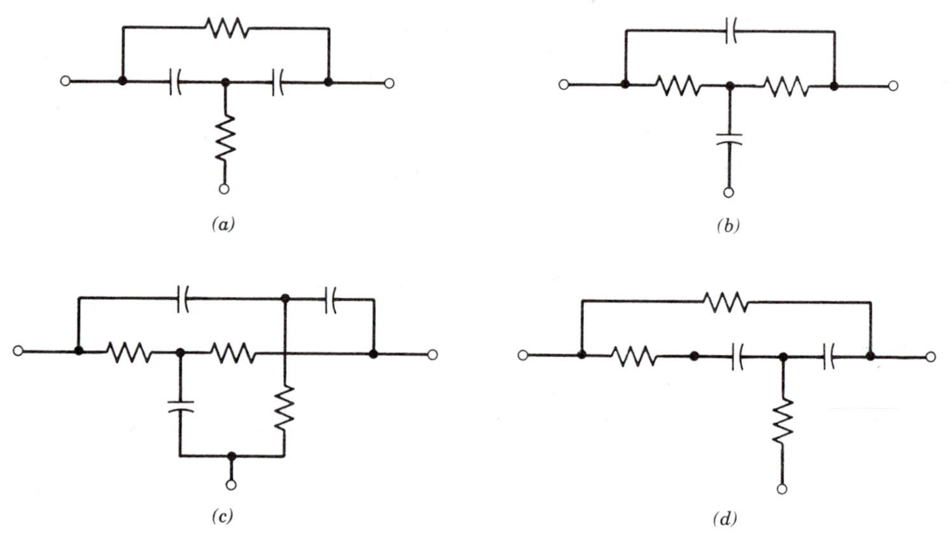

Figure 5.23

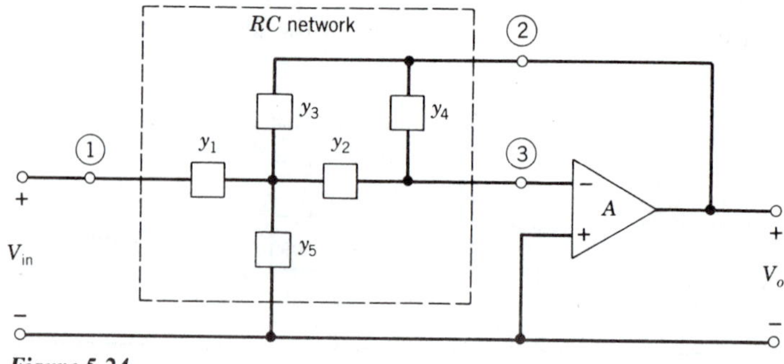

Figure 5.24

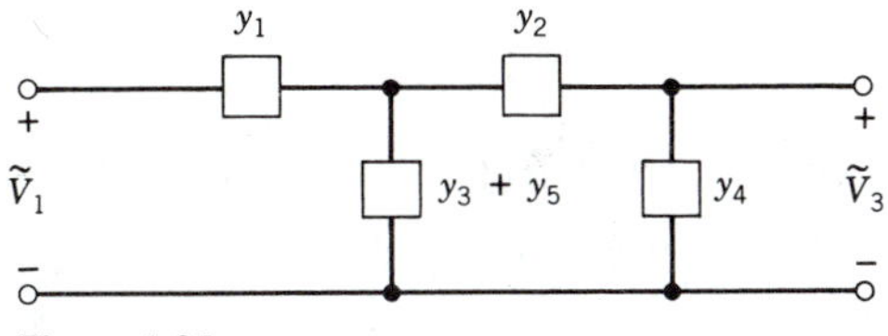

Figure 5.25

For the feedback transfer function, we short-circuit terminal 1 to ground. The resulting network is shown in Figure 5.26. This leads to

$$H_{fb}(s)=\frac{\hat{V}_3(s)}{\hat{V}_2(s)}=\frac{y_4(y_1+y_2+y_3+y_5)+y_2y_3}{(y_2+y_4)(y_3+y_5)+y_4(y_1+y_2)+y_1y_2} \tag{5.48}$$

Substituting (5.47) and (5.48) in (5.44) or (5.46) gives the transfer voltage-ratio function of the *RC* amplifier filter of Figure 5.24 as

$$H(s)=\frac{V_o(s)}{V_{\text{in}}(s)}=-\frac{y_1y_2}{y_4(y_1+y_2+y_3+y_5)+y_2y_3} \tag{5.49}$$

As a specific application, let

$$y_1=1/R_1, \qquad y_2=sC_1, \qquad y_3=sC_2, \qquad y_4=1/R_2, \qquad y_5=0 \tag{5.50}$$

The network of Figure 5.24 becomes that of Figure 5.27 with the resulting transfer

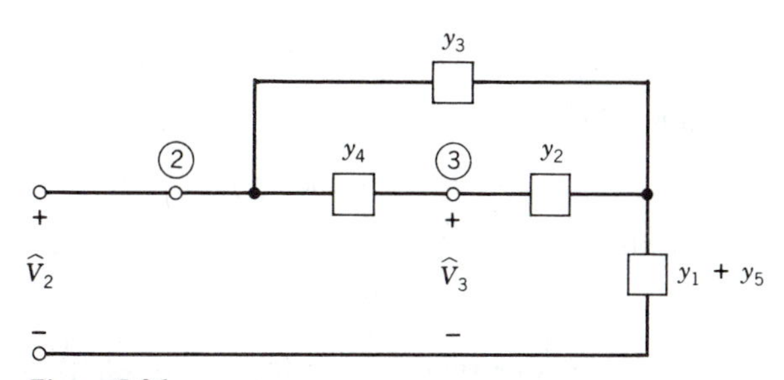

Figure 5.26

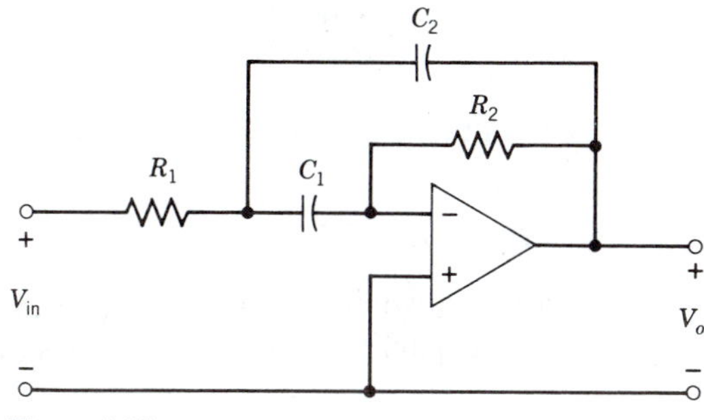

Figure 5.27

voltage-ratio function

$$H(s)=\frac{V_o(s)}{V_{\text{in}}(s)}=-\frac{s/R_1C_2}{s^2+(1/R_2C_1+1/R_2C_2)s+1/R_1R_2C_1C_2} \tag{5.51}$$

Observe that the three-terminal RC network in Figure 5.27 corresponds to the bridged-T network of Figure 5.23(*a*).

Suppose now that we wish to realize the biquadratic function

$$H(s)=\frac{5000s}{s^2+3000s+6\times10^7} \tag{5.52}$$

by the network of Figure 5.27. For this we compare the coefficients of like powers in s in (5.51) and (5.52) and obtain the following relationships:

$$\frac{1}{R_1R_2C_1C_2}=6\times10^7 \tag{5.53a}$$

$$\frac{1}{R_2C_1}+\frac{1}{R_2C_2}=3000 \tag{5.53b}$$

Since there are two equations in four unknowns R_1, R_2, C_1, and C_2, we may fix any two of them prior to solving these equations. However, not all the solutions will lead to physically realizable elements. One simple choice is to let

$$C_1=C_2=1\text{ F} \tag{5.54}$$

Using this in (5.53) we obtain

$$R_1=25\times10^{-6}\,\Omega,\qquad R_2=\tfrac{2}{3}\times10^{-3}\,\Omega \tag{5.55}$$

The gain constant obtained by this realization is

$$K_r=-1/R_1C_2=-4\times10^4 \tag{5.56}$$

whereas the desired constant is

$$K_d=5000 \tag{5.57}$$

Since the desired constant is lower than that realized, the input signal has to be attenuated by the factor

$$\frac{K_r}{K_d}=-\frac{4\times10^4}{5000}=-8 \tag{5.58}$$

To obtain the practical element values, we magnitude-scale the network by a factor of 10^8 to yield

$$C_1=C_2=0.01\ \mu\text{F},\qquad R_1=2.5\text{ k}\Omega,\qquad R_2=66.67\text{ k}\Omega \tag{5.59}$$

The complete network realization is shown in Figure 5.28, where an inverting op-amp network has been used to reduce the gain by a factor of 8 to meet the specifications exactly.

Finally, we mention that since there are more unknowns than the number of

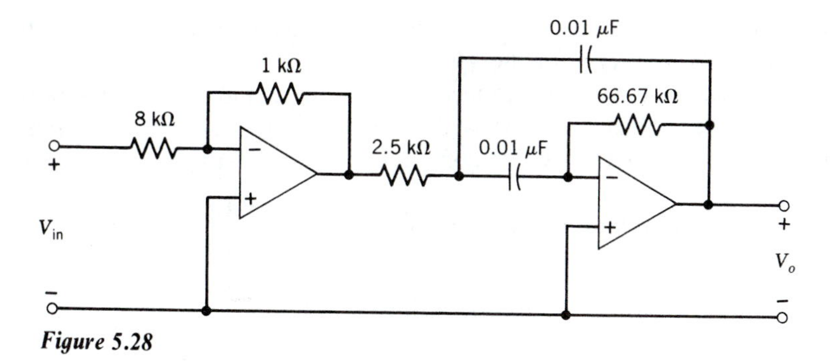

Figure 5.28

equations in (5.53), one might be tempted to set

$$1/R_1C_2 = 5000 \tag{5.60}$$

which when combined with (5.53) yields

$$1/R_2C_2 = -9000 \tag{5.61}$$

This implies that R_2 and C_2 cannot be both positive, showing that the network is not physically realizable. In view of this fact, what is often done is to match only the coefficients that determine the pole and zero locations. The gain constant is adjusted later to any desired value by using the inverting amplifier network of Figure 5.8, as we did in this example, or by input attenuation or gain enhancement to be discussed in Section 5.4. Note that the gain constant affects only the level of the output voltage, not its frequency characteristic. The latter is the most significant character of a filter.

5.3.2 Positive Feedback Topology

The *positive feedback topology* is shown in Figure 5.29, where a feedback path provided by the *RC* network is connected to the positive terminal of the op amp. In addition to this feedback, a part of the output voltage is also fed back to the negative terminal via the resistors r_1 and r_2. Strictly speaking, this is really a mixed feedback

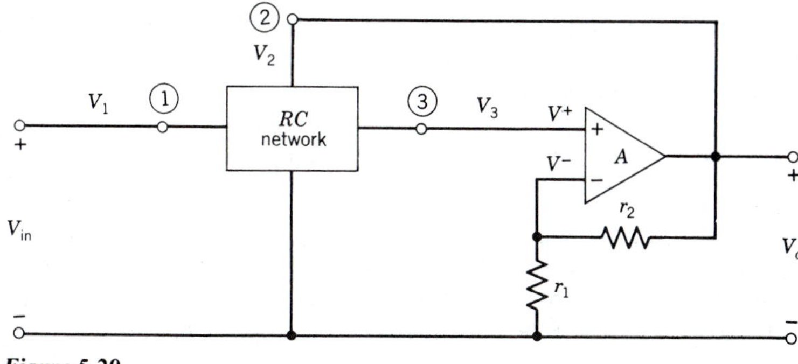

Figure 5.29

topology containing both positive and negative feedback. The negative feedback is used to define a positive gain for the VCVS.

As in the case for the negative feedback topology, the transfer function of the network of Figure 5.29 can be expressed in terms of the *feedforward* and *feedback transfer functions* of the *RC* network, defined by the equations

$$\text{Feedforward transfer function} = H_{ff}(s) = \left. \frac{V_3(s)}{V_1(s)} \right|_{V_2(s)=0} \tag{5.62}$$

$$\text{Feedback transfer function} = H_{fb}(s) = \left. \frac{V_3(s)}{V_2(s)} \right|_{V_1(s)=0} \tag{5.63}$$

From the op-amp circuit of Figure 5.29, the terminal voltages are related by

$$V_3(s) = V^+(s) = H_{ff}(s)V_{\text{in}}(s) + H_{fb}(s)V_o(s) \tag{5.64a}$$

$$V_o(s) = A[V^+(s) - V^-(s)] \tag{5.64b}$$

$$V^-(s) = = \frac{r_1 V_o(s)}{r_1 + r_2} = \frac{V_o(s)}{k} \tag{5.64c}$$

where $k = 1 + r_2/r_1$. Eliminating V^- and V^+ in (5.64) gives the desired transfer voltage-ratio function

$$H(s) = \frac{V_o(s)}{V_{\text{in}}(s)} = \frac{kH_{ff}(s)}{1 - kH_{fb}(s) + k/A} \tag{5.65}$$

In the case of ideal op amp, $A = \infty$ and (5.65) reduces to

$$H(s) = \frac{kH_{ff}(s)}{1 - kH_{fb}(s)} \tag{5.66}$$

As before, write H_{ff} and H_{fb} explicitly as

$$H_{ff}(s) = N_{ff}(s)/D(s), \qquad H_{fb}(s) = N_{fb}(s)/D(s) \tag{5.67}$$

Substituting these in (5.66) yields

$$H(s) = \frac{kN_{ff}(s)}{D(s) - kN_{fb}(s)} \tag{5.68}$$

Equation (5.68) indicates that the zeros of H are determined by the zeros of the feed-forward transfer function of the *RC* network, and therefore they can be complex. The poles of H are determined by the poles of the *RC* network, the zeros of the feed-back transfer function, and the factor $k = 1 + r_2/r_1$, and can be located anywhere in the open LHS. Some of the *RC* networks used in the positive feedback topology are exhibited in Figure 5.30.

Example 5.4 The *RC* amplifier filter configuration shown in Figure 5.31 is of positive feedback topology with a specific form of passive network. The terminals are labeled as those in Figure 5.29. We use (5.68) to compute its transfer voltage-ratio function.

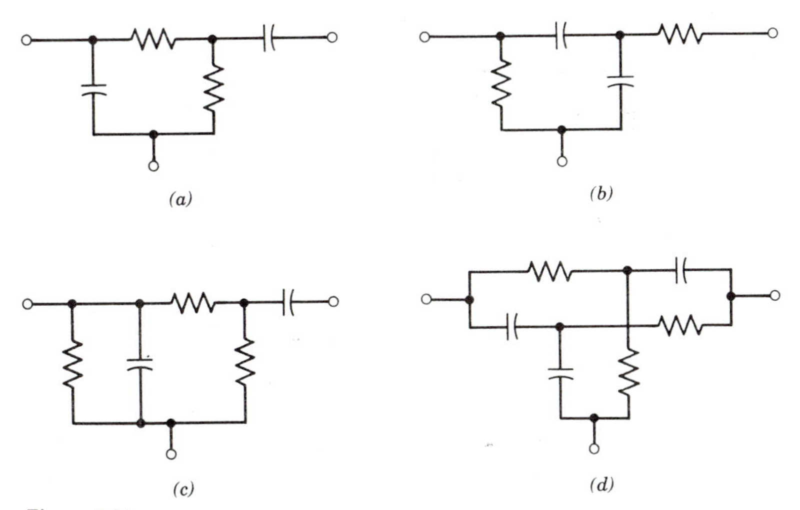

Figure 5.30

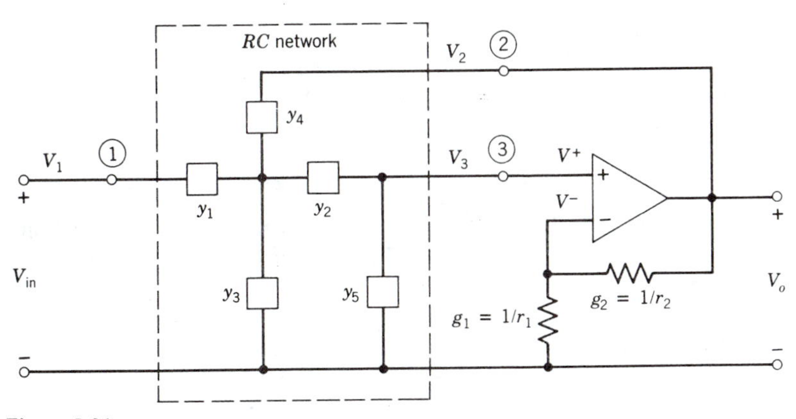

Figure 5.31

To compute the feedforward transfer function, we use the network of Figure 5.32, which is obtained from the *RC* network of Figure 5.31 by short-circuiting terminal 2 to ground, obtaining

$$H_{ff}(s)=\frac{\tilde{V}_3(s)}{\tilde{V}_1(s)}=\frac{y_1 y_2}{(y_3+y_4)(y_2+y_5)+y_5(y_1+y_2)+y_1 y_2} \tag{5.69}$$

For the feedback transfer function, we short-circuit terminal 1 of the *RC* network to ground and obtain the network of Figure 5.33. Using this network, we compute

$$H_{fb}(s)=\frac{\hat{V}_3(s)}{\hat{V}_2(s)}=\frac{y_2 y_4}{(y_1+y_3)(y_2+y_5)+y_5(y_2+y_4)+y_2 y_4} \tag{5.70}$$

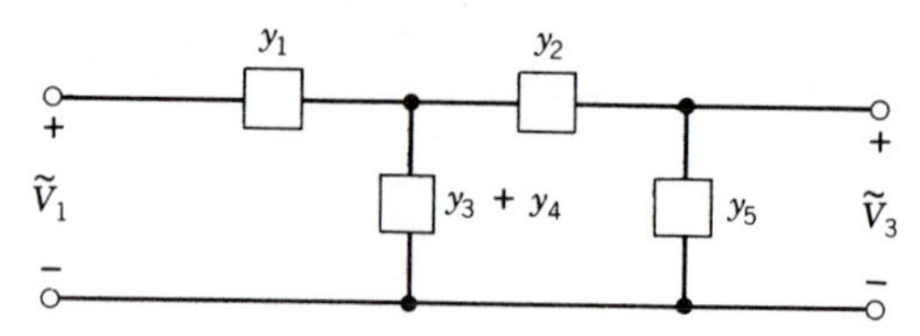

Figure 5.32

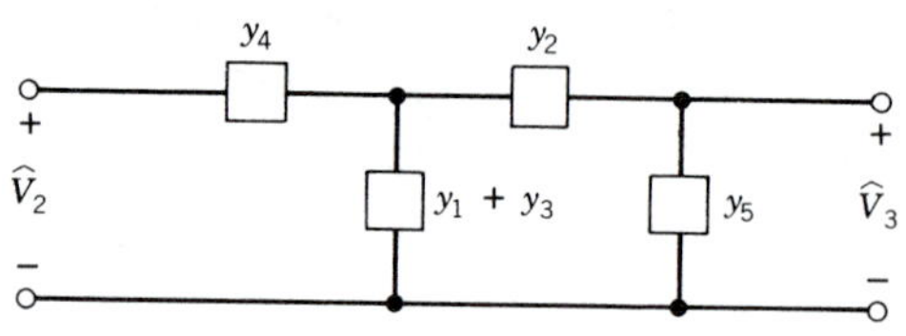

Figure 5.33

Substituting these in (5.68) gives

$$H(s)=\frac{V_o(s)}{V_{in}(s)}=\frac{ky_1y_2}{(y_2+y_5)(y_1+y_3+y_4)+y_2y_5-ky_2y_4} \tag{5.71}$$

where $k=1+r_2/r_1=1+g_1/g_2$, $g_1=1/r_1$, and $g_2=1/r_2$.

As an application of this configuration, let

$$y_1=1/R_1, \qquad y_2=1/R_2, \qquad y_3=0, \qquad y_4=sC_1, \qquad y_5=sC_2 \tag{5.72}$$

Using these in Figure 5.31, the resulting filter configuration is shown in Figure 5.34 and is known as the *low-pass Sallen and Key circuit.*† The corresponding transfer function is from (5.71)

$$H(s)=\frac{k/R_1R_2C_1C_2}{s^2+[1/R_1C_1+1/R_2C_1+(1-k)/R_2C_2]s+1/R_1R_2C_1C_2} \tag{5.73}$$

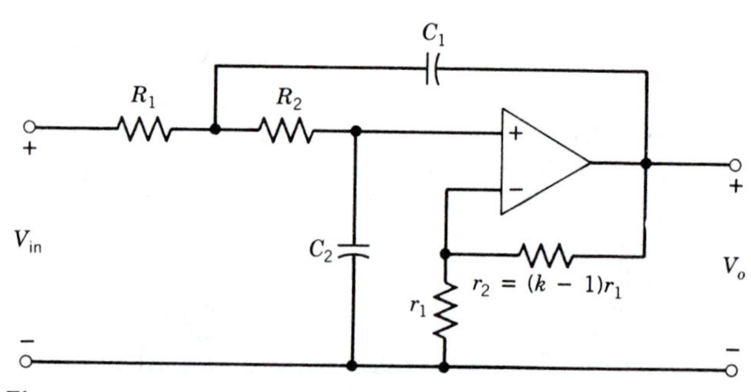

Figure 5.34

†R. P. Sallen and E. L. Key, "A practical method of designing *RC* active filters," *IRE Trans. Circuit Theory*, vol. CT-2, 1955, pp. 74–85.

Suppose now that we wish to realize the biquadratic transfer voltage-ratio function

$$H(s)=\frac{V_o(s)}{V_{in}(s)}=\frac{K}{s^2+as+b} \tag{5.74}$$

For this we equate the coefficients of like powers in s in the denominators of (5.73) and (5.74) and obtain the following relationships:

$$\frac{1}{R_1C_1}+\frac{1}{R_2C_1}+\frac{1-k}{R_2C_2}=a \tag{5.75a}$$

$$\frac{1}{R_1R_2C_1C_2}=b \tag{5.75b}$$

Since there are two equations in five unknowns R_1, R_2, C_1, C_2, and k, a solution may be obtained by specifying

$$R_1=R_2=R \qquad \text{and} \qquad C_1=C_2=C \tag{5.76}$$

Substituting these in (5.75) yields

$$1/RC=\sqrt{b}, \qquad k=3-a/\sqrt{b} \tag{5.77}$$

The realized gain constant is from (5.73)

$$K_r=\frac{k}{R_1R_2C_1C_2}=\frac{k}{R^2C^2}=3b-a\sqrt{b} \tag{5.78}$$

Another possibility in obtaining a unique solution for (5.75) is by specifying

$$C_1=C_2=1\text{ F}, \qquad k=2 \tag{5.79}$$

This gives

$$R_1=1/a, \qquad R_2=a/b \tag{5.80}$$

The realized gain constant becomes

$$K_r=\frac{k}{R_1R_2C_1C_2}=\frac{k}{R_1R_2}=2b \tag{5.81}$$

In the above analysis, we let the gain constant float. Suppose that we wish to fix this constant by introducing another constraint obtained by equating the numerators of (5.73) and (5.74):

$$\frac{k}{R_1R_2C_1C_2}=K \tag{5.82}$$

Combining this with (5.75b) and using $k=1+r_2/r_1$ results in

$$r_2=(K/b-1)r_1 \tag{5.83}$$

showing that r_2 is negative for $K/b<1$. Thus, the realizable range of K is restricted. As mentioned in the previous example, since K affects only the level of the output voltage, not its frequency characteristic, we often match only the coefficients that determine the pole locations of (5.74).

As a numerical example, consider the normalized Butterworth transfer voltage-ratio function

$$H(s)=\frac{2}{s^2+\sqrt{2}s+1} \tag{5.84}$$

giving $a=\sqrt{2}$, $b=1$, and $K=2$. Using the equal-resistance and equal-capacitance design of (5.76), we obtain

$$R_1=R_2=R=\frac{1}{\sqrt{b}C}=\frac{1}{\sqrt{b}}=1\ \Omega \tag{5.85a}$$

$$r_2=\left(2-\frac{a}{\sqrt{b}}\right)r_1=(2-\sqrt{2})\times 10=5.86\ \Omega \tag{5.85b}$$

where we have selected $C=1$ F and $r_1=10\ \Omega$, or

$$C_1=C_2=C=1\ \text{F}, \qquad r_1=10\ \Omega \tag{5.86}$$

The realized gain constant is from (5.78)

$$K_r=3b-a\sqrt{b}=3-\sqrt{2}=1.586 \tag{5.87}$$

Suppose that the radian cutoff frequency of the Butterworth response is 10^4 rad/s. To obtain the final design, we frequency-scale the network by a factor of 10^4 and apply an additional magnitude-scaling of 10^3, resulting in the design values

$$C_1=C_2=0.1\ \mu\text{F}, \qquad R_1=R_2=1\ \text{k}\Omega \tag{5.88a}$$

$$r_1=10\ \text{k}\Omega, \qquad r_2=5.86\ \Omega, \qquad K_r=1.586 \tag{5.88b}$$

The desired low-pass Butterworth filter is shown in Figure 5.35, in which a non-inverting op-amp network has been used to provide a gain of $2/1.586=1.261$ to meet the specifications exactly.

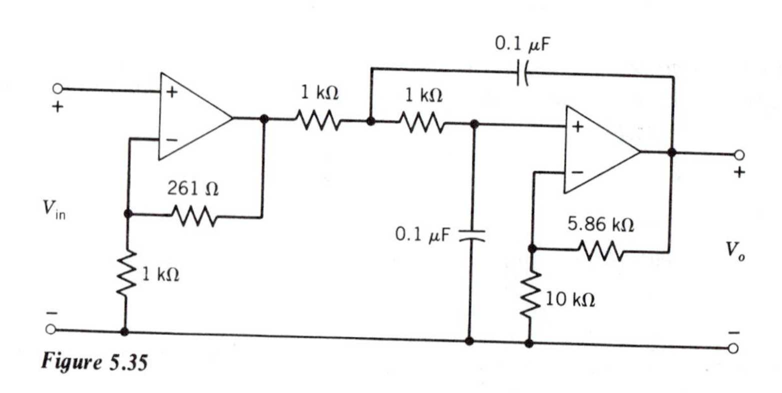

Figure 5.35

5.4 GAIN CONSTANT ADJUSTMENT

In the foregoing discussion, we recognize that in the design of biquads we often match only the coefficients of a biquadratic function that determine the zero and pole locations, leaving the gain constant floating. The resulting gain constant can later be adjusted by the insertion of an inverting amplifier of Figure 5.8 or a noninverting amplifier of Figure 5.13. In this section, we introduce alternate solutions that do not require an additional amplifier. To facilitate our discussion, two cases are distinguished.

Case 1: Input attenuation. In this case, assume that the output voltage of a given network is larger than desired. To attenuate the input in order to attain the desired output voltage, we consider the original input network of Figure 5.36(*a*). Suppose that the desired input voltage to a filter is $k_1 V_{in}$. To obtain an output voltage $k_1 V_{in}$, we use the potential divider network of Figure 5.36(*b*), the output resistance of which is the same as that for the original network in (*a*). This requires that

$$k_1 = \frac{R_3}{R_2 + R_3} \tag{5.89a}$$

$$R_1 = \frac{R_2 R_3}{R_2 + R_3} \tag{5.89b}$$

Solving these for R_2 and R_3 yields

$$R_2 = \frac{R_1}{k_1}, \qquad R_3 = \frac{R_1}{1 - k_1}, \qquad k_1 < 1 \tag{5.90}$$

Example 5.5 Realize the filter of Figure 5.28 without the use of an inverting op-amp network to reduce the gain by a factor of 8. From Figures 5.28 and 5.36, we can make the following identifications:

$$k_1 = \tfrac{1}{8}, \qquad R_1 = 2.5\ \text{k}\Omega \tag{5.91}$$

Substituting these in (5.90) gives

$$R_2 = 20\ \text{k}\Omega, \qquad R_3 = 20/7\ \text{k}\Omega \tag{5.92}$$

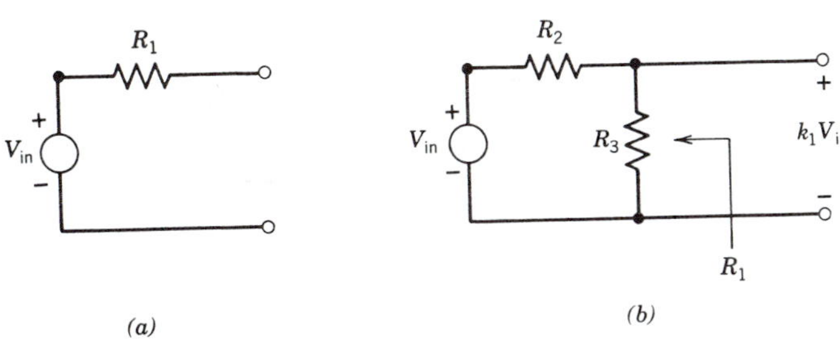

Figure 5.36

The desired network realization is shown in Figure 5.37. Note that the minus sign has not been considered in the realization. This is usually of no concern because neither the magnitude nor the delay of the transfer function is affected by it. It merely causes 180° of phase shift.

Case 2: Gain enhancement. In this case, we assume that the output voltage of the original network is lower than desired. Consider the *RC* amplifier structure of Figure 5.29 or other configurations operating on the same principle. Let $H(s)=V_o/V_{\text{in}}$ be the transfer voltage-ratio function of the original network of Figure 5.29. Now suppose that the output of the op amp is attenuated by k_2 by the introduction of the potential divider network as shown in Figure 5.38, where

$$k_2=\frac{R_b}{R_a+R_b}<1 \tag{5.93}$$

For moderate k_2's, the decrease in the amplifier gain will not affect its performance appreciably. Thus, if we replace the op amp in Figure 5.29 by that of Figure 5.38, the resulting network of Figure 5.39 will have approximately the same transfer function between V_o' and V_{in} as in the original network, or

$$\frac{V_o'(s)}{V_{\text{in}}(s)}\approx H(s)=\frac{V_o(s)}{V_{\text{in}}(s)} \tag{5.94}$$

Since from Figure 5.39, V_o'' and V_o' are related by

$$V_o'(s)=k_2V_o''(s) \tag{5.95}$$

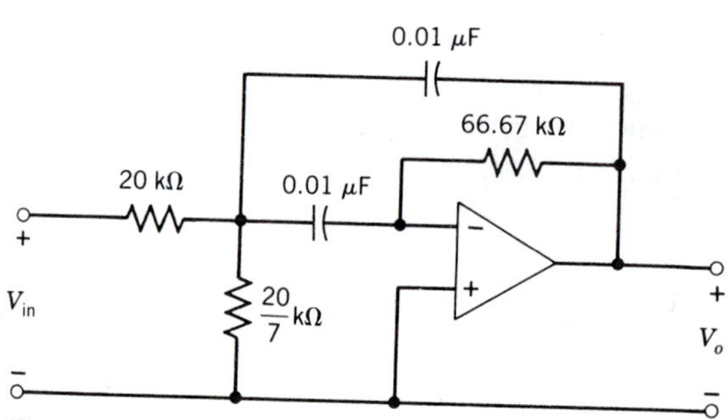

Figure 5.37

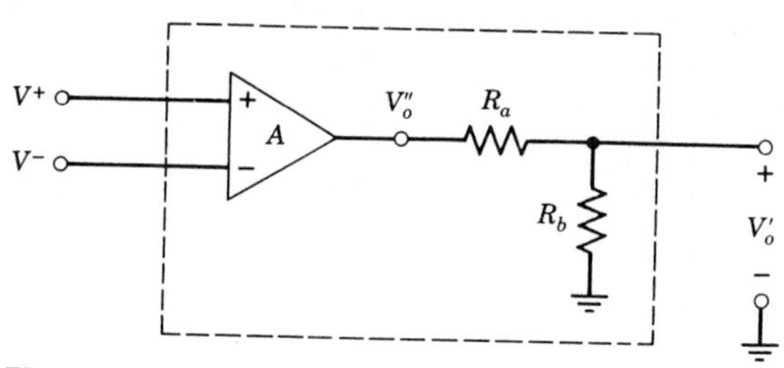

Figure 5.38

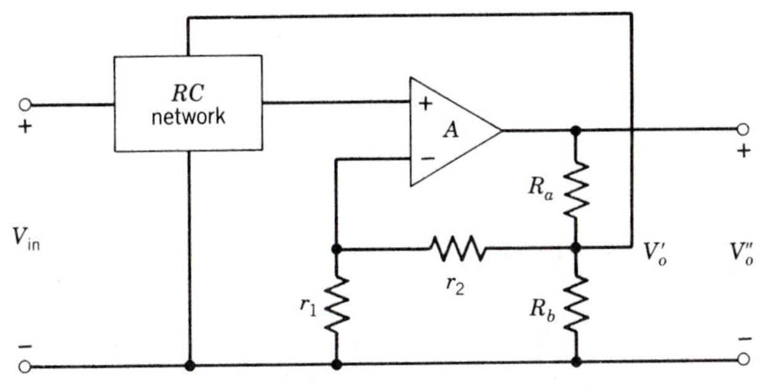

Figure 5.39

the transfer voltage ratio between V_{in} and V_o'' is found to be from (5.94)

$$\frac{V_o''(s)}{V_{in}(s)} \approx \frac{1}{k_2} H(s) \tag{5.96}$$

Thus, the gain constant is enhanced by a factor $1/k_2 > 1$. We recognize that this approximation is valid if the impedance looking into the passive network from the terminals of R_b is large compared to R_b, and is therefore neglected in evaluating k_2. Clearly, the smaller the R_b is the better the approximation will be. However, the minimum resistance $R_a + R_b$ is dictated by the maximum current that the op amp can deliver. A compromise has to be reached between the performance of the op amp and the accuracy of the approximation.

Example 5.6 Realize the filter network of Figure 5.35 without the use of a noninverting amplifier to provide a gain of 1.261 in order to meet the specifications exactly.

From Figure 5.39 and equation (5.93), we obtain

$$\frac{R_b}{R_a + R_b} = k_2 = \frac{1}{1.261} \tag{5.97}$$

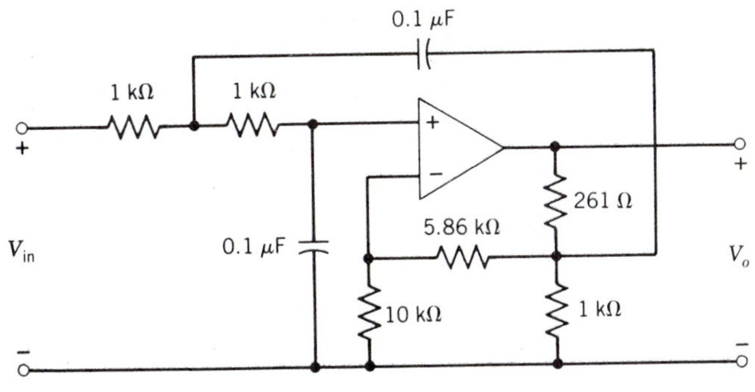

Figure 5.40

Choosing $R_b = 1\ \text{k}\Omega$ yields

$$R_a = 261\ \Omega \tag{5.98}$$

The desired filter network is shown in Figure 5.40.

5.5 THE DIRECT APPROACH

In the foregoing, we discussed the filter realization by cascade approach. The method requires that the transfer function to be realized be first factored into a product of first-order and/or second-order terms. Each term is then individually realized by an active RC network. A cascade connection of these individual networks realizes the overall transfer function. In the present section, we focus on the direct method, where a single network is used to realize the entire transfer function, as was done for the passive filter design discussed in the preceding chapters. A discussion of the direct method will also be given in Chapter 8, where the active filters are realized by means of inductance simulation and variable-frequency scaling.

Most of the direct realization techniques start with a general network configuration that can generate almost any desired transfer function. Each configuration contains RC one-port networks, VCVS, and/or operational amplifiers. The transfer function of the overall network depends explicitly on the driving-point immittances of the RC one-ports. As a result, the problem of realizing a transfer function is reduced to that of realizing RC one-port immittances. In the following, we discuss several representative methods.

5.5.1 Single-Amplifier Realization

Consider the network configuration of Figure 5.41 with node-to-datum voltages V_{in}, V^-, and V_o as labeled. The network consists of one op amp and six RC one-ports. Our first step is to compute the transfer voltage-ratio function of the network in terms of the one-port RC admittances Y_i $(i = 1, 2, \ldots, 6)$. For this we write the nodal

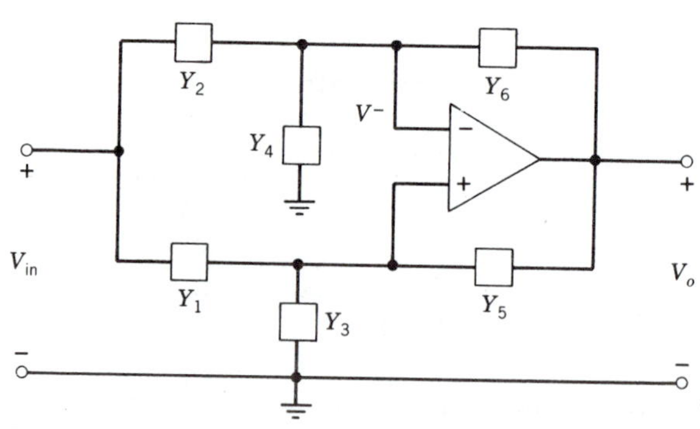

Figure 5.41

equations at the two input terminals of the op amp:

$$Y_2(V^- - V_{\text{in}}) + Y_4 V^- + Y_6(V^- - V_o) = 0 \tag{5.99a}$$

$$Y_1(V^- - V_{\text{in}}) + Y_3 V^- + Y_5(V^- - V_o) = 0 \tag{5.99b}$$

Eliminating V^- from (5.99) yields the desired transfer voltage-ratio function

$$H(s) = \frac{V_o(s)}{V_{\text{in}}(s)} = \frac{Y_1(Y_2 + Y_4 + Y_6) - Y_2(Y_1 + Y_3 + Y_5)}{Y_6(Y_1 + Y_3 + Y_5) - Y_5(Y_2 + Y_4 + Y_6)} \tag{5.100}$$

To simplify the transfer function, choose

$$Y_1 + Y_3 + Y_5 = Y_2 + Y_4 + Y_6 \tag{5.101}$$

Under this condition, (5.100) is reduced to

$$H(s) = \frac{V_o(s)}{V_{\text{in}}(s)} = \frac{Y_1 - Y_2}{Y_6 - Y_5} = \frac{Y_2 - Y_1}{Y_5 - Y_6} \tag{5.102}$$

To realize a given transfer function $H(s)$, we write $H(s)$ explicitly as the ratio of two polynomials

$$H(s) = P(s)/Q(s) \tag{5.103}$$

Recall that in Chapter 3 Theorem 3.2 states that a real rational function is the driving-point impedance of an RC one-port network if and only if the poles and zeros are simple, lie on the negative real axis, and alternate with each other, the first critical frequency being a pole. This is equivalent to stating that a real rational function is the driving-point admittance of an RC one-port network if and only if the poles and zeros are simple, lie on the negative real axis, and alternate with each other, the first critical frequency being a zero. Thus, to proceed with the realization, we divide the numerator and denominator of (5.103) by an appropriate polynomial $D(s)$ of degree n_D having only simple negative real roots, where

$$n_D \geqq \max(n_P, n_Q) - 1 \tag{5.104}$$

and n_P and n_Q are the degrees of the polynomials $P(s)$ and $Q(s)$, respectively, and obtain

$$H(s) = \frac{P(s)/D(s)}{Q(s)/D(s)} \tag{5.105}$$

We remark that the roots of $D(s)$ are unrestricted as long as they are simple, real, and negative. However, to simplify the resulting network realization its roots are frequently chosen to coincide with those real negative roots of $P(s)$ and/or $Q(s)$. Comparing (5.105) with (5.102), we can make the following identifications:

$$\frac{P(s)}{D(s)} = Y_1 - Y_2 \tag{5.106a}$$

$$\frac{Q(s)}{D(s)} = Y_6 - Y_5 \tag{5.106b}$$

or

$$\frac{P(s)}{D(s)} = Y_2 - Y_1 \tag{5.107a}$$

$$\frac{Q(s)}{D(s)} = Y_5 - Y_6 \tag{5.107b}$$

To obtain the realizable one-port RC admittances Y_1, Y_2, Y_5, and Y_6, we expand P/sD and Q/sD in partial fractions. Since all their poles $s = -\sigma_i$ are real, simple, and nonpositive, the residues

$$k_i = (s+\sigma_i)\left.\frac{P(s)}{sD(s)}\right|_{s=-\sigma_i} \quad \text{or} \quad (s+\sigma_i)\left.\frac{Q(s)}{sD(s)}\right|_{s=-\sigma_i} \tag{5.108}$$

evaluated at these poles are real but may be negative. Thus, we can write

$$\frac{P(s)}{D(s)} = \sum_i \frac{k_i s}{s+\sigma_i} - \sum_j \frac{k_j s}{s+\sigma_j} + k_\infty s \tag{5.109a}$$

$$\frac{Q(s)}{D(s)} = \sum_u \frac{k'_u s}{s+\sigma_u} - \sum_v \frac{k'_v s}{s+\sigma_v} + k'_\infty s \tag{5.109b}$$

where k_i, k_j, k'_u, and k'_v are real and positive, and k_∞ and k'_∞ are real. To be specific, if k_∞ and k'_∞ are nonnegative, from (5.106) and (5.107) we can make the following identifications:

$$Y_1(s) = k_\infty s + \sum_i \frac{k_i s}{s+\sigma_i}, \qquad Y_2(s) = \sum_j \frac{k_j s}{s+\sigma_j} \tag{5.110a}$$

$$Y_5(s) = \sum_v \frac{k'_v s}{s+\sigma_v}, \qquad Y_6(s) = k'_\infty s + \sum_u \frac{k'_u s}{s+\sigma_u} \tag{5.110b}$$

or

$$Y_1(s) = \sum_j \frac{k_j s}{s+\sigma_j}, \qquad Y_2(s) = k_\infty s + \sum_i \frac{k_i s}{s+\sigma_i} \tag{5.111a}$$

$$Y_5(s) = k'_\infty s + \sum_u \frac{k'_u s}{s+\sigma_u}, \qquad Y_6(s) = \sum_v \frac{k'_v s}{s+\sigma_v} \tag{5.111b}$$

Since each term in the above expansions corresponds to the series combination of a resistor and a capacitor or a capacitor, the admittances can be realized by the parallel connections of these RC one-ports. If k_∞ or k'_∞ is negative, a similar conclusion can be reached by an appropriate assignment of k_∞ and k'_∞.

Finally, to find Y_3 and Y_4 we appeal to (5.101), which can be rewritten either as

$$Y_3 - Y_4 = (Y_6 - Y_5) - (Y_1 - Y_2) = \frac{Q(s) - P(s)}{D(s)} \tag{5.112}$$

or as

$$Y_4 - Y_3 = (Y_5 - Y_6) - (Y_2 - Y_1) = \frac{Q(s) - P(s)}{D(s)} \tag{5.113}$$

Observe that no matter which form is used, the right-hand side of either (5.112) or (5.113) is similar to the left-hand side of (5.106) or (5.107). Thus, they can be expanded as in (5.109) to give

$$\frac{Q(s)-P(s)}{D(s)}=\sum_m \frac{k''_m s}{s+\sigma_m}-\sum_q \frac{k''_q s}{s+\sigma_q}+k''_\infty s \tag{5.114a}$$

$$=\pm(Y_3-Y_4) \tag{5.114b}$$

where k''_m and k''_q are real and positive, and k''_∞ is real. If $k''_\infty \geqq 0$ and if (5.112) is used, we can identify

$$Y_3(s)=k''_\infty s+\sum_m \frac{k''_m s}{s+\sigma_m}, \qquad Y_4(s)=\sum_q \frac{k''_q s}{s+\sigma_q} \tag{5.115}$$

If we choose (5.113), again with nonnegative k''_∞, we obtain

$$Y_3(s)=\sum_q \frac{k''_q s}{s+\sigma_q}, \qquad Y_4(s)=k''_\infty s+\sum_m \frac{k''_m s}{s+\sigma_m} \tag{5.116}$$

Again, they represent the one-port *RC* admittance functions. If k''_∞ is negative, a similar conclusion can be reached by the reassignment of k''_∞.

Our conclusion is that the *RC* amplifier of Figure 5.41 can generate almost any desired transfer function. The design of such a filter amounts to the realization of the one-port *RC* admittances. Because the realization of complex poles in $H(s)$ is accomplished by the difference of two rational functions, the network is sensitive to the changes of the parameters of the op amp. This is a significant drawback of the structure. We illustrate the above results by the following example.

Example 5.7 Realize the transfer voltage-ratio function

$$H(s)=\frac{600s}{s^2+600s+10^7} \tag{5.117}$$

by the network configuration of Figure 5.41. Comparing this with (5.103) shows that $n_P=1$, $n_Q=2$ and

$$P(s)=600s \tag{5.118a}$$

$$Q(s)=s^2+600s+10^7 \tag{5.118b}$$

For our purposes we choose

$$D(s)=(s+2000)(s+3000) \tag{5.119}$$

and expand P/sD and Q/sD in partial fractions. The results are given by

$$\frac{P(s)}{sD(s)}=\frac{600s}{s(s+2000)(s+3000)}=\frac{0.6}{s+2000}-\frac{0.6}{s+3000} \tag{5.120}$$

$$\frac{Q(s)}{sD(s)}=\frac{s^2+600s+10^7}{s(s+2000)(s+3000)}=\frac{5}{3s}-\frac{6.4}{s+2000}+\frac{86/15}{s+3000} \tag{5.121}$$

Multiplying both sides of (5.120) and (5.121) by s and using (5.110) yields

$$Y_1(s)=\frac{0.6s}{s+2000}, \qquad Y_2(s)=\frac{0.6s}{s+3000} \tag{5.122a}$$

$$Y_5(s)=\frac{6.4s}{s+2000}, \qquad Y_6(s)=\frac{5}{3}+\frac{(86/15)s}{s+3000} \tag{5.122b}$$

Finally, to ascertain Y_3 and Y_4 we perform the partial-fraction expansion of the function

$$\begin{aligned}\frac{Q(s)-P(s)}{sD(s)}&=\frac{s^2+10^7}{s(s+2000)(s+3000)}\\&=\frac{5}{3s}-\frac{7}{s+2000}+\frac{19/3}{s+3000}\end{aligned} \tag{5.123}$$

and obtain from (5.115)

$$Y_3(s)=\frac{5}{3}+\frac{(19/3)s}{s+3000}, \qquad Y_4(s)=\frac{7s}{s+2000} \tag{5.124}$$

By substituting the RC admittance realizations of Y_i $(i=1, 2, \ldots, 6)$ in Figure 5.41, the desired network realization is shown in Figure 5.42. To obtain the practical network element values, we magnitude-scale the network by a factor of 10^3, yielding the final realization of (5.117) as shown in Figure 5.43.

5.5.2 **Two-Amplifier Realization**

A two-amplifier network configuration for the synthesis of a general transfer function is shown in Figure 5.44. The transfer voltage-ratio function of the network is found

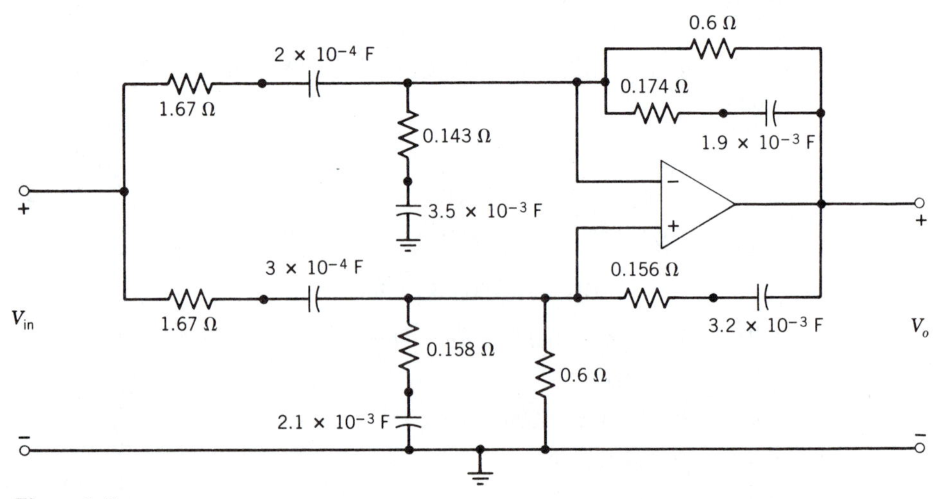

Figure 5.42

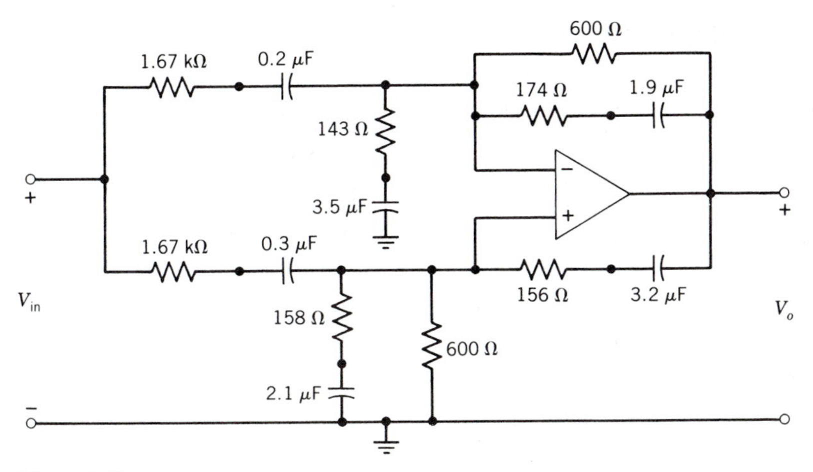

Figure 5.43

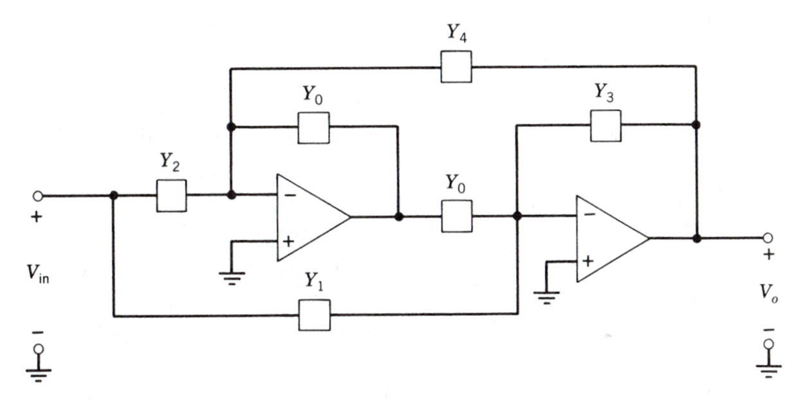

Figure 5.44

to be independent of Y_0 and is given by

$$H(s)=\frac{V_o(s)}{V_{\text{in}}(s)}=\frac{Y_2-Y_1}{Y_3-Y_4}=\frac{Y_1-Y_2}{Y_4-Y_3} \tag{5.125}$$

Comparing this equation with (5.102), we see that they have the same form. Consequently, the realization procedure for the two-amplifier network is similar to that discussed in the preceding section for the single-amplifier structure. It is hardly necessary to go through the same ground again. We merely present the procedure in terms of an illustrative example.

Example 5.8 Realize the transfer voltage-ratio function

$$H(s)=\frac{600s}{s^2+600s+10^7} \tag{5.126}$$

by the two-amplifier network of Figure 5.44.

By choosing the same polynomial $D(s)$ as in (5.119), we obtain (5.120) and (5.121). In terms of (5.125), we can write

$$Y_2 - Y_1 = \frac{P(s)}{D(s)} = \frac{0.6s}{s+2000} - \frac{0.6s}{s+3000} \tag{5.127a}$$

$$Y_3 - Y_4 = \frac{Q(s)}{D(s)} = \frac{5}{3} - \frac{6.4s}{s+2000} + \frac{(86/15)s}{s+3000} \tag{5.127b}$$

identifying

$$Y_1(s) = \frac{0.6s}{s+3000}, \qquad Y_2(s) = \frac{0.6s}{s+2000} \tag{5.128a}$$

$$Y_3(s) = \frac{5}{3} + \frac{(86/15)s}{s+3000}, \qquad Y_4(s) = \frac{6.4s}{s+2000} \tag{5.128b}$$

These admittances can easily be realized by the RC one-ports. By substituting them in the network of Figure 5.44, and then performing the magnitude-scaling of the resulting network by a factor of 10^3, the final realization is shown in Figure 5.45, where Y_0 was chosen to be 1 kΩ after scaling.

The two-amplifier configuration of Figure 5.44 suffers the same drawback as the single-amplifier network of Figure 5.41 in that it is sensitive to the changes of the parameters of the active elements. In addition, basic to all active filters, a small variation in an active element such as the controlling parameter of a VCVS can easily influence a marginally stable network to become unstable. This is particularly true if the partial-fraction decomposition of the denominator ratio gives rise to both positive and negative coefficient terms. Therefore, whenever possible, we should choose $D(s)$ to yield only positive residues in the partial-fraction expansion of $Q(s)/sD(s)$.

We mention that the above two configurations are far from canonic, and, while theoretically general, they are good candidates for some moderate-order ($\leqslant 5$) designs with unchallenging requirements. They are included here for completeness.

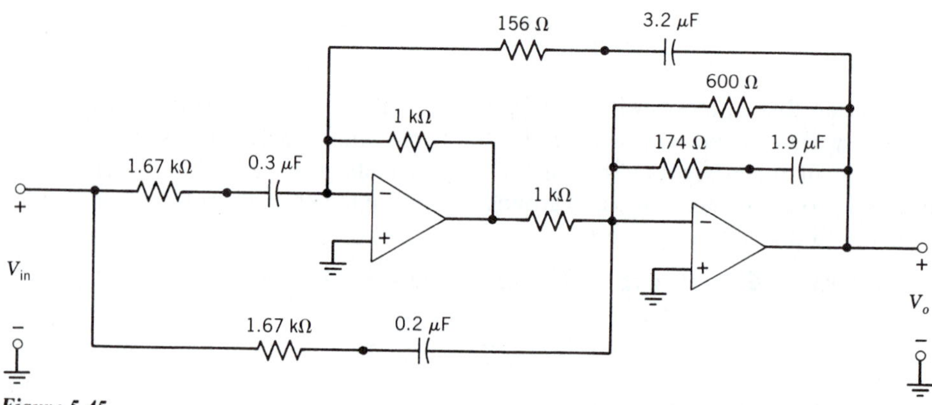

Figure 5.45

5.5.3 **State-Variable Realization**

Another network configuration for the synthesis of a general transfer function is shown in Figure 5.46, and is known as the *state-variable realization*. The node-to-datum voltages V_k $(k=1, 2, \ldots, n)$, V_a, and V_{in} are as indicated in the figure. Also shown are the two possible connections for the resistor R_n, depending on whether n is odd or even. The equations describing the network can be written directly from Figure 5.46 by inspection, as follows:

$$V_k = -RCsV_{k+1} = (-1)^{n-k}(RCs)^{n-k}V_n, \qquad k=1, 2, \ldots, n-1 \tag{5.129a}$$

$$V_a = -\frac{R_a}{R_2}V_2 - \frac{R_a}{R_4}V_4 - \cdots - \frac{R_a}{R_p}V_p \tag{5.129b}$$

$$C_1 sV_1 + \frac{V_a}{R_a} + \frac{V_{in}}{R_0} + \frac{V_1}{R_1} + \frac{V_3}{R_3} + \cdots + \frac{V_q}{R_q} = 0 \tag{5.129c}$$

where $p=n-1$ and $q=n$, n odd, and $p=n$ and $q=n-1$, n even. Substituting (5.129a) and (5.129b) in (5.129c) yields

$$\begin{aligned}(RCs)^n V_n &+ \frac{(RCs)^{n-1}}{R_1}V_n + \frac{(RCs)^{n-2}}{R_2}V_n + \frac{(RCs)^{n-3}}{R_3}V_n \\ &+ \frac{(RCs)^{n-4}}{R_4}V_n + \cdots + \frac{RCs}{R_{n-1}}V_n + \frac{1}{R_n}V_n = (-1)^n \frac{V_{in}}{R_0}\end{aligned} \tag{5.130}$$

The transfer voltage-ratio function of the network can then be written as

$$H_n(s) = \frac{V_n(s)}{V_{in}(s)} = \frac{(-1)^n/R_0}{(RCs)^n + (RCs)^{n-1}/R_1 + (RCs)^{n-2}/R_2 + \cdots + RCs/R_{n-1} + 1/R_n} \tag{5.131}$$

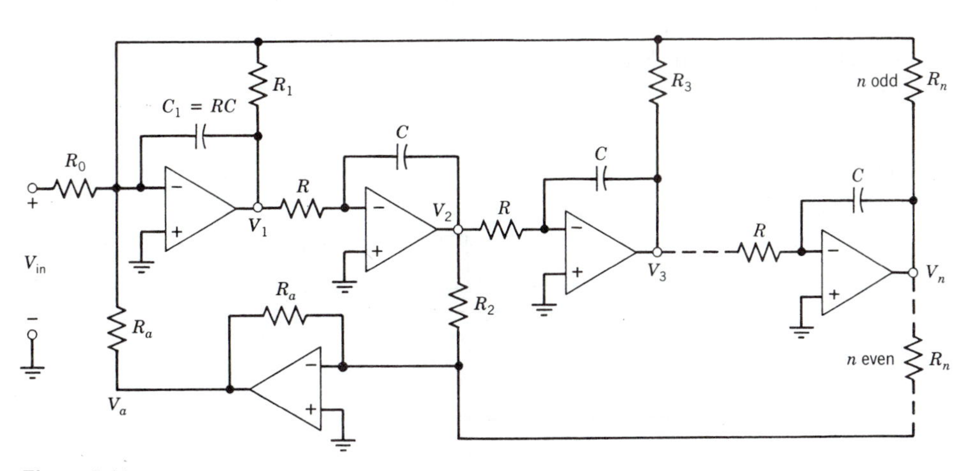

Figure 5.46

If V_k is chosen as the output voltage instead of V_n, the transfer relationships between V_k and V_{in} are found to be

$$H_k(s)=\frac{V_k(s)}{V_{in}(s)}$$

$$=\frac{(-1)^k(RCs)^{n-k}/R_0}{(RCs)^n+(RCs)^{n-1}/R_1+(RCs)^{n-2}/R_2+\cdots+RCs/R_{n-1}+1/R_n},$$

$$k=1, 2, \ldots, n \tag{5.132}$$

Observe that the transfer functions are independent of R_a. Thus, we can choose any convenient value for R_a. From Figure 5.46, we see that the state-variable network configuration requires at least $n+1$ op amps in addition to a large number of resistors. This is costly. However, tuning is easier than in the other two structures because it provides an individual tuning mechanism to the resistor values $R_1, R_2, \ldots, R_n$ to fit the coefficients of a given transfer function.

Finally, to realize a general transfer voltage-ratio function (5.103), we need only to insert a *summing amplifier* such as the one shown in Figure 5.47, the transfer voltage relationships of which are given by

$$V_o=-\sum_{i=1}^{n}\frac{R_f}{R_i'}V_i \tag{5.133}$$

Substituting (5.132) in (5.133) gives the general transfer voltage-ratio function as

$$H(s)=\frac{V_o(s)}{V_{in}(s)}$$

$$=\frac{R_f}{R_0}\cdot\frac{(RCs)^{n-1}/R_1'-(RCs)^{n-2}/R_2'+(RCs)^{n-3}/R_3'+\cdots-(-1)^n/R_n'}{(RCs)^n+(RCs)^{n-1}/R_1+(RCs)^{n-2}/R_2+\cdots+RCs/R_{n-1}+1/R_n} \tag{5.134}$$

Comparing this with (5.103), we can ascertain an appropriate set of element values. Note that the signs associated with the numerator coefficients can be changed by the sign-reversal amplifier of Figure 5.11. We illustrate the above design procedure by the following example.

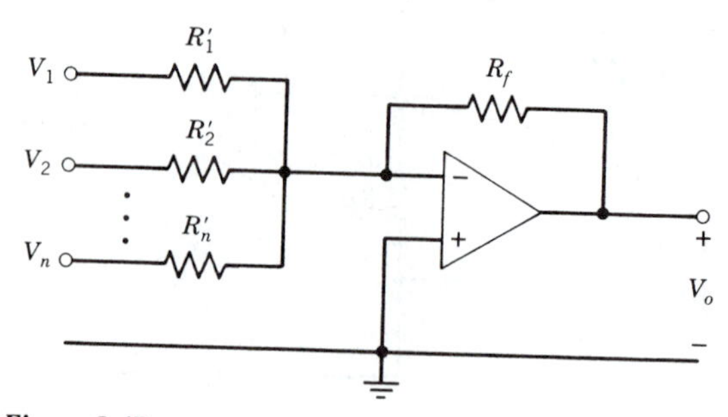

Figure 5.47

Example 5.9 Realize the transfer voltage-ratio function

$$H(s)=\frac{600s}{s^2+600s+10^7} \tag{5.135}$$

by the state-variable network of Figure 5.46.

To put the transfer function (5.135) in the form of (5.134), we divide the numerator and denominator by 10^6 and obtain

$$H(s)=\frac{0.6(s/10^3)}{(s/10^3)^2+0.6(s/10^3)+10} \tag{5.136}$$

Comparing this with (5.134), we can make the following identifications:

$$RC=10^{-3} \tag{5.137a}$$

$$1/R_1=0.6 \quad \text{or} \quad R_1=1.67\ \Omega \tag{5.137b}$$

$$1/R_2=10 \quad \text{or} \quad R_2=0.1\ \Omega \tag{5.137c}$$

$$\frac{R_f}{R_0R'_1}=0.6 \tag{5.137d}$$

$$\frac{R_f}{R_0R'_2}=0 \tag{5.137e}$$

where $n=2$. For our purposes, choose

$$R_a=R=1\ \Omega, \qquad R_f=R'_1=0.5\ \Omega \tag{5.138}$$

giving

$$R_0=R_1=1.67\ \Omega, \qquad C=10^{-3}\text{ F}, \qquad R_2=0.1\ \Omega, \qquad R'_2=\infty \tag{5.139}$$

To obtain the practical element values, we magnitude-scale the network by a factor of 10^4. The resulting network is shown in Figure 5.48 with element values

$$R_0=R_1=16.7\text{ k}\Omega, \qquad R_f=R'_1=5\text{ k}\Omega \tag{5.140a}$$

$$R=R_a=10\text{ k}\Omega, \qquad R_2=1\text{ k}\Omega \tag{5.140b}$$

$$R'_2=\infty \qquad C_1=C=0.1\ \mu\text{F} \tag{5.140c}$$

Before we turn our attention to other aspects of active filter synthesis, we mention that the state-variable technique is the only viable direct approach presented so far. Other direct syntheses based on passive ladder analogs will be discussed in Chapter 8. An important drawback of the single/double amplifier networks is the large sensitivity with respect to the changes of the parameter values of the active elements such as the VCVS and operational amplifiers. These changes can also cause a marginally stable network to become unstable. In many applications, the poor sensitivity performance alone will render these networks impractical. One of the advantages of the direct approach is the ease with which the high-order transfer functions can be implemented with very few op amps. Two such configurations are

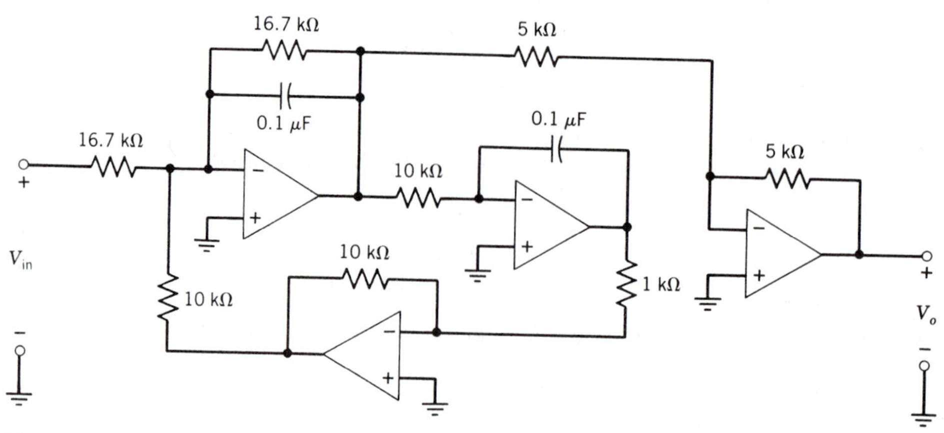

Figure 5.48

shown in Figures 5.41 and 5.44. Although op amps are cheap and physically small, they still consume power and are the predominant sources of noise. Thus, any approach that requires the fewest op amps while not unduly sacrificing performance is welcome.

The most widely used method in industry is the cascade approach because the synthesis procedure needed for determining the element values of a biquad is relatively simple, and the need for the implementation of additional constraints such as the use of standard element values and the minimization of sensitivity is usually easy to attain. In addition, since individual biquads are totally isolated, each active *RC* network may be individually tuned without affecting other networks. It is far easier than trying to tune a network in which all the elements interact, as in the case of passive synthesis and the direct method of realization. For this reason, we shall treat the realization of biquads in greater detail in Chapter 7. There we show that there are general biquads that are capable of realizing various transfer functions with relatively low sensitivities. These structures provide an economical means for implementing a universal kind of building block for realizing filters of any type. Before we do this, we first study the biquadratic function.

5.6 THE BIQUADRATIC FUNCTION

In active cascade synthesis, it is sufficient to consider only the biquadratic function of (5.3):

$$H(s)=K\frac{s^2+cs+d}{s^2+as+b} \tag{5.141}$$

This function is usually put in its standard form as

$$H(s)=K\frac{s^2+\dfrac{\omega_z}{Q_z}s+\omega_z^2}{s^2+\dfrac{\omega_p}{Q_p}s+\omega_p^2} \tag{5.142}$$

where

$$Q_z=\sqrt{d}/c, \qquad \omega_z=\sqrt{d}, \qquad Q_p=\sqrt{b}/a, \qquad \omega_p=\sqrt{b} \tag{5.143}$$

The frequency ω_z is known as the *zero frequency*, and Q_z the *zero* Q. Likewise, the frequency ω_p is called the *pole frequency*, and Q_p the *pole* Q. These quantities play a fundamental role in the characterization of the biquadratic function. Their properties will be studied below.

To obtain the frequency response, we substitute s by $j\omega$ in (5.142) and then consider the logarithm of $H(j\omega)$ instead of $H(j\omega)$. Write

$$20 \log H(j\omega)=\alpha(\omega)+j(20 \log e)\phi(\omega) \tag{5.144}$$

Then from (5.142)

$$\alpha(\omega)=20 \log |K|+10 \log\left[(\omega_z^2-\omega^2)^2+\frac{\omega_z^2}{Q_z^2}\omega^2\right]-10 \log\left[(\omega_p^2-\omega^2)^2+\frac{\omega_p^2}{Q_p^2}\omega^2\right] \tag{5.145}$$

$$\phi(\omega)=\tan^{-1}\frac{\omega_z\omega}{Q_z(\omega_z^2-\omega^2)}-\tan^{-1}\frac{\omega_p\omega}{Q_p(\omega_p^2-\omega^2)} \tag{5.146}$$

where π is to be added in (5.146) in the case of negative K.

Here we see the advantage of using logarithms. Not only has the division been changed to subtraction, but each factor also appears in identical form with the exception of the sign and the constant term. As a result, we need only study the contribution due to a typical term. Once this is accomplished, the overall behavior of the biquadratic function on the $j\omega$-axis is then obtained by merely adding the responses of the components. To this end, we consider the response of the typical term

$$\alpha_1(\omega)=10 \log\left[(\omega_0^2-\omega^2)^2+\frac{\omega_0^2\omega^2}{Q_0^2}\right] \tag{5.147}$$

$$\phi_1(\omega)=\tan^{-1}\frac{\omega_0\omega}{Q_0(\omega_0^2-\omega^2)} \tag{5.148}$$

where $\omega_0=\omega_z$, ω_p and $Q_0=Q_z$, Q_p. At very low frequencies, the asymptote for gain is $40 \log \omega_0$. For large values of ω, the gain is approximated by

$$\alpha_1(\omega)=40 \log \omega \tag{5.149}$$

showing that the slope of the straight line is 40 dB per decade or, equivalently, 12 dB per octave. As to the phase (5.148), it varies from 0° at low frequencies to 180° at high frequencies with the value 90° at $\omega=\omega_0$.

The frequency at which the magnitude achieves a minimum is obtained by equating the derivative of the magnitude to zero:

$$\frac{d}{d\omega}\left[(\omega_0^2-\omega^2)^2+\frac{\omega_0^2\omega^2}{Q_0^2}\right]^{1/2}=0 \tag{5.150}$$

obtaining

$$\omega_{\min}=\omega_0\sqrt{1-1/2Q_0^2}, \qquad 2Q_0^2 \geqq 1 \tag{5.151a}$$

$$=0, \qquad 2Q_0^2<1 \tag{5.151b}$$

For $2Q_0^2 \gg 1$, then $\omega_{\min} \approx \omega_0$ and the minimum value at this frequency is

$$\alpha_{\min}=\alpha(\omega_{\min}) \approx \alpha(\omega_0)=20 \log(\omega_0^2/Q_0) \tag{5.152}$$

Let us next consider the function $\alpha(\omega)$ at the frequencies

$$\omega_1, \omega_2=\omega_0\left(1 \mp \frac{1}{2Q_0}\right) \tag{5.153}$$

Substituting this in (5.147) yields

$$\alpha_1(\omega_i)=20 \log\left|\frac{\omega_0^2}{Q_0}\left(\pm 1-\frac{1}{4Q_0}+j1 \mp j\frac{1}{2Q_0}\right)\right|, \qquad i=1, 2 \tag{5.154}$$

For $2Q_0 \gg 1$, the above equation reduces to

$$\alpha_1(\omega_i) \approx 20 \log\left|\frac{\omega_0^2}{Q_0}(\pm 1+j1)\right|=20 \log\frac{\omega_0^2}{Q_0}+20 \log\sqrt{2}$$

$$=20 \log\frac{\omega_0^2}{Q_0}+3, \qquad i=1, 2 \tag{5.155}$$

Comparing this with (5.152) shows that the magnitude at $s=j\omega_i\,(i=1, 2)$ is 3 dB above the minimum value at the frequency $s=j\omega_0$. These results are sketched in Figure 5.49. The 3-dB *bandwidth* defined as the frequency band between ω_1 and ω_2 is found from (5.153) to be

$$\text{bandwidth}=\omega_2-\omega_1=\omega_0/Q_0 \tag{5.156}$$

Plots of the magnitude (5.147) and phase (5.148) for various values of Q_0 are shown in Figure 5.50. It can be seen that the sharpness of the magnitude and phase plots near the frequency ω_0 increases with Q_0. For Q's greater than 5 the minimum magnitude occurs essentially at the frequency ω_0. For Q's less than 0.707, the function becomes monotonical and does not exhibit a bump. These plots correspond to the contributions by a pair of complex-conjugate zeros. For a pair of complex-conjugate poles, their contributions are simply the negative of those due to the zeros. A sketch of the magnitude response by the poles is shown in Figure 5.51.

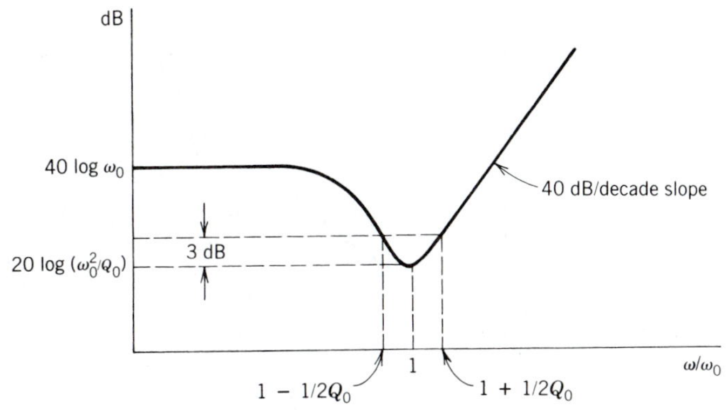

Figure 5.49

As an illustration of identifying the parameters ω_0 and Q_0, consider the low-pass Sallen and Key circuit of Figure 5.34, the transfer function of which was computed earlier in (5.73) and is repeated below:

$$H(s)=\frac{k/R_1R_2C_1C_2}{s^2+[1/R_1C_1+1/R_2C_1+1/R_2C_2-k/R_2C_2]s+1/R_1R_2C_1C_2} \tag{5.157}$$

The standard form for (5.157) is

$$H(s)=\frac{K}{s^2+\dfrac{\omega_p}{Q_p}s+\omega_p^2} \tag{5.158}$$

where

$$K=\frac{k}{R_1R_2C_1C_2} \tag{5.159a}$$

$$\omega_p=\frac{1}{\sqrt{R_1R_2C_1C_2}} \tag{5.159b}$$

$$Q_p=\frac{\omega_p}{B_p}=\frac{1/\sqrt{R_1R_2C_1C_2}}{1/R_1C_1+1/R_2C_1+1/R_2C_2-k/R_2C_2} \tag{5.159c}$$

where B_p is the 3-dB pole bandwidth. For the normalized Butterworth transfer voltage-ratio function

$$H(s)=\frac{2}{s^2+\sqrt{2}s+1} \tag{5.160}$$

the parameters are identified as

$$K=2, \qquad \omega_p=1, \qquad Q_p=1/\sqrt{2}=0.707 \tag{5.161}$$

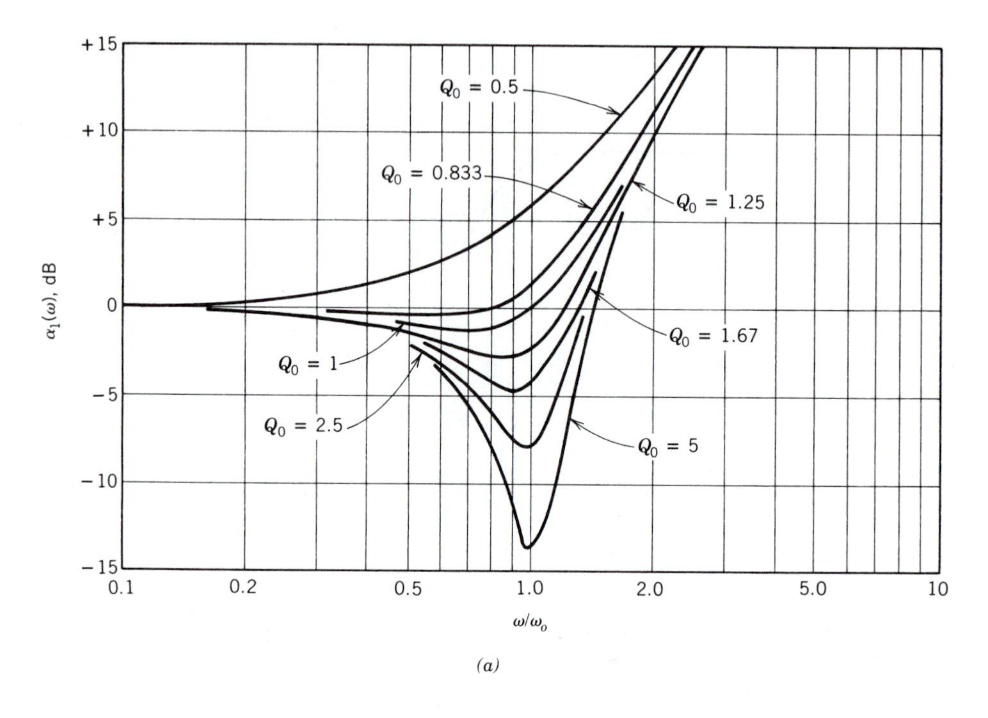

(a)

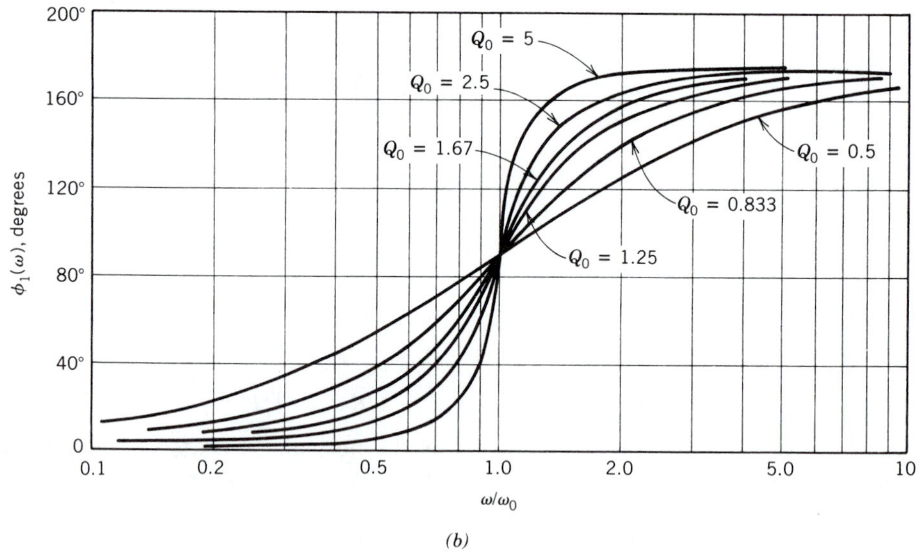

(b)

Figure 5.50

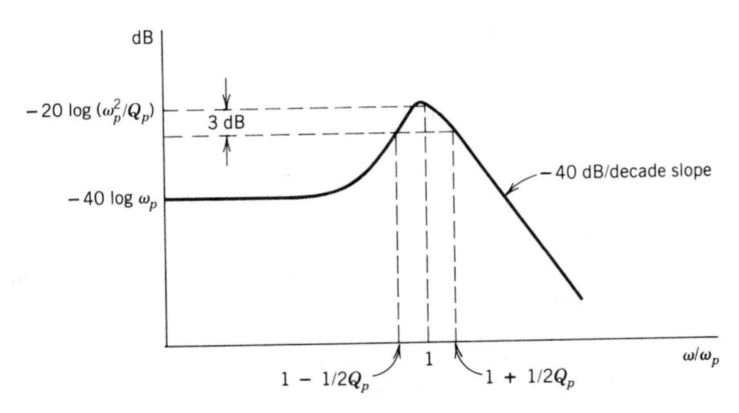

Figure 5.51

Before we turn our attention to sensitivity problems, we mention briefly the classification of biquadratic functions. A biquadratic function that realizes a *low-pass characteristic* takes the general form

$$H(s) = K \frac{b}{s^2 + as + b} = K \frac{\omega_p^2}{s^2 + \dfrac{\omega_p}{Q_p} s + \omega_p^2} \tag{5.162}$$

For a *high-pass characteristic*, the function is given by

$$H(s) = K \frac{s^2}{s^2 + as + b} = K \frac{s^2}{s^2 + \dfrac{\omega_p}{Q_p} s + \omega_p^2} \tag{5.163}$$

A biquadratic function that has a *band-pass characteristic* is

$$H(s) = K \frac{as}{s^2 + as + b} = K \frac{\dfrac{\omega_p}{Q_p} s}{s^2 + \dfrac{\omega_p}{Q_p} s + \omega_p^2} \tag{5.164}$$

A *band-elimination characteristic* is realized by the function

$$H(s) = K \frac{s^2 + b}{s^2 + as + b} = K \frac{s^2 + \omega_z^2}{s^2 + \dfrac{\omega_p}{Q_p} s + \omega_p^2} \tag{5.165}$$

where $\omega_z = \omega_p$. If $\omega_z \gg \omega_p$, the function represents a low-pass characteristic with a null in the stopband. Such a filter characteristic is referred to as a *low-pass notch*. Likewise, if $\omega_z \ll \omega_p$, (5.165) represents a high-pass characteristic with a null in the stopband, and is called a *high-pass-notch* characteristic.

Finally, an *all-pass* function is defined by the equation

$$H(s)=K\frac{s^2-as+b}{s^2+as+b}=K\frac{s^2-\dfrac{\omega_p}{Q_p}s+\omega_p^2}{s^2+\dfrac{\omega_p}{Q_p}s+\omega_p^2} \tag{5.166}$$

The zeros and poles of this function are symmetrical about the $j\omega$-axis, and its $j\omega$-axis magnitude is unity for all ω. Such a characteristic will only cause phase delay, and is often referred to as a *delay* function.

5.7 SUMMARY AND SUGGESTED READINGS

In this chapter, we considered the general subject of active filter synthesis, and presented techniques for the synthesis of transfer functions through the use of both active and passive elements, the latter being restricted exclusively to resistors and capacitors. Such filters are known as *active RC filters.* In general, there are two methods to realize transfer functions using active *RC* networks. The first technique is called the *cascade* method. The transfer function to be realized is first factored into a product of first-order and/or second-order terms. Each term is then individually realized by an active *RC* network. A cascade connection of these individual networks realizes the overall transfer function. This approach is possible provided that the individual networks are isolated from each other in the sense that each successive network does not load the previous one. This condition is readily satisfied in most op-amp network realizations, but is not for the passive networks. Consequently, the realization of higher-order transfer functions is reduced to the much simpler realization of a general second-order function called the *biquadratic function.* The network block that realizes a biquadratic function is commonly referred to as a *biquad.* The second general method is the *direct* method, where a single network is used to realize the entire transfer function, as was done for the passive filter design discussed in the preceding chapters.

Most of the direct realization techniques start with a general network configuration that can generate almost any desired transfer function. For this we presented three representative structures: *single-amplifier realization, two-amplifier realization,* and *state-variable realization.* The first two configurations can realize any high-order transfer functions with one or two op amps, making them efficient in terms of power consumption. However, this efficiency in op amps does not come without sacrifice. The filter sensitivity is large with respect to the changes of the parameter values of the active elements. These changes can cause a marginally stable network to become unstable. In many applications, the poor sensitivity performance alone will render the first two configurations impractical. The state-variable realization is the only viable direct approach. As such, it is not terribly sensitive. In addition, direct syntheses based upon passive ladder analogs to be presented in Chapter 8 will result in networks with a low-sensitivity property. These structures are highly competitive with the cascade approach.

The most widely used method for filter design in industry is the cascade approach because the procedure is relatively simple and the implementation of additional constraints such as the use of standard element values and the minimization of sensitivity is usually easy to attain. In addition, individual biquads can be separately tuned without affecting the others.

The biquad is therefore the basic building block in active filter synthesis. For this we first discussed two simple op-amp networks that realize the first-order factors in the transfer function decomposition. The first configuration is the inverting amplifier structure and the second is the noninverting type. We then introduced the commonly used single-amplifier biquad topologies. These networks consist of an *RC* network and an op amp, and can be used to realize a complex pole-zero pair. There are many networks that can accomplish this, and the majority of them can be classified into one of the two basic categories: the *negative feedback topology* and the *positive feedback topology*, depending on which input terminal of the op amp the *RC* network is connected to. Fundamental properties of these topologies were discussed. We pointed out that in the design of biquads we often match only the coefficients of a biquadratic function that determine the zero and pole locations, leaving the gain constant floating. The resulting gain constant can later be adjusted by the insertion of an amplifier. To avoid the use of an additional amplifier, we presented two techniques: *input attenuation* and *gain enhancement*.

We next discussed the standard form of the biquadratic function. It is shown that the biquadratic function can be characterized by its *zero* and *pole frequencies* and *zero* and *pole* Q's. Depending on these parameters, the biquadratic functions can be classified into *low-pass*, *high-pass*, *band-pass*, *band-elimination*, and *all-pass functions*.

An excellent introduction to various aspects of active filter synthesis can be found in Daryanani [4] and Van Valkenburg [10]. A discussion on direct approach to other filter configurations is given in Ghausi and Laker [5] and Lam [7].

REFERENCES

1. P. Bowron and F. W. Stephenson 1979, *Active Filters for Communications and Instrumentation*, Chapters 5 and 6. New York: McGraw–Hill
2. L. T. Bruton 1980, *RC-Active Circuits: Theory and Design*, Chapters 5, 7, and 8. Englewood Cliffs, NJ: Prentice–Hall.
3. A. Budak 1974, *Passive and Active Network Analysis and Synthesis*, Chapters 10–14. Boston: Houghton Mifflin.
4. G. Daryanani 1976, *Principles of Active Network Synthesis and Design*, Chapters 7–9. New York: John Wiley.
5. M. S. Ghausi and K. R. Laker 1981, *Modern Filter Design: Active RC and Switched Capacitor*, Chapters 4 and 5. Englewood Cliffs, NJ: Prentice–Hall.
6. L. P. Huelsman and P. E. Allen 1980, *Introduction to the Theory and Design of Active Filters*, Chapters 4 and 5. New York: McGraw–Hill.
7. H. Y-F. Lam 1979, *Analog and Digital Filters: Design and Realization*, Chapter 10. Englewood Cliffs, NJ: Prentice–Hall.

8. W. F. Lovering, "Analog computer simulation of transfer functions," *Proc. IEEE*, vol. 53, 1965, p. 306.
9. S. K. Mitra, "Active *RC* filters employing a single operational amplifier as the active element," *Proc. Hawaii International Conference on System Science*, 1968, pp. 433–436.
10. M. E. Van Valkenburg 1982, *Analog Filter Design*, Chapters 4–8 and 10. New York: Holt, Rinehart & Winston.

PROBLEMS

5.1 Realize the transfer voltage-ratio function

$$H(s)=\frac{600s}{(s^2+600s+10^7)(s+2000)} \tag{5.167}$$

5.2 Consider the low-pass Sallen and Key network of Figure 5.34 with the following choice made for the design:

$$k=2, \qquad C_1=C_2=C \tag{5.168}$$

The value for $k=2$ is attractive because it can be realized by the use of equal-valued feedback resistors around an op amp. Determine the design formulas that express the values of R_1, R_2, and $C_2=C$ in terms of the pole frequency and pole Q.

5.3 Repeat Problem 5.2 by choosing

$$k=1, \qquad R_1C_1=R_2C_2, \qquad C_1=1\text{ F} \tag{5.169}$$

5.4 Realize the transfer voltage-ratio function

$$H(s)=\frac{(s+10^3)(s+10^5)}{(s+10^2)(s+10^4)} \tag{5.170}$$

by the cascade approach.

5.5 Realize the transfer function (5.170) by the direct approach, using the network configurations of (a) Figure 5.41, (b) Figure 5.44, and (c) Figure 5.46.

5.6 For the negative feedback biquad network of Figure 5.27, identify the feedback and feedforward transfer functions, and then determine the transfer voltage-ratio function $H=V_o/V_{in}$, assuming an ideal op amp.

5.7 Repeat Problem 5.6 for the op amp with a finite gain A.

5.8 Synthesize the following low-pass transfer voltage-ratio function using the Sallen and Key network of Figure 5.34 as the basic building block:

$$H(s)=\frac{K}{(s^2+2s+100)(s^2+4s+64)} \tag{5.171}$$

5.9 Realize the transfer voltage-ratio function

$$H(s)=\frac{100}{s^2+2s+64} \tag{5.172}$$

by the network configurations of (a) Figure 5.41, (b) Figure 5.44, and (c) Figure 5.46.

5.10 Realize the all-pass function

$$H(s)=\frac{s^2-100s+10^6}{s^2+100s+10^6} \tag{5.173}$$

by the state-variable network of Figure 5.46.

5.11 Realize the transfer function (5.171) by the network configurations of (a) Figure 5.41, (b) Figure 5.44, and (c) Figure 5.46.

5.12 Realize the transfer voltage-ratio function

$$H(s)=\frac{(s+10^3)^2}{(s+10^2)^2} \tag{5.174}$$

by the cascade method.

5.13 Repeat Problem 5.12, using any direct method.

5.14 Using the inverting amplifier of Figure 5.8, realize the transfer function

$$H(s)=-\frac{Ks}{(s+p_1)(s+p_2)} \tag{5.175}$$

chapter 6
SENSITIVITY

In the preceding chapter, we discussed a number of filter configurations that realize a given set of specifications. Among the many possible solutions, the problem is to select what appears to be the best solution based on criteria involving cost, performance, sensitivity, convenience, and engineering judgment. Given this difficult task of selecting a network configuration, it is valuable to identify criteria that can be used to determine the goodness of a network. One of the most important such criteria is sensitivity. This chapter is devoted entirely to this subject.

6.1 SENSITIVITY FUNCTION

Active filters are designed to perform certain functions such as wave shaping or signal processing. Given perfect components, there would be little difference among the many possible designs. In practice, however, real components deviate from their nominal values because of manufacturing tolerances, changes in environmental conditions such as temperature and humidity, or chemical changes due to the aging of the components. As a consequence, the performance of a practical filter differs from the nominal design. This causes the network transfer function to drift away from its nominal value. The cause-and-effect relationship between the network element variations and the resulting changes in the network transfer function is known as the sensitivity. One way to minimize this change or to reduce the sensitivity is to choose components with small manufacturing tolerances, low temperature, aging, and humidity coefficients. However, this approach will usually result in more expensive networks than necessary. A practical solution is to design a network that has a low sensitivity to element changes. This is especially important in active filter design, where active elements such as op amps are much more sensitive to environmental changes. A good understanding of sensitivity is essential to the design of practical active filters. In the present chapter, we study the sensitivity of network functions to changes in the parameters and describe ways of evaluating the sensitivity of a network.

Sensitivity is a measure of the change of the overall network function to the change of a particular parameter in the network. It is formally defined below.

Definition 6.1: *Sensitivity function.* The *sensitivity function* is defined as the ratio of the fractional change in a network function to the fractional change in an element for the situation when all changes concerned are differentially small.

Thus, if H is the network function and x is the element of interest, the sensitivity function, written as S_x^H, is defined by the equation

$$S_x^H = \lim_{\Delta x \to 0} \frac{\Delta H/H}{\Delta x/x} = \frac{x}{H} \cdot \frac{\partial H}{\partial x} \tag{6.1}$$

In terms of logarithms, (6.1) can be simplified to

$$S_x^H = \frac{\partial \ln H}{\partial \ln x} = x \frac{\partial \ln H}{\partial x} \tag{6.2}$$

A partial derivative is used because element values other than x may also be changing. A different version of (6.1) is given by

$$\frac{\partial H}{H} = S_x^H \frac{\partial x}{x} \tag{6.3}$$

which states that the sensitivity function S_x^H is the quantity that multiplies the incremental change in x to indicate the fractional change that will occur in the transfer function H.

As an illustration, consider the inverting amplifier of Figure 6.1 with a finite gain A for the op amp. The transfer function is found to be

$$H(s) = \frac{V_o(s)}{V_{in}(s)} = -\frac{AZ_2}{Z_2 + (1+A)Z_1} \tag{6.4}$$

We compute the sensitivity of H to the change of amplifier gain A. From (6.1) we obtain

$$S_A^H = \frac{A}{H} \cdot \frac{\partial H}{\partial A} = \frac{A}{-\dfrac{AZ_2}{Z_2 + (1+A)Z_1}} \cdot \frac{-Z_2(Z_1 + Z_2)}{[Z_2 + (1+A)Z_1]^2} \tag{6.5}$$

$$= \frac{1 + Z_2/Z_1}{1 + A + Z_2/Z_1} \tag{6.6}$$

If Z_2 is removed from the network so that there is no feedback, then the new transfer function is $\tilde{H} = -A$, and the sensitivity function becomes

$$S_A^{\tilde{H}} = \frac{A}{\tilde{H}} \cdot \frac{\partial \tilde{H}}{\partial A} = \frac{A}{-A}(-1) = 1 \tag{6.7}$$

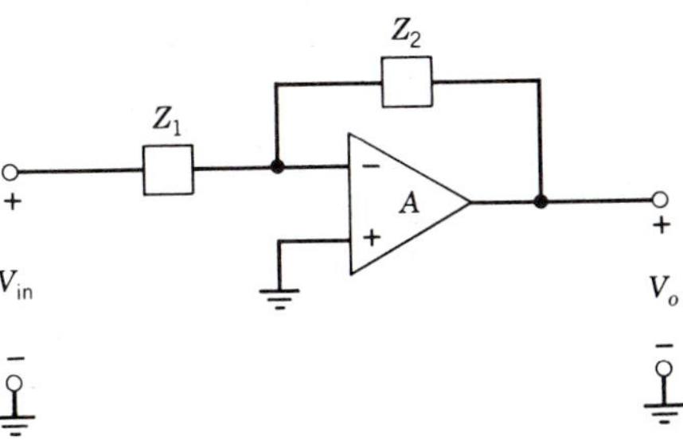

Figure 6.1

As a specific example, let $Z_1 = R_1 = 1\ \text{k}\Omega$, $Z_2 = R_2 = 5\ \text{k}\Omega$, and $A = 10^5$, as shown in Figure 6.2. Then we have

$$S_A^H = 6 \times 10^{-5}, \qquad S_A^{\tilde{H}} = 1 \tag{6.8}$$

meaning that for an amplifier without feedback, a 5% change in amplifier gain will result approximately in a 5% change of the transfer function, whereas for an amplifier with feedback, a 5% change will only result in a 0.0003% change of the transfer function. This is why negative feedback is commonly used in the design of active filters.

We next formulate some general properties of the sensitivity function from its definition (6.1). One form that we will encounter most often is

$$F = f_1^a f_2^b f_3^c \tag{6.9}$$

The natural logarithm of this function is

$$\ln F = a \ln f_1 + b \ln f_2 + c \ln f_3 \tag{6.10}$$

If we differentiate this expression with respect to $\ln f_1$, then from (6.2)

$$S_{f_1}^F = \frac{\partial \ln F}{\partial \ln f_1} = a \tag{6.11}$$

Similarly, we have

$$S_{f_2}^F = b, \qquad S_{f_3}^F = c \tag{6.12}$$

If F is expressed as the product of two functions F_1 and F_2, then

$$S_x^F = \frac{\partial \ln F}{\partial \ln x} = \frac{\partial \ln F_1}{\partial \ln x} + \frac{\partial \ln F_2}{\partial \ln x} \tag{6.13}$$

showing that

$$S_x^{F_1 F_2} = S_x^{F_1} + S_x^{F_2} \tag{6.14}$$

Likewise, we can show that

$$S_x^{F_1/F_2} = S_x^{F_1} - S_x^{F_2} \tag{6.15}$$

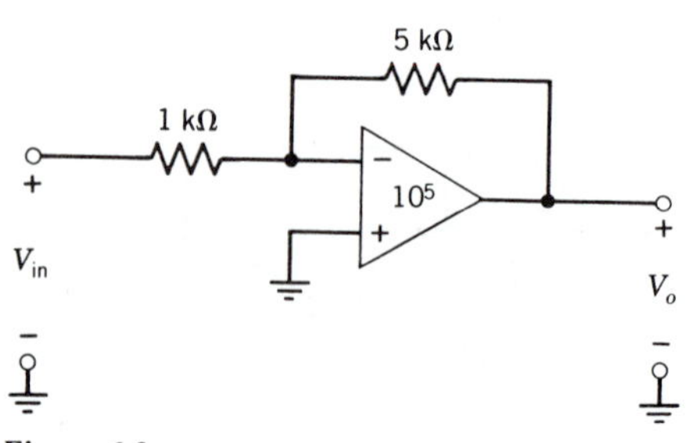

Figure 6.2

Other useful relationships are listed below:

$$S_x^F = -S_x^{1/F} \tag{6.16a}$$

$$S_x^{cF} = S_x^F \qquad (c \text{ is independent of } x) \tag{6.16b}$$

$$S_x^{F_1+F_2} = \frac{F_1 S_x^{F_1} + F_2 S_x^{F_2}}{F_1 + F_2} \tag{6.16c}$$

$$S_x^{F^n} = n S_x^F \tag{6.16d}$$

$$S_{x^n}^F = \frac{1}{n} S_x^F \tag{6.16e}$$

Example 6.1 Perform sensitivity analysis for the low-pass Sallen and Key network of Figure 5.34. From (5.157)–(5.159), we rewrite them as

$$K = kR_1^{-1}R_2^{-1}C_1^{-1}C_2^{-1} \tag{6.17a}$$

$$\omega_p = R_1^{-1/2}R_2^{-1/2}C_1^{-1/2}C_2^{-1/2} \tag{6.17b}$$

$$\frac{\omega_p}{Q_p} = B_p = \frac{1}{R_1 C_1} + \frac{1}{R_2 C_1} + \frac{1-k}{R_2 C_2} \tag{6.17c}$$

where $k = 1 + r_2/r_1$. For convenience, we write

$$S_{x_1, x_2, \ldots, x_q}^F = S_{x_1}^F = S_{x_2}^F = \cdots = S_{x_q}^F \tag{6.18}$$

Then from (6.9) we obtain

$$S_{R_1, R_2, C_1, C_2}^K = -1, \qquad S_k^K = 1 \tag{6.19a}$$

$$S_{R_1, R_2, C_1, C_2}^{\omega_p} = -\tfrac{1}{2}, \qquad S_k^{\omega_p} = S_{r_1}^{\omega_p} = S_{r_2}^{\omega_p} = 0 \tag{6.19b}$$

To compute the Q_p sensitivities, we appeal to (6.15), which states that

$$S_x^{Q_p} = S_x^{\omega_p} - S_x^{B_p} \tag{6.20}$$

where

$$S_x^{B_p} = \frac{x}{B_p} \cdot \frac{\partial B_p}{\partial x} = x \frac{Q_p}{\omega_p} \cdot \frac{\partial B_p}{\partial x} \tag{6.21}$$

We begin by computing Q_p sensitivity with respect to R_1. For this we differentiate (6.17c) with respect to R_1 to give

$$\frac{\partial B_p}{\partial R_1} = -\frac{1}{R_1^2 C_1} \tag{6.22}$$

Substituting this in (6.21) in conjunction with (6.17b) yields

$$S_{R_1}^{B_p} = R_1 Q_p (R_1 R_2 C_1 C_2)^{1/2} \left(\frac{-1}{R_1^2 C_1} \right) = -Q_p R_1^{-1/2} R_2^{1/2} C_1^{-1/2} C_2^{1/2} \tag{6.23}$$

giving from (6.20)

$$S_{R_1}^{Q_p} = -\frac{1}{2} + Q_p \sqrt{\frac{R_2 C_2}{R_1 C_1}} \tag{6.24}$$

Similarly, we obtain

$$S_{R_2}^{Q_p} = -\frac{1}{2} + Q_p \left[\sqrt{\frac{R_1 C_2}{R_2 C_1}} + (1-k) \sqrt{\frac{R_1 C_1}{R_2 C_2}} \right] \tag{6.25a}$$

$$S_{C_1}^{Q_p} = -\frac{1}{2} + Q_p \left(\sqrt{\frac{R_1 C_2}{R_2 C_1}} + \sqrt{\frac{R_2 C_2}{R_1 C_1}} \right) \tag{6.25b}$$

$$S_{C_2}^{Q_p} = -\frac{1}{2} + (1-k) Q_p \sqrt{\frac{R_1 C_1}{R_2 C_2}} \tag{6.25c}$$

$$S_k^{Q_p} = k Q_p \sqrt{\frac{R_1 C_1}{R_2 C_2}} \tag{6.25d}$$

$$S_{r_1}^{Q_p} = -S_{r_2}^{Q_p} = (1-k) Q_p \sqrt{\frac{R_1 C_1}{R_2 C_2}} \tag{6.25e}$$

$$S_{r_1}^{K} = -S_{r_2}^{K} = \frac{1-k}{k} \tag{6.25f}$$

All of these sensitivity functions have been derived for general values of the filter parameters. Once a specific circuit design is chosen, these parameters are known and the numerical values of various sensitivities can be ascertained. As a specific example, consider the normalized Butterworth transfer voltage-ratio function of (5.160),

$$H(s) = \frac{2}{s^2 + \sqrt{2}s + 1} \tag{6.26}$$

giving from (5.161)

$$K = 2, \qquad \omega_p = 1, \qquad Q_p = 0.707 \tag{6.27}$$

As shown in (5.85)–(5.87), this response can be realized by the low-pass Sallen and Key network of Figure 5.34 with

$$R_1 = R_2 = 1\ \Omega, \qquad C_1 = C_2 = 1\ \text{F}, \qquad r_1 = 10\ \Omega, \qquad r_2 = 5.86\ \Omega \tag{6.28}$$

Using these parameter values, we obtain

$$S_{R_1}^{Q_p} = -S_{R_2}^{Q_p} = 0.207, \qquad S_{C_1}^{Q_p} = -S_{C_2}^{Q_p} = 0.914, \qquad S_k^{Q_p} = 1.121 \tag{6.29}$$

$$S_{r_1}^{Q_p} = -S_{r_2}^{Q_p} = -0.414, \qquad S_{r_1}^{K} = -S_{r_2}^{K} = -0.369 \tag{6.30}$$

Others are given in (6.19).

6.2 MAGNITUDE AND PHASE SENSITIVITIES

In this section, we show how to compute the sensitivity functions for the magnitude and phase functions. To begin we express a transfer function in polar form and substitute s by $j\omega$ to give

$$H(j\omega)=|H(j\omega)|e^{j\phi(\omega)} \tag{6.31}$$

Then from (6.1) the sensitivity function becomes

$$S_x^{H(j\omega)}=\frac{x}{H(j\omega)}\cdot\frac{\partial}{\partial x}[|H(j\omega)|e^{j\phi(\omega)}] \tag{6.32}$$

which can be expanded by making use of the product rule for differentiation of a product to give

$$\begin{aligned}S_x^{H(j\omega)}&=\frac{x}{|H(j\omega)|}\cdot\frac{\partial|H(j\omega)|}{\partial x}+jx\frac{\partial\phi(\omega)}{\partial x}\\&=S_x^{|H(j\omega)|}+j\phi(\omega)S_x^{\phi(\omega)}\end{aligned} \tag{6.33}$$

or

$$S_x^{|H(j\omega)|}=\text{Re}\, S_x^{H(j\omega)} \tag{6.34}$$

$$S_x^{\phi(\omega)}=\frac{1}{\phi(\omega)}\,\text{Im}\, S_x^{H(j\omega)} \tag{6.35}$$

They state that the magnitude and phase sensitivities of a transfer function with respect to an element are simply related to the real and imaginary parts of the transfer function sensitivity with respect to the same element.

As an example, consider the low-pass transfer function of (5.162), repeated below as

$$H(s)=K\frac{\omega_p^2}{s^2+(\omega_p/Q_p)s+\omega_p^2} \tag{6.36}$$

Its sensitivity function is found from (6.1) to be

$$S_{Q_p}^{H(s)}=\frac{Q_p}{H(s)}\cdot\frac{\partial H(s)}{\partial Q_p}=\frac{(\omega_p/Q_p)s}{s^2+(\omega_p/Q_p)s+\omega_p^2} \tag{6.37}$$

To compute the sensitivity function of $|H(j\omega)|$ with respect to Q_p, we apply (6.34) by first substituting s by $j\omega$ in (6.37) and then take the real part. The result is given by

$$S_{Q_p}^{|H(j\omega)|}=\text{Re}\, S_{Q_p}^{H(j\omega)}=\frac{(\omega/\omega_p)^2}{Q_p^2[1-(\omega/\omega_p)^2]^2+(\omega/\omega_p)^2} \tag{6.38}$$

At $\omega=\omega_p$, (6.38) reduces to

$$S_{Q_p}^{|H(j\omega)|}(j\omega_p)=1 \tag{6.39}$$

We next compute the sensitivity function of $|H(j\omega)|$ due to the variations in the pole frequency ω_p. Applying (6.1) yields

$$S_{\omega_p}^{H(s)} = \frac{2s^2 + (\omega_p/Q_p)s}{s^2 + (\omega_p/Q_p)s + \omega_p^2} \tag{6.40}$$

giving

$$S_{\omega_p}^{|H(j\omega)|} = \text{Re } S_{\omega_p}^{H(j\omega)} = \frac{(\omega/\omega_p)^2[2(\omega/\omega_p)^2 + 1/Q_p^2 - 2]}{[1-(\omega/\omega_p)^2]^2 + (\omega/Q_p\omega_p)^2} \tag{6.41}$$

At $\omega = \omega_p$, (6.41) reduces to

$$S_{\omega_p}^{|H(j\omega)|}(j\omega_p) = 1 \tag{6.42}$$

As another illustration, consider the band-pass transfer function of (5.164) given by the equation

$$H(s) = K \frac{(\omega_p/Q_p)s}{s^2 + (\omega_p/Q_p)s + \omega_p^2} \tag{6.43}$$

Proceeding as before, we obtain the sensitivity function with respect to Q_p:

$$S_{Q_p}^{H(s)} = -\frac{s^2 + \omega_p^2}{s^2 + (\omega_p/Q_p)s + \omega_p^2} \tag{6.44}$$

Replacing s by $j\omega$ in (6.44) and then taking the real part yields

$$S_{Q_p}^{|H(j\omega)|} = \text{Re } S_{Q_p}^{H(j\omega)} = \frac{-[1-(\omega/\omega_p)^2]^2}{[1-(\omega/\omega_p)^2]^2 + (\omega/Q_p\omega_p)^2} \tag{6.45}$$

At $\omega = \omega_p$, (6.45) becomes

$$S_{Q_p}^{|H(j\omega)|}(j\omega_p) = 0 \tag{6.46}$$

For the sensitivity function of $|H(j\omega)|$ with respect to pole frequency ω_p, we first compute

$$S_{\omega_p}^{H(s)} = \frac{s^2 - \omega_p^2}{s^2 + (\omega_p/Q_p)s + \omega_p^2} \tag{6.47}$$

Substituting $s = j\omega$ and then taking the real part results in

$$S_{\omega_p}^{|H(j\omega)|} = \text{Re } S_{\omega_p}^{H(j\omega)} = \frac{(\omega/\omega_p)^4 - 1}{[1-(\omega/\omega_p)^2]^2 + (\omega/Q_p\omega_p)^2} \tag{6.48}$$

At $\omega = \omega_p$, the function becomes

$$S_{\omega_p}^{|H(j\omega)|}(j\omega_p) = 0 \tag{6.49}$$

This also follows from (6.47), where at $\omega = \omega_p$,

$$S_{\omega_p}^{H(j\omega)}(j\omega_p) = j2Q_p \tag{6.50}$$

Example 6.2 Compute the magnitude sensitivity functions of the normalized Butterworth response of (6.26) due to the variations of the pole frequency and the pole Q, where $\omega_p=1$ and $Q_p=0.707$ as given previously in (6.27).

Substituting $\omega_p=1$ and $Q_p=0.707$ in (6.38) and (6.41) results in

$$S_{Q_p}^{|H(j\omega)|}=\frac{2\omega^2}{1+\omega^4} \tag{6.51}$$

$$S_{\omega_p}^{|H(j\omega)|}=\frac{2\omega^4}{1+\omega^4} \tag{6.52}$$

6.3 THE MULTIPARAMETER SENSITIVITY

So far we have considered the situation where a network function is changed due to a change in a particular network element. In this section, we extend this concept by considering the change of a network function due to the simultaneous variation of many elements in the network.

Let H be the network function, and let x_j ($j=1, 2, \ldots, m$) be the network elements such as resistors, capacitors, inductors, or the parameters describing the active devices that are subject to change in values. Then the change ΔH in H due to the simultaneous variations of all the elements x_j may be obtained by the multivariable Taylor series expansion of H with respect to x_j, as follows:

$$\Delta H=\sum_{j=1}^{m}\frac{\partial H}{\partial x_j}\Delta x_j+\sum_{i=1}^{m}\sum_{j=1}^{m}\frac{\partial^2 H}{\partial x_i\,\partial x_j}\cdot\frac{\Delta x_i\,\Delta x_j}{2!}+\cdots \tag{6.53}$$

If the changes in the elements Δx_j are small, the second- and higher-order terms can be ignored, and the first-order approximation is given by

$$\Delta H\approx\sum_{j=1}^{m}\frac{\partial H}{\partial x_j}\Delta x_j \tag{6.54}$$

To bring the sensitivity functions into evidence, (6.54) is rewritten as

$$\Delta H\approx\sum_{j=1}^{m}\left(\frac{x_j}{H}\cdot\frac{\partial H}{\partial x_j}\right)\left(\frac{\Delta x_j}{x_j}H\right)=H\sum_{j=1}^{m}S_{x_j}^{H}V_{x_j} \tag{6.55}$$

where

$$V_{x_j}=\Delta x_j/x_j \tag{6.56}$$

denotes the fractional change in the element x_j, and is known as the *variability* of x_j. From (6.55) the fractional change in the network function can be approximated by the expression

$$\frac{\Delta H}{H}\approx\sum_{j=1}^{m}S_{x_j}^{H}V_{x_j} \tag{6.57}$$

The multiparameter sensitivity definition above is somewhat simplistic and does not take into account the random element variations. A more realistic and accurate measure is known as the *statistical multiparameter sensitivity*.†

Example 6.3 The transfer function of the active band-pass filter of Figure 6.3 is found to be

$$H(s)=\frac{V_o(s)}{V_{\text{in}}(s)}=\frac{(R_6/R_4R_5C_1)s}{s^2+(1/R_1C_1)s+R_6/R_2R_3R_5C_1C_2} \tag{6.58}$$

Find the worst-case percentage deviation in pole frequency ω_p and pole Q if all the resistances may vary by $\pm r\%$ and the capacitances by $\pm c\%$ from their nominal values.

From (6.58) in conjunction with (5.164), we can make the following identifications:

$$\omega_p=R_2^{-1/2}R_3^{-1/2}R_5^{-1/2}C_1^{-1/2}C_2^{-1/2}R_6^{1/2} \tag{6.59}$$

$$Q_p=R_1C_1R_6^{1/2}R_2^{-1/2}R_3^{-1/2}R_5^{-1/2}C_1^{-1/2}C_2^{-1/2} \tag{6.60}$$

giving from (6.9)–(6.12)

$$S^{\omega_p}_{R_2,R_3,R_5,C_1,C_2}=-\tfrac{1}{2}, \qquad S^{\omega_p}_{R_6}=\tfrac{1}{2}, \qquad S^{\omega_p}_{R_1,R_4}=0 \tag{6.61a}$$

$$S^{Q_p}_{R_2,R_3,R_5,C_2}=-\tfrac{1}{2}, \qquad S^{Q_p}_{R_1}=1, \qquad S^{Q_p}_{C_1,R_6}=\tfrac{1}{2}, \qquad S^{Q_p}_{R_4}=0 \tag{6.61b}$$

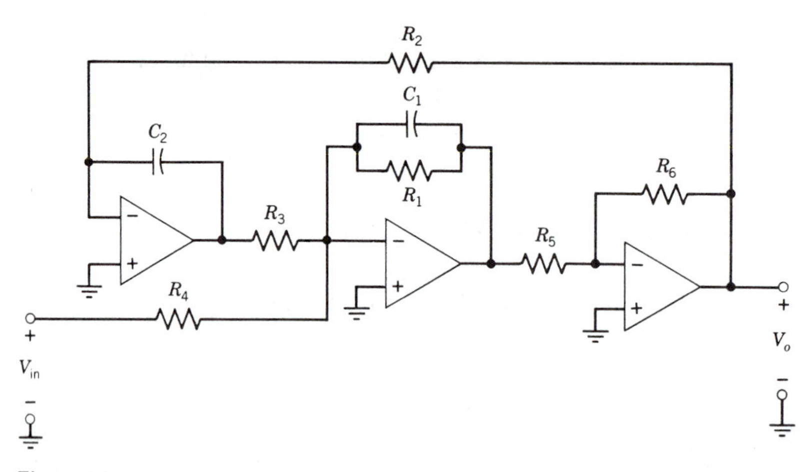

Figure 6.3

†A. L. Rosenblum and M. S. Ghausi, "Multiparameter sensitivity in active *RC* networks," *IEEE Trans. Circuit Theory*, vol. CT-18, 1971, pp. 592–599.

C. Acar and M. Ghausi, "Statistical multiparameter sensitivity measure of gain and phase functions," *Int. J. Circuit Theory Appl.*, vol. 5, 1977, pp. 13–22.

From (6.57), the fractional changes in ω_p and Q_p due to the variations in R's and C's are, respectively, given by the expressions

$$\begin{aligned}\Delta\omega_p/\omega_p &\approx S_{R_2}^{\omega_p}V_{R_2}+S_{R_3}^{\omega_p}V_{R_3}+S_{R_5}^{\omega_p}V_{R_5}+S_{R_6}^{\omega_p}V_{R_6}+S_{C_1}^{\omega_p}V_{C_1}+S_{C_2}^{\omega_p}V_{C_2}\\ &= -\tfrac{1}{2}(V_{R_2}+V_{R_3}+V_{R_5}+V_{C_1}+V_{C_2})+\tfrac{1}{2}V_{R_6}\\ &= \mp 1.5r\% \mp c\% \pm 0.5r\%\end{aligned} \tag{6.62a}$$

$$\begin{aligned}\Delta Q_p/Q_p &\approx S_{R_2}^{Q_p}V_{R_2}+S_{R_3}^{Q_p}V_{R_3}+S_{R_5}^{Q_p}V_{R_5}+S_{R_1}^{Q_p}V_{R_1}+S_{C_1}^{Q_p}V_{C_1}+S_{C_2}^{Q_p}V_{C_2}+S_{R_6}^{Q_p}V_{R_6}\\ &= -\tfrac{1}{2}(V_{R_2}+V_{R_3}+V_{R_5}+V_{C_2})+V_{R_1}+\tfrac{1}{2}(V_{C_1}+V_{R_6})\\ &= \mp 1.5r\% \mp 0.5c\% \pm r\% \pm 0.5c\% \pm 0.5r\%\end{aligned} \tag{6.62b}$$

The worst-case percentage deviations in ω_p and Q_p are given by the expressions

$$\Delta\omega_p/\omega_p = \pm(2r+c)\% \tag{6.63a}$$

$$\Delta Q_p/Q_p = \pm(3r+c)\% \tag{6.63b}$$

Thus, a 5% variation in values of all the elements results in at most a 15% change in values in ω_p and 20% in Q_p.

6.4 GAIN SENSITIVITY

In Section 6.2, we discussed the relationships between the magnitude and phase sensitivities and the transfer function sensitivity. However, in filter design, requirements are frequently stated in terms of the maximum allowable deviation in gain over specified bands of frequencies. In such situations, it is convenient to consider the logarithm of the transfer function of a network operating under the sinusoidal steady state. Thus, from (6.31) the magnitude of the transfer function in dB can be written as

$$\alpha(\omega) = 20 \log |H(j\omega)| \quad \text{dB} \tag{6.64}$$

which is known as the *gain function.*

Definition 6.2: ***Gain sensitivity.*** The *gain sensitivity* is defined as the ratio of the change in gain in dB to the fractional change in an element for the situation when all changes concerned are differentially small.

Thus, following (6.1), if α is the gain function, the gain sensitivity with respect to an element x, written as $\mathcal{S}_x^{\alpha}$, is defined by the equation

$$\mathcal{S}_x^{\alpha(\omega)} = \lim_{\Delta x\to 0} \frac{\Delta\alpha}{\Delta x/x} = x\frac{\partial\alpha(\omega)}{\partial x} \tag{6.65}$$

A comparison of (6.65) with (6.2) shows that the gain sensitivity is similar to the sensitivity function when ln H is under consideration. Thus, for a small change in x, the change in gain can be approximated by

$$\Delta\alpha(\omega) \approx \mathcal{S}_x^{\alpha(\omega)}(\Delta x/x) \tag{6.66}$$

Refer to the magnitude sensitivity in (6.33), which can be rewritten as

$$S_x^{|H(j\omega)|} = \frac{x}{|H(j\omega)|} \cdot \frac{\partial |H(j\omega)|}{\partial x} = x\, \frac{\partial[\alpha(\omega)/(20 \log e)]}{\partial x}$$

$$= \frac{1}{(20 \log e)} \mathscr{S}_x^{\alpha(\omega)} = 0.115 \mathscr{S}_x^{\alpha(\omega)} \tag{6.67}$$

showing that the magnitude sensitivity and gain sensitivity are related by the constant 0.115.

We next compute the gain sensitivity of the biquadratic function of (5.142) with respect to ω_p, ω_z, Q_p, and Q_z. Appealing to formula (6.54), the change in gain in dB due to the changes in ω_p, ω_z, Q_p, and Q_z can be approximated by the equation

$$\Delta\alpha \approx \frac{\partial\alpha}{\partial\omega_p}\Delta\omega_p + \frac{\partial\alpha}{\partial Q_p}\Delta Q_p + \frac{\partial\alpha}{\partial\omega_z}\Delta\omega_z + \frac{\partial\alpha}{\partial Q_z}\Delta Q_z$$

$$= \mathscr{S}_{\omega_p}^{\alpha}\frac{\Delta\omega_p}{\omega_p} + \mathscr{S}_{Q_p}^{\alpha}\frac{\Delta Q_p}{Q_p} + \mathscr{S}_{\omega_z}^{\alpha}\frac{\Delta\omega_z}{\omega_z} + \mathscr{S}_{Q_z}^{\alpha}\frac{\Delta Q_z}{Q_z} \tag{6.68}$$

If the biquadratic parameters are subject to changes of the network elements, from (6.57) we obtain

$$\Delta\alpha \approx \sum_{j=1}^{m} \left(\mathscr{S}_{\omega_p}^{\alpha} S_{x_j}^{\omega_p} V_{x_j} + \mathscr{S}_{Q_p}^{\alpha} S_{x_j}^{Q_p} V_{x_j} + \mathscr{S}_{\omega_z}^{\alpha} S_{x_j}^{\omega_z} V_{x_j} + \mathscr{S}_{Q_z}^{\alpha} S_{x_j}^{Q_z} V_{x_j}\right) \tag{6.69}$$

This equation gives the change in the gain in dB due to the simultaneous variations of the network element values realizing the biquadratic function of (5.142). To apply (6.69) we must compute the gain sensitivities with respect to the biquadratic parameters.

Refer to (5.145). The gain sensitivity with respect to ω_z can be evaluated from its definition, as follows:

$$\mathscr{S}_{\omega_z}^{\alpha(\omega)} = \omega_z \frac{\partial\alpha(\omega)}{\partial\omega_z} = \frac{\partial\alpha(\omega)}{\partial \ln \omega_z} = \frac{10\omega_z(\log e)[4\omega_z(\omega_z^2 - \omega^2) + 2\omega_z\omega^2/Q_z^2]}{(\omega_z^2 - \omega^2)^2 + (\omega_z\omega/Q_z)^2}$$

$$= 8.686\, \frac{2[1 - (\omega/\omega_z)^2] + (\omega/\omega_z Q_z)^2}{[1 - (\omega/\omega_z)^2]^2 + (\omega/\omega_z Q_z)^2} \tag{6.70}$$

Likewise, we can show that

$$\mathscr{S}_{Q_z}^{\alpha(\omega)} = -8.686\, \frac{(\omega/\omega_z Q_z)^2}{[1 - (\omega/\omega_z)^2]^2 + (\omega/\omega_z Q_z)^2} \tag{6.71}$$

$$\mathscr{S}_{\omega_p}^{\alpha(\omega)} = -8.686\, \frac{2[1 - (\omega/\omega_p)^2] + (\omega/\omega_p Q_p)^2}{[1 - (\omega/\omega_p)^2]^2 + (\omega/\omega_p Q_p)^2} \tag{6.72}$$

$$\mathscr{S}_{Q_p}^{\alpha(\omega)} = 8.686\, \frac{(\omega/\omega_p Q_p)^2}{[1 - (\omega/\omega_p)^2]^2 + (\omega/\omega_p Q_p)^2} \tag{6.73}$$

At the zero and pole frequencies, (6.70)–(6.73) reduce to

$$\mathscr{S}^{\alpha}_{\omega_z}(\omega_z) = -\mathscr{S}^{\alpha}_{\omega_p}(\omega_p) = 8.686 \text{ dB} \tag{6.74a}$$

$$\mathscr{S}^{\alpha}_{Q_z}(\omega_z) = -\mathscr{S}^{\alpha}_{Q_p}(\omega_p) = -8.686 \text{ dB} \tag{6.74b}$$

At the 3-dB band-edge frequencies

$$\omega_{zi} = \omega_z(1 \mp 1/2Q_z), \qquad \omega_{pi} = \omega_p(1 \mp 1/2Q_p), \qquad i = 1, 2 \tag{6.75}$$

and for $Q_z \gg 1$ and $Q_p \gg 1$, these sensitivities can be approximated by the following expressions in dB

$$\mathscr{S}^{\alpha}_{\omega_z}(\omega_{zi}) \approx \pm 8.686 Q_z, \qquad \mathscr{S}^{\alpha}_{\omega_p}(\omega_{pi}) \approx \mp 8.686 Q_p \tag{6.76a}$$

$$\mathscr{S}^{\alpha}_{Q_z}(\omega_{zi}) \approx -4.343 \text{ dB}, \qquad \mathscr{S}^{\alpha}_{Q_p}(\omega_{pi}) \approx 4.343 \text{ dB} \tag{6.76b}$$

To see the relative magnitude of the biquadratic gain sensitivities, (6.72) and (6.73) are plotted as a function of the normalized frequency ω/ω_p for various values of pole Q. The results are shown in Figure 6.4. As can be seen from these plots, in the neighborhood of the pole frequency, the dominant sensitivity term is $\mathscr{S}^{\alpha}_{\omega_p}$. This sensitivity increases with the pole Q and achieves the maximum value of approximately $\pm 8.686 Q_p$ at the 3-dB band-edge frequencies. Similar conclusion can be reached for the zero frequency and zero Q. We illustrate the concept of gain sensitivity by the following example.

Example 6.4 Find the approximate change in the gain in dB at the 3-dB band-edge frequency of the band-pass transfer function

$$H(s) = \frac{800s}{s^2 + 5000s + 10^{10}} \tag{6.77}$$

for a 3% increase in the pole Q and a 5% increase in the pole frequency.

From the denominator of (6.77), the pole frequency and pole Q are found to be

$$\omega_p = 10^5 \text{ rad/s}, \qquad Q_p = 20 \tag{6.78}$$

The 3-dB band-edge frequencies are given by

$$\omega_{p1}, \omega_{p2} = \omega_p(1 \mp 1/2Q_p) = 0.975 \times 10^5 \text{ rad/s}, 1.025 \times 10^5 \text{ rad/s} \tag{6.79}$$

Substituting these in (6.72) and (6.73) gives

$$\mathscr{S}^{\alpha}_{\omega_p}(\omega_{p1}) = -182.45 \text{ dB}, \qquad \mathscr{S}^{\alpha}_{\omega_p}(\omega_{p2}) = 165 \text{ dB} \tag{6.80}$$

$$\mathscr{S}^{\alpha}_{Q_p}(\omega_{p1}) = 4.29 \text{ dB}, \qquad \mathscr{S}^{\alpha}_{Q_p}(\omega_{p2}) = 4.4 \text{ dB} \tag{6.81}$$

These are to be compared with the approximation solutions (6.76) given as

$$\mathscr{S}^{\alpha}_{\omega_p}(\omega_{pi}) \approx \mp 8.686 Q_p = \mp 173.72 \text{ dB}, \qquad i = 1, 2 \tag{6.82a}$$

$$\mathscr{S}^{\alpha}_{Q_p}(\omega_{pi}) \approx 4.343 \text{ dB} \qquad i = 1, 2 \tag{6.82b}$$

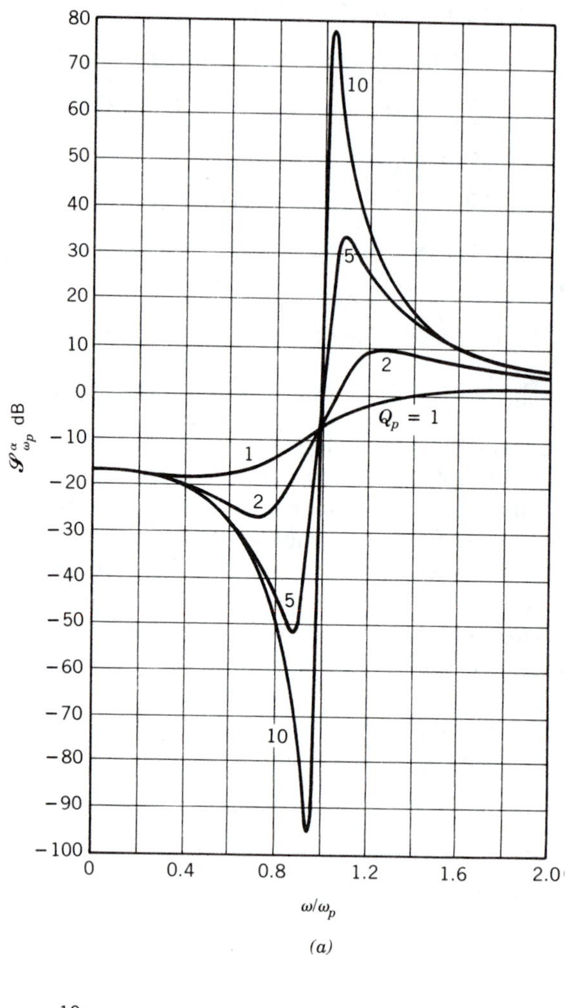
80
70
60
50
40
30
20
10
0
−10
−20
−30
−40
−50
−60
−70
−80
−90
−100
$\mathscr{S}^{\alpha}_{\omega_p}$ dB
10
5
2
$Q_p = 1$
1
2
5
10
0
0.4
0.8
1.2
1.6
2.0
ω/ω_p

(a)

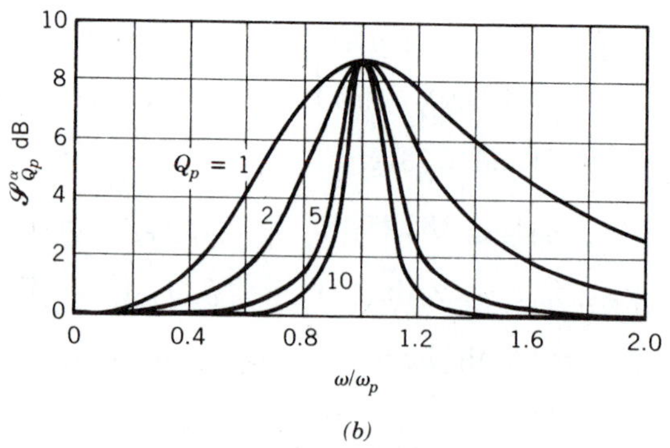
10
8
6
4
2
0
$\mathscr{S}^{\alpha}_{Q_p}$ dB
$Q_p = 1$
2
5
10
0
0.4
0.8
1.2
1.6
2.0
ω/ω_p

(b)

Figure 6.4

The gain deviations at these frequencies are from (6.68)

$$\Delta\alpha(\omega_{pi}) = \mathcal{S}^{\alpha}_{\omega_p}(\omega_{pi}) \frac{\Delta\omega_p}{\omega_p} + \mathcal{S}^{\alpha}_{Q_p}(\omega_{pi}) \frac{\Delta Q_p}{Q_p} = -8.99 \text{ dB}, 8.4 \text{ dB} \tag{6.83}$$

for $i=1, 2$, where $\Delta Q_p/Q_p = 0.03$ and $\Delta\omega_p/\omega_p = 0.05$.

6.5 ROOT SENSITIVITY

The location of the poles of an active filter determines the stability of the network. Therefore, it is important to know the manner in which these poles vary as some of the network elements change. Since poles of a network function are roots of a polynomial, it suffices to examine how the roots of a polynomial change as the value of an element changes.

Let $P(s)$ be a polynomial of interest, and let s_j be a root of $P(s)$ with multiplicity n. Then $P(s)$ can be written as

$$P(s) = (s - s_j)^n P_1(s) \tag{6.84}$$

where $P_1(s)$ represents the polynomial formed by the remaining factors.

Definition 6.3: ***Root sensitivity.*** The *root sensitivity* is defined as the ratio of the change in a root to the fractional change in an element for the situation when all changes concerned are differentially small.

Thus, if s_j is a root of the polynomial $P(s)$, the root sensitivity of s_j with respect to an element x is defined by the equation

$$\hat{S}^{s_j}_x \triangleq \frac{\partial s_j}{\partial x/x} = x\frac{\partial s_j}{\partial x} \tag{6.85}$$

Conceptually, this definition is easy to apply; it involves only a partial differentiation. In reality, to compute the partial differentiation we must first find the functional relationship between the root s_j and the parameter x. In a large network, this is a difficult task. In addition, if s_j is a root of high multiplicity, the differentiation in (6.85) may encounter some computational difficulties. Even if a digital computer is employed to do the computation, the derivative still presents some numerical problems. To circumvent this difficulty, in the later part of this section we demonstrate a procedure that avoids the use of the derivative.

When $P(s)$ is the numerator polynomial of a network function, (6.85) defines the *zero sensitivity*. Likewise, if $P(s)$ is the denominator polynomial, (6.85) is referred to as the *pole sensitivity*. The zero and pole sensitivities of a network function may be used to relate to the function sensitivity as discussed in Section 6.1. For this let $F(s)$ be the network function of interest, and express it explicitly as the ratio of two polynomials $P(s)$ and $Q(s)$:

$$F(s) = K\frac{P(s)}{Q(s)} = K\frac{\prod_{i=1}^{m}(s - z_i)^{m_i}}{\prod_{j=1}^{k}(s - p_j)^{k_j}} \tag{6.86}$$

In terms of the logarithms, (6.86) becomes

$$\ln F(s)=\ln K+\sum_{i=1}^{m} m_i \ln(s-z_i)-\sum_{j=1}^{k} k_j \ln(s-p_j) \tag{6.87}$$

Taking partial derivatives with respect to x on both sides yields

$$\frac{\partial \ln F(s)}{\partial x}=\frac{1}{K}\cdot\frac{\partial K}{\partial x}-\sum_{i=1}^{m}\frac{m_i\dfrac{\partial z_i}{\partial x}}{s-z_i}+\sum_{j=1}^{k}\frac{k_j\dfrac{\partial p_j}{\partial x}}{s-p_j} \tag{6.88}$$

which can be put in the form

$$\frac{x}{F(s)}\cdot\frac{\partial F(s)}{\partial x}=\frac{x}{K}\cdot\frac{\partial K}{\partial x}-\sum_{i=1}^{m}\frac{m_i x\dfrac{\partial z_i}{\partial x}}{s-z_i}+\sum_{j=1}^{k}\frac{k_j x\dfrac{\partial p_j}{\partial x}}{s-p_j} \tag{6.89}$$

This gives a relationship between the function sensitivity and the zero and pole sensitivities:

$$S_x^F=S_x^K-\sum_{i=1}^{m}\frac{m_i\hat{S}_x^{z_i}}{s-z_i}+\sum_{j=1}^{k}\frac{k_j\hat{S}_x^{p_j}}{s-p_j} \tag{6.90}$$

Observe that apart from a constant determined by S_x^K, the zero and pole sensitivities determine the function sensitivity. For simple zeros z_i and poles p_j, $\hat{S}_x^{z_i}$ and $\hat{S}_x^{p_j}$ are the residues of the function sensitivity S_x^F evaluated at its poles z_i and p_j. Thus, the contribution of $\hat{S}_x^{z_i}$ and $\hat{S}_x^{p_j}$ to the overall network function sensitivity is most significant in the neighborhood of z_i or p_j.

Example 6.5 Consider the impedance function of the *RLC* network of Figure 6.5, which is given by

$$Z(s)=\frac{LCs^2+RCs+1}{Cs} \tag{6.91}$$

The impedance has a pole at $s=0$ and a pair of zeros at

$$z_1, z_2=\frac{1}{2L}\left(-R\pm\sqrt{R^2-\frac{4L}{C}}\right) \tag{6.92}$$

From (6.85) the zero and pole sensitivities are found to be

$$\hat{S}_R^{z_1}, \hat{S}_R^{z_2}=R\frac{\partial z_i}{\partial R}=-\frac{R}{2L}\pm\frac{R^2}{2L\sqrt{R^2-4L/C}} \tag{6.93a}$$

$$\hat{S}_L^{z_1}, \hat{S}_L^{z_2}=L\frac{\partial z_i}{\partial L}=\frac{R}{2L}\mp\frac{R^2-2L/C}{2L\sqrt{R^2-4L/C}} \tag{6.93b}$$

$$\hat{S}_C^{z_1}, \hat{S}_C^{z_2}=C\frac{\partial z_i}{\partial C}=\pm\frac{1}{C\sqrt{R^2-4L/C}} \tag{6.93c}$$

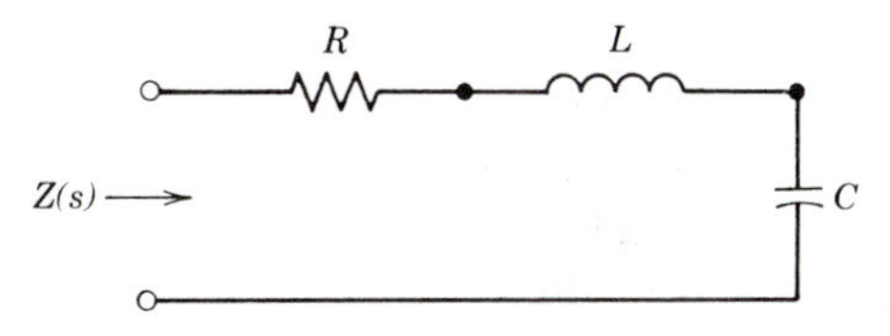

Figure 6.5

$$\hat{S}_R^{p_1} = \hat{S}_L^{p_1} = \hat{S}_C^{p_1} = 0 \tag{6.93d}$$

where $p_1 = 0$ is the pole of $Z(s)$. Comparing (6.91) with (6.86), we can identify $K = L$. This gives from (6.1) the sensitivities

$$S_L^K = 1, \qquad S_R^K = S_C^K = 0 \tag{6.94}$$

To compute the function sensitivities, we use (6.90) and obtain

$$\begin{aligned} S_R^Z &= S_R^K - \frac{\hat{S}_R^{z_1}}{s - z_1} - \frac{\hat{S}_R^{z_2}}{s - z_2} + \frac{\hat{S}_R^{p_1}}{s - p_1} \\ &= 0 - \frac{-R/2L + R^2/(2L\sqrt{R^2 - 4L/C})}{s + R/2L - (1/2L)\sqrt{R^2 - 4L/C}} - \frac{-R/2L - R^2/(2L\sqrt{R^2 - 4L/C})}{s + R/2L + (1/2L)\sqrt{R^2 - 4L/C}} + 0 \\ &= \frac{RCs}{LCs^2 + RCs + 1} \end{aligned} \tag{6.95a}$$

Likewise, the sensitivities of Z with respect to L and C are found to be

$$\begin{aligned} S_L^Z &= S_L^K - \frac{\hat{S}_L^{z_1}}{s - z_1} - \frac{\hat{S}_L^{z_2}}{s - z_2} + \frac{\hat{S}_L^{p_1}}{s - p_1} \\ &= \frac{LCs^2}{LCs^2 + RCs + 1} \end{aligned} \tag{6.95b}$$

$$\begin{aligned} S_C^Z &= S_C^K - \frac{\hat{S}_C^{z_1}}{s - z_1} - \frac{\hat{S}_C^{z_2}}{s - z_2} + \frac{\hat{S}_C^{p_1}}{s - p_1} \\ &= -\frac{1}{LCs^2 + RCs + 1} \end{aligned} \tag{6.95c}$$

6.5.1 **Root Sensitivity Computations**

The root sensitivity is defined as the ratio of the change in a root to the fractional change in an element for the situation when all changes concerned are differentially small. For a small variation in an element value x, the change in a root s_j can be approximated from (6.85) as

$$\Delta s_j \approx \hat{S}_x^{s_j}(\Delta x / x) \tag{6.96}$$

In the present section, we show how to compute the root change Δs_j without actually computing the root sensitivity $\hat{S}_x^{s_j}$, thus avoiding the necessity of first obtaining the functional relationship between the root s_j and the parameter x and the resulting use of the derivative.

Let s_j be a root of a polynomial $P(s)$ with multiplicity n. Write $P(s)$ as in (6.84). Suppose that the element value of interest is changed from x to $x+\Delta x$, the polynomial will change from $P(s)$ to $\hat{P}(s)$ or

$$\hat{P}(s)=P(s)+\Delta P(s) \tag{6.97}$$

where $\Delta P(s)$ denotes the change of $P(s)$ due to the change of x. As a result, the root will change from s_j to $\hat{s}_j$. This requires that

$$P(s_j)=0 \tag{6.98}$$

$$\hat{P}(\hat{s}_j)=0 \tag{6.99}$$

Denote the root displacement of s_j by

$$\Delta s_j=\hat{s}_j-s_j \tag{6.100}$$

To find Δs_j, we can first compute $\hat{s}_j$ by solving (6.99) and then apply (6.100). However, solving (6.99) is generally difficult. In the following, we introduce a technique to compute Δs_j approximately.

From (6.84) the Laurent series expansion of $1/P(s)$ about the point s_j can be expressed as

$$\frac{1}{P(s)}=\frac{K_{jn}}{(s-s_j)^n}+\frac{K_{j(n-1)}}{(s-s_j)^{n-1}}+\cdots+\frac{K_{j1}}{s-s_j}+L_0+L_1(s-s_j)+L_2(s-s_j)^2+\cdots \tag{6.101}$$

In the neighborhood of s_j, the function is approximated by

$$\frac{1}{P(s)}\approx\frac{K_{jn}}{(s-s_j)^n} \tag{6.102}$$

where

$$K_{jn}=\left.\frac{(s-s_j)^n}{P(s)}\right|_{s=s_j} \tag{6.103}$$

is called the *Laurent constant* of $1/P(s)$ at s_j. When s_j is a simple root of $P(s)$, $n=1$ and the constant K_{jn} is also known as the *residue* of $1/P(s)$ at the pole s_j. To proceed, we substitute $s=\hat{s}_j$ in (6.102) and obtain

$$\frac{1}{P(\hat{s}_j)}\approx\frac{K_{jn}}{(\hat{s}_j-s_j)^n}=\frac{K_{jn}}{(\Delta s_j)^n} \tag{6.104}$$

or

$$\Delta s_j=[K_{jn}P(\hat{s}_j)]^{1/n} \tag{6.105}$$

Combining (6.97) and (6.99) yields

$$\hat{P}(\hat{s}_j)=P(\hat{s}_j)+\Delta P(\hat{s}_j)=0 \tag{6.106}$$

or

$$P(\hat{s}_j) = -\Delta P(\hat{s}_j) \tag{6.107}$$

We next consider the Taylor series expansion of $\Delta P(s)$ about s_j, which gives

$$\Delta P(s) = \Delta P(s_j) + \left.\frac{\partial \Delta P(s)}{\partial s}\right|_{s=s_j} (s - s_j) + \cdots \tag{6.108}$$

At $s = \hat{s}_j$, (6.108) reduces to

$$\Delta P(\hat{s}_j) = \Delta P(s_j) + \left.\frac{\partial \Delta P(s)}{\partial s}\right|_{s=s_j} \Delta s_j + \cdots \approx \Delta P(s_j) \tag{6.109}$$

if Δs_j is indeed small. This means that we essentially ignore all second- or higher-order variations of the parameter s_j. Finally, substituting (6.107) in (6.105) in conjunction with (6.109) gives

$$\Delta s_j = \hat{s}_j - s_j \approx [-K_{jn} \Delta P(s_j)]^{1/n} \tag{6.110}$$

Equation (6.110) shows that as a result of small parameter change, each root of $P(s)$ of multiplicity n becomes n roots located equidistant and equiangle from each other on a circle of radius $|\Delta s_j|$, center at the point s_j. For $n = 1$, (6.110) gives the direction and distance of root changes. We illustrate the above procedure by the following examples.

Example 6.6 Consider again the impedance function of Figure 6.5 as given by

$$Z(s) = \frac{LCs^2 + RCs + 1}{Cs} \tag{6.111}$$

with $R = 1\ \Omega$, $L = 1$ H, and $C = 1$ F. The zeros of the function are

$$z_1, z_2 = -\frac{R}{2L} \pm \frac{1}{2L}\sqrt{R^2 - 4L/C} = -\frac{1}{2} \pm j\frac{\sqrt{3}}{2} \tag{6.112}$$

Assume that the element values R, L, and C are changed by ΔR, ΔL, and ΔC, respectively. The resultant polynomial is found to be

$$\begin{aligned} \hat{P}(s) &= (L + \Delta L)(C + \Delta C)s^2 + (R + \Delta R)(C + \Delta C)s + 1 \\ &\approx LCs^2 + (C\Delta L + L\Delta C)s^2 + (RC + C\Delta R + R\Delta C)s + 1 \end{aligned} \tag{6.113}$$

where the second-order variational terms have been ingored, and

$$P(s) = LCs^2 + RCs + 1 \tag{6.114}$$

giving the incremental change in $P(s)$ as

$$\begin{aligned} \Delta P(s) = \hat{P}(s) - P(s) &= (C\Delta L + L\Delta C)s^2 + (C\Delta R + R\Delta C)s \\ &= (\Delta L + \Delta C)s^2 + (\Delta R + \Delta C)s \end{aligned} \tag{6.115}$$

Evaluating this at the nominal zero locations leads to

$$\Delta P(z_1)=(\Delta L+\Delta C)\left(-\tfrac{1}{2}-j\frac{\sqrt{3}}{2}\right)+(\Delta R+\Delta C)\left(-\tfrac{1}{2}+j\frac{\sqrt{3}}{2}\right) \tag{6.116a}$$

$$\Delta P(z_2)=(\Delta L+\Delta C)\left(-\tfrac{1}{2}+j\frac{\sqrt{3}}{2}\right)+(\Delta R+\Delta C)\left(-\tfrac{1}{2}-j\frac{\sqrt{3}}{2}\right) \tag{6.116b}$$

To find the possible zero locations, we need to compute the Laurent constants of $1/P(s)$ at the zeros z_1 and z_2. With $n=1$, (6.103) yields

$$K_{11}=\left.\frac{s-z_1}{P(s)}\right|_{s=z_1}=\frac{1}{z_1-z_2}=-j\frac{1}{\sqrt{3}} \tag{6.117a}$$

$$K_{21}=\left.\frac{s-z_2}{P(s)}\right|_{s=z_2}=\frac{1}{z_2-z_1}=j\frac{1}{\sqrt{3}} \tag{6.117b}$$

Finally, substituting (6.116) and (6.117) in (6.110) gives the displacements of the zeros as

$$\begin{aligned}\Delta z_1&=\hat{z}_1-z_1\approx -K_{11}\Delta P(z_1)\\&=\frac{j}{2\sqrt{3}}[(-1+j\sqrt{3})\Delta R-(1+j\sqrt{3})\Delta L-2\Delta C]\end{aligned} \tag{6.118a}$$

$$\begin{aligned}\Delta z_2&=\hat{z}_2-z_2\approx -K_{21}\Delta P(z_2)\\&=-\frac{j}{2\sqrt{3}}[(-1-j\sqrt{3})\Delta R-(1-j\sqrt{3})\Delta L-2\Delta C]\end{aligned} \tag{6.118b}$$

where $\hat{z}_1$ and $\hat{z}_2$ are the zero locations after element variations.

As a specific example, assume that all element values are increased by 10% or $\Delta L=0.1$, $\Delta R=0.1$, and $\Delta C=0.1$. Using these values in (6.118) gives

$$\Delta z_1\approx -j(0.2/\sqrt{3})=-j0.115 \tag{6.119a}$$

$$\Delta z_2\approx j(0.2/\sqrt{3})=j0.115 \tag{6.119b}$$

For illustrative purposes, we compute the actual zero locations after variations by finding the roots of the polynomial (6.113) given by the equation

$$\hat{P}(s)=1.21s^2+1.21s+1 \tag{6.120}$$

They are found to be

$$\hat{z}_1,\hat{z}_2=-0.5\pm j0.759 \tag{6.121}$$

The actual root increments are obtained as

$$\Delta z_1=\hat{z}_1-z_1=-j0.107, \qquad \Delta z_2=\hat{z}_2-z_2=j0.107 \tag{6.122}$$

Comparing these with (6.119) indicates an error of about 7.4% of actual root displacements by using the approximation formula (6.110).

Now suppose that the element values R and L are both increased by 10%, whereas C is decreased by 10%. This is equivalent to setting $\Delta R=0.1$, $\Delta L=0.1$, and $\Delta C=-0.1$. Substituting these in (6.118) gives

$$\Delta z_1=\Delta z_2 \approx 0 \tag{6.123}$$

The actual zero locations are determined from the polynomial

$$\hat{P}(s)=0.99s^2+0.99s+1 \tag{6.124}$$

yielding

$$\hat{z}_1, \hat{z}_2=-0.5\pm j0.872 \tag{6.125}$$

$$\Delta z_1, \Delta z_2=\pm j0.006 \tag{6.126}$$

These results are to be compared with the roots obtained by approximation, which are found from (6.110) as

$$\hat{z}_1, \hat{z}_2=z_i+\Delta z_i=z_i=-0.5\pm j0.866 \qquad (i=1, 2) \tag{6.127}$$

We see an excellent agreement between the approximation and the exact solutions. This agreement is valid in general if the network is subject to small variations of the element values so that all the second- or higher-order variations can be ignored. For large variations, additional terms in the series expansions must be included.

Example 6.7 Consider the low-pass Sallen and Key filter of Figure 5.34, the transfer voltage-ratio function of which is given in (5.73). For our purposes, assume that the nominal values of the elements are given by

$$R_1=R_2=1\,\Omega, \qquad C_1=C_2=1\text{ F} \tag{6.128}$$

Substituting these in (5.73), the resulting transfer function is

$$H(s)=\frac{k}{s^2+(3-k)s+1} \tag{6.129}$$

where $k=1+r_2/r_1$. The poles of the transfer function are the roots of the polynomial

$$P(s)=s^2+(3-k)s+1 \tag{6.130}$$

and are located at

$$p_1, p_2=-\tfrac{1}{2}(3-k)\pm\tfrac{1}{2}\sqrt{k^2-6k+5} \tag{6.131}$$

From (6.85), the pole sensitivities with respect to k are found to be

$$\hat{S}_k^{p_i}=k\frac{\partial p_i}{\partial k}=\frac{k}{2}\pm\frac{k(k-3)}{2\sqrt{k^2-6k+5}} \qquad (i=1, 2) \tag{6.132}$$

Assume that the nominal value for k is 3. The nominal pole locations and pole sensitivities are from (6.131) and (6.132)

$$p_1, p_2=\pm j1 \tag{6.133}$$

$$\hat{S}_k^{p_1}, \hat{S}_k^{p_2}=1.5 \tag{6.134}$$

From (6.96) we obtain the pole displacements

$$\Delta p_i \approx \hat{S}_k^{p_i} \frac{\Delta k}{k} = 1.5 \frac{\Delta k}{k} \qquad (i=1, 2) \tag{6.135}$$

This means that if k is increased by any amount, say, by 1%, from 3 to 3.03, the new pole locations will be at

$$\hat{p}_i = p_i + \Delta p_i = 0.015 \pm j1 \qquad (i=1, 2) \tag{6.136}$$

which are in the right-half of the s-plane. As a result of this small change in k, the filter becomes unstable.

Alternately, we can use (6.110) to estimate the pole movement without first determining the functional relationship between poles and k as in (6.131). Using (6.103) the Laurent constants at the poles p_1 and p_2 are obtained as

$$K_{11} = \left.\frac{s-p_1}{P(s)}\right|_{s=p_1} = \frac{1}{p_1 - p_2} = \frac{1}{\sqrt{k^2 - 6k + 5}} = -j\tfrac{1}{2} \tag{6.137a}$$

$$K_{21} = \left.\frac{s-p_2}{P(s)}\right|_{s=p_2} = \frac{1}{p_2 - p_1} = -\frac{1}{\sqrt{k^2 - 6k + 5}} = j\tfrac{1}{2} \tag{6.137b}$$

where $k=3$. When k is increased from its nominal value to $k+\Delta k$, the resultant polynominal becomes

$$\hat{P}(s) = s^2 + (3 - k - \Delta k)s + 1 = s^2 - 0.03s + 1 \tag{6.138}$$

where $\Delta k = 0.03$, giving

$$\Delta P(s) = \hat{P}(s) - P(s) = -\Delta k s \tag{6.139}$$

Evaluating this at the poles p_i, we obtain

$$\begin{aligned} \Delta P(p_i) &= -\Delta k p_i = -\tfrac{1}{2}\Delta k(k - 3 \pm \sqrt{k^2 - 6k + 5}) \\ &= \mp j0.03 \qquad (i=1, 2) \end{aligned} \tag{6.140}$$

Finally, substituting (6.137) and (6.140) in (6.110) yields the pole displacements as

$$\Delta p_i \approx -K_{i1} \Delta P(p_i) = \pm j\tfrac{1}{2}(\mp j0.03) = 0.015 \qquad (i=1, 2) \tag{6.141}$$

The new poles will be located approximately at

$$\hat{p}_i \approx p_i + \Delta p_i = 0.015 \pm j1 \qquad (i=1, 2) \tag{6.142}$$

This is to be compared with the exact pole locations determined by the polynomial (6.138):

$$\hat{p}_1, \hat{p}_2 = 0.015 \pm j0.9999 \tag{6.143}$$

We see an excellent agreement between the exact solutions and the approximation.

6.5.2 Multiparameter Root Sensitivity

In the foregoing, we have considered the situation where the roots of a polynomial are changed due to a change in a particular element. In practice, a network contains many nonideal elements each of which is subject to variation. In this section, we extend the preceding concept by introducing the notion of multiparameter root sensitivity.

Let s_i be the root of a polynomial $P(s)$ and let $x_1, x_2, \ldots, x_m$ be the elements of interest. Because s_i depends not only on s but also on the values of $x_1, x_2, \ldots, x_m$, we denote this relationship by writing s_i as a function of $x_1, x_2, \ldots, x_m$:

$$s_i = s_i(\mathbf{x}) = s_i(x_1, x_2, \ldots, x_m) \tag{6.144}$$

where the row vector $\mathbf{x}$ is defined by

$$\mathbf{x} = [x_1, x_2, \ldots, x_m] \tag{6.145}$$

Suppose that the parameter vector $\mathbf{x}$ changes from its nominal value to $\mathbf{x} + \Delta\mathbf{x}$. Then the new root location is at

$$\hat{s}_i = s_i(\mathbf{x} + \Delta\mathbf{x}) \tag{6.146}$$

The multiparameter Taylor series expansion about the point $\mathbf{x}$ is given by

$$\hat{s}_i = s_i(\hat{\mathbf{x}}) = s_i(\mathbf{x} + \Delta\mathbf{x}) = s_i(\mathbf{x}) + \sum_{j=1}^{m} \frac{\partial s_i}{\partial x_j} \Delta x_j + \sum_{u=1}^{m} \sum_{v=1}^{m} \frac{\partial^2 s_i}{\partial x_u \, \partial x_v} \cdot \frac{\Delta x_u \, \Delta x_v}{2!} + \cdots \tag{6.147}$$

where $\hat{\mathbf{x}} = \mathbf{x} + \Delta\mathbf{x}$ and

$$\Delta\mathbf{x} = [\Delta x_1, \Delta x_2, \ldots, \Delta x_m] \tag{6.148}$$

If the changes in the elements of $\mathbf{x}$ are small, the second- and higher-order terms in (6.147) can be ignored, and the first-order approximation is

$$\hat{s}_i = s_i(\hat{\mathbf{x}}) \approx s_i(\mathbf{x}) + \sum_{j=1}^{m} \frac{\partial s_i}{\partial x_j} \Delta x_j \tag{6.149}$$

To bring the sensitivity functions into evidence, (6.149) is rewritten as

$$\begin{aligned} \Delta s_i &= \hat{s}_i - s_i(\mathbf{x}) \\ &= \sum_{j=1}^{m} \frac{\partial s_i}{\partial x_j} \Delta x_j = \sum_{j=1}^{m} \left(x_j \frac{\partial s_i}{\partial x_j} \right) \frac{\Delta x_j}{x_j} \end{aligned} \tag{6.150}$$

or

$$\Delta s_i = \sum_{j=1}^{m} \hat{S}_{x_j}^{s_i} V_{x_j} \tag{6.151}$$

where

$$V_{x_j} = \Delta x_j / x_j \tag{6.152}$$

denotes the fractional change in the element x_j and is called the variability of x_j as previously given (6.56).

Example 6.8 Consider the impedance function of the RLC network of Figure 6.5, which is given by

$$Z(s)=\frac{LCs^2+RCs+1}{Cs} \tag{6.153}$$

The zeros of the function are located at

$$z_1, z_2=\frac{1}{2L}\left(-R\pm\sqrt{R^2-\frac{4L}{C}}\right) \tag{6.154}$$

The zero sensitivities were computed earlier in Example 6.5 and are given by (6.93). Assume that the nominal values for the elements are

$$R=1\ \Omega, \qquad L=1\ \text{H}, \qquad C=1\ \text{F} \tag{6.155}$$

From (6.93) the zero sensitivities are found to be

$$\hat{S}_R^{z_1}, \hat{S}_R^{z_2}=-\frac{1}{2}\mp j\frac{1}{2\sqrt{3}} \tag{6.156a}$$

$$\hat{S}_L^{z_1}, \hat{S}_L^{z_2}=\frac{1}{2}\mp j\frac{1}{2\sqrt{3}} \tag{6.156b}$$

$$\hat{S}_C^{z_1}, \hat{S}_C^{z_2}=\mp j\frac{1}{\sqrt{3}} \tag{6.156c}$$

Substituting these in (6.151) yields the approximate zero displacements as

$$\Delta z_i=\hat{S}_R^{z_i}V_R+\hat{S}_L^{z_i}V_L+\hat{S}_C^{z_i}V_C \qquad (i=1, 2) \tag{6.157}$$

If all the element values are increased by 10% or

$$V_R=V_L=V_C=0.1 \tag{6.158}$$

the zero displacements are found from (6.157) as

$$\Delta z_i=0.1(\hat{S}_R^{z_i}+\hat{S}_L^{z_i}+\hat{S}_C^{z_i})=\mp j0.115 \qquad (i=1, 2) \tag{6.159}$$

This is to be compared with the results previously obtained in (6.119).

Suppose that the element values for R and L are both increased by 10% whereas C is decreased by 10%. This is equivalent to setting

$$V_R=V_L=0.1, \qquad V_C=-0.1 \tag{6.160}$$

and from (6.157) we obtain

$$\Delta z_i=0.1(\hat{S}_R^{z_i}+\hat{S}_L^{z_i}-\hat{S}_C^{z_i})=0.1\times 0=0 \tag{6.161}$$

We see an excellent agreement between this approximation and that of (6.123). Note that the approximation formula (6.151) is valid under the assumption of small variations in element values so that all the second- and higher-order terms in the Taylor series expansion (6.147) can be ignored. For large variations, additional terms in the series expansion must be included.

6.5.3 General Relationships Among the Root Sensitivities

Consider an active network containing resistors R_i, inductors L_i, capacitors C_i, voltage-controlled voltage sources (VCVS) $V_i = \mu_j V_j$, current-controlled current sources (CCCS) $I_i = \alpha_j I_j$, current-controlled voltage sources (CCVS) $V_i = r_j I_j$, and voltage-controlled current sources (VCCS) $I_i = g_j V_j$. Our objective is to establish relationships among the root sensitivities of a network function with respect to the variations of the network elements.

Let s_k be a root of either the numerator or the denominator polynomial of a network function. Since s_k is a function of all the network parameters, write s_k explicitly as

$$s_k = s_k(R_i, L_i, C_i, \mu_i, \alpha_i, r_i, g_i) \tag{6.162}$$

As mentioned in Section 1.9, if we raise the impedance level of the network by a factor of b, the transfer voltage-ratio function will not change, and the driving-point impedance will be increased by a factor of b. Consequently, the root s_k will remain unaltered. This leads to the expression

$$s_k(bR_i, bL_i, C_i/b, \mu_i, \alpha_i, br_i, g_i/b) = s_k(R_i, L_i, C_i, \mu_i, \alpha_i, r_i, g_i) \tag{6.163}$$

Differentiating this equation with respect to b yields

$$\sum_{R_i} \frac{\partial s_k(\hat{\mathbf{x}})}{\partial bR_i}\frac{\partial bR_i}{\partial b} + \sum_{L_i} \frac{\partial s_k(\hat{\mathbf{x}})}{\partial bL_i}\frac{\partial bL_i}{\partial b} + \sum_{C_i} \frac{\partial s_k(\hat{\mathbf{x}})}{\partial C_i/b}\frac{\partial C_i/b}{\partial b}$$

$$+ \sum_{r_i} \frac{\partial s_k(\hat{\mathbf{x}})}{\partial br_i}\frac{\partial br_i}{\partial b} + \sum_{g_i} \frac{\partial s_k(\hat{\mathbf{x}})}{\partial g_i/b}\frac{\partial g_i/b}{\partial b} = 0 \tag{6.164}$$

or

$$\sum_{R_i} R_i \frac{\partial s_k(\hat{\mathbf{x}})}{\partial bR_i} + \sum_{L_i} L_i \frac{\partial s_k(\hat{\mathbf{x}})}{\partial bL_i} - \sum_{C_i} \frac{C_i}{b^2}\frac{\partial s_k(\hat{\mathbf{x}})}{\partial C_i/b}$$

$$+ \sum_{r_i} r_i \frac{\partial s_k(\hat{\mathbf{x}})}{\partial br_i} - \sum_{g_i} \frac{g_i}{b^2}\frac{\partial s_k(\hat{\mathbf{x}})}{\partial g_i/b} = 0 \tag{6.165}$$

where

$$s_k(\hat{\mathbf{x}}) = s_k(bR_i, bL_i, C_i/b, \mu_i, \alpha_i, br_i, g_i/b) \tag{6.166}$$

and $\sum_W$ denotes that the sum is taken over all elements of type W. Since (6.165) is valid for all values of b, for our purposes we set $b = 1$ and obtain

$$\sum_{R_i} R_i \frac{\partial s_k}{\partial R_i} + \sum_{L_i} L_i \frac{\partial s_k}{\partial L_i} - \sum_{C_i} C_i \frac{\partial s_k}{\partial C_i} + \sum_{r_i} r_i \frac{\partial s_k}{\partial r_i} - \sum_{g_i} g_i \frac{\partial s_k}{\partial g_i} = 0 \tag{6.167}$$

In terms of root sensitivities as defined in (6.85), the above equation can be rewritten as

$$\sum_{R_i} \hat{S}_{R_i}^{s_k} + \sum_{L_i} \hat{S}_{L_i}^{s_k} - \sum_{C_i} \hat{S}_{C_i}^{s_k} + \sum_{r_i} \hat{S}_{r_i}^{s_k} - \sum_{g_i} \hat{S}_{g_i}^{s_k} = 0 \tag{6.168}$$

This is a desired relation among the root sensitivities. In the special situation of an active RC network with op amps, which are considered as VCVS, the general relation

(6.168) is simplified to

$$\sum_{R_i} \hat{S}_{R_i}^{s_k} = \sum_{C_i} \hat{S}_{C_i}^{s_k} \tag{6.169}$$

Example 6.9 Consider again the impedance function

$$Z(s) = \frac{LCs^2 + RCs + 1}{Cs} \tag{6.170}$$

of the RLC network of Figure 6.5, the zeros of which are located at

$$z_1, z_2 = \frac{1}{2L}\left(-R \pm \sqrt{R^2 - \frac{4L}{C}}\right) \tag{6.171}$$

The zero sensitivities were previously computed in Example 6.5, and are repeated below:

$$\hat{S}_R^{z_1}, \hat{S}_R^{z_2} = -\frac{R}{2L} \pm \frac{R^2}{2L\sqrt{R^2 - 4L/C}} \tag{6.172a}$$

$$\hat{S}_L^{z_1}, \hat{S}_L^{z_2} = \frac{R}{2L} \mp \frac{R^2 - 2L/C}{2L\sqrt{R^2 - 4L/C}} \tag{6.172b}$$

$$\hat{S}_C^{z_1}, \hat{S}_C^{z_2} = \pm \frac{1}{C\sqrt{R^2 - 4L/C}} \tag{6.172c}$$

For the network of Figure 6.5, the zero sensitivities of the impedance $Z(s)$ are related from (6.168) by the equation

$$\hat{S}_R^{z_i} + \hat{S}_L^{z_i} - \hat{S}_C^{z_i} = 0 \qquad (i = 1, 2) \tag{6.173}$$

This can easily be confirmed by substituting (6.172) in (6.173).

In deriving the identity (6.168), we first raised the impedance level of the network by a factor of b. Then we performed the differentiation of the root with respect to b, and finally set $b = 1$ to obtain the desired formula. A similar result on root sensitivities can be obtained by frequency scaling.

As before, let s_k be a root of the polynomial $P(s)$ of interest. By frequency-scaling the network by a factor of a, the root s_k becomes as_k. Thus, we can write

$$s_k(R_i, L_i/a, C_i/a, \mu_i, \alpha_i, r_i, g_i) = as_k(R_i, L_i, C_i, \mu_i, \alpha_i, r_i, g_i) \tag{6.174}$$

or, more compactly,

$$s_k(\mathbf{x}_a) = as_k(\mathbf{x}) \tag{6.175}$$

where

$$\mathbf{x}_a = [R_i, L_i/a, C_i/a, \mu_i, \alpha_i, r_i, g_i] \tag{6.176a}$$

$$\mathbf{x} = [R_i, L_i, C_i, \mu_i, \alpha_i, r_i, g_i] \tag{6.176b}$$

Differentiating (6.174) with respect to a yields

$$-\sum_{L_i} \frac{L_i}{a^2} \frac{\partial s_k(\mathbf{x}_a)}{\partial L_i/a} - \sum_{C_i} \frac{C_i}{a^2} \frac{\partial s_k(\mathbf{x}_a)}{\partial C_i/a} = s_k(\mathbf{x}) \tag{6.177}$$

Setting $a=1$ results in

$$\sum_{L_i} L_i \frac{\partial s_k}{\partial L_i} + \sum_{C_i} C_i \frac{\partial s_k}{\partial C_i} = -s_k(\mathbf{x}) = -s_k \tag{6.178}$$

In terms of root sensitivity (6.85), (6.178) becomes

$$\sum_{L_i} \hat{S}_{L_i}^{s_k} + \sum_{C_i} \hat{S}_{C_i}^{s_k} = -s_k \tag{6.179}$$

In particular, for the active RC networks (6.179) is simplified to

$$\sum_{C_i} \hat{S}_{C_i}^{s_k} = -s_k \tag{6.180}$$

Combining this with (6.169) gives

$$\sum_{R_i} \hat{S}_{R_i}^{s_k} = \sum_{C_i} \hat{S}_{C_i}^{s_k} = -s_k \tag{6.181}$$

As an illustration, consider the results in Example 6.9. From (6.172) and (6.171), we have

$$\hat{S}_L^{z_k} + \hat{S}_C^{z_k} = \frac{R}{2L} \mp \frac{\sqrt{R^2 - 4L/C}}{2L} = -z_k \qquad (k=1, 2) \tag{6.182}$$

confirming (6.179).

Example 6.10 The transfer voltage-ratio function of the low-pass Sallen and Key network of Figure 6.6 was computed in Example 5.4, and is repeated below:

$$H(s) = \frac{k/R_1 R_2 C_1 C_2}{s^2 + [1/R_1 C_1 + 1/R_2 C_1 + (1-k)/R_2 C_2]s + 1/R_1 R_2 C_1 C_2} \tag{6.183}$$

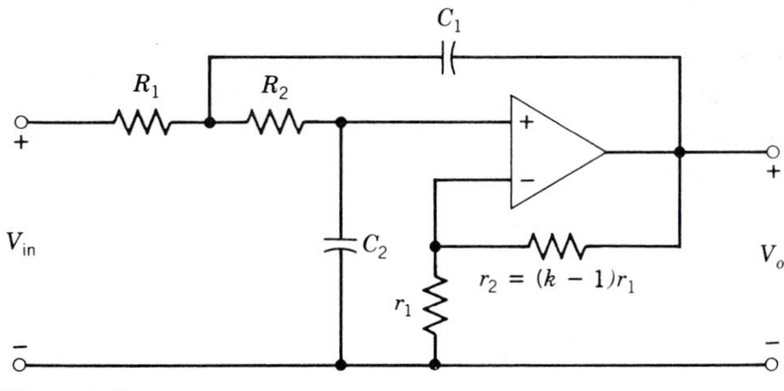

Figure 6.6

The poles of the function are the roots of the polynomial

$$P(s)=s^2+\left(\frac{1}{R_1C_1}+\frac{1}{R_2C_1}+\frac{1-k}{R_2C_2}\right)s+\frac{1}{R_1R_2C_1C_2} \tag{6.184}$$

and are given by

$$p_1, p_2=-\frac{M}{2}\pm\sqrt{\frac{M^2}{4}-\frac{1}{R_1R_2C_1C_2}} \tag{6.185}$$

where

$$M=\frac{1}{R_1C_1}+\frac{1}{R_2C_1}+\frac{1-k}{R_2C_2} \tag{6.186}$$

From (6.85) the pole sensitivities are found to be (Problem 6.11)

$$\hat{S}_{R_1}^{p_1}, \hat{S}_{R_1}^{p_2}=\frac{1}{2R_1C_1}\mp\frac{M-2/R_2C_2}{2R_1C_1\sqrt{M^2-4/R_1R_2C_1C_2}} \tag{6.187a}$$

$$\hat{S}_{R_2}^{p_1}, \hat{S}_{R_2}^{p_2}=\frac{1}{2R_2C_1}+\frac{1-k}{2R_2C_2}\mp\frac{M(M-1/R_1C_1)-2/R_1R_2C_1C_2}{2\sqrt{M^2-4/R_1R_2C_1C_2}} \tag{6.187b}$$

$$\hat{S}_{C_1}^{p_1}, \hat{S}_{C_1}^{p_2}=\frac{1}{2R_1C_1}+\frac{1}{2R_2C_1}\mp\frac{M(1/R_1C_1+1/R_2C_1)-2/R_1R_2C_1C_2}{2\sqrt{M^2-4/R_1R_2C_1C_2}} \tag{6.187c}$$

$$\hat{S}_{C_2}^{p_1}, \hat{S}_{C_2}^{p_2}=\frac{1-k}{2R_2C_2}\mp\frac{M(1-k)/R_2C_2-2/R_1R_2C_1C_2}{2\sqrt{M^2-4/R_1R_2C_1C_2}} \tag{6.187d}$$

giving

$$\hat{S}_{R_1}^{p_1}+\hat{S}_{R_2}^{p_1}=\hat{S}_{C_1}^{p_1}+\hat{S}_{C_2}^{p_1}=\tfrac{1}{2}M-\sqrt{\tfrac{1}{4}M^2-1/R_1R_2C_1C_2}=-p_1 \tag{6.188a}$$

$$\hat{S}_{R_1}^{p_2}+\hat{S}_{R_2}^{p_2}=\hat{S}_{C_1}^{p_2}+\hat{S}_{C_2}^{p_2}=\tfrac{1}{2}M+\sqrt{\tfrac{1}{4}M^2-1/R_1R_2C_1C_2}=-p_2 \tag{6.188b}$$

confirming (6.181).

6.6 GENERAL RELATIONS OF NETWORK FUNCTION SENSITIVITIES

In the preceding section, we derived general relationships (6.168) and (6.179) among the root sensitivities and the root. These relationships will now be extended to the network function sensitivity as defined in (6.2).

Let $H(s)$ be a network function of an active network containing R, L, C, and all four types of controlled sources. Then $H(s)$ can be represented explicitly as

$$H=H(R_i, L_i, C_i, \mu_i, \alpha_i, r_i, g_i, s) \tag{6.189}$$

where the parameters are defined the same as in (6.162). As before, if we raise the impedance level of the network by a factor of b, the resulting network function becomes

$$H(bR_i, bL_i, C_i/b, \mu_i, \alpha_i, br_i, g_i/b, s) = wH(R_i, L_i, C_i, \mu_i, \alpha_i, r_i, g_i, s) \quad (6.190)$$

or, more compactly,

$$H(\mathbf{x}_b) = wH(\mathbf{x}) \quad (6.191)$$

where

$$w = b \text{ if } H \text{ is a driving-point impedance} \quad (6.192a)$$

$$= 1/b \text{ if } H \text{ is a driving-point admittance} \quad (6.192b)$$

$$= 1 \text{ if } H \text{ is a transfer voltage-ratio function} \quad (6.192c)$$

$$\mathbf{x}_b = [bR_i, bL_i, C_i/b, \mu_i, \alpha_i, br_i, g_i/b, s] \quad (6.193)$$

$$\mathbf{x} = [R_i, L_i, C_i, \mu_i, \alpha_i, r_i, g_i, s] \quad (6.194)$$

As in (6.164), we differentiate (6.190) with respect to b and obtain

$$\sum_{R_i} R_i \frac{\partial H(\mathbf{x}_b)}{\partial bR_i} + \sum_{L_i} L_i \frac{\partial H(\mathbf{x}_b)}{\partial bL_i} - \sum_{C_i} \frac{C_i}{b^2} \frac{\partial H(\mathbf{x}_b)}{\partial C_i/b} + \sum_{r_i} r_i \frac{\partial H(\mathbf{x}_b)}{\partial br_i} - \sum_{g_i} \frac{g_i}{b^2} \frac{\partial H(\mathbf{x}_b)}{\partial g_i/b} = \frac{\partial w}{\partial b} H(\mathbf{x}) \quad (6.195)$$

Dividing both sides by H, setting $b = 1$, and using (6.2) yields

$$\sum_{R_i} S_{R_i}^H + \sum_{L_i} S_{L_i}^H - \sum_{C_i} S_{C_i}^H + \sum_{r_i} S_{r_i}^H - \sum_{g_i} S_{g_i}^H = \left.\frac{\partial w}{\partial b}\right|_{b=1} \quad (6.196)$$

where

$$\left.\frac{\partial w}{\partial b}\right|_{b=1} = 1 \quad \text{if } H \text{ is a driving-point impedance} \quad (6.197a)$$

$$= -1 \quad \text{if } H \text{ is a driving-point admittance} \quad (6.197b)$$

$$= 0 \quad \text{if } H \text{ is a transfer voltage-ratio function} \quad (6.197c)$$

For an active RC network with op amps, (6.196) is simplified to

$$\sum_{R_i} S_{R_i}^H - \sum_{C_i} S_{C_i}^H = \left.\frac{\partial w}{\partial b}\right|_{b=1} \quad (6.198)$$

If, in addition, H is a transfer voltage-ratio function, we have

$$\sum_{R_i} S_{R_i}^H = \sum_{C_i} S_{C_i}^H \quad (6.199)$$

The implication of (6.199) is that if all resistances and all capacitances are varied in the same absolute fractional value but with opposite signs, the transfer voltage-ratio function remains essentially unaltered. This follows directly from (6.57). Thus, in constructing an active RC filter with op amps, the resistive and capacitive materials should have the same absolute temperature coefficient values but with opposite signs to ensure that temperature variations will not change the filter performance.

Likewise, by performing frequency-scaling by a factor of a, formulas similar to (6.179)–(6.181) can be obtained. Because frequency-scaling will only affect inductors and capacitors, (6.189) can be rewritten as

$$H(R_i, L_i/a, C_i/a, \mu_i, \alpha_i, r_i, g_i, s) = H(R_i, L_i, C_i, \mu_i, \alpha_i, r_i, g_i, s/a) \tag{6.200}$$

Differentiating (6.200) with respect to a results in

$$-\sum_{L_i} \frac{L_i}{a^2} \frac{\partial H(\mathbf{x}_\alpha)}{\partial L_i/a} - \sum_{C_i} \frac{C_i}{a^2} \frac{\partial H(\mathbf{x}_\alpha)}{\partial C_i/a} = -\frac{s}{a^2} \frac{\partial H(\mathbf{x}_\beta)}{\partial s/a} \tag{6.201}$$

where

$$\mathbf{x}_\alpha = [R_i, L_i/a, C_i/a, \mu_i, \alpha_i, r_i, g_i, s] \tag{6.202a}$$

$$\mathbf{x}_\beta = [R_i, L_i, C_i, \mu_i, \alpha_i, r_i, g_i, s/a] \tag{6.202b}$$

Dividing both sides of (6.201) by H, setting $a = 1$, and using (6.2) gives

$$\sum_{L_i} S_{L_i}^H + \sum_{C_i} S_{C_i}^H = S_s^H \tag{6.203}$$

In particular, for an active RC network with op amps, (6.203) is simplified to

$$\sum_{C_i} S_{C_i}^H = S_s^H \tag{6.204}$$

If, in addition, H is a transfer voltage-ratio function, (6.204) can be combined with (6.199) to yield

$$\sum_{R_i} S_{R_i}^H = \sum_{C_i} S_{C_i}^H = S_s^H \tag{6.205}$$

Example 6.11 Consider the low-pass active filter of Figure 6.6, the transfer voltage-ratio function of which is given by

$$H(s) = \frac{V_o}{V_{\text{in}}} = \frac{k}{R_1 R_2 C_1 C_2 s^2 + [(R_1 + R_2)C_2 + (1-k)R_1 C_1]s + 1} \tag{6.206}$$

Using (6.2) the network function sensitivities are found to be (Problem 6.2)

$$S_{R_1}^H = -R_1 \frac{C_1 C_2 R_2 s^2 + [C_1(1-k) + C_2]s}{D(s)} \tag{6.207a}$$

$$S_{R_2}^H = -R_2 \frac{R_1 C_1 C_2 s^2 + C_2 s}{D(s)} \tag{6.207b}$$

$$S_{C_1}^H = -C_1 \frac{R_1 R_2 C_2 s^2 + (1-k)R_1 s}{D(s)} \tag{6.207c}$$

$$S_{C_2}^H = -C_2 \frac{R_1 R_2 C_1 s^2 + (R_1 + R_2)s}{D(s)} \tag{6.207d}$$

where

$$D(s) = R_1 R_2 C_1 C_2 s^2 + [(R_1 + R_2)C_2 + (1-k)R_1 C_1]s + 1 \tag{6.208}$$

It is easy to confirm that

$$S_{R_1}^H + S_{R_2}^H = S_{C_1}^H + S_{C_2}^H \tag{6.209}$$

To verify (6.205), we compute the sensitivity of H with respect to the complex frequency s:

$$S_s^H = \frac{s}{H}\frac{\partial H}{\partial s} = -\frac{2R_1R_2C_1C_2s^2 + [(R_1+R_2)C_2 + (1-k)R_1C_1]s}{D(s)} \tag{6.210}$$

From (6.207) we see that

$$S_{R_1}^H + S_{R_2}^H = S_{C_1}^H + S_{C_2}^H = S_s^H \tag{6.211}$$

indeed holds.

6.7 SUMMARY AND SUGGESTED READINGS

The performance of the built filters differs from the nominal design in that real components may deviate from their nominal values because of changes in environmental conditions such as temperature and humidity. A practical way to minimize this difference is to design a network that has a low sensitivity to the component variations. This is especially important in active filter design because active elements such as op amps are much more sensitive to the changes of the environment. A good understanding of sensitivity is essential to the design of practical active filters. For this we introduced the concept of *sensitivity function*, which is defined as the ratio of the fractional change in a network function to the fractional change in an element for the situation when all changes concerned are differentially small. Useful relationships among the sensitivity functions were derived. Specifically, we showed that the *magnitude sensitivity* of a network function with respect to an element is equal to the real part of the network function sensitivity with respect to the same element. This result is useful in computing the sensitivity function for a magnitude function. The single-parameter sensitivity concept was extended to the *multiparameter sensitivity* where the change of a network function is due to the simultaneous variations of many elements in the network. We also considered the notion of *gain sensitivity*, which is defined as the ratio of the change in gain in dB to the fractional change in an element for the situation when all changes concerned are differentially small. This concept is useful in that in filter design, requirements are frequently stated in terms of the maximum allowable deviation in gain over specified bands of frequencies. The gain sensitivity would be a convenient measure of these requirements. Explicit formulas for gain sensitivities with respect to the biquadratic parameters were derived.

The location of the poles of an active filter determines the stability of the filter. Therefore, it is important to know the manner in which these poles vary as some of the network elements change. Since poles of a network function are roots of a polynomial, it suffices to examine how the roots of a polynomial change as the value of an element changes. For this we defined the term *root sensitivity* as the ratio of the change in a root to the fractional change in an element for the situation when all changes concerned are differentially small. When the polynomial represents the numerator of a network function, root sensitivity is referred to as the *zero sensitivity*. When the polynomial represents the denominator, it is called the *pole sensitivity*. We showed that apart from a constant, the zero and pole sensitivities determine the function sensitivity. In computing the root sensitivity, it is necessary to first obtain the functional relationship between a root and the parameter of interest. This is usually difficult for large networks. To circumvent this difficulty, we introduced a computational technique that estimates the root change without actually computing the root sensitivity.

A network contains many nonideal elements each of which is subject to variation. The root variation due to the simultaneous changes of many elements in the network is described by the *multiparameter root sensitivity*. An approximation formula was derived that estimates the root change due to the simultaneous small variations of the elements. Finally, we obtained several relationships among the root sensitivities and the root, and also the relationships among the network function sensitivities. They are useful in filter design and in checking correctness of the sensitivity computations.

A discussion of other aspects of sensitivity can be found in Moschytz [9]. An introduction to statistical sensitivity is given by Daryanani [5], Ghausi and Laker [6], and Huelsman and Allen [7].

REFERENCES

1. P. Bowron and F. W. Stephenson 1979, *Active Filters for Communications and Instrumentation*, Chapter 4. New York: McGraw–Hill.
2. L. T. Bruton 1980, *RC-Active Circuits: Theory and Design*, Chapters 6 and 7. Englewood Cliffs, NJ: Prentice–Hall.
3. A. Budak 1974, *Passive and Active Network Analysis and Synthesis*, Chapters 10–14. Boston: Houghton Mifflin.
4. W. K. Chen 1980, *Active Network and Feedback Amplifier Theory*, Chapter 5. New York: McGraw–Hill.
5. G. Daryanani 1976, *Principles of Active Network Synthesis and Design*, Chapter 5. New York: John Wiley.
6. M. S. Ghausi and K. R. Laker 1981, *Modern Filter Design: Active RC and Switched Capacitor*, Chapter 3. Englewood Cliffs, NJ: Prentice–Hall.
7. L. P. Huelsman and P. E. Allen 1980, *Introduction to the Theory and Design of Active Filters*, Chapter 3. New York: McGraw–Hill.
8. C. S. Lindquist 1977, *Active Network Design with Signal Filtering Applications*, Chapter 1. Long Beach, CA: Steward & Sons.
9. G. S. Moschytz 1974, *Linear Integrated Networks: Fundamentals*, Chapter 4. New York: Van Nostrand–Reinhold.
10. A. S. Sedra and P. O. Brackett 1978, *Filter Theory and Design: Active and Passive*, Chapter 8. Champaign, IL: Matrix Publishers.
11. M. E. Van Valkenburg 1982, *Analog Filter Design*, Chapter 9. New York: Holt, Rinehart & Winston.

PROBLEMS

6.1 For the network of Figure 6.7, determine the pole frequency ω_p and pole Q. Also, compute the sensitivities of ω_p and Q_p with respect to the three network elements.

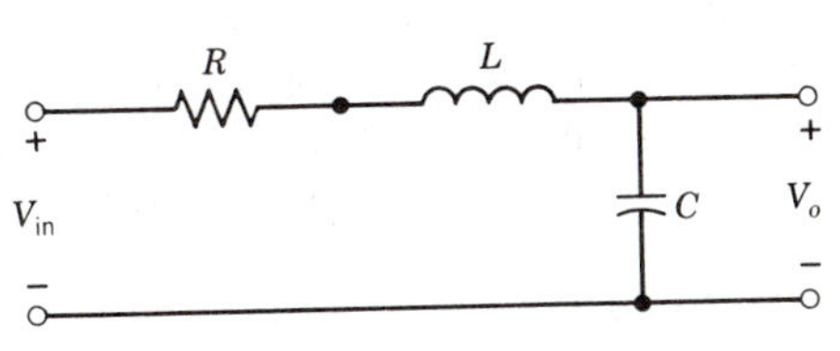

Figure 6.7

6.2 For the low-pass Sallen and Key network of Figure 6.6, verify the following sensitivity functions:

$$S_k^{H(s)} = \frac{R_1R_2C_1C_2s^2 + (R_1C_1 + R_1C_2 + R_2C_2)s + 1}{D(s)} \tag{6.212a}$$

$$S_{R_1}^{H(s)} = -R_1\frac{R_2C_1C_2s^2 + (C_1 + C_2 - kC_1)s}{D(s)} \tag{6.212b}$$

$$S_{R_2}^{H(s)} = -R_2\frac{R_1C_1C_2s^2 + C_2s}{D(s)} \tag{6.212c}$$

$$S_{C_1}^{H(s)} = -C_1\frac{R_1R_2C_2s^2 + (1-k)R_1s}{D(s)} \tag{6.212d}$$

$$S_{C_2}^{H(s)} = -C_2\frac{R_1R_2C_1s^2 + (R_1 + R_2)s}{D(s)} \tag{6.212e}$$

where

$$D(s) = R_1R_2C_1C_2s^2 + (R_1C_2 + R_2C_2 + R_1C_1 - kR_1C_1)s + 1 \tag{6.213}$$

6.3 Using the definition of sensitivity, prove the identities

$$S_x^{f+c} = \frac{f}{f+c}S_x^f \tag{6.214}$$

$$S_x^f = S_y^f S_x^y \tag{6.215}$$

6.4 Verify the following sensitivity relations:

$$S_x^{\cos y} = y(\tan y)S_x^y \tag{6.216}$$

$$S_x^{\sinh y} = y(\coth y)S_x^y \tag{6.217}$$

6.5 Compute the approximate change in gain in dB at $\omega = 9$ rad/s for the low-pass biquadratic transfer function

$$H(s) = \frac{100}{s^2 + 2s + 64} \tag{6.218}$$

for a 3% increase in the pole Q and a 5% increase in the pole frequency.

6.6 Compute the approximate change in gain in dB at the pole frequency of the low-pass biquadratic function

$$H(s)=\frac{1}{s^2+2s+100} \tag{6.219}$$

for a 2% increase in the pole Q and a 3% increase in the pole frequency.

6.7 Refer to the problem considered in Example 6.3. If it is desired to limit both the pole frequency and pole Q deviations to within $\pm 6\%$, find the maximum tolerances on the resistors and the capacitors, assuming all resistors have the same tolerance and all capacitors have the same tolerance.

6.8 An active filter has the band-pass transfer function

$$H(s)=-\frac{(1/R_4C_1)s}{s^2+(1/R_1C_1)s+1/R_2R_3C_1C_2} \tag{6.220}$$

Due to an increase in surrounding temperature, suppose that the resistances are all increased by $r\%$ and the capacitances are increased by $c\%$. Show that there is no change in pole Q and that the fractional change in the pole frequency is $-(r+c)\%$.

6.9 Perform sensitivity analysis as in Example 6.1 for the band-pass transfer voltage-ratio function of (5.51).

6.10 Perform sensitivity analysis as in Example 6.1 for the band-pass transfer voltage-ratio function of (6.220).

6.11 Verify that the pole sensitivities for the transfer voltage-ratio function (6.183) are those given in (6.187).

6.12 As in (5.1b), the general biquadratic function is sometimes written in the form

$$H(s)=\frac{b_2s^2+b_1s+b_0}{a_2s^2+a_1s+a_0} \tag{6.221}$$

Show that the coefficient sensitivities are related to the transfer-function sensitivity by the relation

$$S_x^{H(s)}=\frac{\sum_{i=0}^{2} S_x^{b_i}b_is^i}{b_2s^2+b_1s+b_0}-\frac{\sum_{i=0}^{2} S_x^{a_i}a_is^i}{a_2s^2+a_1s+a_0} \tag{6.222}$$

6.13 Extend identity (6.222) to any transfer function expressed as a ratio of two polynomials in s.

6.14 Verify that the coefficient sensitivities are related to the transfer-function sensitivity by the relation (6.222) for the active RC network of Figure 6.8.

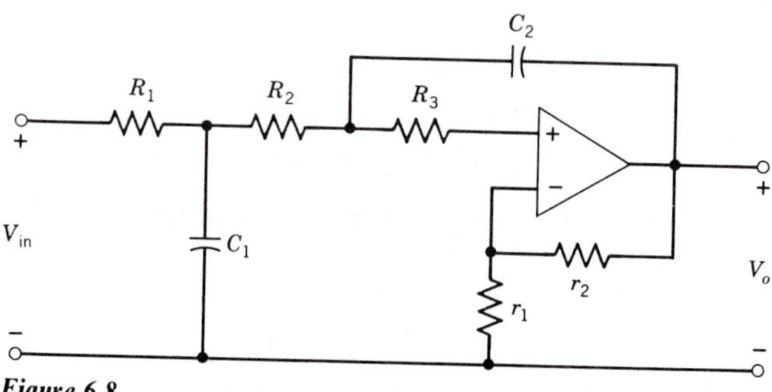

Figure 6.8

6.15 For the band-pass transfer function (6.220), compute the pole sensitivities with respect to the elements R_1, R_2, R_3, R_4, C_1, and C_2. Using these results, verify the identity (6.169).

6.16 Consider the transfer function (6.220) with $R_1=R_2=R_3=R_4=1\,\Omega$ and $C_1=C_2=2\,\text{F}$. Using the technique discussed in Section 6.5.1, estimate the pole displacements if all the resistances and capacitances are increased by 5%.

6.17 Repeat Problem 6.16, using formula (6.151).

6.18 Using the transfer function (6.220), verify the identity (6.205).

6.19 The transfer voltage-ratio function for the band-pass filter of Figure 6.9 is found to be

$$H(s)=\frac{V_o(s)}{V_{in}(s)}=\frac{-s/R_1C_2}{s^2+(1/R_2C_1+1/R_2C_2)s+1/R_1R_2C_1C_2} \tag{6.223}$$

Compute its pole sensitivities with respect to the elements R_1, R_2, C_1, and C_2. Using these results, verify the identity (6.169).

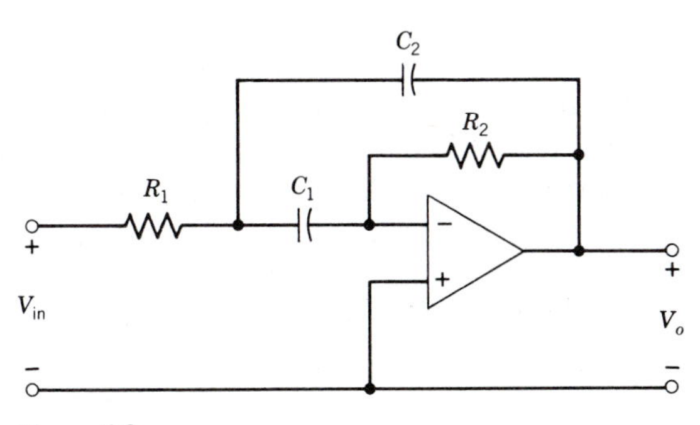

Figure 6.9

6.20 Consider the transfer function (6.223) with $R_1=R_2=2\,\Omega$ and $C_1=C_2=1\,\text{F}$. Using the technique discussed in Section 6.5.1, estimate the pole displacements if all the resistances are increased by 5% and all the capacitances are decreased by 5%.

6.21 Repeat Problem 6.20, using formula (6.151).

6.22 The transfer voltage-ratio function of a low-pass active filter is found to be

$$H(s)=\frac{1/R_2R_3C_1C_2}{s^2+(1/R_1C_1+1/R_2C_2-1/AR_2C_2)s+1/R_2R_3C_1C_2} \tag{6.224}$$

Identify its pole frequency and pole Q. Compute the sensitivities of the pole frequency and pole Q to the network elements R_1, R_2, R_3, C_1, and C_2 and also to the amplifier gain A.

chapter 7
THE ACTIVE BIQUAD

In Chapter 5, we introduced two commonly used single-amplifier biquad topologies and discussed some of their realizations. In this chapter, we continue this discussion by presenting selected biquads. In particular, we show that there are general biquads that are capable of realizing the high-pass, band-pass, band-elimination, and all-pass functions with relatively low sensitivities. These networks provide an economical means for implementing a universal building block for realizing filters of virtually any order and type.

7.1 SINGLE-AMPLIFER BAND-PASS BIQUADS

A specific negative feedback biquad is shown in Figure 5.27. This network, as indicated in (5.51), realizes the band-pass biquadratic function of (5.164). In the following, we present a single-amplifier band-pass biquad using the positive feedback topology of Figure 5.29.

Refer to the network of Figure 7.1. This network has the specific form of Figure 5.31. Comparing the elements of Figure 7.1 with those of Figure 5.31, we can make the

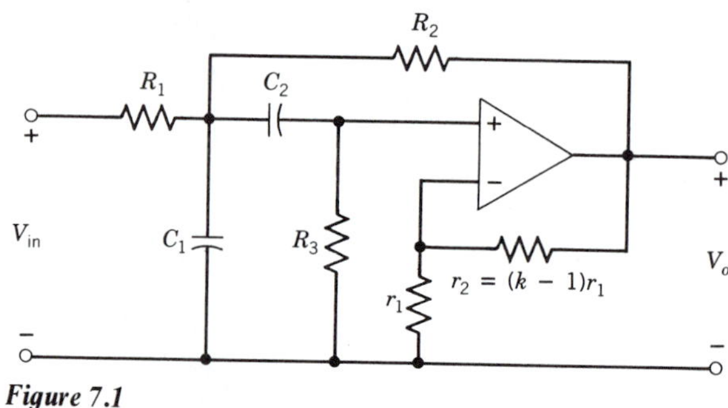

Figure 7.1

following identifications:

$$y_1=1/R_1, \qquad y_2=C_2 s, \qquad y_3=C_1 s, \qquad y_4=1/R_2, \qquad y_5=1/R_3 \tag{7.1}$$

Substituting these in (5.71), we obtain the transfer voltage-ratio function

$$H(s)=\frac{V_o(s)}{V_{\text{in}}(s)}=\frac{ks/R_1C_1}{s^2+\left(\dfrac{1}{R_1C_1}+\dfrac{1}{R_3C_2}+\dfrac{1}{R_3C_1}+\dfrac{1-k}{R_2C_1}\right)s+\dfrac{R_1+R_2}{R_1R_2R_3C_1C_2}} \tag{7.2}$$

The network of Figure 7.1 is known as the *Sallen and Key band-pass network.* To proceed with the design, we compare (7.2) with the standard band-pass transfer function

$$H(s)=K\frac{(\omega_p/Q_p)s}{s^2+(\omega_p/Q_p)s+\omega_p^2} \tag{7.3}$$

obtaining

$$\omega_p=\sqrt{\frac{R_1+R_2}{R_1R_2R_3C_1C_2}} \tag{7.4a}$$

$$Q_p=\frac{\sqrt{(R_1+R_2)/R_1R_2R_3C_1C_2}}{1/R_1C_1+1/R_3C_2+1/R_3C_1+1/R_2C_1-k/R_2C_1} \tag{7.4b}$$

$$K=\frac{k/R_1C_1}{1/R_1C_1+1/R_3C_2+1/R_3C_1+1/R_2C_1-k/R_2C_1} \tag{7.4c}$$

Since there are three constraints and six unknowns, three of the unknowns can be fixed. However, by fixing any three of the unknowns, the resulting solution may not be physically realizable. One simple physical solution is achieved by selecting an equal-valued-resistor and equal-valued-capacitor design:

$$R_1=R_2=R_3=R, \qquad C_1=C_2=C \tag{7.5}$$

In this case, equations (7.4) reduce to

$$\omega_p=\frac{\sqrt{2}}{RC}, \qquad Q_p=\frac{\sqrt{2}}{4-k}, \qquad K=\frac{k}{4-k} \tag{7.6}$$

Solving for RC and k yields

$$RC=\frac{\sqrt{2}}{\omega_p}, \qquad k=4-\frac{\sqrt{2}}{Q_p} \tag{7.7}$$

Observe that for positive k, Q_p must be greater than $\sqrt{2}/4$. The resulting gain constant of this realization is found to be

$$K_r=\frac{k}{R_1C_1}=\omega_p\left(2\sqrt{2}-\frac{1}{Q_p}\right) \tag{7.8}$$

This constant can be adjusted by using the techniques of input attenuation or gain enhancement, as outlined in Section 5.4.

Example 7.1 Realize the biquadratic band-pass function

$$H(s)=\frac{5000s}{s^2+3000s+6\times 10^7} \tag{7.9}$$

From (7.9) the biquadratic parameters are identified as

$$\omega_p=\sqrt{60}\times 10^3 \text{ rad/s}, \qquad Q_p=\sqrt{20/3} \tag{7.10}$$

The required element values are found from (7.5) and (7.7) by choosing $C=0.1\ \mu$F:

$$C_1=C_2=C=0.1\ \mu\text{F} \tag{7.11a}$$

$$R_1=R_2=R_3=R=\frac{\sqrt{2}}{\sqrt{60}\times 10^3\times 10^{-7}}=1.826\text{ k}\Omega \tag{7.11b}$$

$$k=4-\frac{\sqrt{2}}{\sqrt{20/3}}=3.452 \tag{7.11c}$$

With this choice of element values, the gain constant realized by the network is from (7.8)

$$K_r=\sqrt{60}\times 10^3\left(2\sqrt{2}-\frac{1}{\sqrt{20/3}}\right)=18{,}909 \tag{7.12}$$

Since the desired gain constant is 5000, the input will have to be attenuated by a factor of $18{,}909/5000=3.78$. Using the input attenuation scheme described in Section 5.4, the input resistance $R_1=R$ is replaced by two resistors R_a and R_b, as shown in Figure 7.2, having values obtained from (5.90) of

$$R_a=\frac{1.826\times 10^3}{1/3.78}=6.90\text{ k}\Omega \tag{7.13a}$$

$$R_b=\frac{1.826\times 10^3}{1-1/3.78}=2.48\text{ k}\Omega \tag{7.13b}$$

Since $k=1+r_2/r_1=3.452$, making $r_1=1$ kΩ gives $r_2=2.452$ kΩ. The final network realization of the transfer function of (7.9) is shown in Figure 7.2.

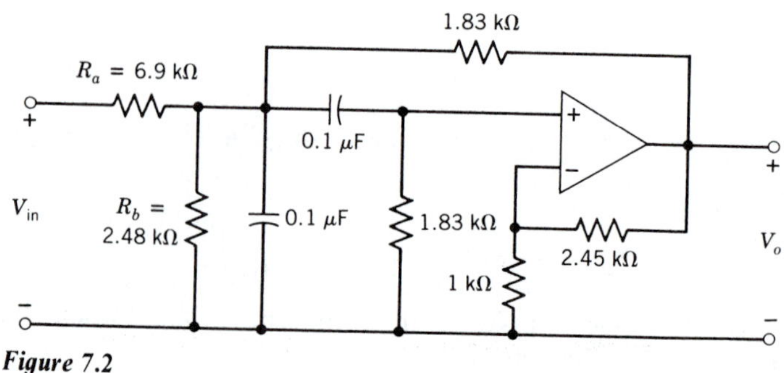

Figure 7.2

Another very practical design for the network of Figure 7.1 is obtained by specifying

$$R_1 = 1\ \Omega, \qquad C_1 = C_2 = 1\ \text{F}, \qquad k = 2 \tag{7.14}$$

In this case, $r_1 = r_2$ and we can use two equal-valued precision resistors as feedback around the op amp, resulting in a network with great stability. Using the above values in (7.4) gives

$$\omega_p^2 = \frac{1}{R_3}\left(1 + \frac{1}{R_2}\right) \tag{7.15a}$$

$$\frac{\omega_p}{Q_p} = 1 + \frac{2}{R_3} - \frac{1}{R_2} \tag{7.15b}$$

These nonlinear equations can be solved to yield R_2 and R_3.

Example 7.2 We realize the band-pass function of (7.9) using (7.15):

$$6 \times 10^7 = \frac{1}{R_3}\left(1 + \frac{1}{R_2}\right) \tag{7.16a}$$

$$3000 = 1 + \frac{2}{R_3} - \frac{1}{R_2} \tag{7.16b}$$

Solving the second equation for $1/R_2$ and substituting the result in the first equation, we obtain

$$6 \times 10^7 R_3^2 + 2998 R_3 - 2 = 0 \tag{7.17}$$

giving $R_3 = 1.593 \times 10^{-4}$, -2.093×10^{-4}. Choose $R_3 = 1.593 \times 10^{-4}\ \Omega$. The other resistance R_2 is found to be

$$R_2 = 1.046 \times 10^{-4}\ \Omega \tag{7.18a}$$

To get the practical element values, we magnitude-scale the network by a factor of 10^6 and obtain

$$R_1 = 1\ \text{M}\Omega, \qquad R_2 = 104.6\ \Omega, \qquad R_3 = 159.3\ \Omega \tag{7.18b}$$

$$C_1 = C_2 = 1\ \mu\text{F}, \qquad r_1 = r_2 = 1\ \text{k}\Omega \tag{7.18c}$$

where r_1 and r_2 were arbitrarily chosen as 1 kΩ.

7.2 *RC–CR* TRANSFORMATION

The general subject of frequency transformation was discussed in Section 2.5. There we showed that the results obtained for the low-pass approximation can readily be adapted to other cases such as high-pass, band-pass, and band-elimination by means of transformations of the frequency variable.

Recall that, given a normalized low-pass function with unity passband edge frequency, the transformation

$$s' = \omega_0 / s \tag{7.19}$$

will convert a low-pass characteristic to a high-pass characteristic with passband edge at $\omega = \omega_0$. This is equivalent to replacing any capacitor having impedance $1/Cs'$ in the low-pass structure by a one-port network having impedance $s/\omega_0 C$. The resistors in the low-pass structure are not affected. In this section we show that by a slight modification of this procedure, high-pass active *RC* filters can be obtained from those of the low-pass type.

A simple extension of the transformation (7.19) to convert a low-pass function with passband edge at $\omega = \omega_p$ to a high-pass function with passband edge at $\omega = \omega_p$ is given by the expression

$$s' = \omega_p^2 / s \tag{7.20}$$

We first apply this transformation to the elements of a low-pass network, and then perform an impedance normalization of ω_p / s. The latter operation will not affect the transfer voltage-ratio function because it is dimensionless, being the ratio of two voltages. The above procedure can be depicted symbolically as follows:

$$R \to R \to \frac{\omega_p R}{s} \tag{7.21a}$$

$$\frac{1}{Cs'} \to \frac{s}{\omega_p^2 C} \to \frac{1}{\omega_p C} \tag{7.21b}$$

Thus, the result of the two transformations can be effected by replacing the resistors of value R in the low-pass structure by capacitors of value $1/\omega_p R$, and capacitors of value C in the low-pass network by resistors of value $1/\omega_p C$, as shown in Figure 7.3. The op amp that is modeled as a VCVS is not affected by these transformations, and so K, a gain factor used in this and preceding chapters, is not changed. Also, the transformations need not be applied to the resistors r_1 and r_2 used to set the gain for the noninverting op-amp network because they always appear as a ratio r_2/r_1 in the transfer voltage-ratio function.

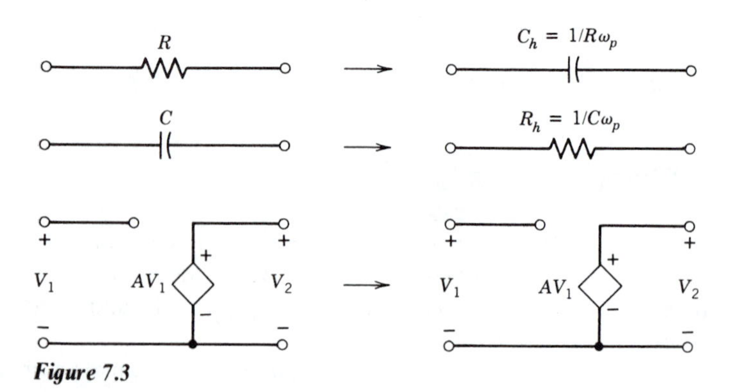

***Figure** 7.3*

To see that the transformation (7.20) will indeed convert a low-pass biquadratic function of the general form

$$H_{LP}(s') = \frac{\omega_p^2}{s'^2 + (\omega_p/Q_p)s' + \omega_p^2} \tag{7.22}$$

to a high-pass function, we substitute (7.20) in (7.22) and obtain

$$H_{HP}(s) = \frac{s^2}{s^2 + (\omega_p/Q_p)s + \omega_p^2} \tag{7.23}$$

a standard high-pass characteristic. The process as depicted in Figure 7.3 is known as the *RC–CR transformation* for converting an active *RC* low-pass filter to a high-pass filter, and cannot be applied to networks in which the active element is not dimensionless, such as FDNR to be discussed in Chapter 8.

Example 7.3 Consider the low-pass Sallen and Key network of Figure 5.34, which is redrawn in Figure 7.4. The transfer function was computed earlier in (5.73) and is given by

$$H(s) = \frac{k/R_1R_2C_1C_2}{s^2 + [1/R_1C_1 + 1/R_2C_1 + 1/R_2C_2 - k/R_2C_2]s + 1/R_1R_2C_1C_2} \tag{7.24}$$

The standard form for (7.24) is

$$H(s) = \frac{K}{s^2 + (\omega_p/Q_p)s + \omega_p^2} \tag{7.25}$$

where

$$K = \frac{k}{R_1R_2C_1C_2} \tag{7.26a}$$

$$\omega_p = \frac{1}{\sqrt{R_1R_2C_1C_2}} \tag{7.26b}$$

$$Q_p = \frac{1/\sqrt{R_1R_2C_1C_2}}{1/R_1C_1 + 1/R_2C_1 + 1/R_2C_2 - k/R_2C_2} \tag{7.26c}$$

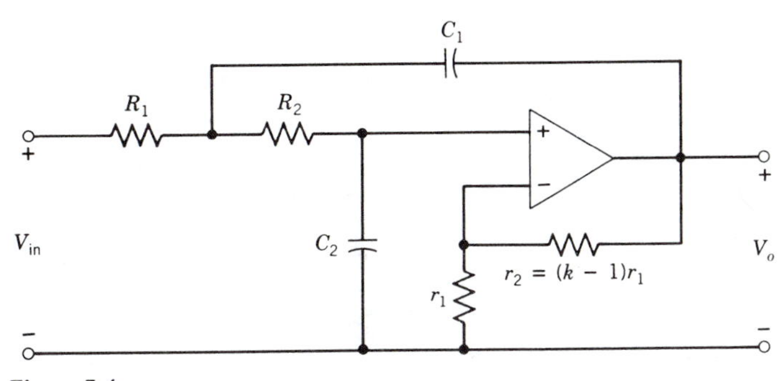

Figure 7.4

Following the design of (5.76) with

$$R_1 = R_2 = R, \qquad C_1 = C_2 = C \tag{7.27}$$

we obtain from (5.77)

$$1/RC = \omega_p, \qquad k = 3 - 1/Q_p \tag{7.28}$$

To convert the low-pass filter of Figure 7.4 to a high-pass filter, we apply the transformations of Figure 7.3 to the network of Figure 7.4. The resulting high-pass network is shown in Figure 7.5 with the following element values:

$$C_1' = C_2' = \frac{1}{R\omega_p} = C' \tag{7.29a}$$

$$R_1' = R_2' = \frac{1}{C\omega_p} = R' \tag{7.29b}$$

$$\frac{r_2}{r_1} = k - 1 = 2 - \frac{1}{Q_p} \tag{7.29c}$$

As a numerical example, consider the normalized Butterworth transfer voltage-ratio function

$$H(s) = \frac{2}{s^2 + \sqrt{2}s + 1} \tag{7.30}$$

giving $\omega_p = 1$, $Q_p = 1/\sqrt{2}$, and $K = 2$. Substituting these in (7.29) with $C = 1$ F and $R = 1\ \Omega$ yields

$$C_1' = C_2' = C' = 1\text{ F}, \qquad R_1' = R_2' = R' = 1\ \Omega \tag{7.31a}$$

$$r_2 = (2 - \sqrt{2})r_1 = 0.586r_1 = 5.86\ \Omega \tag{7.31b}$$

where r_1 was selected to be 10 Ω. The resulting high-pass network is shown in Figure 7.6, the transfer function of which can be obtained directly from (7.24) by substituting

$$R_1 = R_2 = 1/C'\omega_p = 1, \qquad C_1 = C_2 = 1/R'\omega_p = 1 \tag{7.32}$$

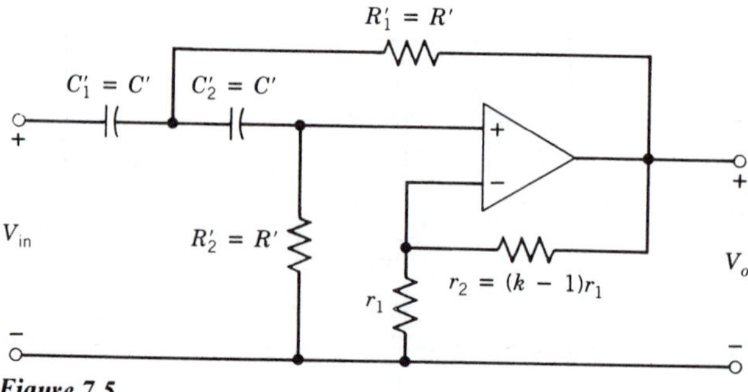

Figure 7.5

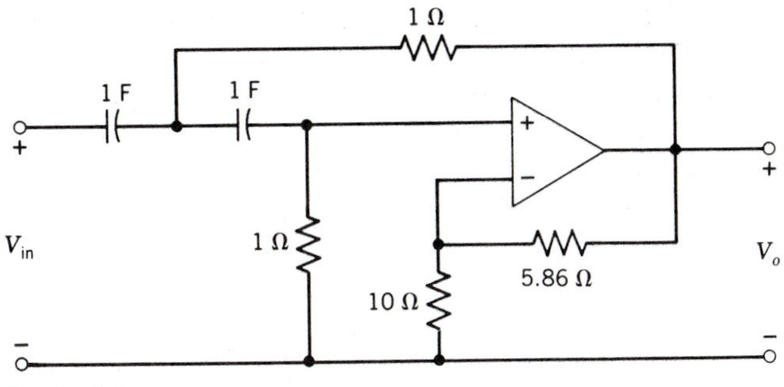

Figure 7.6

along with replacing s by ω_p^2/s. The result is

$$H(s) = \frac{(3-\sqrt{2})s^2}{s^2+\sqrt{2}s+1} \tag{7.33}$$

The realized gain constant for the high-pass network of Figure 7.6 is

$$K_r = 3-\sqrt{2} = 1.586 \tag{7.34}$$

which is the same as that for the low-pass filter obtained in (5.87).

7.3 SINGLE-AMPLIFIER GENERAL BIQUAD

In this section, we consider a negative feedback RC amplifier network that is capable of realizing a general biquadratic transfer function. This network as shown in Figure 7.7 is due to Friend, Harris, and Hilberman,† and is often considered as an extension of the network proposed by Delyiannis.‡ It is highly suited to thin-film, hybrid

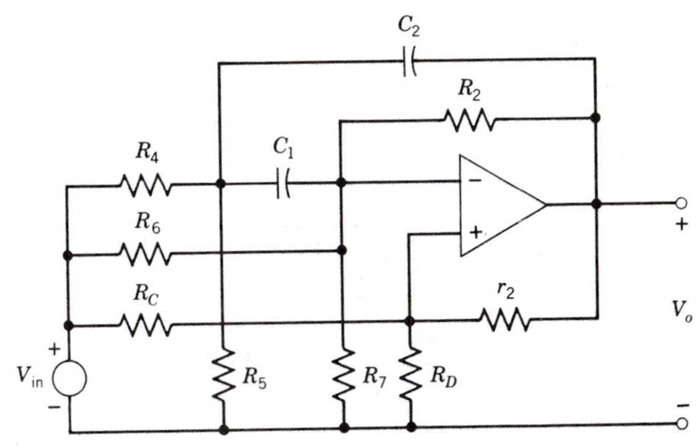

Figure 7.7

†J. J. Friend, C. A. Harris, and D. Hilberman, "STAR: An active biquadratic filter section," *IEEE Trans. Circuits and Systems*, vol. CAS-22, 1975, pp. 115–121.

‡T. Delyiannis, "High-Q factor circuit with reduced sensitivity," *Electronics Letters*, vol. 4, 1968, p. 577.

integrated-circuit fabrication, and is known as the *single-amplifier biquad* (SAB) or as the *standard tantalum active resonator* (STAR), and has been used extensively for audio- and voice-frequency signal-processing applications in the Bell Telephone System.

To proceed with the analysis, we consider the equivalent network of Figure 7.8, where the constants K_1, K_2, and K_3 are related to the potential dividers of Figure 7.7 by the expressions

$$K_1 = \frac{R_5}{R_4 + R_5}, \qquad R_1 = \frac{R_4 R_5}{R_4 + R_5} \tag{7.35a}$$

$$K_2 = \frac{R_D}{R_C + R_D}, \qquad r_1 = \frac{R_C R_D}{R_C + R_D} \tag{7.35b}$$

$$K_3 = \frac{R_7}{R_6 + R_7}, \qquad R_3 = \frac{R_6 R_7}{R_6 + R_7} \tag{7.35c}$$

Application of the principle of superposition shows that the output voltage V_o due to the three sources $K_1 V_{\text{in}}$, $K_2 V_{\text{in}}$, and $K_3 V_{\text{in}}$ equals the sum of the voltages due to the individual sources applied separately. Thus, we can write

$$V_o(s) = H_1(s) K_1 V_{\text{in}}(s) + H_2(s) K_2 V_{\text{in}}(s) + H_3(s) K_3 V_{\text{in}}(s) \tag{7.36}$$

giving

$$\frac{V_o(s)}{V_{\text{in}}(s)} = H(s) = K_1 H_1(s) + K_2 H_2(s) + K_3 H_3(s) \tag{7.37}$$

Writing H in the form

$$H(s) = \frac{V_o(s)}{V_{\text{in}}(s)} = K \frac{b_2 s^2 + b_1 s + b_0}{s^2 + a_1 s + a_0} \tag{7.38}$$

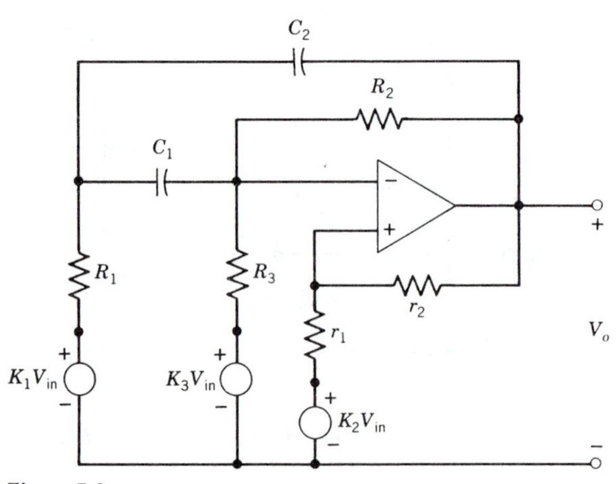

Figure 7.8

the coefficients are found to be $K=1$ and

$$b_2=K_2 \tag{7.39a}$$

$$b_1=\frac{K_2}{C_2}\left(\frac{1}{R_1}+\frac{1}{R_2}+\frac{1}{R_3}\right)+\frac{K_2}{C_1}\left(\frac{1}{R_2}+\frac{1}{R_3}\right)-\left(1+\frac{r_1}{r_2}\right)\left[\frac{K_1}{R_1C_2}+\frac{K_3}{R_3}\left(\frac{1}{C_1}+\frac{1}{C_2}\right)\right] \tag{7.39b}$$

$$b_0=\frac{1}{C_1C_2}\left[\frac{K_2}{R_1}\left(\frac{1}{R_2}+\frac{1}{R_3}\right)-\frac{K_3}{R_1R_3}\left(1+\frac{r_1}{r_2}\right)\right] \tag{7.39c}$$

$$a_1=\frac{C_1+C_2}{C_1C_2}\left(\frac{1}{R_2}-\frac{r_1}{R_3r_2}\right)-\frac{r_1}{R_1r_2C_2} \tag{7.39d}$$

$$a_0=\frac{1}{R_1C_1C_2}\left(\frac{1}{R_2}-\frac{r_1}{R_3r_2}\right) \tag{7.39e}$$

Observe that there are five equations in nine unknown variables C_1, C_2, R_1, R_2, R_3, K_1, K_2, K_3, and r_1/r_2. Therefore, four of these variables and either r_1 or r_2 can be specified independently by the designer. However, fixing any four of these variables may result in nonphysical elements. As in the previous situations, the capacitances are usually made equal. The other fixed parameters are K_3 and r_1/r_2. The scaling factor K_3 is chosen to ensure that the element values are all nonnegative, while the ratio r_1/r_2 is selected to yield a low-sensitivity circuit with a reasonable spread of element values. Thus, our objective is to express R_1, R_2, R_3, and K_1 in terms of C_1, C_2, K_3, and r_1/r_2. After some algebraic manipulation, equations (7.39) can be solved to yield

$$R_1=\frac{2r_1}{r_2C_2[-a_1+\sqrt{a_1^2+4a_0(1+C_1/C_2)r_1/r_2}]} \tag{7.40a}$$

$$K_1=\frac{b_2+b_0(1+C_1/C_2)R_1^2C_2^2-b_1R_1C_2}{1+r_1/r_2} \tag{7.40b}$$

$$R_3=\frac{(1+r_1/r_2)(b_2-K_3)}{R_1C_1C_2(b_0-a_0b_2)} \tag{7.40c}$$

$$R_2=\frac{R_3}{R_1R_3C_1C_2a_0+r_1/r_2} \tag{7.40d}$$

For our design, we set

$$C_1=C_2=1\text{ F},\qquad 0\leqslant K_3\leqslant 1,\qquad \gamma=r_1/r_2 \tag{7.41}$$

Substituting these in (7.40) gives

$$R_1=\frac{2\gamma}{-a_1+\sqrt{a_1^2+8a_0\gamma}} \tag{7.42a}$$

$$K_1=\frac{b_2+2b_0R_1^2-b_1R_1}{1+\gamma} \tag{7.42b}$$

$$R_3 = \frac{(1+\gamma)(b_2 - K_3)}{R_1 a_0(b_0/a_0 - b_2)} \tag{7.42c}$$

$$R_2 = \frac{R_3}{R_1 R_3 a_0 + \gamma} \tag{7.42d}$$

As can be seen from (7.35), K_1, K_2, and K_3 are bounded between 0 and 1. If in the solution of (7.42b) K_1 is found to be greater than unity, we must first scale the numerator coefficients b_2, b_1, and b_0 down by a common factor, say, k_1, so that K_1 will lie between 0 and 1. This reduced gain constant can be restored by increasing K by the same factor of k_1. In other words, we need only consider the modified transfer function

$$H(s) = Kk_1 \frac{(b_2/k_1)s^2 + (b_1/k_1)s + b_0/k_1}{s^2 + a_1 s + a_0} \tag{7.43}$$

The constant K_3 should be chosen so that the resulting resistance R_3 in (7.42c) is nonnegative, using the scaling technique as outlined in (7.43) if necessary. Finally, in the band-pass case $b_2 = b_0 = 0$, K_1 becomes negative. To circumvent this difficulty, we simply change the sign of the numerator coefficient b_1. This is usually of no concern because neither the magnitude nor the delay of the transfer function is affected by it.

For the band-pass realization, the SAB of Figure 7.7 reduces to that shown in Figure 7.9, because in this case $K_2 = b_2 = 0$ and $R_3 = \infty$. For a low-pass-notch network, we set $b_1 = 0$ and $b_0 > 0$. To this end, we choose $K_3 = 0$ and set b_1 of (7.39b) to zero, obtaining

$$\left(1 + \frac{r_1}{r_2}\right)\frac{R_2 K_1}{R_1 K_2} = \frac{R_2}{R_1} + \left(1 + \frac{C_2}{C_1}\right)\left(1 + \frac{R_2}{R_3}\right) \tag{7.44}$$

or K_1 and K_2 are related by the expression

$$\frac{K_1}{K_2} = \frac{1 + (R_1/R_2)(1 + C_2/C_1)(1 + R_2/R_3)}{1 + r_1/r_2} \tag{7.45}$$

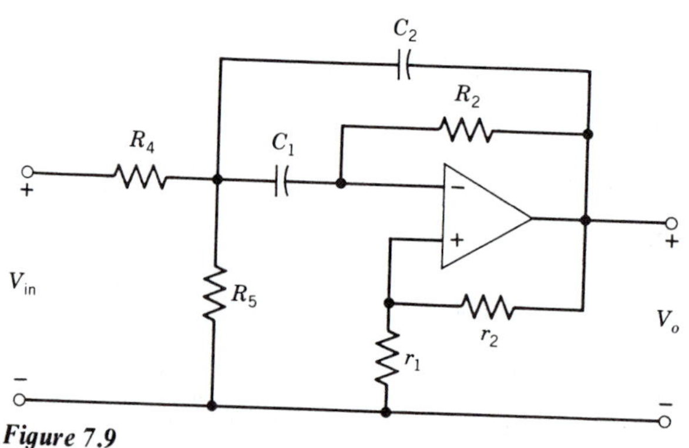

Figure 7.9

With $K_3 = 0$, from (7.35c) $R_6 = \infty$, and the SAB of Figure 7.7 reduces to the low-pass-notch filter of Figure 7.10, the transfer function of which is found from (7.39) to be

$$H(s) = \frac{K_2(s^2 + \omega_z^2)}{s^2 + (\omega_p/Q_p)s + \omega_p^2} \tag{7.46}$$

where

$$\omega_z = \sqrt{\frac{R_2 + R_3}{R_1 R_2 R_3 C_1 C_2}} \tag{7.47a}$$

$$\omega_p = \sqrt{\frac{R_3 - R_2 r_1/r_2}{R_1 R_2 R_3 C_1 C_2}} \tag{7.47b}$$

$$Q_p = \frac{\sqrt{(R_3 - R_2 r_1/r_2)/R_1 R_2 R_3 C_1 C_2}}{1/R_2 C_1 + 1/R_2 C_2 - (r_1/r_2)(1/R_3 C_1 + 1/R_3 C_2 + 1/R_1 C_2)} \tag{7.47c}$$

Observe that for H to be realizable, it is necessary that

$$\frac{R_2}{R_3} \cdot \frac{r_1}{r_2} < 1 \tag{7.48}$$

This also guarantees that $\omega_z > \omega_p$, a required property for a low-pass-notch filter. Finally, if the design parameters are chosen in accordance with (7.41), the biquadratic parameters of (7.47) reduce to

$$\omega_z = \sqrt{\frac{R_2 + R_3}{R_1 R_2 R_3}}, \qquad \omega_p = \sqrt{\frac{R_3 - \gamma R_2}{R_1 R_2 R_3}} \tag{7.49a}$$

$$Q_p = \frac{\sqrt{R_1 R_2 R_3 (R_3 - \gamma R_2)}}{2R_1(R_3 - \gamma R_2) - \gamma R_2 R_3} \tag{7.49b}$$

Similar procedures can be used to realize a high-pass-notch response, a band-elimination response, or an all-pass response. The only response that cannot be

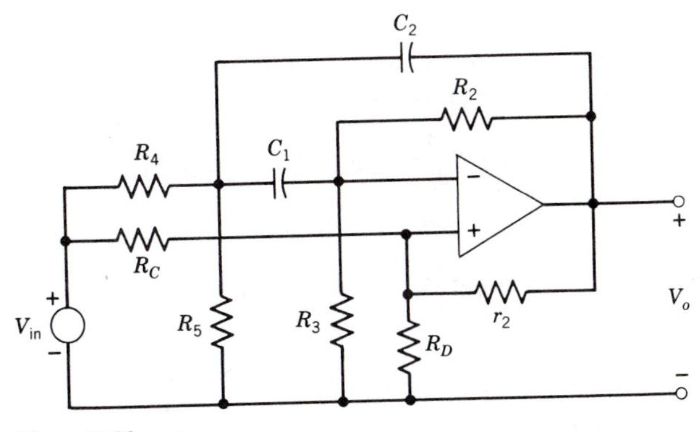

Figure 7.10

realized directly with the SAB is the low-pass response. However, the low-pass Sallen and Key network can be used for this purpose.

All the circuits discussed in this and preceding chapters can be realized with the tiny thin-film hybrid integrated circuit (HIC). This HIC can be conveniently packaged in a standard 16-pin dual-in-line package (DIP). The SAB, therefore, provides an economical means for implementing a universal type of building blocks for active filter design. We illustrate the above results by the following examples.

Example 7.4 Realize the all-pass transfer function

$$H(s)=\frac{s^2-20s+4000}{s^2+20s+4000} \tag{7.50}$$

Comparing this equation with (7.38), we can make the following identifications:

$$K=1, \quad b_2=1, \quad b_1=-20, \quad b_0=4000, \quad a_1=20, \quad a_0=4000 \tag{7.51}$$

We shall follow the equal-capacitance design of (7.41)–(7.42). For our purposes, the fixed design parameters are chosen as

$$C_1=C_2=1\text{ F}, \quad \gamma=0.1, \quad 0\leqslant K_3\leqslant 1 \tag{7.52}$$

Substituting these in (7.42a) yields

$$R_1=\frac{0.2}{-20+\sqrt{400+3200}}=5\times10^{-3}\,\Omega \tag{7.53a}$$

Using this R_1 in (7.42b), we obtain

$$K_1=\frac{1+2\times4\times10^3\times25\times10^{-6}+20\times5\times10^{-3}}{1+0.1}=1.182>1 \tag{7.53b}$$

Since K_1 cannot exceed unity, we must scale the coefficients b_2, b_1, and b_0 down by a factor k_1 of at least 1.182, and correspondingly increase K by the same factor. Choose $k_1=1.182$. The new coefficients become

$$b_2=0.846, \quad b_1=-16.92, \quad b_0=3384 \tag{7.54a}$$

$$K=1.182, \quad a_1=20, \quad a_0=4000 \tag{7.54b}$$

Substituting these again in (7.42) gives

$$R_1=5\times10^{-3}\,\Omega, \quad R_3=\infty, \quad R_2=0.05\,\Omega \tag{7.55}$$

where $K_1=1$, as expected. Using $K_1=1$ and $K_2=b_2=0.846$ in (7.35) gives

$$R_4=5\times10^{-3}\,\Omega, \quad R_5=\infty, \quad R_C=1.182r_1, \quad R_D=6.494r_1 \tag{7.56}$$

To obtain the practical element values, we magnitude-scale the network by a factor of 10^6, yielding

$$R_1=5\text{ k}\Omega, \quad R_2=50\text{ k}\Omega, \quad R_3=\infty \tag{7.57a}$$

$$C_1=C_2=1\,\mu\text{F}, \quad R_4=5\text{ k}\Omega, \quad R_5=\infty \tag{7.57b}$$

A practical choice for the resistors r_1 and r_2 to realize the ratio $r_1/r_2=0.1$ is

$$r_1=1\text{ k}\Omega, \qquad r_2=10\text{ k}\Omega \tag{7.58}$$

obtaining from (7.56)

$$R_C=1.18\text{ k}\Omega, \qquad R_D=6.49\text{ k}\Omega \tag{7.59}$$

Finally, to realize the gain constant $K=1.182$ by using the gain enhancement technique as discussed in Section 5.4, we use the potential divider resistors R_a and R_b, where

$$K=1+R_a/R_b=1.182 \tag{7.60}$$

A practical choice for R_a and R_b is given by

$$R_a=182\ \Omega, \qquad R_b=1\text{ k}\Omega \tag{7.61}$$

The final network realization for the all-pass function of (7.50) is shown in Figure 7.11.

Example 7.5 Realize the high-pass-notch transfer voltage-ratio function

$$H(s)=\frac{s^2+10^6}{s^2+100s+10^8} \tag{7.62}$$

Comparing this equation with (7.38), we can make the following identifications:

$$K=1, \qquad b_2=1, \qquad b_1=0, \qquad b_0=10^6, \qquad a_1=100, \qquad a_0=10^8 \tag{7.63}$$

We shall follow the equal-capacitance design of (7.41)–(7.42) with $\gamma=0.1$. Substituting these in (7.42) gives

$$R_1=\frac{2\times 0.1}{-100+\sqrt{10^4+8\times 10^8\times 0.1}}=2.26\times 10^{-5}\ \Omega \tag{7.64a}$$

$$K_1=\frac{1+2\times 10^6\times 5.1076\times 10^{-10}-0}{1+0.1}=0.91 \tag{7.64b}$$

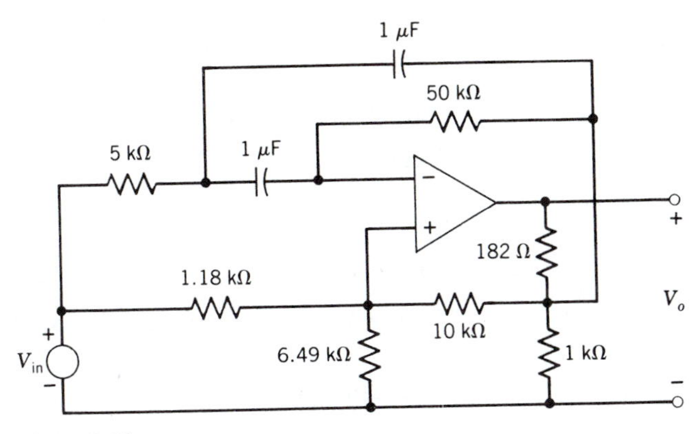

Figure 7.11

$$R_3 = \frac{(1+0.1)(1-K_3)}{2.26\times10^{-5}\times10^8(10^6/10^8-1)} < 0, \qquad K_3 < 1,$$

$$= 0, \qquad K_3 = 1 \tag{7.64c}$$

To circumvent the difficulty encountered in (7.64c), we consider the function

$$H(s) = 2\,\frac{0.5s^2 + 0.5\times10^6}{s^2 + 100s + 10^8} \tag{7.65}$$

having the following new coefficients:

$$K=2, \qquad b_2=0.5, \qquad b_1=0, \qquad b_0=0.5\times10^6, \qquad a_1=100, \qquad a_0=10^8 \tag{7.66}$$

Substituting these again in (7.42) yields

$$R_1 = 2.26\times10^{-5}\ \Omega \tag{7.67a}$$

$$K_1 = 0.455 \tag{7.67b}$$

$$R_3 = \frac{(1+0.1)(0.5-K_3)}{2.26\times10^{-5}\times10^8(0.5\times10^6/10^8-0.5)} = 4.91\times10^{-4}\ \Omega \tag{7.67c}$$

where K_3 was chosen to be 1, and

$$R_2 = \frac{4.91\times10^{-4}}{2.26\times10^{-5}\times4.91\times10^{-4}\times10^8+0.1} = 4.06\times10^{-4}\ \Omega \tag{7.67d}$$

Using

$$K_1 = 0.455, \qquad K_2 = b_2 = 0.5, \qquad K_3 = 1 \tag{7.68}$$

in (7.35), the other resistances are obtained as follows:

$$R_4 = \frac{R_1}{K_1} = 4.97\times10^{-5}\ \Omega, \qquad R_5 = \frac{R_1}{1-K_1} = 4.15\times10^{-5}\ \Omega \tag{7.69a}$$

$$R_6 = \frac{R_3}{K_3} = 4.91\times10^{-4}\ \Omega, \qquad R_7 = \frac{R_3}{1-K_3} = \infty \tag{7.69b}$$

$$R_C = \frac{r_1}{K_2} = 2r_1, \qquad R_D = \frac{r_1}{1-K_2} = 2r_1 \tag{7.69c}$$

A practical choice for the resistors r_1 and r_2 to realize the ratio $\gamma = r_1/r_2 = 0.1$ is

$$r_1 = 1\ \text{k}\Omega, \qquad r_2 = 10\ \text{k}\Omega \tag{7.70}$$

obtaining from (7.69c)

$$R_C = R_D = 2\ \text{k}\Omega \tag{7.71}$$

To obtain the practical element values, we magnitude-scale the network by a factor of 10^8 and obtain

$$C_1 = C_2 = 0.01\ \mu\text{F}, \qquad R_1 = 2.26\ \text{k}\Omega, \qquad R_2 = 40.6\ \text{k}\Omega \tag{7.72a}$$

$$R_3=49.1\text{ k}\Omega, \qquad R_4=4.97\text{ k}\Omega, \qquad R_5=4.15\text{ k}\Omega \tag{7.72b}$$

$$R_6=49.1\text{ k}\Omega, \qquad R_7=\infty, \qquad R_C=R_D=2\text{ k}\Omega \tag{7.72c}$$

Finally, to realize the gain constant $K=2$ by using the gain enhancement technique as discussed in Section 5.4, we use the potential divider resistors R_a and R_b, where

$$K=1+R_a/R_b=2 \tag{7.73}$$

A practical choice for R_a and R_b is given by

$$R_a=R_b=1\text{ k}\Omega \tag{7.74}$$

The final network realizing the high-pass-notch function of (7.62) is shown in Figure 7.12.

Example 7.6 Realize the band-pass transfer voltage-ratio function

$$H(s)=\frac{-600s}{s^2+600s+10^7} \tag{7.75}$$

As before, we compare the coefficients of this equation with (7.38) and make the following identifications:

$$K=1, \qquad b_2=0, \qquad b_1=-600, \qquad b_0=0, \qquad a_1=600, \qquad a_0=10^7 \tag{7.76}$$

We shall again follow the equal-capacitance design of (7.41)–(7.42) with $\gamma=0.1$. Substituting these in (7.42) gives

$$R_1=8.73\times 10^{-5}\ \Omega \tag{7.77a}$$

$$K_1=0.0476 \tag{7.77b}$$

$$R_3=\infty \tag{7.77c}$$

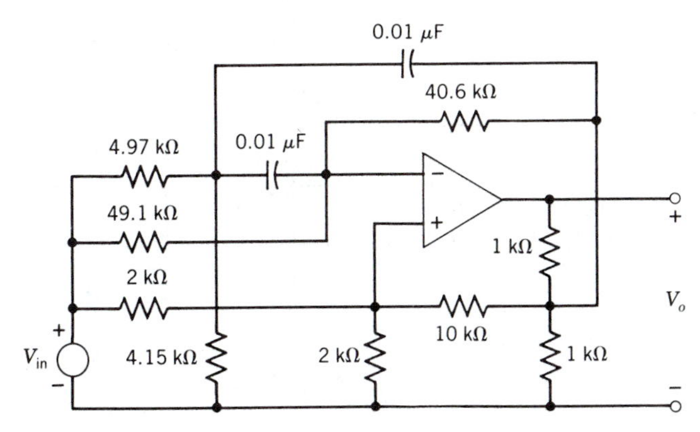

Figure 7.12

$$R_2 = \frac{R_3}{R_1 R_3 a_0 + \gamma} = \frac{1}{R_1 a_0} = 1.15 \times 10^{-3}\ \Omega \tag{7.77d}$$

Using $K_1 = 0.0476$ and $K_2 = b_2 = 0$ in (7.35) yields

$$R_4 = \frac{R_1}{K_1} = 1.83 \times 10^{-3}\ \Omega, \qquad R_5 = \frac{R_1}{1 - K_1} = 9.16 \times 10^{-5}\ \Omega \tag{7.78a}$$

$$R_C = \frac{r_1}{K_2} = \infty, \qquad R_D = \frac{r_1}{1 - K_2} = r_1 \tag{7.78b}$$

Choose again

$$r_1 = 1\ \text{k}\Omega, \qquad r_2 = 10\ \text{k}\Omega \tag{7.79}$$

to realize the resistance ratio $\gamma = r_1/r_2 = 0.1$, obtaining from (7.78b)

$$R_D = r_1 = 1\ \text{k}\Omega \tag{7.80}$$

Finally, we magnitude-scale the network by a factor of 10^8 to yield practical element values. The results are given by

$$C_1 = C_2 = 0.01\ \mu\text{F}, \qquad R_1 = 8.73\ \text{k}\Omega, \qquad R_2 = 115\ \text{k}\Omega \tag{7.81a}$$

$$R_3 = \infty, \qquad R_4 = 183\ \text{k}\Omega, \qquad R_5 = 9.16\ \text{k}\Omega \tag{7.81b}$$

Since $R_3 = \infty$ there is no need to compute R_6 and R_7. The final network realization for the band-pass transfer voltage-ratio function of (7.75) is shown in Figure 7.13.

Before we turn our attention to the design of multiple-amplifier biquads, we consider the realization of the general band-pass function

$$H(s) = \frac{(\omega_p/Q_p)s}{s^2 + (\omega_p/Q_p)s + \omega_p^2} \tag{7.82}$$

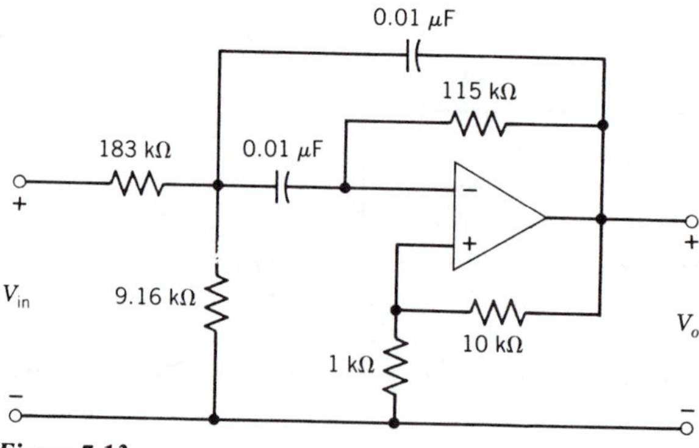

Figure 7.13

by the SAB. Comparing this equation with (7.38), we can make the following identifications:

$$K=1, \qquad b_2=K_2=0, \qquad b_0=0 \tag{7.83a}$$

$$b_1=\omega_p/Q_p, \qquad a_1=\omega_p/Q_p, \qquad a_0=\omega_p^2 \tag{7.83b}$$

obtaining from (7.40c), (7.40d), and (7.39d)

$$R_3=\infty, \qquad R_2=\frac{1}{a_0 R_1 C_1 C_2}=\frac{1}{R_1 C_1 C_2 \omega_p^2} \tag{7.84a}$$

$$a_1=\frac{1}{R_2 C_1}+\frac{1}{R_2 C_2}-\frac{r_1}{R_1 C_2 r_2}=\frac{\omega_p}{Q_p} \tag{7.84b}$$

or

$$\omega_p=\frac{1}{\sqrt{R_1 R_2 C_1 C_2}} \tag{7.85a}$$

$$Q_p=\frac{1/\sqrt{R_1 R_2 C_1 C_2}}{1/R_2 C_1+1/R_2 C_2-r_1/R_1 C_2 r_2} \tag{7.85b}$$

With $K_2=0$, we obtain from (7.35b)

$$R_C=\infty, \qquad R_D=r_1 \tag{7.86}$$

The resulting band-pass filter is shown in Figure 7.14, the transfer function of which is determined from (7.85) in conjunction with (7.39b) as

$$H(s)=\frac{V_o(s)}{V_{\text{in}}(s)}=\frac{-\dfrac{K_1(1+r_1/r_2)}{R_1 C_2}s}{s^2+(1/R_2 C_1+1/R_2 C_2-r_1/R_1 C_2 r_2)s+1/R_1 R_2 C_1 C_2} \tag{7.87}$$

Let us next consider the synthesis of an equal-capacitance band-pass filter. Let

$$C_1=C_2=1\text{ F}, \qquad \gamma=r_1/r_2 \tag{7.88}$$

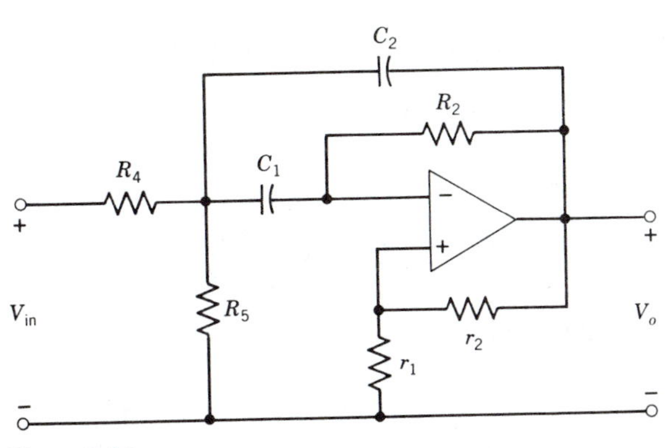

Figure 7.14

Then from (7.42)

$$R_1 = \frac{2\gamma Q_p}{\omega_p(-1+\sqrt{1+8\gamma Q_p^2})} \tag{7.89a}$$

$$R_2 = \frac{1}{R_1\omega_p^2} \tag{7.89b}$$

$$K_1 = \frac{-b_1 R_1}{1+\gamma} \tag{7.89c}$$

Thus, for the band-pass filter of Figure 7.14 to be physically realizable, the coefficient b_1 has to be negative. In the case of positive b_1, we simply change the sign of b_1 and proceed with the synthesis, because in doing so neither the magnitude nor the delay of the transfer function is affected. The parameter γ can be chosen to minimize sensitivity. To properly select γ, we must first evaluate the sensitivities of the biquadratic parameters to the passive elements. These sensitivities are readily obtained from (7.85), as follows:

$$S^{\omega_p}_{R_1, R_2, C_1, C_2} = -\tfrac{1}{2}, \qquad S^{\omega_p}_{r_1, r_2} = 0 \tag{7.90}$$

For Q_p sensitivities, we appeal to (6.15)

$$S^{Q_p}_x = S^{\omega_p}_x - S^{B_p}_x \tag{7.91}$$

where as in (5.156) the bandwidth B_p is defined by the equation

$$B_p = \frac{\omega_p}{Q_p} = \frac{1}{R_2 C_1} + \frac{1}{R_2 C_2} - \frac{\gamma}{R_1 C_2} \tag{7.92}$$

obtaining

$$S^{Q_p}_{R_1} = -S^{Q_p}_{R_2} = -\tfrac{1}{2} - \gamma Q_p \sqrt{\frac{R_2 C_1}{R_1 C_2}} \tag{7.93a}$$

$$S^{Q_p}_{C_1} = -S^{Q_p}_{C_2} = -\tfrac{1}{2} + Q_p \sqrt{\frac{R_1 C_2}{R_2 C_1}} \tag{7.93b}$$

To compute the Q_p sensitivities to r_1 and r_2, we apply the identity

$$S^{Q_p}_x = S^{Q_p}_\gamma S^\gamma_x \tag{7.94}$$

and obtain

$$S^{Q_p}_{r_1} = -S^{Q_p}_{r_2} = \gamma Q_p \sqrt{\frac{R_2 C_1}{R_1 C_2}} \tag{7.95}$$

The realized gain constant is from (7.87)

$$K_r = -\frac{K_1(1+\gamma)}{R_1 C_2} \tag{7.96}$$

Its sensitivities to the element changes are

$$S_{R_1,C_2}^{K_r} = -1, \qquad S_{R_2,C_1}^{K_r} = 0, \qquad S_{r_1}^{K_r} = -S_{r_2}^{K_r} = \frac{\gamma}{1+\gamma} \tag{7.97}$$

Observe that R_1 and R_2 and C_1 and C_2 appear in (7.93) and (7.95) in the form of ratios R_2/R_1 and C_2/C_1. It is convenient to introduce

$$\delta = R_2/R_1, \qquad \lambda = C_2/C_1 \tag{7.98}$$

Substituting these in (7.93) and (7.95) yields

$$S_{R_1}^{Q_p} = -S_{R_2}^{Q_p} = -\tfrac{1}{2} - \gamma Q_p \sqrt{\frac{\delta}{\lambda}} = \tfrac{1}{2} - Q_p \left(1 + \frac{1}{\lambda}\right) \sqrt{\frac{\lambda}{\delta}} \tag{7.99a}$$

$$S_{C_1}^{Q_p} = -S_{C_2}^{Q_p} = -\tfrac{1}{2} + Q_p \sqrt{\frac{\lambda}{\delta}} = \tfrac{1}{2} + Q_p \left(\gamma - \frac{1}{\delta}\right) \sqrt{\frac{\delta}{\lambda}} \tag{7.99b}$$

$$S_{r_1}^{Q_p} = -S_{r_2}^{Q_p} = \gamma Q_p \sqrt{\frac{\delta}{\lambda}} = \frac{(\lambda+1)Q_p}{\sqrt{\lambda\delta}} - 1 \tag{7.99c}$$

In the case of equal-capacitance design of (7.88), $\lambda = 1$ and the above sensitivity functions reduce to

$$S_{R_1}^{Q_p} = -S_{R_2}^{Q_p} = -\tfrac{1}{2} - \gamma Q_p \sqrt{\delta} = \tfrac{1}{2} - \frac{2Q_p}{\sqrt{\delta}} \tag{7.100a}$$

$$S_{C_1}^{Q_p} = -S_{C_2}^{Q_p} = -\tfrac{1}{2} + \frac{Q_p}{\sqrt{\delta}} = \tfrac{1}{2} + \left(\gamma - \frac{1}{\delta}\right) Q_p \sqrt{\delta} \tag{7.100b}$$

$$S_{r_1}^{Q_p} = -S_{r_2}^{Q_p} = \gamma Q_p \sqrt{\delta} = \frac{2Q_p}{\sqrt{\delta}} - 1 \tag{7.100c}$$

An inspection of the sensitivity equations (7.90), (7.97), and (7.100) shows that the pole frequency ω_p is minimally sensitive to the change of the passive elements and that the Q_p sensitivities are directly proportional to Q_p. For a fixed resistance ratio $\delta = R_2/R_1$, the Q_p sensitivities can be reduced by decreasing $\gamma = r_1/r_2$. Likewise, for a fixed γ, the Q_p sensitivities can be reduced by increasing δ. In other words, the low Q_p sensitivities can be traded for large resistance spreads. Also, a check of (7.99) reveals that as the capacitance ratio $\lambda = C_2/C_1$ increases, some Q_p sensitivities increase while others decrease. For this reason, little design freedom is lost by setting $\lambda = 1$ or using the equal-capacitance design.

As an alternative to the design formulas (7.89) where γ is fixed, suppose that we fixed δ instead of γ, so that δ can be chosen to minimize the network sensitivity. Then γ cannot be chosen independently. We shall express R_1, R_2, γ, and K_1 in terms of ω_p, Q_p, b_1, and δ. To this end, we solve (7.89) and (7.98) for R_1, R_2, γ, and K_1. The results are given by

$$R_1 = \frac{1}{\sqrt{\delta}\,\omega_p}, \qquad R_2 = \frac{\sqrt{\delta}}{\omega_p} \tag{7.101a}$$

$$\gamma = \frac{2}{\delta} - \frac{1}{\sqrt{\delta Q_p}} \tag{7.101b}$$

$$K_1 = \frac{-b_1\sqrt{\delta Q_p}}{\omega_p(\delta Q_p - \sqrt{\delta} + 2Q_p)} \tag{7.101c}$$

To complete our analysis, we should also determine the effect of the finite op-amp gain-bandwidth on the sensitivities. First, we compute the transfer function of the network of Figure 7.14 by assuming finite gain A for the op amp. This analysis results in the more general transfer function

$$H(s) = \frac{-\dfrac{K_1 A(s)(\gamma+1)s}{R_1C_2[A(s)+\gamma+1]}}{s^2 + \left(\dfrac{1}{R_2C_1} + \dfrac{1}{R_2C_2} + \dfrac{\gamma - \gamma A(s) + 1}{R_1C_2[A(s)+\gamma+1]}\right)s + \dfrac{1}{R_1R_2C_1C_2}} \tag{7.102}$$

Consider the terms containing A. For large amplifier gain, they can be approximated by retaining only the first two terms in the binomial expansions, as follows:

$$\frac{A(s)}{A(s)+\gamma+1} = \left[1 + \frac{\gamma+1}{A(s)}\right]^{-1} = 1 - \frac{\gamma+1}{A(s)} + \cdots \tag{7.103a}$$

$$\frac{-\gamma A(s)+\gamma+1}{A(s)+\gamma+1} = \left[-\gamma + \frac{\gamma+1}{A(s)}\right]\left[1 + \frac{\gamma+1}{A(s)}\right]^{-1}$$

$$= \left[-\gamma + \frac{\gamma+1}{A(s)}\right]\left[1 - \frac{\gamma+1}{A(s)} + \cdots\right] \tag{7.103b}$$

Many op amps have a gain characteristic that can be approximated by a single pole. Therefore, the gain function can be represented by the equation

$$A(s) = \frac{A_0 s_0}{s + s_0} \tag{7.104}$$

where s_0 is a low-frequency pole typically around the frequency of 10 Hz, and A_0 is the dc gain. At $s = jA_0s_0$ the gain of the amplifier is approximately unity. Thus, the term A_0s_0 is known as the *gain-bandwidth product*. In active filter applications where frequency is above 100 Hz, the amplifier gain characteristic (7.104) is more conveniently approximated by the expression

$$A(s) = A_0 s_0 / s \tag{7.105}$$

Substituting this in (7.103), using only two terms in the binomial expansions, and applying the resulting expressions in (7.102) yields the approximation

$$H(s) \approx \frac{-\dfrac{K_1 s}{R_1C_2}(\gamma+1)\left[1 - \dfrac{(\gamma+1)s}{B_0}\right]}{\left[1 + \dfrac{(\gamma+1)^2}{B_0R_1C_2}\right]s^2 + \left(\dfrac{1}{R_2C_1} + \dfrac{1}{R_2C_2} - \dfrac{\gamma}{R_1C_2}\right)s + \dfrac{1}{R_1R_2C_1C_2}} \tag{7.106}$$

where $B_0 = A_0 s_0$. Dividing the numerator and denominator by the s^2 coefficient in the denominator and using the expansion

$$\left[1+\frac{(\gamma+1)^2}{B_0 R_1 C_2}\right]^{-1} = 1-\frac{(\gamma+1)^2}{B_0 R_1 C_2}+\cdots \tag{7.107}$$

in the resulting equation, the transfer function can be approximated by

$$H(s) \approx \frac{-\dfrac{K_1 s}{R_1 C_2}(\gamma+1)\left[1-\dfrac{(\gamma+1)s}{B_0}-\dfrac{(\gamma+1)^2}{B_0 R_1 C_2}\right]}{s^2+\left(\dfrac{1}{R_2 C_1}+\dfrac{1}{R_2 C_2}-\dfrac{\gamma}{R_1 C_2}\right)\left[1-\dfrac{(\gamma+1)^2}{B_0 R_1 C_2}\right]s+\dfrac{1}{R_1 R_2 C_1 C_2}\left[1-\dfrac{(\gamma+1)^2}{B_0 R_1 C_2}\right]} \tag{7.108}$$

From this the pole frequency and pole Q are identified as

$$\omega_p = \frac{1}{\sqrt{R_1 R_2 C_1 C_2}}\sqrt{1-\frac{(\gamma+1)^2}{B_0 R_1 C_2}} \tag{7.109a}$$

$$Q_p = \frac{(1/\sqrt{R_1 R_2 C_1 C_2})\sqrt{1-(\gamma+1)^2/B_0 R_1 C_2}}{(1/R_2 C_1 + 1/R_2 C_2 - \gamma/R_1 C_2)[1-(\gamma+1)^2/B_0 R_1 C_2]} \tag{7.109b}$$

From these expressions, the sensitivities of the pole frequency and pole Q to the gain-bandwidth B_0 can be approximated by the equations

$$S_{B_0}^{\omega_p} \approx -S_{B_0}^{Q_p} \approx \frac{\omega_p(\gamma+1)^2}{2B_0}\sqrt{\frac{R_2 C_1}{R_1 C_2}} = \frac{\omega_p(\gamma+1)^2}{2B_0}\sqrt{\frac{\delta}{\lambda}} \tag{7.110}$$

Observe that the sensitivities to B_0 are directly proportional to the pole frequency. They also increase with increasing δ or decreasing in λ. In contrast, the Q_p sensitivities to the passive elements decrease with increasing δ. In the case of the equal-capacitance design, (7.110) reduces to

$$S_{B_0}^{\omega_p} \approx -S_{B_0}^{Q_p} \approx \frac{\omega_p(\gamma+1)^2\sqrt{\delta}}{2B_0} \tag{7.111}$$

Example 7.7 Using the design formulas (7.101), realize the band-pass transfer function

$$H(s) = \frac{-600s}{s^2+600s+10^7} \tag{7.112}$$

The pole frequency and pole Q are found to be

$$\omega_p = 3.16\times 10^3 \text{ rad/s}, \qquad Q_p = 5.27 \tag{7.113a}$$

For our purposes, choose $R_2/R_1 = 30$ or from (7.98)

$$\delta = 30 \tag{7.113b}$$

Substituting these in (7.101) gives

$$R_1 = 5.77 \times 10^{-5}\ \Omega, \qquad R_2 = 1.73 \times 10^{-3}\ \Omega \tag{7.114a}$$

$$\gamma = 0.032, \qquad K_1 = 0.03356 \tag{7.114b}$$

where $b_1 = -600$. The capacitances were chosen to be

$$C_1 = C_2 = 1\ \text{F} \tag{7.115}$$

Magnitude-scaling the network by a factor of 10^8 yields

$$C_1 = C_2 = 0.01\ \mu\text{F}, \qquad R_1 = 5.77\ \text{k}\Omega, \qquad R_2 = 173\ \text{k}\Omega \tag{7.116}$$

The other element values are found from (7.35) and (7.114b) to be

$$R_4 = 172\ \text{k}\Omega, \qquad R_5 = 5.97\ \text{k}\Omega, \qquad r_1 = 320\ \Omega \tag{7.117}$$

where $r_2 = 10\ \text{k}\Omega$, as in (7.79).

The sensitivities of the network to the passive elements are determined from (7.90), (7.95), (7.97), and (7.100) as

$$S_{R_1,R_2,C_1,C_2}^{\omega_p} = -\tfrac{1}{2}, \qquad S_{r_1,r_2}^{\omega_p} = 0 \tag{7.118a}$$

$$S_{R_1}^{Q_p} = -S_{R_2}^{Q_p} = -1.42, \qquad S_{C_1}^{Q_p} = -S_{C_2}^{Q_p} = 0.46 \tag{7.118b}$$

$$S_{r_1}^{Q_p} = -S_{r_2}^{Q_p} = 0.92, \qquad S_{R_1,C_2}^{K_r} = -1 \tag{7.118c}$$

$$S_{R_2,C_1}^{K_r} = 0, \qquad S_{r_1}^{K_r} = -S_{r_2}^{K_r} = 0.03 \tag{7.118d}$$

For the sensitivities to the gain-bandwidth product, we apply (7.111) and obtain

$$S_{B_0}^{\omega_p} \approx -S_{B_0}^{Q_p} = \frac{3.16 \times 10^3 (0.032 + 1)^2 \sqrt{30}}{2 \times 2\pi \times 10^6} = 0.0015 \tag{7.119}$$

where the gain-bandwidth product of a typical op amp is taken to be†

$$B_0 = A_0 s_0 = 10^5 \times 20\pi = 2\pi \times 10^6 \tag{7.120}$$

with $A_0 = 10^5$ and $s_0 = 20\pi$ rad/s.

These sensitivities are to be compared with the sensitivities of the network of Figure 7.13, which are found to be

$$S_{R_1}^{Q_p} = -S_{R_2}^{Q_p} = -2.41, \qquad S_{C_1}^{Q_p} = -S_{C_2}^{Q_p} = 0.96 \tag{7.121a}$$

$$S_{r_1}^{Q_p} = -S_{r_2}^{Q_p} = 1.91, \qquad S_{B_0}^{\omega_p} = -S_{B_0}^{Q_p} = 0.011 \tag{7.121b}$$

We see a reduction in Q_p sensitivities of more than 48% with respect to all the passive elements, while the active sensitivities are increased slightly. Our conclusion is that by a proper choice of the resistance ratio $R_2/R_1 = \delta$, we may minimize the sensitivity of the network. However, in many situations, this choice of resistance ratio may be

†Typical parameter values of A_0 and s_0 for the op amps used in active filters are given as follows: 10 Hz $\leqslant s_0/2\pi \leqslant 100$ Hz, $A_0 > 10^4$.

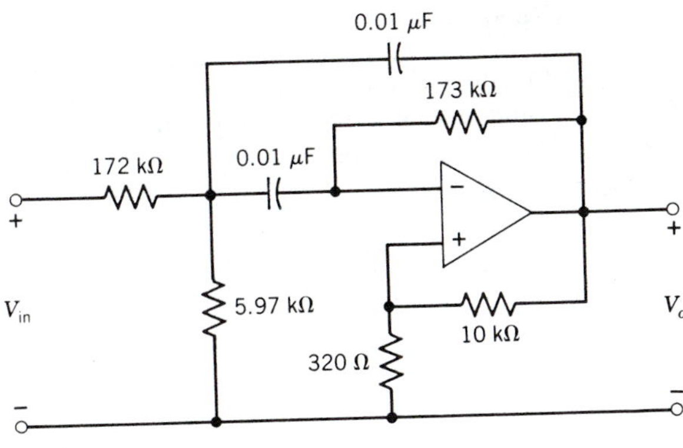

Figure 7.15

difficult. This difficulty is resolved by choosing δ so that the total gain deviation is minimized.

Finally, the realized gain constant is from (7.96)

$$K_r = -600 \tag{7.122}$$

as expected. The complete network realization is shown in Figure 7.15.

7.4 MULTIPLE-AMPLIFIER GENERAL BIQUADS

In this section, we introduce general biquads that use three or more op amps for the realization of the biquadratic function. In spite of its requiring more op amps than the single-amplifier configuration discussed in the preceding section, the structure is attractive because we can derive several benefits, often simultaneously. It can realize the general biquadratic function having low-pass, band-pass, high-pass, band-elimination, or all-pass characteristic with no exception. The network generally has low sensitivity and can easily be tuned to match the nominal requirements, allowing the pole frequency and pole Q to be independently adjusted. In addition, it permits the simultaneous realization of a variety of filter characteristics with the minimum number of topological changes. However, the price paid for these features is the increased power dissipation and noise. This added power dissipation may be critical in some applications. The noise generated in the op amps effectively limits the minimum signal level. To obtain large values of signal attenuation in the stopband of a filter characteristic, low noise levels are necessary, requiring the use of fewer op amps. It is always a good practice to carefully evaluate the filter performance and power and noise requirements before a particular topology is chosen. For a very high pole Q design, where Q_p is of order of 50 or higher, the SAB realization is impractical. As we show in this section, the use of multiple op-amp networks can lead to biquads with low sensitivities, very large pole Q, and easier-to-fabricate components.

7.4.1 Basic Building Blocks

The basic building blocks for the multiple-amplifier biquads are the inverting amplifier of Figure 7.16(*a*), the noninverting amplifier of Figure 7.16(*b*), and the summing amplifier of Figure 7.17. The transfer functions of the networks of Figures 7.16(*a*) and (*b*) are given, respectively, by

$$H(s)=\frac{V_o(s)}{V_{\text{in}}(s)}=-\frac{Z_2(s)}{Z_1(s)} \tag{7.123a}$$

$$H(s)=\frac{V_o(s)}{V_{\text{in}}(s)}=1+\frac{r_2}{r_1} \tag{7.123b}$$

The output voltage of the summing amplifier can be expressed in terms of the input voltages V_i and E_j by the equation

$$V_o(s)=\sum_{i=1}^{n}\frac{1+R_fG}{r_ig}E_i(s)-\sum_{i=1}^{m}\frac{R_f}{R_i}V_i(s) \tag{7.124}$$

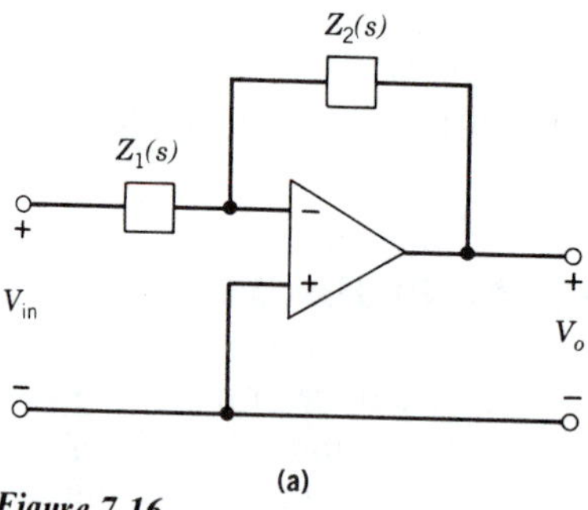

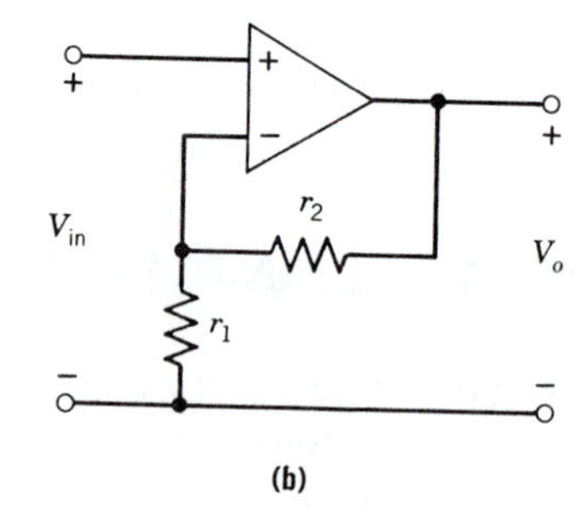

Figure 7.16

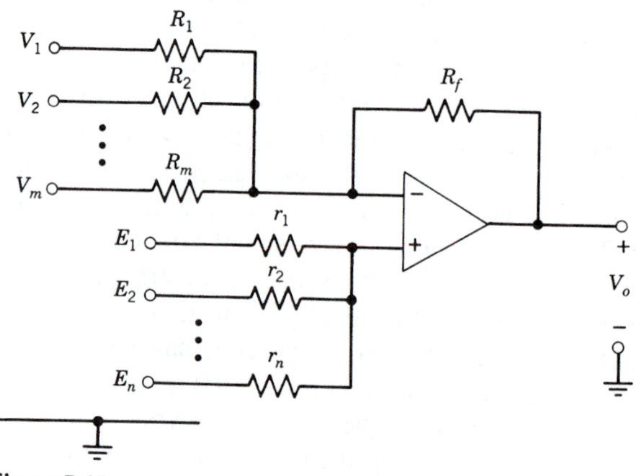

Figure 7.17

where

$$G=\sum_{i=1}^{m}\frac{1}{R_i},\qquad g=\sum_{i=1}^{n}\frac{1}{r_i} \tag{7.125}$$

These blocks can be used to generate a variety of biquadratic functions. Consider, for example, the network of Figure 7.18. Applying (7.123), the nodal voltages V_1, V_2, and V_3 are related by the expressions

$$\frac{V_3(s)}{V_2(s)}=-\frac{R_6}{R_5} \tag{7.126a}$$

$$\frac{V_2(s)}{V_1(s)}=-\frac{1/R_2C_2}{s} \tag{7.126b}$$

$$V_1(s)=\frac{-1/R_4C_1}{s+1/R_1C_1}V_{\text{in}}(s)+\frac{-1/R_3C_1}{s+1/R_1C_1}V_3(s) \tag{7.126c}$$

Eliminating V_2 and V_3 in these equations yields

$$\left[1+\frac{1/R_3C_1}{s+1/R_1C_1}\cdot\frac{R_6}{R_2R_5C_2s}\right]V_1(s)=\frac{-1/R_4C_1}{s+1/R_1C_1}V_{\text{in}}(s) \tag{7.127}$$

or

$$\frac{V_1(s)}{V_{\text{in}}(s)}=\frac{-(1/R_4C_1)s}{s^2+(1/R_1C_1)s+R_6/R_2R_3R_5C_1C_2}\triangleq H_{BP}(s) \tag{7.128}$$

This equation is recognized to possess the band-pass characteristic.

On the other hand, if we eliminate V_1 in (7.126b) and (7.128), the resulting function exhibits the low-pass characteristic:

$$\frac{V_2(s)}{V_{\text{in}}(s)}=-\frac{1/R_2C_2}{s}\cdot\frac{V_1(s)}{V_{\text{in}}(s)}=\frac{1/R_2R_4C_1C_2}{s^2+(1/R_1C_1)s+R_6/R_2R_3R_5C_1C_2}\triangleq H_{LP}(s) \tag{7.129}$$

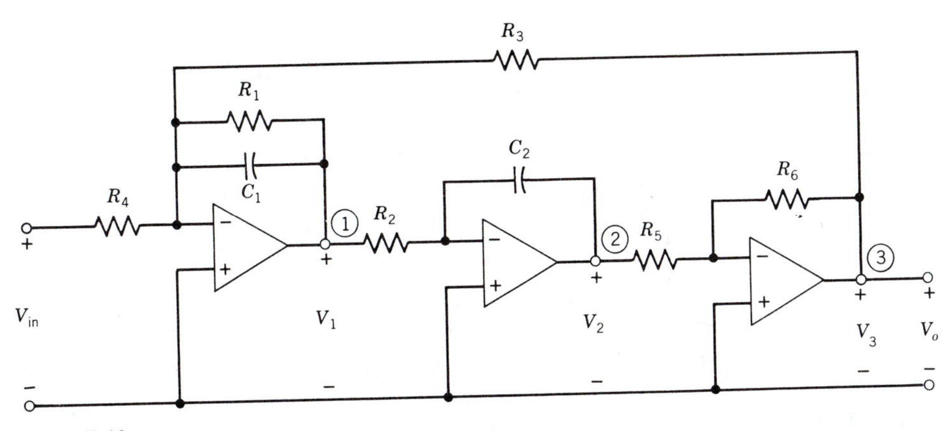

Figure 7.18

Finally, combining (7.126a) and (7.129) gives the transfer function between V_3 and V_{in} as

$$\frac{V_3(s)}{V_{in}(s)} = -\frac{R_6}{R_5} \cdot \frac{V_2(s)}{V_{in}(s)} = \frac{-R_6/R_2R_4R_5C_1C_2}{s^2+(1/R_1C_1)s+R_6/R_2R_3R_5C_1C_2} \tag{7.130}$$

which again possesses the low-pass characteristic except with the reversal of sign in comparison with (7.129).

7.4.2 **The Summing Four-Amplifier Biquad**

In this section, we show that with the addition of a summing amplifier of Figure 7.17, the basic configuration of Figure 7.18 can be used to realize the general biquadratic function

$$H(s) = -K\frac{s^2+b_1s+b_0}{s^2+a_1s+a_0} \tag{7.131}$$

Refer to the network of Figure 7.19, which is the same as that of Figure 7.18 except that a summing amplifier is inserted, where the inputs to the summing amplifier are the nodal voltages in the basic circuit of Figure 7.18. The output of the summing amplifier is found to be

$$V_o(s) = -\frac{R_{10}}{R_7}V_1(s) - \frac{R_{10}}{R_8}V_3(s) - \frac{R_{10}}{R_9}V_{in}(s) \tag{7.132}$$

The resulting transfer function obtained by substituting (7.128) and (7.130) in (7.132) is given by the equation

$$H(s) = \frac{V_o(s)}{V_{in}(s)} = \frac{R_6R_{10}/R_2R_4R_5R_8C_1C_2+(R_{10}/R_4R_7C_1)s}{s^2+(1/R_1C_1)s+R_6/R_2R_3R_5C_1C_2} - \frac{R_{10}}{R_9} \tag{7.133}$$

We next manipulate (7.131) into the form of (7.133), as follows:

$$H(s) = K\frac{(a_0-b_0)-(b_1-a_1)s}{s^2+a_1s+a_0} - K \tag{7.134}$$

Comparing this with (7.133), we can make the following identifications:

$$a_1 = \frac{1}{R_1C_1}, \qquad a_0 = \frac{R_6}{R_2R_3R_5C_1C_2} \tag{7.135a}$$

$$K = \frac{R_{10}}{R_9}, \qquad a_1 - b_1 = \frac{R_9}{R_4R_7C_1} \tag{7.135b}$$

$$a_0 - b_0 = \frac{R_6R_9}{R_2R_4R_5R_8C_1C_2} \tag{7.135c}$$

There are five equations in twelve unknown variables, ten resistors and two capacitors. Thus, we can fix any seven of them and solve the remainder of the variables in terms of these. The resulting solution, however, may not be physically realizable.

One simple realizable choice is the following:

$$C_1 = C_2 = 1\text{ F}, \qquad R_2 = R_3 = R_5 = R_6 = R_7 = R_{10} = R \tag{7.136}$$

Substituting these in (7.135) and then solving the resulting equations for R_1, R_4, R_8, R_9, and R in terms of the known coefficients a_0, a_1, b_0, b_1, and K, the results are given by

$$R_1 = \frac{1}{a_1}, \qquad R_4 = \frac{1}{K(a_1 - b_1)} \tag{7.137a}$$

$$R_2 = R_3 = R_5 = R_6 = R_7 = R_{10} = R = \frac{1}{\sqrt{a_0}} \tag{7.137b}$$

$$R_8 = \frac{a_1 - b_1}{a_0 - b_0}, \qquad R_9 = \frac{1}{K\sqrt{a_0}} \tag{7.137c}$$

These design equations will give nonnegative element values provided that

$$a_1 \geqq b_1, \qquad a_0 \geqq b_0 \tag{7.138}$$

The first condition requires that the magnitude of the real part of the zeros be smaller than that of the poles. This constraint is satisfied for all the approximation functions discussed in Chapter 2. The second constraint requires that the pole frequency be larger than the zero frequency. This restriction can be removed by using V_2 as the input to the summing amplifier instead of V_3 as shown in Figure 7.19, with the dashed

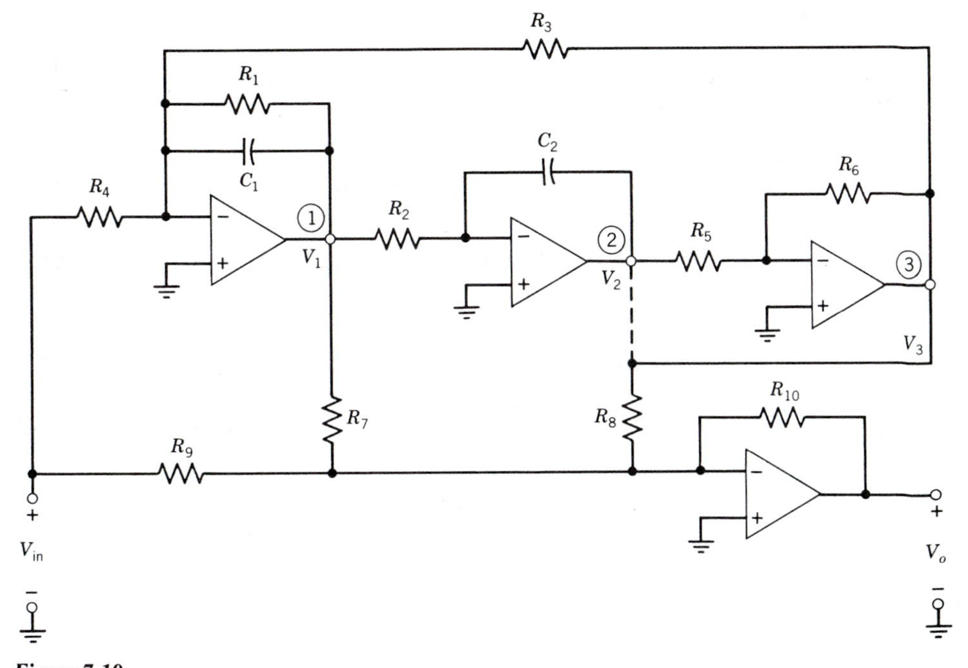

Figure 7.19

line indicating the new connection. Under this condition, the output of the summing amplifier becomes

$$V_o(s) = -\frac{R_{10}}{R_7}V_1(s) - \frac{R_{10}}{R_8}V_2(s) - \frac{R_{10}}{R_9}V_{in}(s) \tag{7.139}$$

Since V_2 is related to V_3 by (7.126a), (7.139) can be written as

$$V_o(s) = -\frac{R_{10}}{R_7}V_1(s) + \frac{R_5R_{10}}{R_6R_8}V_3(s) - \frac{R_{10}}{R_9}V_{in}(s) \tag{7.140}$$

For our choice of the resistances shown in (7.136), $R_5 = R_6$, (7.140) is identical to that of (7.132) except that the sign of the second term on the right-hand side is changed from minus to plus. This is equivalent to replacing $a_0 - b_0$ in (7.137c) by $b_0 - a_0$, everything else being the same. In fact, if $a_1 < b_1$ a simple modification of the network can also accommodate such a situation. The four-amplifier network realization of the general biquadratic function of Figure 7.19 is due to Tow† and Thomas.‡ We illustrate the above procedure by the following examples.

Example 7.8 Realize the following all-pass function

$$H(s) = -\frac{s^2 - 100s + 10^6}{s^2 + 100s + 10^6} \tag{7.141}$$

Comparing this equation with (7.131), we can make the following identifications:

$$K = 1, \qquad b_1 = -100, \qquad b_0 = 10^6, \qquad a_1 = 100, \qquad a_0 = 10^6 \tag{7.142}$$

Following the design of (7.136) by letting

$$C_1 = C_2 = 1\text{ F}, \qquad R_2 = R_3 = R_5 = R_6 = R_7 = R_{10} = R \tag{7.143}$$

the other element values are found from (7.137), as follows:

$$R_1 = \frac{1}{100} = 0.01\ \Omega, \qquad R_4 = \frac{1}{1 \times (100 + 100)} = 0.005\ \Omega \tag{7.144a}$$

$$R_2 = R_3 = R_5 = R_6 = R_7 = R_{10} = R = \frac{1}{\sqrt{10^6}} = 0.001\ \Omega \tag{7.144b}$$

$$R_8 = \frac{100 + 100}{10^6 - 10^6} = \infty, \qquad R_9 = \frac{1}{1\sqrt{10^6}} = 0.001\ \Omega \tag{7.144c}$$

To obtain the practical element values, we magnitude-scale the network by a factor of 10^7, giving

$$C_1 = C_2 = 0.1\ \mu\text{F}, \qquad R_1 = 100\text{ k}\Omega \tag{7.145a}$$

†J. Tow, "Design formulas for active *RC* filters using operational-amplifier biquad," *Electronics Letters*, vol. 5, 1969, pp. 339–341.

‡L. C. Thomas, "The Biquad: Part I—Some practical design considerations," *IEEE Trans. Circuit Theory*, vol. CT-18, 1971, pp. 350–357.

$$R_2 = R_3 = R_5 = R_6 = R_7 = R_{10} = 10\ \text{k}\Omega \tag{7.145b}$$

$$R_4 = 50\ \text{k}\Omega, \qquad R_8 = \infty, \qquad R_9 = 10\ \text{k}\Omega \tag{7.145c}$$

The final network realization is shown in Figure 7.20.

Example 7.9 Realize the following two transfer functions simultaneously:

$$H_1(s) = -\frac{s^2 + 10^6}{s^2 + 100s + 10^6} \tag{7.146}$$

$$H_2(s) = \frac{400s}{s^2 + 100s + 10^6} \tag{7.147}$$

We first realize the transfer function H_1. Comparing (7.146) with (7.131), we can make the following identifications:

$$K = 1, \qquad b_1 = 0, \qquad b_0 = 10^6, \qquad a_1 = 100, \qquad a_0 = 10^6 \tag{7.148}$$

We again follow the design of (7.136) by letting

$$C_1 = C_2 = 1\ \text{F}, \qquad R_2 = R_3 = R_5 = R_6 = R_7 = R_{10} = R \tag{7.149}$$

Applying formulas (7.137) gives

$$R_1 = 0.01\ \Omega, \qquad R_4 = 0.01\ \Omega, \qquad R = 10^{-3}\ \Omega \tag{7.150a}$$

$$R_8 = \infty, \qquad R_9 = 10^{-3}\ \Omega \tag{7.150b}$$

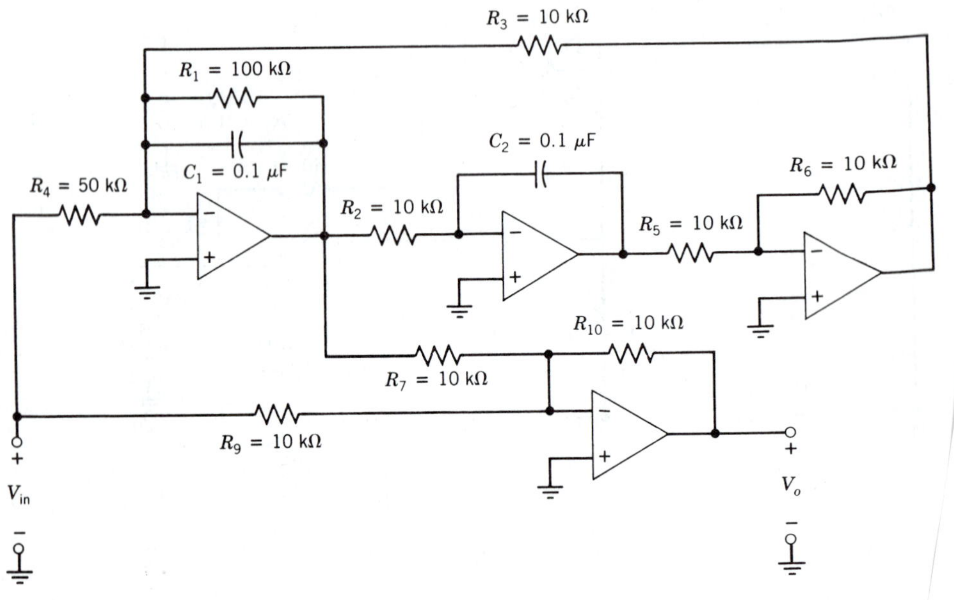

Figure 7.20

Magnitude-scaling the network by a factor of 10^7 yields

$$C_1 = C_2 = 0.1\ \mu\text{F}, \qquad R_1 = R_4 = 100\ \text{k}\Omega, \qquad R_8 = \infty \tag{7.151a}$$

$$R_2 = R_3 = R_5 = R_6 = R_7 = R_9 = R_{10} = 10\ \text{k}\Omega \tag{7.151b}$$

The network for the realization of the band-elimination function H_1 is shown in Figure 7.21 with V_{o1} as the output.

As indicated in (7.128), the network of Figure 7.21 gives the band-pass function between V_{in} and V_1. Its transfer function is obtained from (7.128) as

$$\frac{V_1(s)}{V_{\text{in}}(s)} = \frac{-100s}{s^2 + 100s + 10^6} \tag{7.152}$$

The desired gain constant for H_2 is 400. An inverting amplifier is required to provide a gain of $-400/100 = -4$. The transfer function between V_{in} and the output V_{o2} of the inverting amplifier gives the desired band-pass function H_2. The final network realization is shown in Figure 7.21.

We next consider the sensitivity of the summing four-amplifier biquad of Figure 7.19, the transfer function of which is given in (7.133). Since the dominant terms in effecting the sensitivity of the biquad are the pole frequency and pole Q, we shall

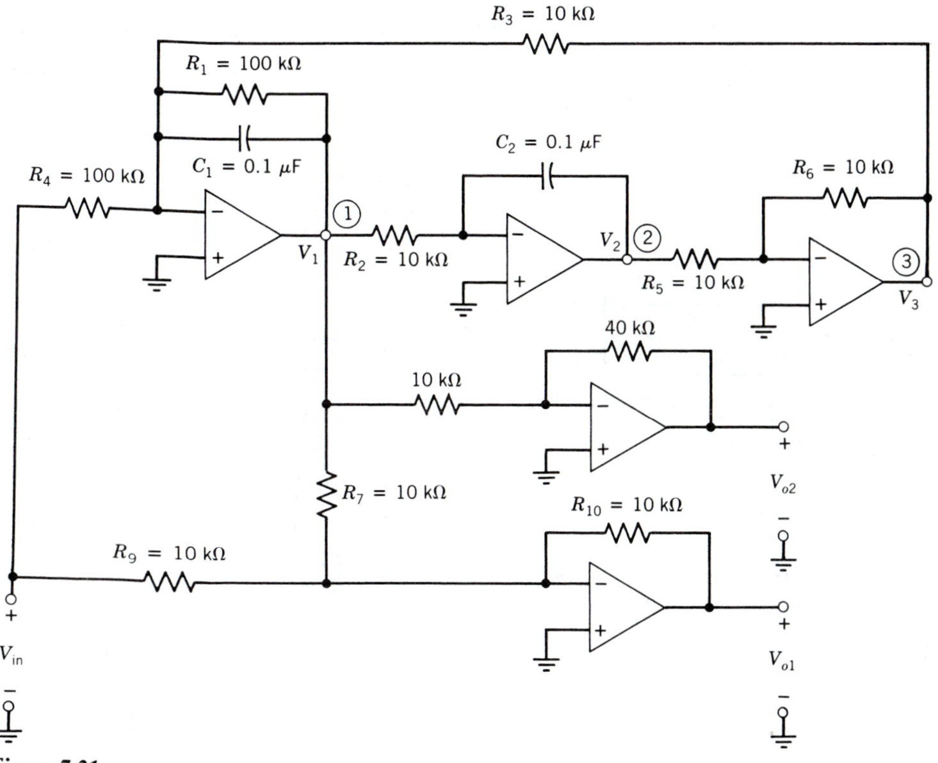

Figure 7.21

evaluate their sensitivities to the passive elements. First, the pole frequency ω_p and pole Q_p are determined from (7.133) as

$$\omega_p = R_2^{-1/2} R_3^{-1/2} R_5^{-1/2} R_6^{1/2} C_1^{-1/2} C_2^{-1/2} \tag{7.153}$$

$$Q_p = R_1 R_2^{-1/2} R_3^{-1/2} R_5^{-1/2} R_6^{1/2} C_1^{1/2} C_2^{-1/2} \tag{7.154}$$

obtaining

$$S^{\omega_p}_{R_2,R_3,R_5,C_1,C_2} = -\tfrac{1}{2}, \qquad S^{\omega_p}_{R_6} = \tfrac{1}{2} \tag{7.155a}$$

$$S^{Q_p}_{R_1} = 1, \qquad S^{Q_p}_{R_2,R_3,R_5} = -S^{Q_p}_{R_6} = -\tfrac{1}{2} \tag{7.155b}$$

$$S^{Q_p}_{C_1} = -S^{Q_p}_{C_2} = \tfrac{1}{2} \tag{7.155c}$$

To complete our analysis, we should also determine the effect of the finite op-amp gain-bandwidth on the sensitivities. First, we compute the transfer function of the network by assuming that all the op amps are identical with finite gain as in (7.102). Then we assume that the op amp can be modeled by a single pole at the origin as in (7.105). We next identify the approximate pole frequency and pole Q as in (7.109). After considerable computation, it can be shown† that the ω_p and Q_p sensitivities to the gain-bandwidth B_0 can be approximated by the expressions

$$S^{\omega_p}_{B_0} = \frac{3\omega_p}{2B_0}, \qquad S^{Q_p}_{B_0} = -\frac{4\omega_p Q_p}{B_0} \tag{7.156}$$

Observe that both ω_p and Q_p are minimally sensitive to the change of the passive elements and that the ω_p sensitivity to B_0 for very high pole Q design is much less than the Q_p sensitivity to B_0. This is in contrast to the situation in the single-amplifier biquad networks, where these two sensitivities had the same magnitude, as indicated in (7.111).

Example 7.10 Compute the active sensitivities of the networks of Figures 7.20 and 7.21. The pole frequency and pole Q are found to be

$$\omega_p = 10^3 \text{ rad/s}, \qquad Q_p = 10 \tag{7.157}$$

Assume that the gain-bandwidth of a typical op amp is given by

$$B_0 = 2\pi \times 10^6 \tag{7.158}$$

as in (7.120). Then from (7.156) the active sensitivities are obtained as

$$S^{\omega_p}_{B_0} = 0.00024, \qquad S^{Q_p}_{B_0} = -0.0064 \tag{7.159}$$

showing that Q_p sensitivity is much more than the ω_p sensitivity.

†D. Åkerberg and K. Mossberg," A versatile active *RC* building block with inherent compensation for the finite bandwidth of the amplifier," *IEEE Trans. Circuits Syst.*, vol. CAS-21, 1974, pp. 75–78.

7.4.3 The Three-Amplifier Biquad

In the preceding section, we showed how to insert a summing amplifier to realize the general biquadratic function. To avoid the use of an additional amplifier, we may feed fractions of the input forward into the input of each op amp as shown in Figure 7.22. To compute the transfer function of this network, we write the nodal equations at the three inverting terminals of the op amps, obtaining

$$-\left(\frac{1}{R_1}+C_1 s\right) V_1(s)-\frac{1}{R_3} V_3(s)-\frac{1}{R_4} V_{\text{in}}(s)=0 \tag{7.160a}$$

$$-\frac{1}{R_7} V_1(s)-\frac{1}{R_8} V_2(s)-\frac{1}{R_6} V_{\text{in}}(s)=0 \tag{7.160b}$$

$$-\frac{1}{R_2} V_2(s)-C_2 s V_3(s)-\frac{1}{R_5} V_{\text{in}}(s)=0 \tag{7.160c}$$

These equations are solved to yield the transfer function

$$H(s)=\frac{V_o(s)}{V_{\text{in}}(s)}=-\frac{(R_8/R_6)s^2+(R_8/R_1R_6C_1-R_8/R_4R_7C_1)s+R_8/R_3R_5R_7C_1C_2}{s^2+(1/R_1C_1)s+R_8/R_2R_3R_7C_1C_2} \tag{7.161}$$

This is to be compared with the biquadratic function

$$H(s)=K\frac{b_2s^2+b_1s+b_0}{s^2+a_1s+a_0} \tag{7.162}$$

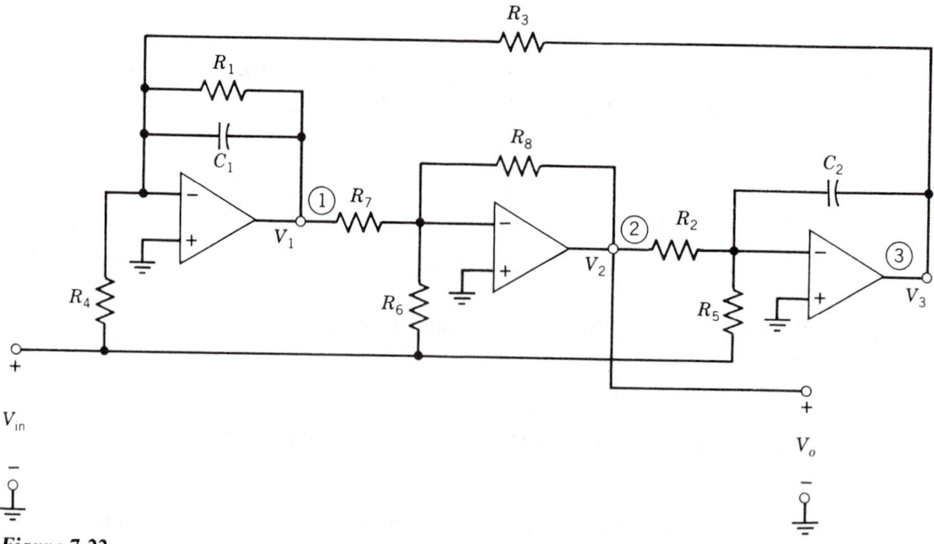

Figure 7.22

identifying

$$K = -1 \tag{7.163a}$$

$$b_2 = \frac{R_8}{R_6} \tag{7.163b}$$

$$b_1 = \frac{R_8}{R_1 R_6 C_1} - \frac{R_8}{R_4 R_7 C_1} \tag{7.163c}$$

$$b_0 = \frac{R_8}{R_3 R_5 R_7 C_1 C_2} \tag{7.163d}$$

$$a_1 = \frac{1}{R_1 C_1} \tag{7.163e}$$

$$a_0 = \frac{R_8}{R_2 R_3 R_7 C_1 C_2} \tag{7.163f}$$

Excluding (7.163a), there are five equations in ten unknowns. Therefore, five of the variables may be fixed. For our purposes, we choose C_1, C_2, R_8, and

$$k_1 = \sqrt{\frac{R_2 R_8 C_2}{R_3 R_7 C_1}}, \qquad k_2 = \frac{R_7}{R_8} \tag{7.164}$$

as the free parameters. Solving the other variables in terms of these and a's and b's yields the design formulas, as follows:

$$R_1 = \frac{1}{a_1 C_1} \tag{7.165a}$$

$$R_2 = \frac{k_1}{C_2 \sqrt{a_0}} \tag{7.165b}$$

$$R_3 = \frac{1}{k_1 k_2 C_1 \sqrt{a_0}} \tag{7.165c}$$

$$R_4 = \frac{1}{k_2 (b_2 a_1 - b_1) C_1} \tag{7.165d}$$

$$R_5 = \frac{k_1 \sqrt{a_0}}{b_0 C_2} \tag{7.165e}$$

$$R_6 = \frac{R_8}{b_2} \tag{7.165f}$$

Using these values in the network of Figure 7.22, the transfer functions at the output of the three op amps due to V_{in} can be expressed as

$$\frac{V_1(s)}{V_{in}(s)} = -k_2 \frac{(b_2 a_1 - b_1)s + (b_2 a_0 - b_0)}{s^2 + a_1 s + a_0} \tag{7.166a}$$

$$\frac{V_3(s)}{V_{in}(s)} = -\frac{(b_0 - b_2 a_0)s + (a_1 b_0 - a_0 b_1)}{k_1 \sqrt{a_0}(s^2 + a_1 s + a_0)} \tag{7.166b}$$

and $V_2/V_{in} = V_o/V_{in}$ is as given in (7.161) or (7.162) with $K = -1$.

To compute the sensitivities of the biquadratic parameters to the passive elements, we first determine the biquadratic parameters of the network. These are found from (7.161) as

$$\omega_p = \sqrt{\frac{R_8}{R_2 R_3 R_7 C_1 C_2}} = R_2^{-1/2} R_3^{-1/2} R_7^{-1/2} R_8^{1/2} C_1^{-1/2} C_2^{-1/2} \tag{7.167a}$$

$$Q_p = R_1 \sqrt{\frac{R_8 C_1}{R_2 R_3 R_7 C_2}} = R_1 R_2^{-1/2} R_3^{-1/2} R_7^{-1/2} R_8^{1/2} C_1^{1/2} C_2^{-1/2} \tag{7.167b}$$

$$\omega_z = \sqrt{\frac{R_6}{R_3 R_5 R_7 C_1 C_2}} = R_3^{-1/2} R_5^{-1/2} R_7^{-1/2} C_1^{-1/2} C_2^{-1/2} R_6^{1/2} \tag{7.167c}$$

$$Q_z = \frac{\sqrt{R_6/R_3 R_5 R_7 C_1 C_2}}{1/R_1 C_1 - R_6/R_4 R_7 C_1} \tag{7.167d}$$

Thus, from (6.9)–(6.12) we obtain

$$S^{\omega_p}_{R_2, R_3, R_7, C_1, C_2} = -\tfrac{1}{2}, \qquad S^{\omega_p}_{R_8} = \tfrac{1}{2} \tag{7.168a}$$

$$S^{Q_p}_{R_2, R_3, R_7, C_2} = -\tfrac{1}{2}, \qquad S^{Q_p}_{R_1} = 1, \qquad S^{Q_p}_{C_1, R_8} = \tfrac{1}{2} \tag{7.168b}$$

$$S^{\omega_z}_{R_3, R_5, R_7, C_1, C_2} = -\tfrac{1}{2}, \qquad S^{\omega_z}_{R_6} = \tfrac{1}{2} \tag{7.168c}$$

For the Q_z sensitivities, we appeal to (6.15)

$$S^{Q_z}_x = S^{\omega_z}_x - S^{B_z}_x \tag{7.169}$$

where

$$B_z = \frac{1}{R_1 C_1} - \frac{R_6}{R_4 R_7 C_1} \tag{7.170}$$

and obtain

$$S^{Q_z}_{R_3, R_5, C_2} = -S^{Q_z}_{C_1} = -\frac{1}{2} \tag{7.171a}$$

$$S^{Q_z}_{R_1} = \frac{R_4 R_7}{R_4 R_7 - R_1 R_6} \tag{7.171b}$$

$$S^{Q_z}_{R_4} = \frac{R_1 R_6}{R_1 R_6 - R_4 R_7} \tag{7.171c}$$

$$S^{Q_z}_{R_6} = \frac{1}{2} + \frac{R_1 R_6}{R_4 R_7 - R_1 R_6} \tag{7.171d}$$

$$S^{Q_z}_{R_7} = -\frac{1}{2} - \frac{R_1 R_6}{R_4 R_7 - R_1 R_6} \tag{7.171e}$$

Finally, we mention that since we have freedom to choose C_1, C_2, k_1, k_2, and R_8, one set of synthesis equations for realizing the general biquadratic function of (7.162) is obtained by setting

$$C_1=C_2=1\text{ F},\qquad k_1=1,\qquad k_2=1,\qquad R_8=R_2 \tag{7.172}$$

Then the remaining element values are determined from (7.165) as

$$R_1=\frac{1}{a_1},\qquad R_2=\frac{1}{\sqrt{a_0}},\qquad R_3=\frac{1}{\sqrt{a_0}} \tag{7.173a}$$

$$R_4=\frac{1}{b_2a_1-b_1},\qquad R_5=\frac{\sqrt{a_0}}{b_0},\qquad R_6=\frac{1}{b_2\sqrt{a_0}} \tag{7.173b}$$

We note that in the band-pass case, $b_2=b_0=0$ and R_4 becomes negative for $b_1>0$. To circumvent this difficulty, we simply change the sign of b_1, resulting neither in a change in magnitude nor in the delay.

Example 7.11 Realize the all-pass function

$$H(s)=-\frac{s^2-100s+10^6}{s^2+100s+10^6} \tag{7.174}$$

Comparing this equation with (7.162), we can make the following identifications:

$$K=-1,\qquad b_2=1,\qquad b_1=-100,\qquad b_0=10^6,\qquad a_1=100,\qquad a_0=10^6 \tag{7.175}$$

Following the design of (7.172) by letting

$$C_1=C_2=1\text{ F},\qquad k_1=k_2=1,\qquad R_8=R_2 \tag{7.176}$$

the other element values are found from (7.173), as follows:

$$R_1=0.01\ \Omega,\qquad R_2=10^{-3}\ \Omega,\qquad R_3=10^{-3}\ \Omega \tag{7.177a}$$

$$R_4=5\times10^{-3}\ \Omega,\qquad R_5=10^{-3}\ \Omega,\qquad R_6=10^{-3}\ \Omega \tag{7.177b}$$

From (7.164) we obtain

$$R_7=k_2R_8=R_8=R_2=10^{-3}\ \Omega \tag{7.178}$$

To obtain the practical element values, we magnitude-scale the network by a factor of 10^7, yielding

$$R_1=100\text{ k}\Omega,\qquad R_2=10\text{ k}\Omega,\qquad R_3=10\text{ k}\Omega \tag{7.179a}$$

$$R_4=50\text{ k}\Omega,\qquad R_5=10\text{ k}\Omega,\qquad R_6=10\text{ k}\Omega \tag{7.179b}$$

$$R_7=10\text{ k}\Omega,\qquad R_8=10\text{ k}\Omega,\qquad C_1=C_2=0.1\ \mu\text{F} \tag{7.179c}$$

The final network realization is shown in Figure 7.23.

We next compute the sensitivities of the biquadratic parameters to the passive elements. The ω_p, Q_p, and ω_z sensitivities are listed in (7.168). For the Q_z sensitivities,

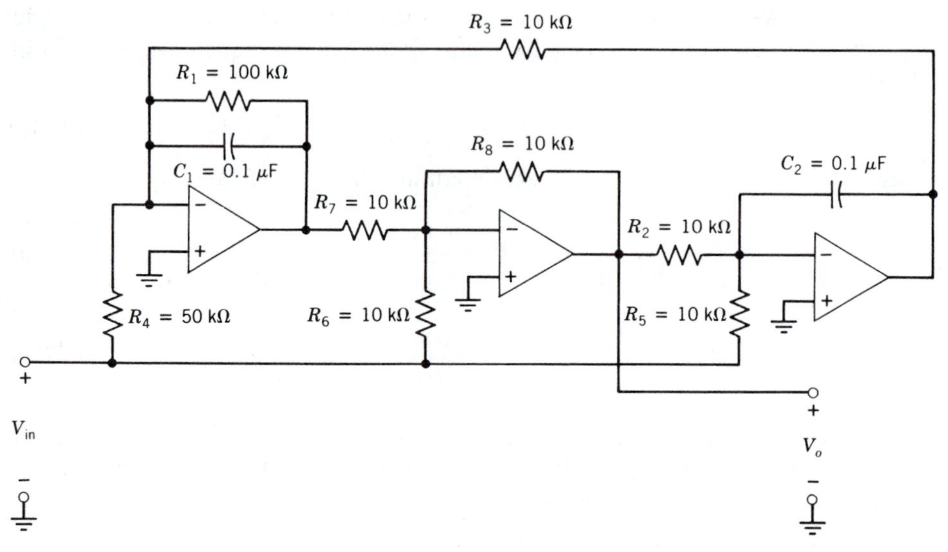

Figure 7.23

we apply (7.171) and obtain

$$S^{Q_z}_{R_3,R_5,C_2} = -S^{Q_z}_{C_1} = -\tfrac{1}{2} \tag{7.180a}$$

$$S^{Q_z}_{R_1} = -1, \qquad S^{Q_z}_{R_4} = 2 \tag{7.180b}$$

$$S^{Q_z}_{R_6} = -\tfrac{3}{2}, \qquad S^{Q_z}_{R_7} = \tfrac{3}{2} \tag{7.180c}$$

7.4.4 **The State-Variable Biquad**

In this section, we introduce the state-variable biquad. The name *state variable* is derived from the fact that the state-variable technique was used in the development of this network.

Consider the network of Figure 7.24, the nodal equations of which are given by

$$V_1(s) = \frac{R_2(R_{10}+R_3)}{R_3(R_1+R_2)} V_{in}(s) + \frac{R_1(R_{10}+R_3)}{R_3(R_1+R_2)} V_2(s) - \frac{R_{10}}{R_3} V_3(s) \tag{7.181a}$$

$$V_2(s) = -\frac{V_1(s)}{sC_1R_8} \tag{7.181b}$$

$$V_3(s) = -\frac{V_2(s)}{sC_2R_9} \tag{7.181c}$$

From these equations, we obtain

$$\frac{V_3(s)}{V_{in}(s)} = K_1 \frac{1/R_8R_9C_1C_2}{s^2 + K_1(R_1/R_2R_8C_1)s + R_{10}/R_3R_8R_9C_1C_2} \tag{7.182a}$$

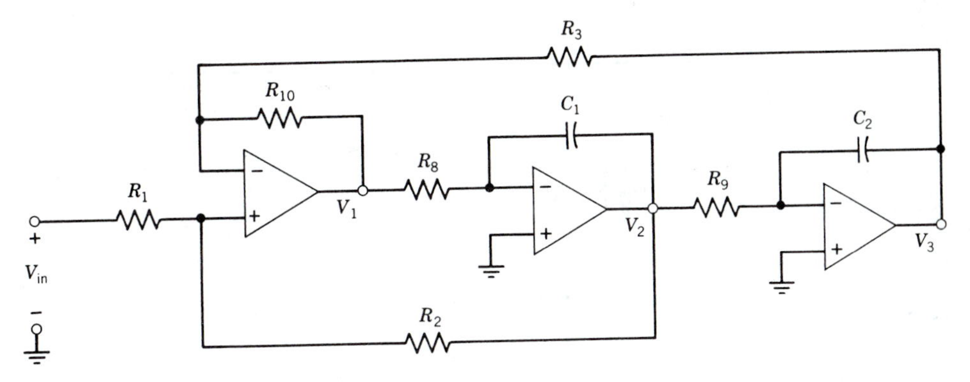

Figure 7.24

$$\frac{V_2(s)}{V_{in}(s)}=K_1\frac{-s/R_8C_1}{s^2+K_1(R_1/R_2R_8C_1)s+R_{10}/R_3R_8R_9C_1C_2} \tag{7.182b}$$

$$\frac{V_1(s)}{V_{in}(s)}=K_1\frac{s^2}{s^2+K_1(R_1/R_2R_8C_1)s+R_{10}/R_3R_8R_9C_1C_2} \tag{7.182c}$$

where

$$K_1=\frac{R_2(R_3+R_{10})}{R_3(R_1+R_2)} \tag{7.183}$$

To realize the general biquadratic function

$$H(s)=K\frac{s^2+b_1s+b_0}{s^2+a_1s+a_0} \tag{7.184}$$

we insert a summing amplifier of Figure 7.17 in Figure 7.24. The resulting network is shown in Figure 7.25, where the output of this summing amplifier is a weighted sum of V_1, V_2, and V_3:

$$V_o(s)=\frac{R_5(R_6+R_7)}{R_7(R_4+R_5)}V_1(s)+\frac{R_4(R_6+R_7)}{R_7(R_4+R_5)}V_3(s)-\frac{R_6}{R_7}V_2(s) \tag{7.185}$$

Substituting (7.182) in (7.185) gives

$$H(s)=\frac{V_o(s)}{V_{in}(s)}=K_1\frac{R_5(R_6+R_7)}{R_7(R_4+R_5)}\cdot\frac{s^2+\dfrac{R_6(R_4+R_5)}{R_5R_8C_1(R_6+R_7)}s+\dfrac{R_4}{R_5R_8R_9C_1C_2}}{s^2+\dfrac{K_1R_1}{R_2R_8C_1}s+\dfrac{R_{10}}{R_3R_8R_9C_1C_2}} \tag{7.186}$$

Comparing this equation with (7.184), we can make the following identifications:

$$K=K_1\frac{R_5(R_6+R_7)}{R_7(R_4+R_5)}=\frac{R_2R_5(R_3+R_{10})(R_6+R_7)}{R_3R_7(R_1+R_2)(R_4+R_5)} \tag{7.187a}$$

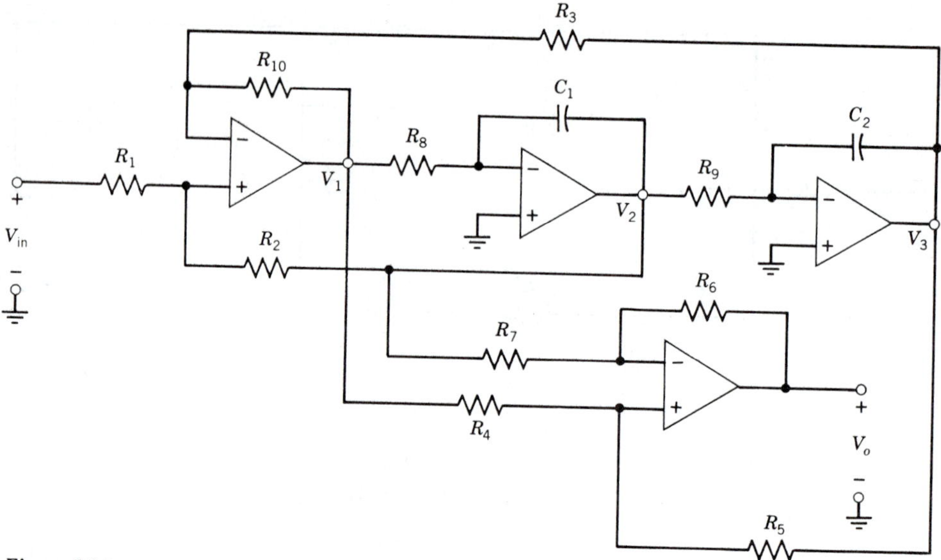

Figure 7.25

$$b_0 = \frac{R_4}{R_5 R_8 R_9 C_1 C_2} \tag{7.187b}$$

$$b_1 = \frac{R_6(R_4 + R_5)}{R_5 R_8 C_1 (R_6 + R_7)} \tag{7.187c}$$

$$a_0 = \frac{R_{10}}{R_3 R_8 R_9 C_1 C_2} \tag{7.187d}$$

$$a_1 = \frac{K_1 R_1}{R_2 R_8 C_1} = \frac{R_1(R_3 + R_{10})}{R_3 R_8 C_1 (R_1 + R_2)} \tag{7.187e}$$

Observe that there are five equations in twelve unknown variables. Therefore, seven of them can be fixed, and we solve the remaining variables in terms of these. The seven chosen variables can be used to satisfy additional constraints such as to minimize the sensitivity, to reduce the spread of resistor and capacitor values, and to equalize the input and output impedance matching.

To complete our analysis, we next compute the biquadratic parameter sensitivities to the passive elements. From (7.186) the biquadratic parameters are found to be

$$\omega_p = \sqrt{\frac{R_{10}}{R_3 R_8 R_9 C_1 C_2}} = R_3^{-1/2} R_8^{-1/2} R_9^{-1/2} R_{10}^{1/2} C_1^{-1/2} C_2^{-1/2} \tag{7.188a}$$

$$Q_p = \frac{\sqrt{R_{10}/R_3 R_8 R_9 C_1 C_2}}{\dfrac{R_1(R_3 + R_{10})}{R_3 R_8 C_1 (R_1 + R_2)}} = \frac{(R_1 + R_2)}{R_1(R_3 + R_{10})} \sqrt{\frac{R_3 R_8 R_{10} C_1}{R_9 C_2}} \tag{7.188b}$$

$$\omega_z = \sqrt{\frac{R_4}{R_5 R_8 R_9 C_1 C_2}} \tag{7.188c}$$

$$Q_z = \frac{\sqrt{R_4/R_5 R_8 R_9 C_1 C_2}}{\dfrac{R_6(R_4+R_5)}{R_5 R_8 C_1(R_6+R_7)}} = \frac{(R_6+R_7)}{R_6(R_4+R_5)}\sqrt{\frac{R_4 R_5 R_8 C_1}{R_9 C_2}} \tag{7.188d}$$

From these equations we obtain

$$S^{\omega_p}_{R_3,R_8,R_9,C_1,C_2} = -S^{\omega_p}_{R_{10}} = -\tfrac{1}{2} \tag{7.189a}$$

$$S^{Q_p}_{R_8,C_1} = \tfrac{1}{2}, \qquad S^{Q_p}_{R_9,C_2} = -\tfrac{1}{2} \tag{7.189b}$$

$$S^{Q_p}_{R_2} = -S^{Q_p}_{R_1} = \frac{R_2}{R_1+R_2}, \qquad S^{Q_p}_{R_3} = -S^{Q_p}_{R_{10}} = \frac{R_{10}-R_3}{2(R_{10}+R_3)} \tag{7.189c}$$

$$S^{\omega_z}_{R_5,R_8,R_9,C_1,C_2} = -S^{\omega_z}_{R_4} = -\tfrac{1}{2} \tag{7.189d}$$

$$S^{Q_z}_{R_8,C_1} = \tfrac{1}{2}, \qquad S^{Q_z}_{R_9,C_2} = -\tfrac{1}{2} \tag{7.189e}$$

$$S^{Q_z}_{R_7} = -S^{Q_z}_{R_6} = \frac{R_7}{R_6+R_7}, \qquad S^{Q_z}_{R_4} = -S^{Q_z}_{R_5} = \frac{R_5-R_4}{2(R_4+R_5)} \tag{7.189f}$$

Example 7.12 Realize the all-pass function

$$H(s) = \frac{s^2-5s+100}{s^2+5s+100} \tag{7.190}$$

Comparing (7.190) and (7.184), we can make the following identifications:

$$K=1, \qquad b_1=-5, \qquad b_0=100, \qquad a_1=5, \qquad a_0=100 \tag{7.191}$$

Substituting $b_1=-5$ in (7.187c) gives

$$-5 = \frac{R_6(R_4+R_5)}{R_5 R_8 C_1(R_6+R_7)} \tag{7.192}$$

This equation cannot be satisfied without resulting in the negative element values. To circumvent this difficulty, we reconnect the inputs to the summing amplifier as shown in Figure 7.26. The summing amplifier output becomes

$$V_o(s) = \frac{1+R_8'/R_7}{1/R_4+1/R_5+1/R_6}\left[\frac{V_1(s)}{R_4}+\frac{V_2(s)}{R_5}+\frac{V_3(s)}{R_6}\right] \tag{7.193}$$

Substituting (7.182) in (7.193), the new transfer function can be written as

$$H(s) = \frac{K_1 R_4 R_5 R_6(R_7+R_8')}{R_4 R_7(R_5 R_6+R_4 R_5+R_4 R_6)} \cdot \frac{s^2-(R_4/R_5 R_8 C_1)s+R_4/R_6 R_8 R_9 C_1 C_2}{s^2+(K_1 R_1/R_2 R_8 C_1)s+R_{10}/R_3 R_8 R_9 C_1 C_2} \tag{7.194}$$

Comparing this equation with (7.190), we see that there are five constraints in thirteen variables. Therefore, we can fix eight of them. Let

$$C_1=C_2=1\text{ F}, \qquad R_2=R_4=R_8=R_8'=R_9=R_{10}=1\ \Omega \tag{7.195}$$

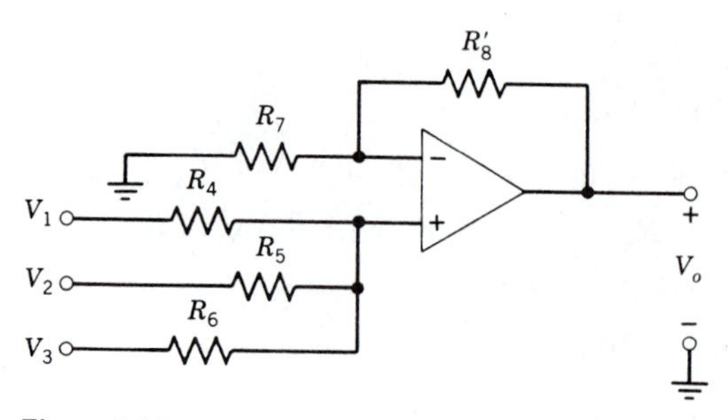

Figure 7.26

Using these in (7.194) and comparing the resulting equation with (7.190) yields

$$R_5 = 0.2\ \Omega, \qquad R_6 = 0.01\ \Omega \tag{7.196a}$$

$$K_1 R_1 = 5 \qquad R_3 = 0.01\ \Omega \tag{7.196b}$$

$$\frac{K_1(1+1/R_7)}{1+1/R_5+1/R_6} = 1, \qquad K_1 = \frac{101}{1+R_1} \tag{7.196c}$$

giving

$$R_1 = 0.052\ \Omega, \qquad K_1 = 96, \qquad R_7 = 9.6\ \Omega \tag{7.197}$$

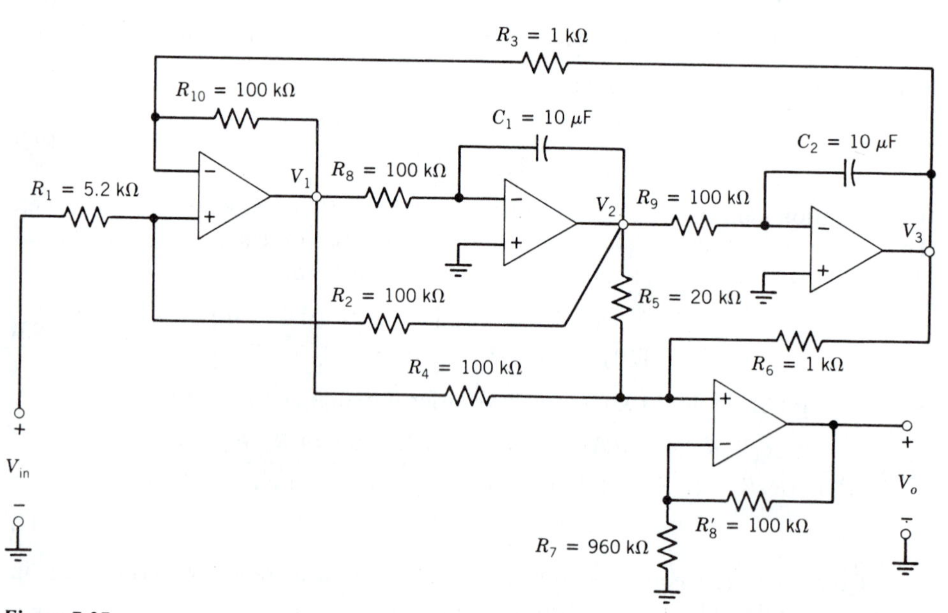

Figure 7.27

To obtain practical element values, we magnitude-scale the network by a factor of 10^5, yielding

$$R_1 = 5.2\ \text{k}\Omega, \qquad R_2 = R_4 = R_8 = R_8' = R_9 = R_{10} = 100\ \text{k}\Omega \tag{7.198a}$$

$$R_3 = 1\ \text{k}\Omega, \qquad R_5 = 20\ \text{k}\Omega, \qquad R_6 = 1\ \text{k}\Omega \tag{7.198b}$$

$$R_7 = 960\ \text{k}\Omega, \qquad C_1 = C_2 = 10\ \mu\text{F} \tag{7.198c}$$

The final network is shown in Figure 7.27.

The pole frequency and pole Q sensitivities are determined directly from (7.189) as

$$S^{\omega_p}_{R_3,R_8,R_9,C_1,C_2} = -S^{\omega_p}_{R_{10}} = -\tfrac{1}{2}, \qquad S^{Q_p}_{R_8,C_1} = \tfrac{1}{2} \tag{7.199a}$$

$$S^{Q_p}_{R_9,C_2} = -\tfrac{1}{2}, \qquad S^{Q_p}_{R_2} = -S^{Q_p}_{R_1} = 0.95 \tag{7.199b}$$

$$S^{Q_p}_{R_3} = -S^{Q_p}_{R_{10}} = 0.49 \tag{7.199c}$$

7.5 **Summary and Suggested Readings**

This chapter is a continuation of our study on the general subject of active filter synthesis using RC amplifier circuits. We showed that there are general biquads that are capable of realizing the low-pass, high-pass, band-pass, band-elimination, and all-pass functions with relatively low sensitivities. These networks provide an economical means for implementing a universal kind of building block for realizing filters of virtually any order and of any type.

We began this chapter by studying a single-amplifier band-pass biquad using the positive feedback topology. We then showed that by a slight modification of the frequency-transformation procedure, high-pass active RC filters can be obtained from those of the low-pass type. The result of this transformation can be effected by replacing the resistors in the low-pass structure by capacitors of appropriate values, and capacitors in the low-pass network by resistors of proper values. The op amp that is modeled as a VCVS is not affected by the transformation, and neither are the gain constant nor the resistance ratios. This frequency transformation is known as the *RC–CR transformation*, which may also be used to convert a high-pass structure to a low-pass type.

For the single-amplifier general biquad, we introduced a negative feedback RC amplifier network known as the SAB or the STAR, which is highly suited to thin-film, hybrid integrated-circuit fabrication, and has been used extensively for audio- and voice-frequency signal-processing applications. This network is capable of realizing all types of biquadratic response except the low-pass response. However, the low-pass Sallen and Key network can bé used for these purposes.

To demonstrate how the finite op-amp gain-bandwidth affects the network sensitivity, we computed the pole frequency and pole Q sensitivities of the band-pass SAB network to the gain-bandwidth product. The general approach to the problem is as follows: First, compute the transfer function of the network by assuming a finite op-amp gain. The op-amp characteristic is then approximated by a single pole at the origin. Terms are then expanded by binomial expansions and the resulting equation will yield the first-order approximation of the transfer function, from which the sensitivity functions are computed. We found that these sensitivities are directly proportional to the pole frequency. They also depend on the spread of the resistances and the capacitances.

One of the limitations of the SAB network is that for a very high pole Q design, where Q_p is of order of 50 or higher, its realization is impractical. To circumvent this difficulty, we introduced general biquads that use three or more op amps. In spite of using more op amps, the multiple-amplifier biquads are attractive because the networks generally have low sensitivities and can be tuned easily to match the nominal requirements. In addition, it realizes the general biquadratic function with no exception, and permits the simultaneous realization of a variety of filter characteristics with the minimum number of topological changes. However, the price paid in achieving this is the increased power dissipation and noise. This added power dissipation may be critical in some applications.

Three multiple-amplifier biquads were discussed. The *summing four-amplifier biquads* uses three basic blocks to generate the low-pass and band-pass characteristics. A summing amplifier is then inserted to produce the desired biquadratic function. To avoid the use of this additional summing amplifier, we may feed the input forward into the input of each op amp and produce the general biquadratic function, resulting in a *three-amplifier biquad.* Finally, we studied the *state-variable biquad*, which was developed by using the state-variable technique. The state-variable biquad is a four-amplifier biquad composed of three basic blocks and one summing amplifier, is generally flexible, and has good performance and low sensitivities.

An excellent introduction to various aspects of the biquad can be found in Daryanani [2] and Ghausi and Laker [5]. A more detailed discussion on a number of biquads is given by Huelsman and Allen [6]. A step-by-step active filter design is outlined by Tow [14]. For a more detailed study on various biquads introduced in this chapter, see appropriate papers listed in the references [4], [7], and [13].

REFERENCES

1. D. Åkerberg and K. Mossberg, "A versatile active *RC* building block with inherent compensation for the finite bandwidth of the amplifier," *IEEE Trans. Circuits Syst.*, vol. CAS-21, 1974, pp. 75–78.
2. G. Daryanani 1976, *Principles of Active Network Synthesis and Design*," Chapters 8–10. New York: John Wiley.
3. P. E. Fleischer and J. Tow, "Design formulas for biquad active filters using three operational amplifiers," *Proc. IEEE*, vol. 61, 1973, pp. 662–663.
4. J. J. Friend, C. A. Harris, and D. Hilberman, "STAR: An active biquadratic filter section," *IEEE Trans. Circuits Syst.*, vol. CAS-22, 1975, pp. 115–121.

5. M. S. Ghausi and K. R. Laker 1981, *Modern Filter Design: Active RC and Switched Capacitor*, Chapter 4. Englewood Cliffs, NJ: Prentice–Hall.
6. L. P. Huelsman and P. E. Allen 1980, *Introduction to the Theory and Design of Active Filters*, Chapters 4 and 5. New York: McGraw–Hill.
7. W. J. Kerwin, L. P. Huelsman, and R. W. Newcomb, "State-variable synthesis for insensitive integrated circuit transfer functions," *IEEE J. Solid-State Circuits*, vol. SC-2, 1967, pp. 87–92.
8. H. Y-F. Lam 1979, *Analog and Digital Filters: Design and Realization*, Chapter 10. Englewood Cliffs, NJ: Prentice–Hall.
9. W. B. Mikhael and B. B. Bhattacharyya, "A practical design for insensitive *RC*-active filters," *IEEE Trans. Circuits Syst.*, vol. CAS-22, 1975, pp. 407–415.
10. R. Tarmy and M. S. Ghausi, "Very high-*Q* insensitive active *RC* networks," *IEEE Trans. Circuit Theory*, vol. CT-17, 1970, pp. 358–366.
11. L. C. Thomas, "The biquad: Pt. I. Some practical design considerations," *IEEE Trans. Circuit Theory*, vol. CT-18, 1971, pp. 350–357.
12. L. C. Thomas, "The biquad: Pt. II. Multipurpose active filtering system," *IEEE Trans. Circuit Theory*, vol. CT-18, 1971, pp. 358–361.
13. J. Tow, "Design formulas for active *RC* filters using operational-amplifier biquad," *Electronics Letters*, vol. 5, 1969, pp. 339–341.
14. J. Tow, "A step-by-step active-filter design," *IEEE Spectrum*, vol. 6, 1969, pp. 64–68.
15. P. W. Vogel, "Method for phase correction in active *RC* circuits using two integrators," *Electronics Letters*, vol. 10, 1971, pp. 273–275.
16. W. Worobey and J. Rutkiewicz, "Tantalum thin-film *RC* circuit technology for a universal active filter," *IEEE Trans. Parts, Hybrids Packag.*, vol. PHP-12, 1976, pp. 276–282.

PROBLEMS

7.1 Synthesize the following transfer functions using the single-amplifier biquad of Figure 7.10:

(a) $$H(s)=\frac{s^2+10^6}{s^2+100s+10^6} \tag{7.200}$$

(b) $$H(s)=\frac{20s^2}{s^2+100s+10^6} \tag{7.201}$$

(c) $$H(s)=\frac{s^2-100s+10^6}{s^2+100s+10^6} \tag{7.202}$$

(d) $$H(s)=\frac{50s(s^2+10^6)}{(s^2+100s+10^6)(s^2+50s+10^6)} \tag{7.203}$$

(e) $$H(s)=\frac{s^2+10^8}{s^2+100s+10^6} \tag{7.204}$$

(f) $$H(s)=\frac{s^2+10^4}{s^2+100s+10^6} \tag{7.205}$$

7.2 Synthesize the transfer functions in Problem 7.1 using the summing four-amplifier biquad of Figure 7.19.

7.3 Synthesize the transfer functions in Problem 7.1 using the three-amplifier biquad of Figure 7.22.

7.4 Synthesize the transfer functions in Problem 7.1 using the state-variable biquad of Figure 7.25.

7.5 Using the Sallen and Key band-pass network of Figure 7.1, synthesize the band-pass transfer function

$$H(s)=\frac{600s}{s^2+600s+3\times 10^8} \tag{7.206}$$

7.6 By introducing an additional input to the noninverting terminal of the op amp of the network as shown in Figure 7.28, verify that the transfer function of the network is given by

$$H(s)=\frac{V_o(s)}{V_{\text{in}}(s)}=\frac{s^2+\dfrac{1}{R_2C_1}\left(1+\dfrac{C_1(R_1+R_2)}{R_1C_2}-\dfrac{1}{K_1}\right)s+\dfrac{1}{R_1R_2C_1C_2}}{s^2+\left(\dfrac{R_1+R_2}{R_1R_2C_2}\right)s+\dfrac{1}{R_1R_2C_1C_2}} \tag{7.207}$$

Show that this network can be used to realize a general all-pass function. Derive the necessary design formulas.

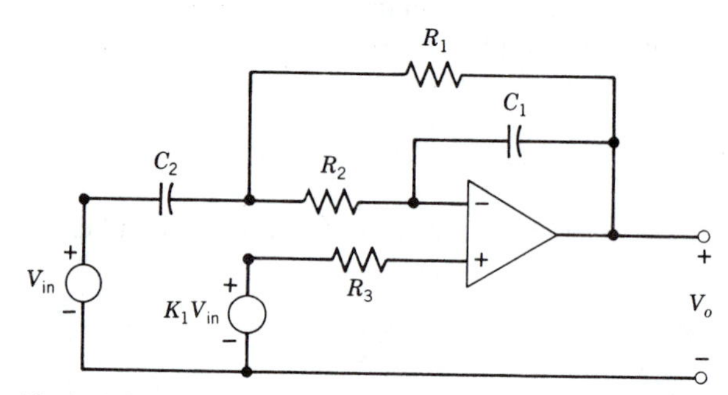

Figure 7.28

7.7 Verify that the transfer function of the network of Figure 7.29 can be expressed in the general form of a biquadratic function as

$$H(s)=\frac{V_o(s)}{V_{\text{in}}(s)}=-\frac{R_8}{R_6}\cdot\frac{s^2+\left(\dfrac{1}{R_1C_1}-\dfrac{R_6}{R_5C_2}\right)s+\left(\dfrac{R_6}{R_2R_4R_7C_1C_2}-\dfrac{R_6}{R_1R_5R_7C_1C_2}\right)}{s^2+\left(\dfrac{1}{R_1C_1}\right)s+\dfrac{R_8}{R_2R_3R_7C_1C_2}} \tag{7.208}$$

7.8 Refer to the filter network of Figure 7.29, the transfer function of which is shown in (7.208). Using this equation, derive design formulas for a band-elimination function. Compute the pole frequency and pole Q sensitivities.

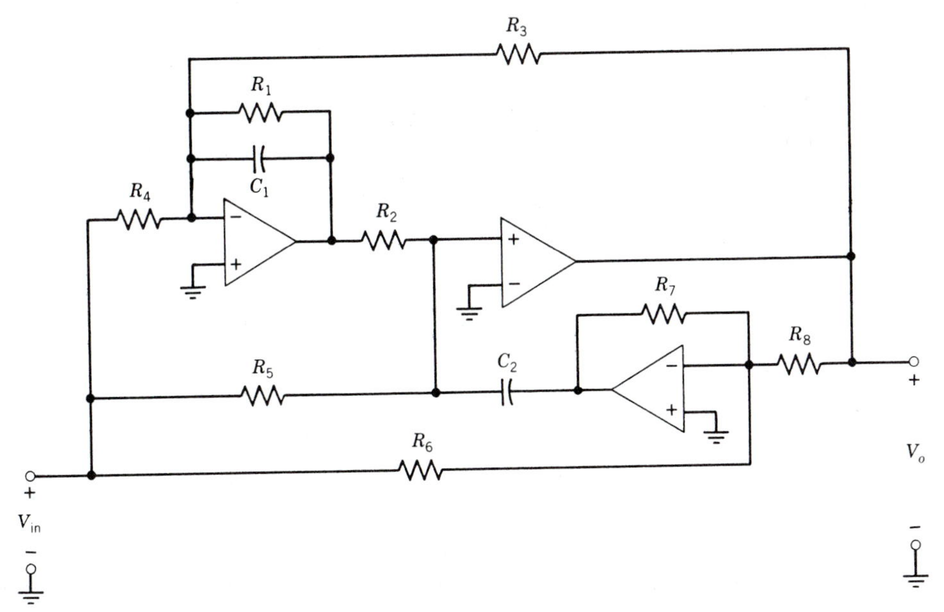

Figure 7.29

7.9 Show that if the biquad of Figure 7.19 is used to realize the high-pass function

$$H(s) = -\frac{b_2 s^2}{s^2 + a_1 s + a_0} \tag{7.209}$$

the design formulas can be expressed as

$$R_1 = \frac{1}{a_1 C_1}, \qquad R_2 = \frac{k_1}{\sqrt{a_0} C_2}, \qquad R_3 = \frac{1}{k_1 \sqrt{a_0} C_1} \tag{7.210a}$$

$$R_4 = \frac{1}{k_2 b_2 a_1 C_1}, \qquad R_5 = R_6, \qquad R_7 = k_2 R_{10} \tag{7.210b}$$

$$R_8 = \frac{k_2 a_1 R_{10}}{k_1 \sqrt{a_0}}, \qquad R_9 = \frac{R_{10}}{b_2} \tag{7.210c}$$

where k_1 and k_2 are arbitrary nonnegative numbers.

7.10 If the biquad of Figure 7.19 is used to realize the all-pass function

$$H(s) = -K \frac{s^2 - a_1 s + a_0}{s^2 + a_1 s + a_0} \tag{7.211}$$

show that the design formulas can be expressed as

$$R_1 = \frac{1}{a_1 C_1}, \qquad R_2 = \frac{k_1}{\sqrt{a_0} C_2} \qquad R_3 = \frac{1}{k_1 \sqrt{a_0} C_1} \tag{7.212a}$$

$$R_4 = \frac{1}{2 k_2 a_1 C_1}, \qquad R_5 = R_6, \qquad R_7 = \frac{k_2 R_{10}}{K} \tag{7.212b}$$

$$R_8=\infty, \qquad R_9=\frac{R_{10}}{K} \tag{7.212c}$$

where k_1 and k_2 are arbitrary nonnegative numbers.

7.11 Show that if the biquad of Figure 7.19 is used to realize the notch filter function

$$H(s)=-\frac{b_2s^2+b_0}{s^2+a_1s+a_0} \tag{7.213}$$

the design formulas can be expressed as

$$R_1=\frac{1}{a_1C_1}, \qquad R_2=\frac{k_1}{\sqrt{a_0}C_2}, \qquad R_3=\frac{1}{k_1\sqrt{a_0}C_1} \tag{7.214a}$$

$$R_4=\frac{1}{k_2b_2a_1C_1}, \qquad R_5=R_6, \qquad R_7=k_2R_{10} \tag{7.214b}$$

$$R_8=\frac{k_2b_2a_1\sqrt{a_0}R_{10}}{k_1(b_2a_0-b_0)}, \qquad R_9=\frac{R_{10}}{b_2} \tag{7.214c}$$

where k_1 and k_2 are arbitrary nonnegative numbers.

7.12 Design a band-pass filter for which the pole frequency is 10^5 rad/s with a bandwidth of 10^3 rad/s. Scale the network so that all the element values are in a practical range.

7.13 Design a low-pass active *RC* filter using the specifications of Problem 2.3.

7.14 Design a high-pass active *RC* filter using the specifications of Problem 2.4.

7.15 Design a band-pass active *RC* filter using the specifications of Problem 2.7.

7.16 Design a band-elimination filter using the SAB and satisfying the specifications of Problem 2.6.

7.17 Design a band-elimination filter using the summing four-amplifier biquad satisfying the specifications of Problem 2.10.

7.18 Repeat Problem 7.16 by using the state-variable biquad.

7.19 Repeat Problem 7.17 by using the three-amplifier biquad.

7.20 Using the state-variable biquad realize a second-order all-pass function having pole and zero frequency $\omega_p=\omega_z=5\times10^4$ rad/s and pole and zero Q to be $Q_p=-Q_z=10$.

chapter 8

REALIZATION OF ACTIVE TWO-PORT NETWORKS

In Chapter 5 we indicated that there are two general methods for realizing a transfer function using active RC networks. The first such method was the cascade approach, in which active RC networks were used to realize first- and second-order functions. These individual realizations were connected in cascade to yield higher-order functions. The second general method was the direct approach, where a single network is used to realize the entire transfer function, as was done for passive filter design discussed in Chapters 3 and 4. A brief discussion of the direct approach was given in Section 5.5. In this chapter, we shall continue our discussion for using the direct method to realize network functions. Specifically, we show that based on passive prototype networks, certain portions of the passive realizations can be simulated by using active RC networks. Such a technique is referred to as a *passive network simulation method.*

One of the important advantages of an active RC simulated structure is that it emulates the low-sensitivity property of the passive prototype network. In addition, there is a large body of knowledge and experience with the passive networks, especially passive ladders, where tables of element values are readily available. The synthesis procedures for active RC simulated networks become very straightforward. Other advantages and disadvantages of this approach will be discussed when specific filter structures are introduced.

8.1 LADDER NETWORKS

The doubly terminated passive ladder structure of Figure 8.1 is attractive from the engineering viewpoint in that it is unbalanced, so that all the shunt branches can be grounded. Furthermore, the structure has very low sensitivity, which can be emulated by using active RC networks.

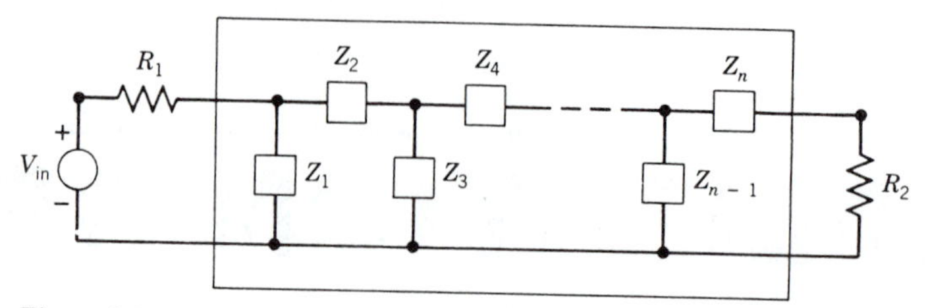

Figure 8.1

To establish this low-sensitivity property, consider the general doubly terminated LC network of Figure 8.2, the input impedance of which, as seen from input terminals 1–1′, may be expressed as a function of $s=j\omega$ as

$$Z_{11}(j\omega)=R_{11}(\omega)+jX_{11}(\omega) \tag{8.1}$$

From the input impedance and the input current, the input power to the LC network is found to be

$$\begin{aligned} P_1 &= |I_1(j\omega)|^2 \operatorname{Re} Z_{11}(j\omega) \\ &= \frac{|V_{\text{in}}|^2 R_{11}}{(R_1+R_{11})^2+X_{11}^2} \end{aligned} \tag{8.2}$$

Since the LC two-port network is lossless, the input power P_1 equals the output power P_2 or

$$P_2=\frac{|V_{\text{in}}|^2 R_{11}}{(R_1+R_{11})^2+X_{11}^2} \tag{8.3}$$

To calculate the sensitivity of P_2 to a network element x, we appeal to (6.1) and obtain

$$S_x^{P_2}=\frac{x}{P_2}\cdot\frac{\partial P_2}{\partial x}=\frac{x}{P_2}\left(\frac{\partial P_2}{\partial R_{11}}\cdot\frac{\partial R_{11}}{\partial x}+\frac{\partial P_2}{\partial X_{11}}\cdot\frac{\partial X_{11}}{\partial x}\right) \tag{8.4}$$

where from (8.3)

$$\frac{\partial P_2}{\partial R_{11}}=\frac{R_1^2-R_{11}^2+X_{11}^2}{[(R_1+R_{11})^2+X_{11}^2]^2}|V_{\text{in}}|^2 \tag{8.5a}$$

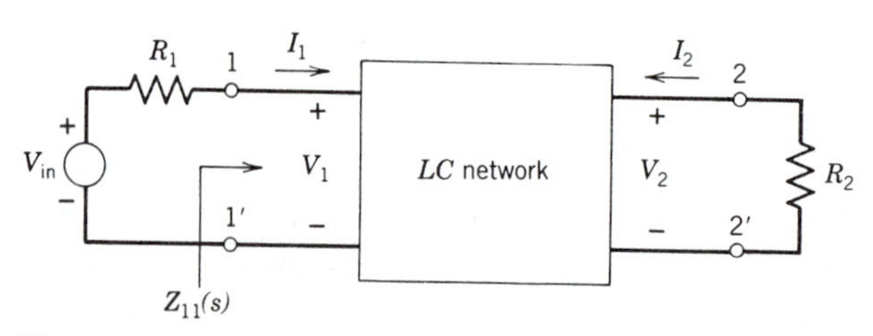

Figure 8.2

$$\frac{\partial P_2}{\partial X_{11}}=\frac{-2R_{11}X_{11}}{[(R_1+R_{11})^2+X_{11}^2]^2}|V_{in}|^2 \tag{8.5b}$$

A filter is normally designed to match the source to the load so that the maximum average power is transferred from the source to the load within the passband. It is well known that this condition of maximum power transfer is achieved when the terminating impedances of a two-port network are conjugate-matched to the input and output impedances of the two-port network. From Figure 8.2, we see that for maximum power transfer we must have

$$Z_{11}(j\omega)=R_1 \tag{8.6}$$

or

$$R_{11}=R_1 \qquad \text{and} \qquad X_{11}=0 \tag{8.7}$$

At the frequencies where these conditions are satisfied, the power transmission of the filter is maximum. Substituting (8.7) in (8.5) gives

$$\frac{\partial P_2}{\partial R_{11}}=0 \qquad \text{and} \qquad \frac{\partial P_2}{\partial X_{11}}=0 \tag{8.8}$$

Equation (8.8) states in effect that the sensitivity in (8.4) is zero at the frequencies of maximum power transfer:

$$S_x^{P_2}=0 \tag{8.9}$$

These frequencies will, of course, lie within the filter passband. Since the power P_2 is a smooth continuous function, it is reasonable to expect that the sensitivity in the vicinity of these frequencies will remain very small. Also, the frequencies of maximum power transfer are normally well distributed within the passband. Such a passive realization should therefore have a low sensitivity in the entire passband. Since the output power is $|V_2(j\omega)|^2/R_2$, the output voltage is maximum at the frequencies of power maxima. Consequently, the magnitude of the transfer voltage-ratio function will be maximum at these power maxima. Any change in a component value, whether it be an increase or a decrease, can only result in a decrease in the gain. At each gain maximum in the passband, the sensitivity is seen to be zero. Filters designed for flat-magnitude response such as the Butterworth or Chebyshev response achieve maximum power transfer over most of the passband frequencies. Thus, low sensitivity over much of the passband is typical of these filters. However, for filters that achieve near maximum power transfer over a small portion of the passband such as the Bessel response, the in-band sensitivity will be larger than that for the flat-magnitude type of filter. The above argument, however, does not apply to the sensitivity in the stopband or the transitional band. This is not of great concern, because in most filter applications the requirements in these bands are not very stringent. Moreover, if response variations outside the passband are of concern, optimized lossy ladder networks can be designed to provide lower sensitivity than their lossless counterparts.

The low-sensitivity property of the doubly terminated LC ladder structures motivates us to search for active RC realizations based on these passive prototype ladders. In the following, we develop different approaches for realizing such active *RC* networks, and show that the low-sensitivity character of the passive prototype networks is preserved.

8.2 INDUCTANCE SIMULATION

In this section we consider the realization of the active *RC* filters from the passive *RLC* prototype structures by direct replacement of the inductors with equivalent active *RC* networks. The procedure consists of first obtaining the resistively terminated *LC* network and then replacing each inductor by an active *RC* equivalent. The resulting filter is expected to have the same low sensitivity as its passive counterpart except for the imperfections in the realization of the inductor using active *RC* network.

In Section 1.3.3 of Chapter 1, we introduced a useful secondary building block called the gyrator represented by the symbols shown in Figure 1.25. A realization of the gyrator with two op amps is shown in Figure 1.26. The gyrator is a device with the terminal voltages and currents related by

$$\begin{bmatrix} V_1 \\ V_2 \end{bmatrix} = \begin{bmatrix} 0 & -r \\ r & 0 \end{bmatrix} \begin{bmatrix} I_1 \\ I_2 \end{bmatrix} \tag{8.10}$$

for Figure 1.25(a), where r is the gyration resistance, and

$$\begin{bmatrix} V_1 \\ V_2 \end{bmatrix} = \begin{bmatrix} 0 & r \\ -r & 0 \end{bmatrix} \begin{bmatrix} I_1 \\ I_2 \end{bmatrix} \tag{8.11}$$

for Figure 1.25(b). An important application of the gyrator in active filter synthesis is its use in the simulation of an inductor. The network of Figure 1.27, repeated in Figure 8.3, is a schematic realization of an inductor. As shown in (1.25), the input impedance of the terminated gyrator is

$$Z_{\text{in}} = (Cr^2)s = Ls \tag{8.12}$$

where $L = Cr^2$, which is equivalent to a self-inductor of inductance L.

Another network realization of an inductor is shown in Figure 8.4 with two op amps. The three nodal equations at the input terminals of the op amps are found

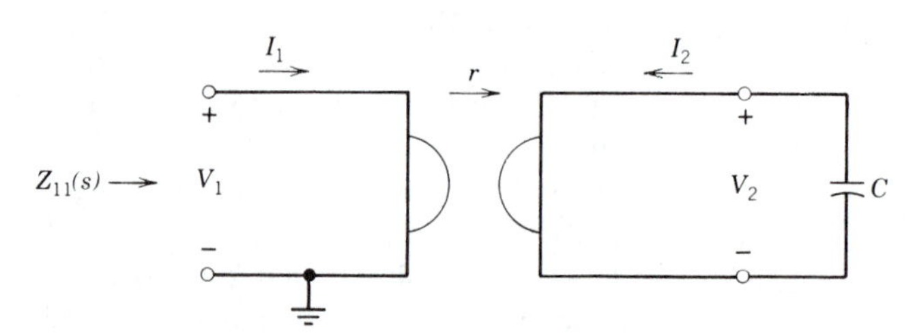

Figure 8.3

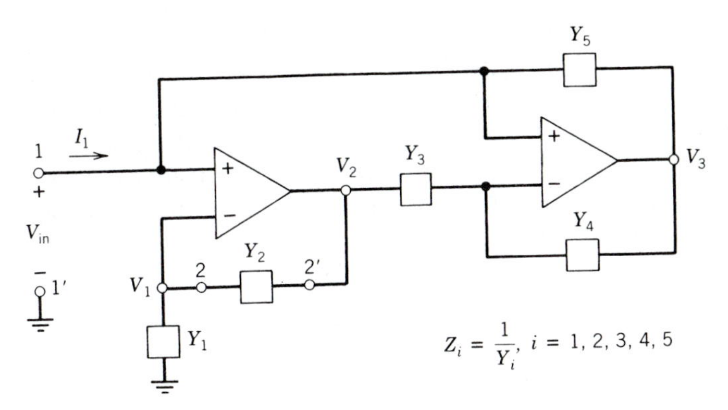

Figure 8.4

to be

$$Y_2(V_{in} - V_2) + Y_1 V_{in} = 0 \tag{8.13a}$$

$$Y_4(V_{in} - V_3) + Y_3(V_{in} - V_2) = 0 \tag{8.13b}$$

$$Y_5(V_{in} - V_3) = I_1 \tag{8.13c}$$

where $Y_i = 1/Z_i$ for $i = 1, 2, 3, 4, 5$. Solving these equations for I_1 in terms of V_{in} yields the input impedance facing voltage source V_{in} as

$$Z_{in} = \frac{V_{in}}{I_1} = \frac{Y_2 Y_4}{Y_1 Y_3 Y_5} = \frac{Z_1 Z_3 Z_5}{Z_2 Z_4} \tag{8.14}$$

In particular, if all the elements are resistive except Z_2, which represents a capacitor,

$$Z_1 = Z_3 = Z_4 = Z_5 = R, \qquad Z_2 = 1/sC \tag{8.15}$$

the input impedance becomes

$$Z_{in} = (R^2 C)s = Ls \tag{8.16}$$

where $L = R^2 C$. Equation (8.16) represents the impedance of an inductor of inductance $L = R^2 C$, the network of which is shown in Figure 8.5.

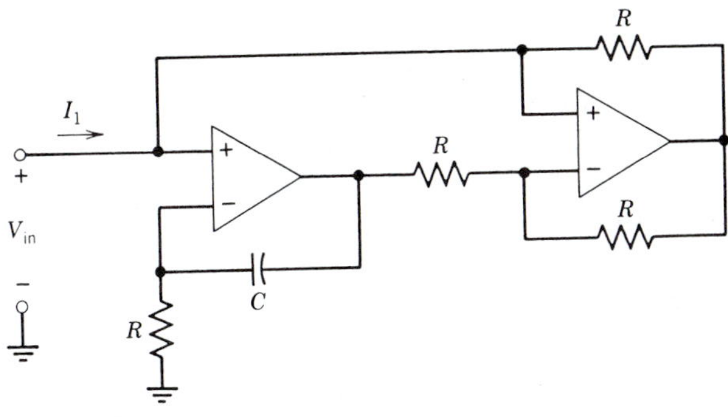

Figure 8.5

Observe that the networks of Figures 8.3 and 8.4 have one terminal of the input port grounded. Thus, these networks can only realize grounded inductors. To realize a floating inductor, we connect two gyrators in cascade as shown in Figure 8.6.† An analysis of this network will show that it is equivalent to a floating inductor of inductance $L=r^2C$ of Figure 8.7. If the gyrators are not identical with gyration resistances r_1 and r_2, as shown in Figure 8.8, parasitic elements will appear and the equivalent network becomes that of Figure 8.9 with

$$L_1=\frac{r_1^2r_2C}{r_2-r_1} \tag{8.17a}$$

$$L_2=\frac{r_2^2r_1C}{r_1-r_2} \tag{8.17b}$$

$$L=r_1r_2C \tag{8.17c}$$

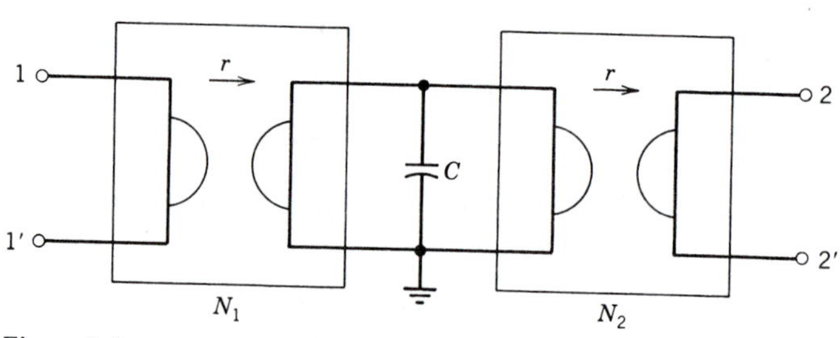

Figure 8.6

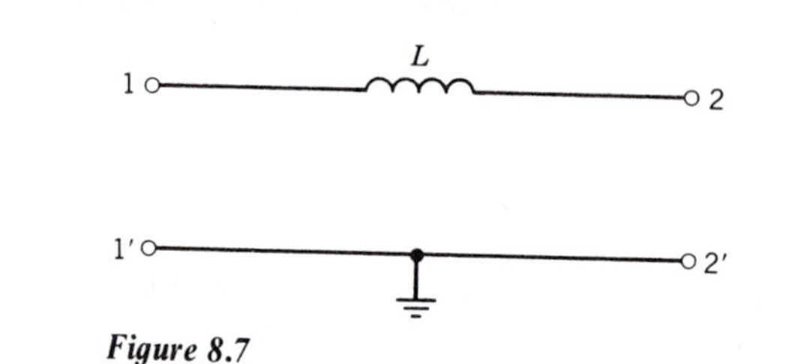

Figure 8.7

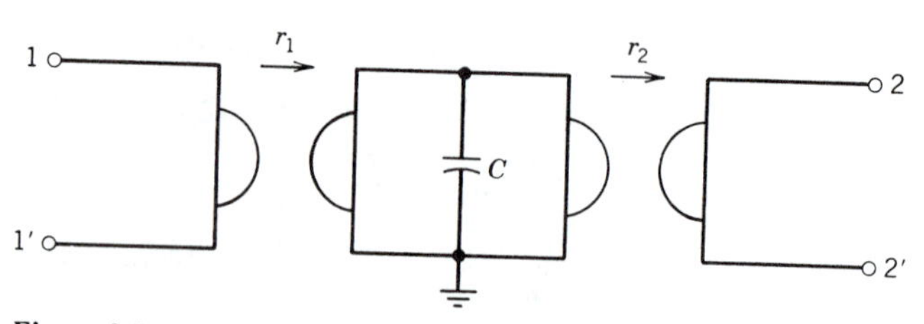

Figure 8.8

†The configuration is due to A. G. J. Holt and J. Taylor, "Method of replacing ungrounded inductors by grounded gyrators," *Electronics Letters*, vol. 1, 1965, p. 105.

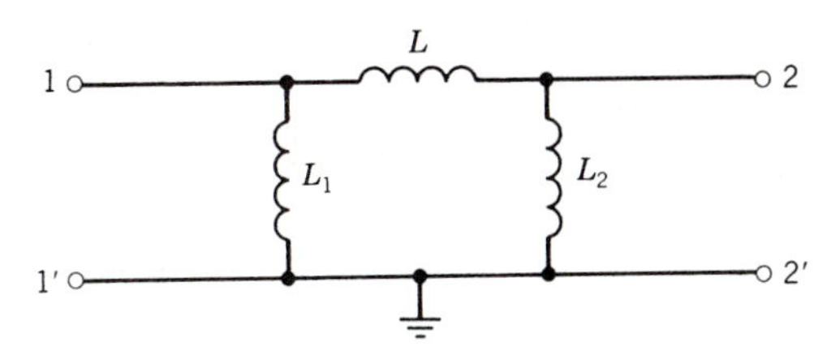

Figure 8.9

For identical gyrators, $r_1 = r_2 = r$ and the network of Figure 8.9 reduces to that of Figure 8.7.

Another solution to the realization of a floating inductor is to connect two identical networks of Figure 8.5 back-to-back, as shown in Figure 8.10. It is straightforward to show that the input impedance at port 1–2 is

$$Z_{\text{in}} = R^2 Cs \tag{8.18}$$

Thus, it is equivalent to a floating inductor of Figure 8.7 with inductance

$$L = R^2 C \tag{8.19}$$

The inductance simulation techniques described above lead to a class of active network elements called the generalized immittance converter. A *generalized immittance converter* (GIC) is a two-port device, as shown in Figure 8.11, capable of making the input immittance (impedance or admittance) of one of its two ports the product of the immittance terminating at the other port and some internal immittances of the

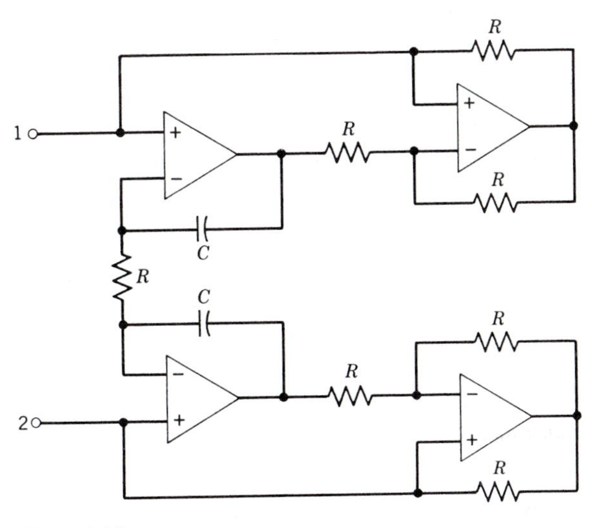

Figure 8.10

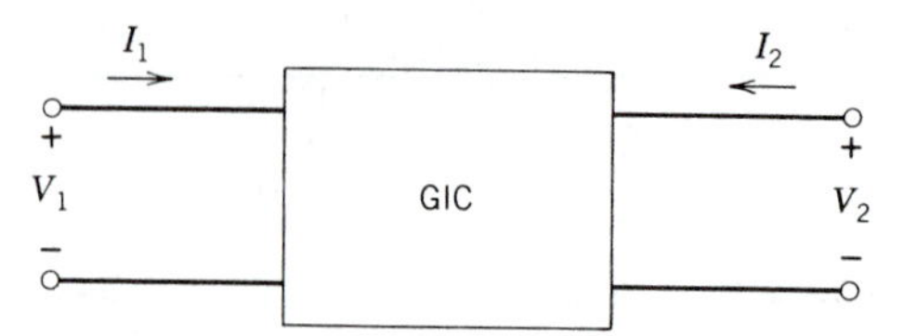

Figure 8.11

device. It is characterized by the following transmission matrix:

$$\begin{bmatrix} V_1 \\ I_1 \end{bmatrix} = \begin{bmatrix} k & 0 \\ 0 & \dfrac{1}{f(s)} \end{bmatrix} \begin{bmatrix} V_2 \\ -I_2 \end{bmatrix} \tag{8.20}$$

where $f(s)$ is called the *impedance transformation function* and k is usually normalized to unity. The generalized immittance converter is a secondary building block and can be realized by RC networks and op amps. The realization shown in Figure 8.12 is probably the most used and practical one, and is available today as a hybrid integrated circuit.† With Z_5 considered as the load impedance, the transmission matrix of the two-port network of Figure 8.12 is given by

$$\begin{bmatrix} A & B \\ C & D \end{bmatrix} = \begin{bmatrix} 1 & 0 \\ 0 & \dfrac{Z_2 Z_4}{Z_1 Z_3} \end{bmatrix} \tag{8.21}$$

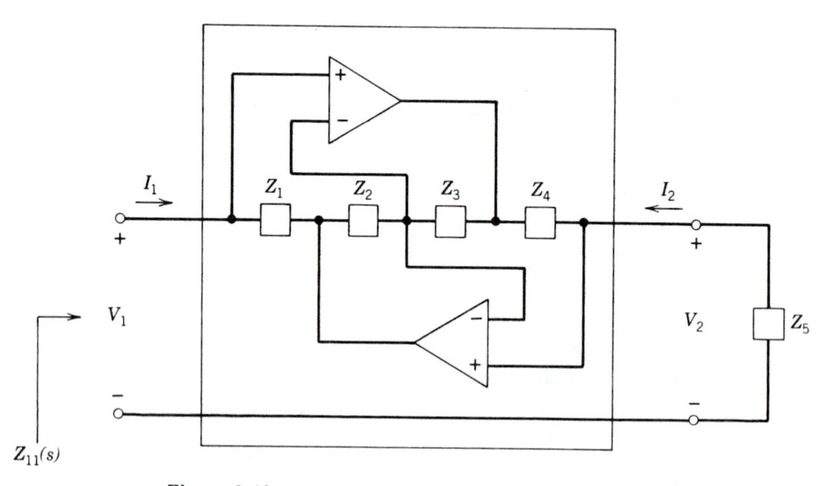

Figure 8.12

†ATF 431 Gyrator, Amperex Electronic Corp., Slatesville, RI. TCA 580 Gyrator, Signetics Corp., Sunnyvale, CA.

From (8.20) we can make the following identifications:

$$k=1, \qquad f(s)=\frac{Z_1Z_3}{Z_2Z_4} \tag{8.22}$$

Thus, the two-port device of Figure 8.12 is a GIC.

If, instead, Z_4 is considered as the load and Z_5 becomes part of internal impedances, as depicted in Figure 8.13, the transmission matrix of the resulting two-port network becomes

$$\begin{bmatrix} A & B \\ C & D \end{bmatrix}=\begin{bmatrix} 0 & Z_5 \\ \dfrac{Z_2}{Z_1Z_3} & 0 \end{bmatrix} \tag{8.23}$$

If only the input impedance is of concern, both networks of Figures 8.12 and 8.13 are identical with

$$Z_{\text{in}}=\frac{Z_1Z_3Z_5}{Z_2Z_4} \tag{8.24}$$

This equation is the same as (8.14). Thus, both networks can be used to simulate an inductor. For the network of Figure 8.12, the input impedance can be expressed as

$$Z_{\text{in}}=\frac{Z_1Z_3}{Z_2Z_4}Z_5=f(s)Z_5 \tag{8.25}$$

showing that Z_{in} is the product of $f(s)$ and the terminating impedance Z_5. Likewise, for the network of Figure 8.13, (8.24) can be rewritten as

$$Z_{\text{in}}=\frac{1}{(Z_2/Z_1Z_3Z_5)Z_4} \tag{8.26}$$

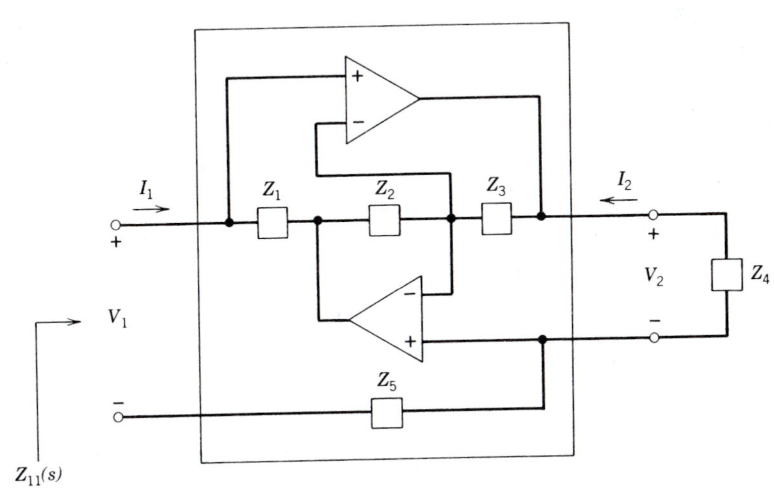

Figure 8.13

In other words, the input impedance of the network of Figure 8.13 equals the reciprocal of the product of the terminating impedance Z_4 and an impedance transformation function $Z_2/Z_1Z_3Z_5$. Such a device is termed a *generalized immittance inverter* (GII) as opposed to GIC, where no inversion occurs.

Let us now examine how the simulated inductors are used to realize active *RC* filters. The synthesis begins with a selection of an appropriate passive prototype network. Typically, the selected prototype network is a doubly terminated *LC* ladder, for which explicit formulas for element values and design tables are widely available.† We then replace each passive inductor with an active simulated inductor. Observe that each grounded inductor is replaced by an active network using two op amps, four resistors, and one capacitor, while for each floating inductor the network requires four op amps, seven resistors, and two capacitors if Figure 8.10 is used. The resulting structure has the same topology as the passive prototype network. Thus, we would expect that the sensitivity of the active *RC* realization is equivalent to that of the passive prototype, except for this increase in the total number of components required to simulate the inductors. However, since the op amps are not ideal, they will cause some increase in sensitivity in the active realizations. We illustrate the above procedure by the following examples.

Example 8.1 It is desired to design a high-pass filter with a radian cutoff frequency of $\omega_c = 10^3$ rad/s. The filter is required to possess the fourth-order Chebyshev response with peak-to-peak ripple in the passband not to exceed 1 dB, and is to be operated between a resistive generator of internal resistance 400 Ω and a 100-Ω load.

The problem is essentially the same as that solved in Example 2.9 except for a frequency-scaling. Multiplying the element values of the inductors and capacitors of the network of Figure 2.24 by a factor of 100 yields the desired high-pass filter as shown in Figure 8.14. Replacing each grounded inductor by its equivalent of Figure 8.5, we obtain the active *RC* high-pass filter of Figure 8.15 with $R = 1$ kΩ.

Example 8.2 Design a band-pass filter with a passband that extends from $\omega = 10^3$ rad/s to $\omega = 4 \times 10^3$ rad/s. The filter is required to possess the fourth-order

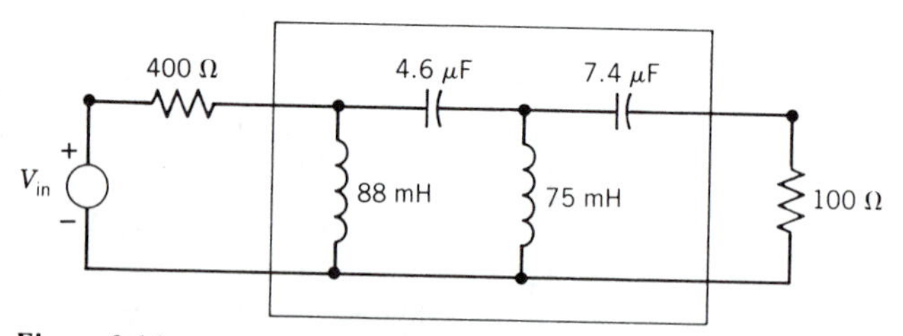

Figure 8.14

†See, for example, W. K. Chen 1976, *Theory and Design of Broadband Matching Networks*, Chapter 3. New York: Pergamon Press; and L. Weinberg 1962, *Network Analysis and Synthesis*. New York: McGraw-Hill. Reissued by R. E. Krieger Publishing Co., Melbourne, Fl, 1975, Chapter 13.

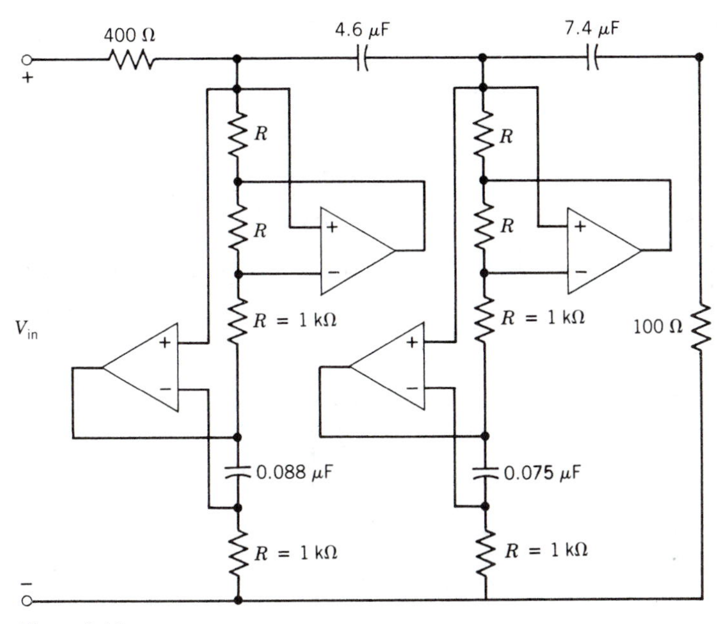

Figure 8.15

Chebyshev response with peak-to-peak ripple in the passband not to exceed 1 dB, and is to be operated between a resistive generator of internal resistance 400 Ω and a 100-Ω load.

The problem was solved in Example 2.10 except for a scaling factor. Multiplying the element values of the inductors and capacitors of the network of Figure 2.28 by a factor of 100 yields the desired band-pass filter as shown in Figure 8.16. By replacing each grounded inductor by its equivalent of Figure 8.5 and each floating inductor by its equivalent of Figure 8.10, we obtain the active *RC* band-pass filter of Figure 8.17 with $R = 1$ kΩ. Alternately, we can use the gyrator-simulated inductances of Figures 8.3 and 8.6, and obtain the active band-pass filter of Figure 8.18.

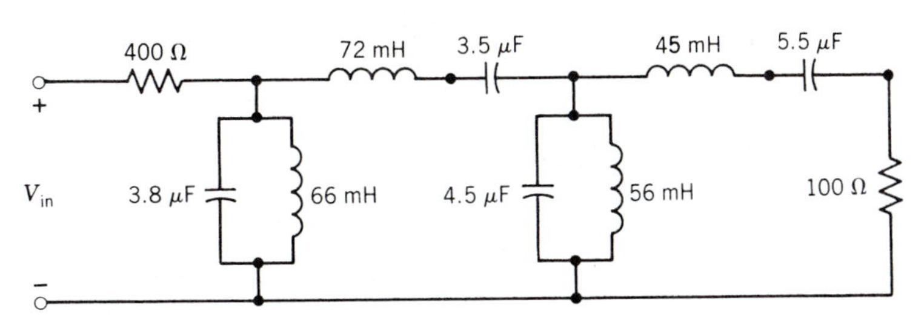

Figure 8.16

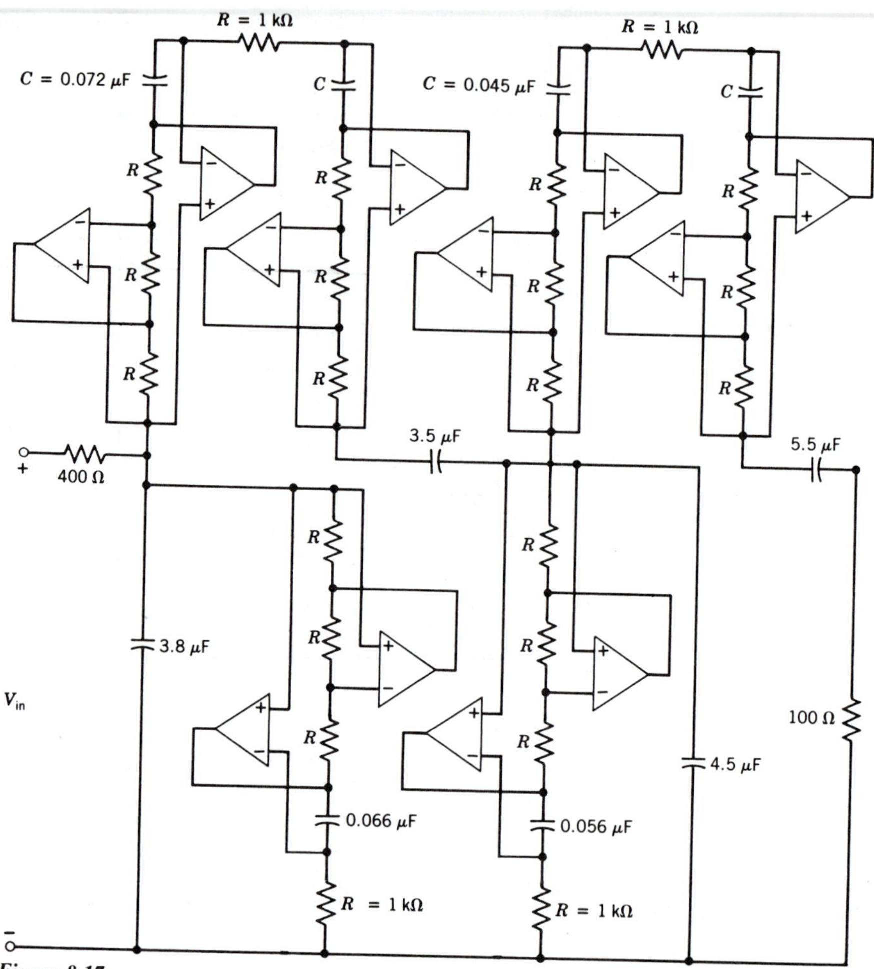

Figure 8.17

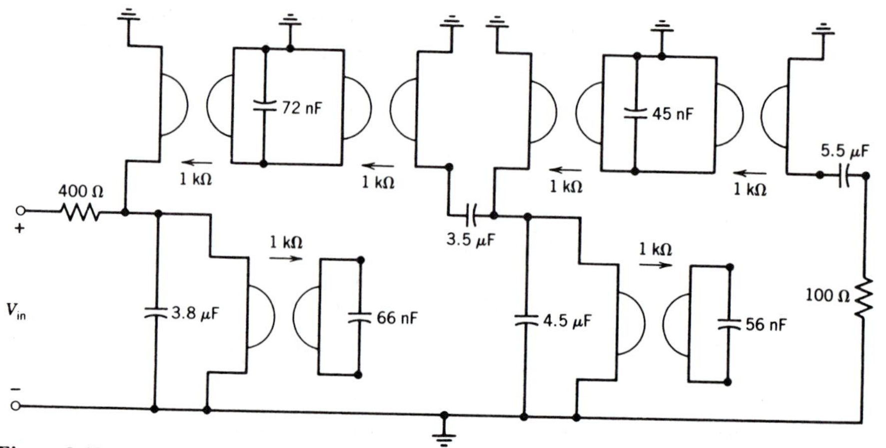

Figure 8.18

8.3 FREQUENCY-DEPENDENT NEGATIVE RESISTORS

In the preceding section, we showed how to realize an active RC filter from its passive RLC prototype by directly replacing the inductors with equivalent active RC networks. In this section, we present an alternate procedure for obtaining an active RC equivalent by using a new secondary building block known as the *frequency-dependent negative resistor* (FDNR), which can be realized by an active RC network.

Recall that in impedance scaling of a network, if the impedance of each branch is multiplied by b, the input impedance of the scaled network is increased by the same factor b. However, the transfer voltage-ratio function, being the ratio of two determinants of the same order, is unaffected by the process. If, instead of scaling by a real positive constant b, the impedances were scaled by the factor b/s, it is clear that the transfer voltage-ratio function would remain unaltered. Thus, by multiplying each impedance of an RLC network by b/s, the resistors are transformed into capacitors, inductors into resistors, and capacitors into devices with impedances inversely proportional to s^2. These transformations are summarized in Figure 8.19. A device whose impedance is inversely proportional to s^2 or

$$Z(s)=1/Ds^2 \tag{8.27}$$

where D is a positive real constant, is called a *frequency-dependent negative resistor*, abbreviated as FDNR. The unit for the constant D is farad-second (Fs). The reason for this name is that on the $j\omega$-axis

$$Z(j\omega)=-1/D\omega^2 \tag{8.28}$$

which is a negative real frequency-dependent function, and possesses the characteristic of a resistor. The symbolic representation of the FDNR is shown in Figure 8.19, and is explicitly presented in Figure 8.20.

An FDNR can be realized using the terminated generalized immittance converter (GIC) of Figure 8.12, the input impedance of which is from (8.24)

$$Z_{\text{in}}(s)=\frac{Z_1Z_3Z_5}{Z_2Z_4} \tag{8.29}$$

To realize an FDNR, we make any two of the impedances Z_1, Z_3, and Z_5 capacitive

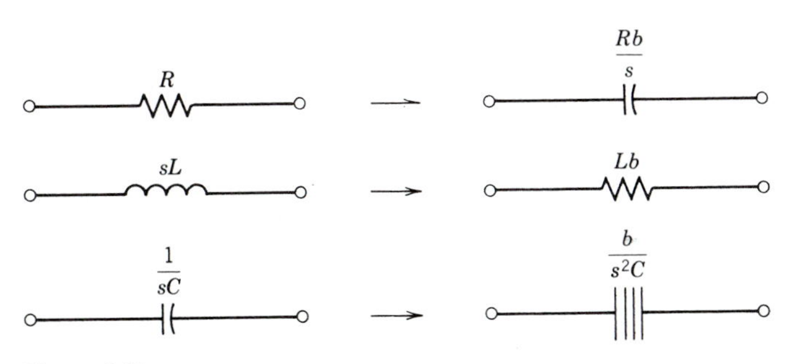

Figure 8.19

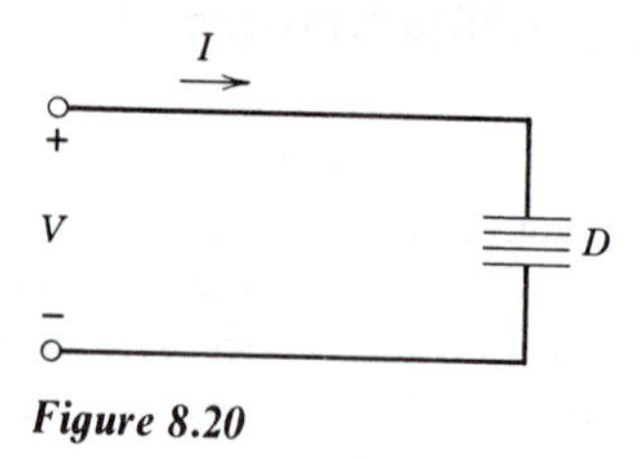

Figure 8.20

and the remaining impedances resistive. It can be shown† that if we make the following specific choice

$$Z_1 = Z_3 = 1/sC, \qquad Z_2 = Z_4 = Z_5 = R \tag{8.30}$$

the resulting network of Figure 8.21 will yield the lowest sensitivity to the active elements. Equation (8.29) then reduces to

$$Z_{in}(s) = \frac{1}{RC^2 s^2} = \frac{1}{Ds^2} \tag{8.31a}$$

where

$$D = RC^2 \tag{8.31b}$$

This is the impedance of an FDNR. Although other network configurations for the FDNR are possible, the active RC realization of Figure 8.21 is of good performance, and is commercially available as the ATF 431 Gyrator.‡

To apply the FDNR to the synthesis of active filters, we first realize a given approximation function by an LC doubly terminated ladder or other suitable LC prototypes. The equivalent active RC network is then obtained by replacing each

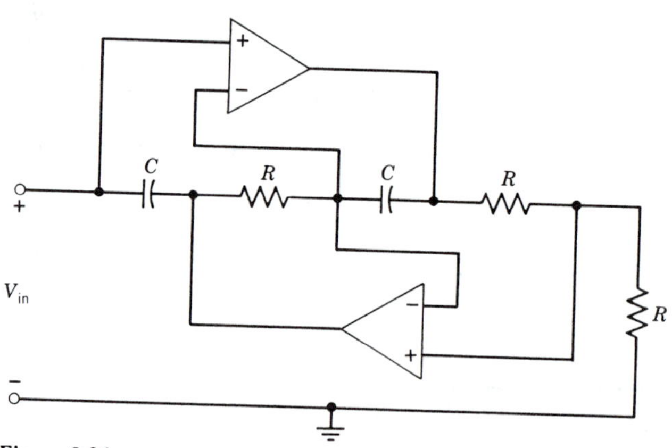

Figure 8.21

†L. T. Bruton and J. T. Lim, "High-frequency comparison of GIC-simulated inductance circuits," *Int. J. Circuit Theory and Applications*, vol. 2, 1974, pp. 401–404.
‡Amperex Electronic Corp., Slatesville, RI.

resistor of resistance R by a capacitor of capacitance $1/Rb$, each inductor of inductance L by a resistor of resistance Lb, and each capacitor of capacitance C by an FDNR whose D value is C/b, where b is any appropriate positive-real constant.

Since one of the input terminals of the active network of Figure 8.21 is grounded, the network can only realize a grounded FDNR. The realization of a floating FDNR is achieved by placing two of these networks back-to-back as shown in Figure 8.22, as in the realization of a floating inductor. As in the case of inductance simulation, the above transformation does not affect the topology of its prototype, nor does it affect the transfer voltage-ratio function realized. Therefore, we expect the transformed network to possess the same low sensitivity as its passive prototype except for the increase in the number of elements used and the imperfections of the op amps employed to realize the FDNR.

Observe that for an FDNR realization the total number of op amps required equals twice the number of grounded capacitors plus four times the number of floating capacitors in the passive prototype. For an inductance-simulated realization, the number of op amps needed is twice the number of grounded inductors plus four times the number of floating inductors in the passive prototype network. As a result, in low-pass realizations the inductance-simulated realizations will always require more op amps than the corresponding FDNR realizations. On the other hand, in high-pass realizations the reverse is true. For band-pass and band-elimination filters, each realization will have to be evaluated individually. We illustrate the above procedure by the following examples.

Example 8.3 Consider the normalized low-pass filter of Figure 8.23. Applying the transformation b/s to the filter of Figure 8.23 produces the FDNR realization of

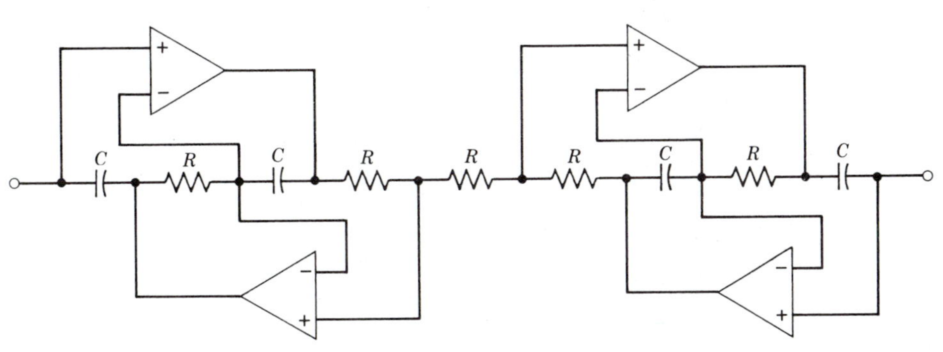

Figure 8.22

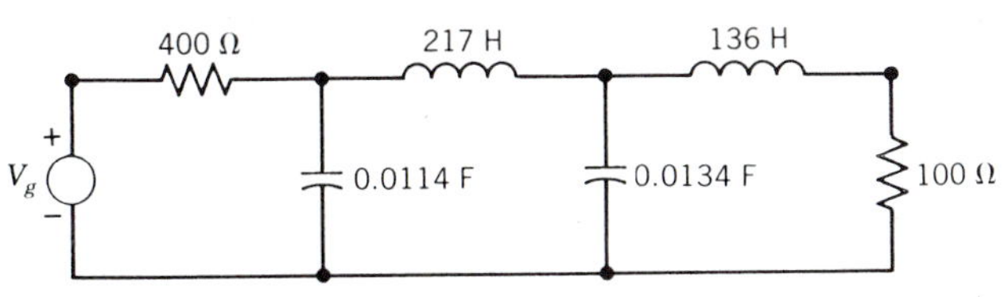

Figure 8.23

Figure 8.24 where $b=10^3$. Suppose now that we select the active RC network of Figure 8.21 for the realization of the FDNR by choosing $R=10\ \text{k}\Omega$. The required capacitances are shown in the final realization of Figure 8.25.

Observe that this realization needs four op amps, eight resistors, and six capacitors. In contrast, if Figure 8.23 is realized by inductance simulation of Figure 8.10, a total of eight op amps, sixteen resistors, and six capacitors is required. Therefore, for low-pass filters the FDNR approach is superior to the inductance simulation.

Example 8.4 Consider the high-pass filter of Figure 8.14. Applying the transformation b/s to Figure 8.14 yields the FDNR realization of Figure 8.26, where b

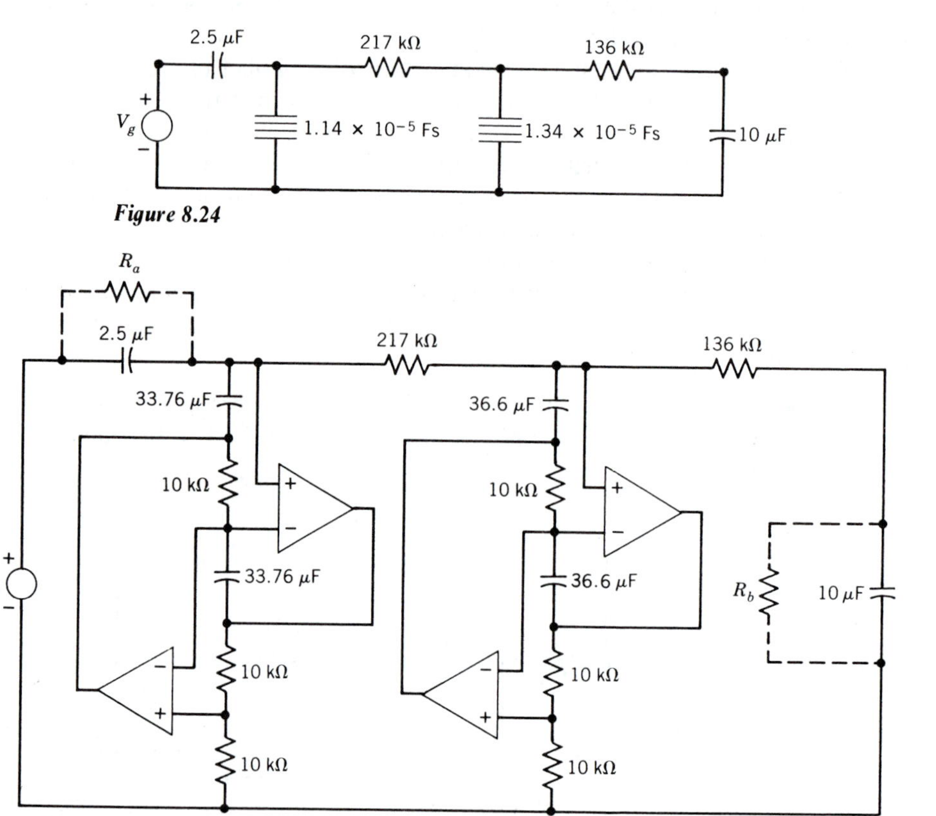

Figure 8.24

Figure 8.25

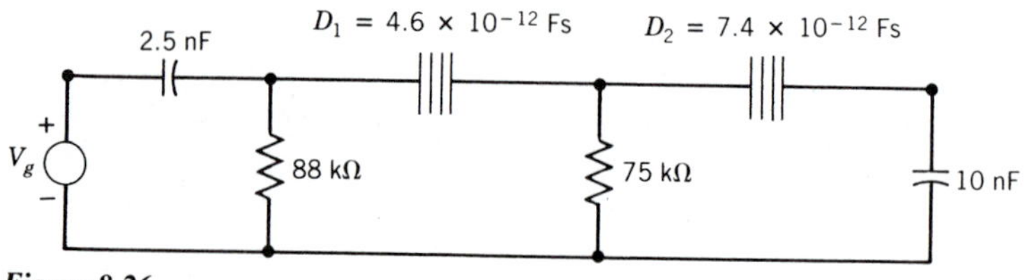

Figure 8.26

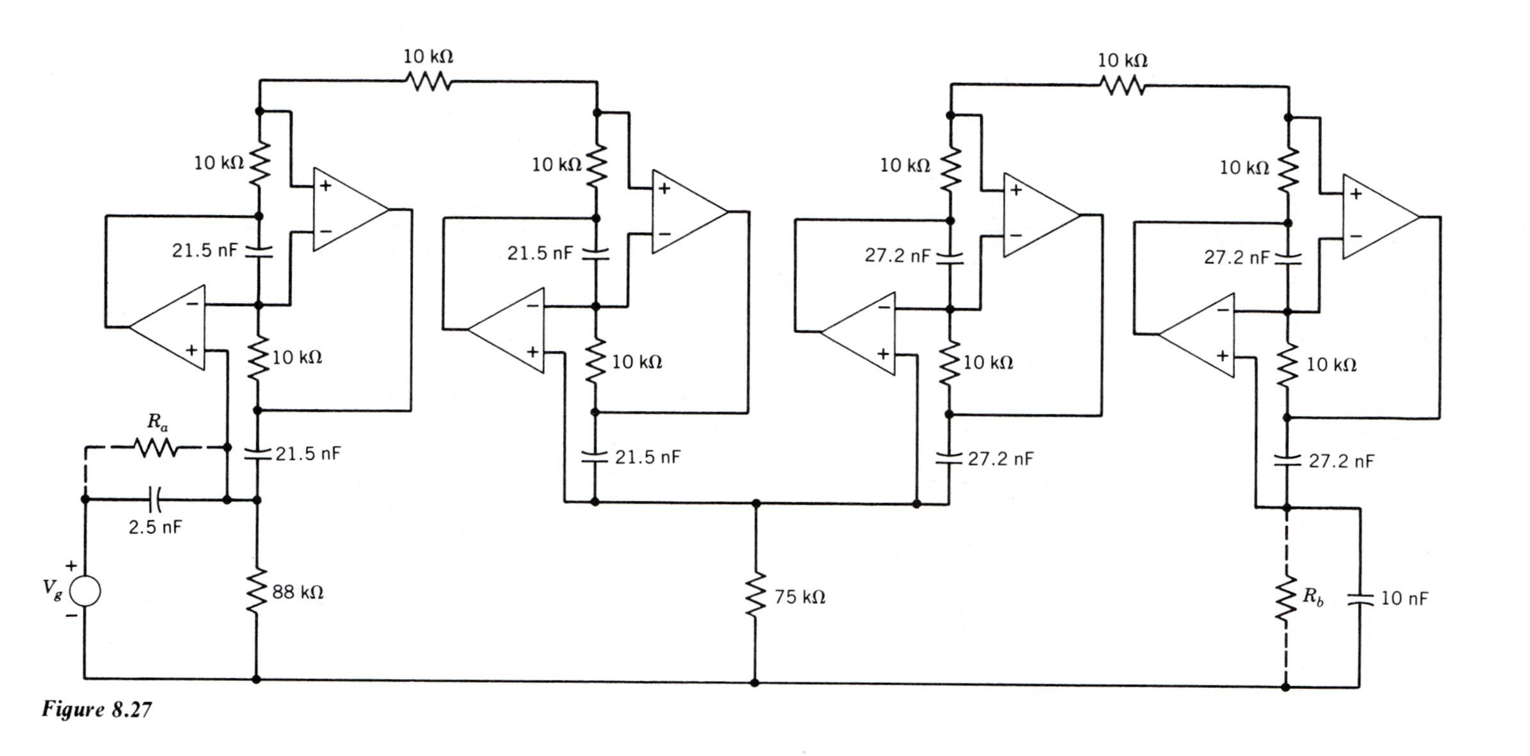

Figure 8.27

was chosen to be 10^6. Using the floating FDNR of Figure 8.22 in Figure 8.26 with $R = 10\,\text{k}\Omega$, we obtain the active realization of Figure 8.27.

Observe that in this realization it requires eight op amps, ten capacitors, and twelve resistors. This is to be compared with the inductance-simulated realization of Figure 8.15, where four op amps, four capacitors, and ten resistors are used. Therefore, for high-pass filters the inductance-simulated realization is better than the FDNR realization.

Before we proceed to other topics, we reexamine the FDNR realizations of Figures 8.25 and 8.27. Observe that in both networks the noninverting terminals of some op amps do not have a dc path to ground. Such a path is needed for the biasing of op amps. One simple solution is to place large resistors R_a and R_b across the blocking capacitors as indicated in Figures 8.25 and 8.27, thereby providing the necessary dc paths to ground for the biasing circuitry. This, of course, will introduce some error in the function. Another problem associated with these networks is that the source and load impedances are required to be capacitive. But active filters are frequently embedded in other electronic circuitry that may require resistive source and load. In such cases, buffering the input and output with unity-gain amplifier of Figure 1.23 will overcome this difficulty.

8.4 LEAPFROG SIMULATION OF LADDERS

In the foregoing discussion, we use either the simulated inductance or the FDNR in the realizations. In both cases, only one particular element type was realized actively using RC and op-amp network. In the present section, we introduce a different approach in which the active network is obtained by simulating equations describing the passive ladder structure with analog operations such as summing, differencing, or integration. The use of analog computer simulation for the filter appears to be impractical. However, such is not the case at all, and the filters that result are very competitive with those discussed in the previous chapters.

The network we wish to simulate is the ladder structure shown in Figure 8.28, in which all series elements or combinations of elements are represented by their admittances and all shunt elements or combinations of elements are represented by impedances. Using the reference directions and symbols indicated, the branch relationships describing the ladder are

$$I_1 = Y_1(V_{\text{in}} - V_2) \tag{8.32a}$$

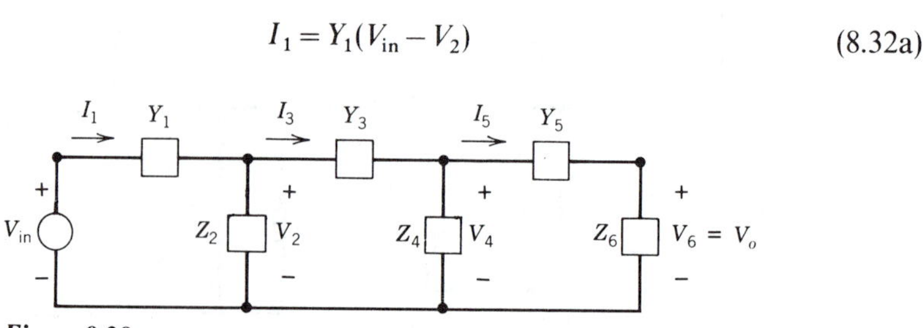

Figure 8.28

$$V_2 = Z_2(I_1 - I_3) \tag{8.32b}$$

$$I_3 = Y_3(V_2 - V_4) \tag{8.32c}$$

$$V_4 = Z_4(I_3 - I_5) \tag{8.32d}$$

$$I_5 = Y_5(V_4 - V_6) \tag{8.32e}$$

$$V_o = V_6 = Z_6 I_5 \tag{8.32f}$$

We recognize that these equations are not unique in describing the ladder network, but the pattern of writing successive voltage and current equations is general and will be useful in simulation. To obtain the transfer voltage-ratio function V_o/V_{in}, we simply eliminate the intermediate variables I_1, V_2, I_3, V_4, and I_5 among the equations (8.32). However, this is not our objective. Our objective is to simulate these equations by analog method. To this end, we first represent the equations (8.32) by the block diagram as depicted in Figure 8.29, in which the output of each block is fed back to the input of the preceding block. As a result, the individual blocks are not isolated from one another and any change in one block affects the voltages and currents in all other blocks. The pattern suggests the children's game called *leapfrog*, and the filters that result are referred to as the *leapfrog realizations*. The method is known as the *leapfrog technique*.†

As in the previous cases of passive simulation, our motivation in simulating the passive ladder structure is that we wish to exploit its low sensitivity to parameter changes as compared to the cascaded topologies. Because of the coupling effect between the blocks of the leapfrog realizations, it will make the tuning of the complete network more difficult. This is in contrast with the cascaded topology, where the sections are totally isolated, so that any change in one biquad will not affect any other one, thereby making tuning of the filter's performance at the time of manufacture relatively simple.

Our next objective is to convert the block diagram representation of the ladder into a form that can be implemented by active *RC* networks. In the block diagram of Figure 8.29, we recognize that we cannot realize currents. Therefore, we have to

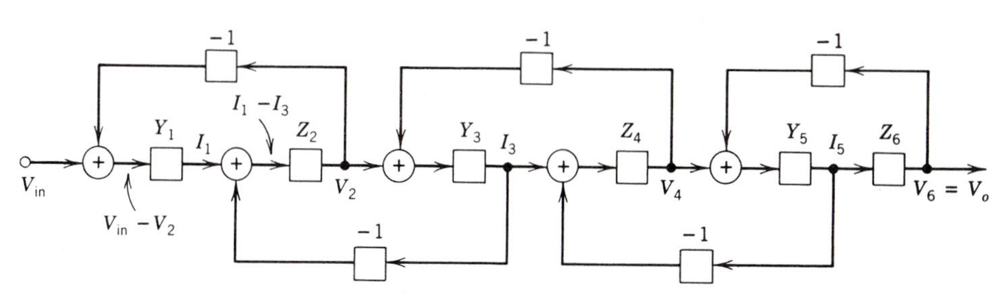

Figure 8.29

†F. E. J. Girling and E. F. Good, "Active filters 12: The leapfrog or active-ladder synthesis," *Wireless World*, vol. 76, 1970, pp. 341–345. The technique was originally reported in RRE Memo No. 1177, Sept. 1955, Royal Radar Establishment, Malvern, U.K.

simulate the currents as voltages. In addition, the blocks with specified immittances will have to be implemented with active RC networks characterized by transfer functions. These difficulties are circumvented by considering the hypothetical network shown in Figure 8.30. The topology of this block representation is identical to that of Figure 8.29 except that all the intermediate variables are voltages. The equations describing the branch relationships are found to be

$$E_1 = Y_1(V_{\text{in}} - V_2) \tag{8.33a}$$

$$V_2 = Z_2(E_1 - E_3) \tag{8.33b}$$

$$E_3 = Y_3(V_2 - V_4) \tag{8.33c}$$

$$V_4 = Z_4(E_3 - E_5) \tag{8.33d}$$

$$E_5 = Y_5(V_4 - V_6) \tag{8.33e}$$

$$V_o = V_6 = Z_6 E_5 \tag{8.33f}$$

where E_1, E_3, and E_5 are voltage variables. Observe that since the variables on both sides of the equations (8.33) are voltages, the Y's and Z's are now the transfer voltage-ratio functions, even though we still use the immittance symbol Y or Z. If the intermediate variables E_1, V_2, E_3, V_4, and E_5 are eliminated among the equations (8.33), the transfer voltage-ratio function V_o/V_{in} obtained will be identical to that obtained from the equations (8.32) by eliminating I_1, V_2, I_3, V_4, and I_5. In other words, the network of Figure 8.30 has the same transfer voltage-ratio function as the original ladder structure of Figure 8.28. Therefore, the simulation of the passive ladder of Figure 8.28 by the active RC networks to yield a prescribed transfer voltage-ratio function can be achieved by the realization of the block diagram of Figure 8.30 with transfer voltage-ratio functions Y_i and Z_i for the blocks.

Before we turn our attention to the actual realization, we reexamine the block diagram of Figure 8.30. Observe that the block diagram requires five unity-gain inverting amplifiers, one for each feedback path. A simpler alternative realization can be obtained by a slight modification of the defining equations. This is accomplished by rewriting (8.33) as

$$E_1 = Y_1(V_{\text{in}} - V_2) \tag{8.34a}$$

$$-V_2 = -Z_2(E_1 - E_3) \tag{8.34b}$$

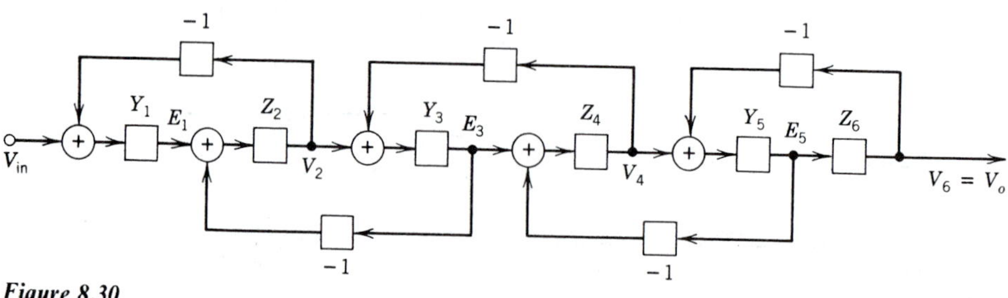

Figure 8.30

$$-E_3 = Y_3(-V_2 + V_4) \tag{8.34c}$$

$$V_4 = -Z_4(-E_3 + E_5) \tag{8.34d}$$

$$E_5 = Y_5(V_4 - V_6) \tag{8.34e}$$

$$-V_6 = -Z_6 E_5 \tag{8.34f}$$

A block diagram realization of these equations is shown in Figure 8.31, in which the output voltage is $-V_6$ instead of V_6 as previously given in Figure 8.30. Note that no unity-gain inverters are required except possibly one at the output. The inverted output $-V_6$ is usually of no concern in practice. In addition, all the Z_i transfer voltage-ratio functions in (8.34) have minus signs preceding them, whereas the Y_i functions do not. It will turn out that the inverting transfer functions are easier to realize than the noninverting ones. Similarly, (8.33) may be modified in a somewhat different form as

$$-E_1 = -Y_1(V_{in} - V_2) \tag{8.35a}$$

$$-V_2 = Z_2(-E_1 + E_3) \tag{8.35b}$$

$$E_3 = -Y_3(-V_2 + V_4) \tag{8.35c}$$

$$V_4 = Z_4(E_3 - E_5) \tag{8.35d}$$

$$-E_5 = -Y_5(V_4 - V_6) \tag{8.35e}$$

$$-V_6 = Z_6(-E_5) \tag{8.35f}$$

These equations are realized by the block diagram of Figure 8.32. In this configuration, all the Y_i transfer voltage ratios are inverting, whereas the Z_i functions are not. It is interesting to note that in both block diagrams the signs of the transfer functions alternate as we progress from input to output, and that around each loop there is one negative transfer function providing the required negative feedback.

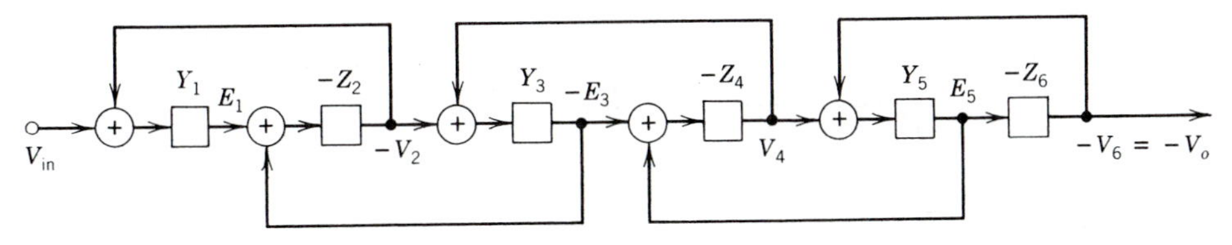

Figure 8.31

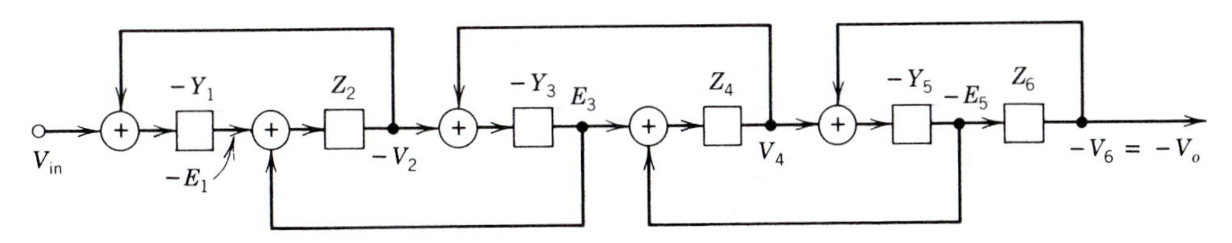

Figure 8.32

We now illustrate the above procedure by applying the leapfrog technique to the synthesis of low-pass filters in which all the transmission zeros are at infinity. To be specific, we consider the sixth-order low-pass ladder structure of Figure 8.33. The source and load resistors R_1 and R_2 can be included in the initial and final elements, respectively, as explicitly shown in Figure 8.34. Comparing this network with that of Figure 8.28, we can make the following identifications:

$$H_1(s)=Y_1(s)=\frac{1/L_1}{s+R_1/L_1} \tag{8.36a}$$

$$H_2(s)=-Z_2(s)=-\frac{1}{sC_2} \tag{8.36b}$$

$$H_3(s)=Y_3(s)=\frac{1}{sL_3} \tag{8.36c}$$

$$H_4(s)=-Z_4(s)=-\frac{1}{sC_4} \tag{8.36d}$$

$$H_5(s)=Y_5(s)=\frac{1}{sL_5} \tag{8.36e}$$

$$H_6(s)=-Z_6(s)=-\frac{1/C_6}{s+1/R_2C_6} \tag{8.36f}$$

where H_i and $-H_i$ are the required transfer functions for the individual blocks of Figures 8.31 and 8.32, respectively. Among all these transfer functions (8.36), H_2 and H_4 can be realized by the inverting op-amp integrator of Figure 8.35, the transfer

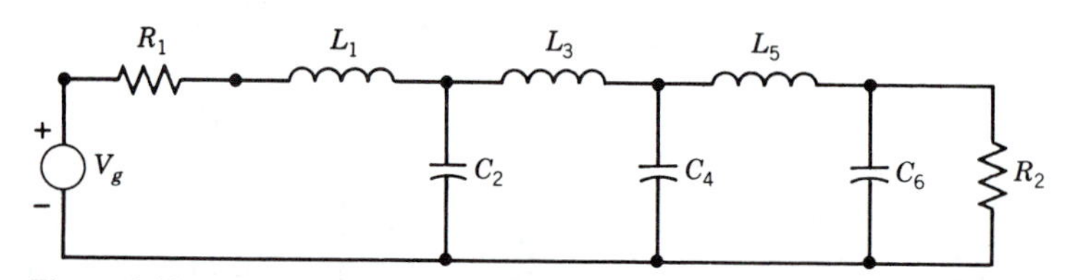

Figure 8.33

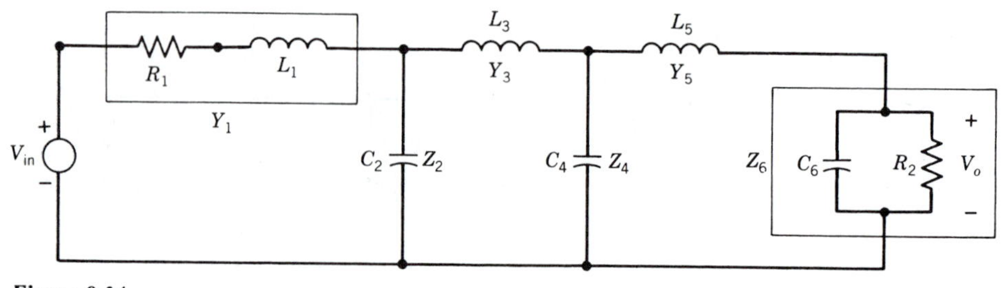

Figure 8.34

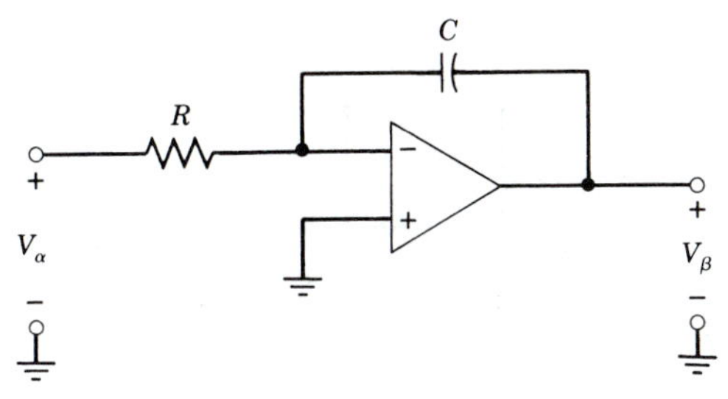

Figure 8.35

voltage-ratio function of which is

$$H(s)=\frac{V_\beta}{V_\alpha}=-\frac{1}{RCs} \tag{8.37}$$

and H_3 and H_5 may be realized by such an integrator in cascade with a unity-gain inverter, or by the noninverting integrator of Figure 8.36, the transfer voltage-ratio function of which is

$$H(s)=\frac{V_\beta}{V_\alpha}=\frac{2}{RCs} \tag{8.38}$$

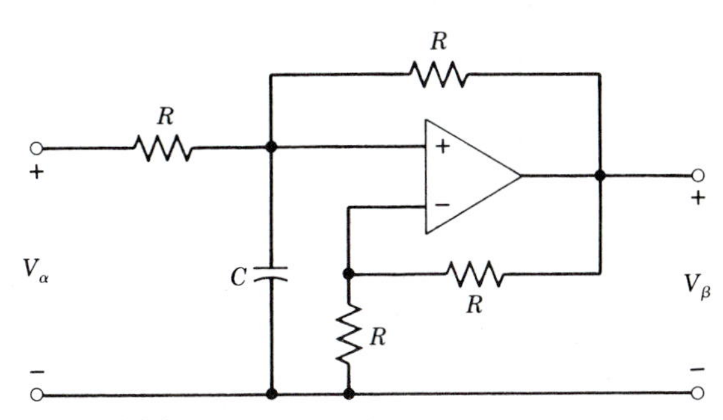

Figure 8.36

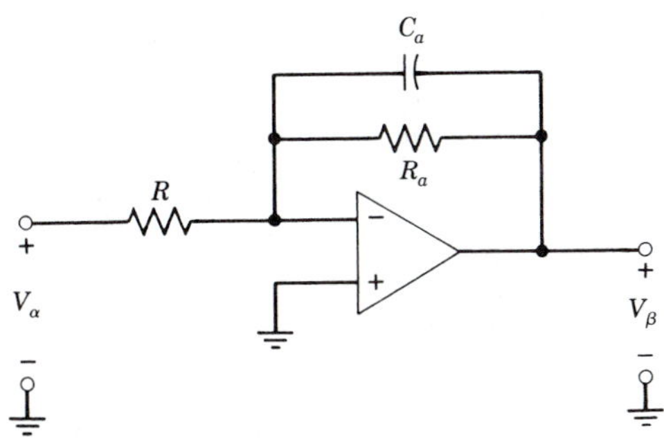

Figure 8.37

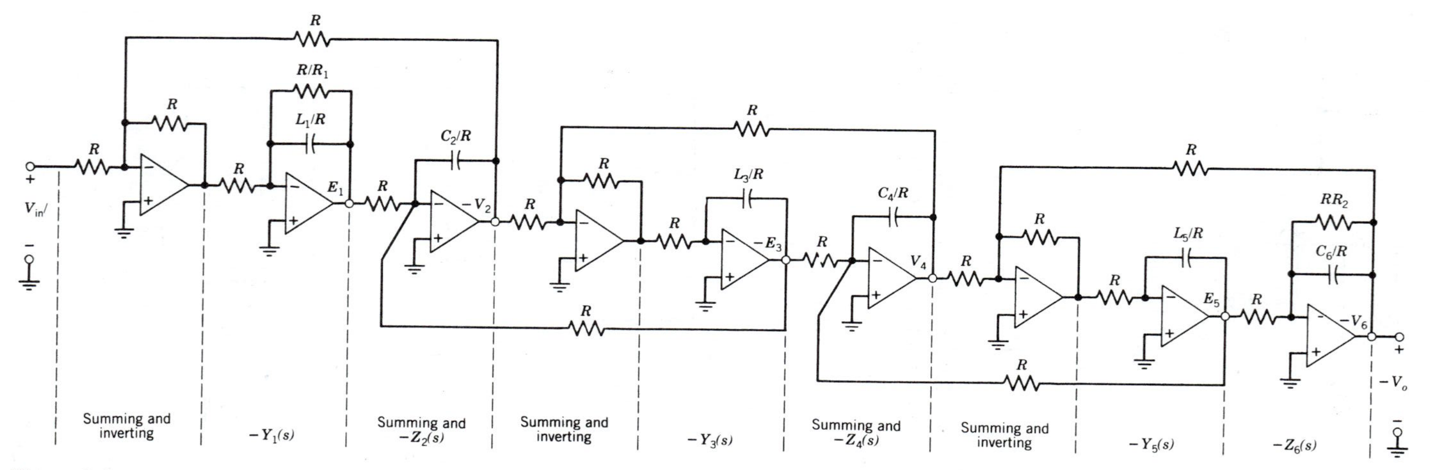

R
R/R1
R
L1/R
C2/R
R
R
R
L3/R
C4/R
R
RR2
R
L5/R
C6/R
R
R
R
R
R
R
R
E1
−V2
−E3
V4
E5
−V6
+
Vin/
−
+
−Vo
−
Summing and inverting
−Y1(s)
Summing and −Z2(s)
Summing and inverting
−Y3(s)
Summing and −Z4(s)
Summing and inverting
−Y5(s)
−Z6(s)

Figure 8.38

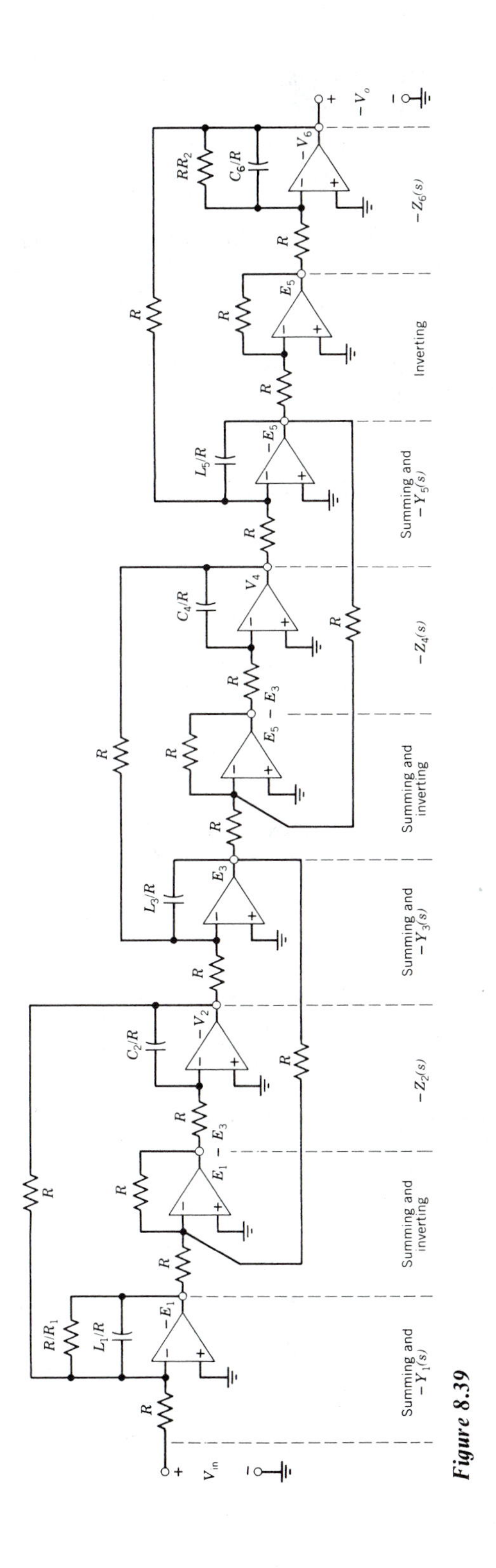

V_{in}
R
R/R_1
L_1/R
$-E_1$
R
R
$E_1 - E_3$
R
C_2/R
$-V_2$
R
R
L_3/R
E_3
R
R
$E_5 - E_3$
R
C_4/R
V_4
R
R
L_5/R
$-E_5$
R
R
E_5
R
R
RR_2
C_6/R
$-V_6$
$-V_o$
Summing and $-Y_1(s)$
Summing and inverting
$-Z_2(s)$
Summing and $-Y_3(s)$
Summing and inverting
$-Z_4(s)$
Summing and $-Y_5(s)$
Inverting
$-Z_6(s)$

Figure 8.39

Finally, the transfer functions H_1 and H_6 may be realized by the inverting lossy integrator of Figure 8.37 for which

$$H(s)=\frac{V_\beta}{V_\alpha}=-\frac{1/RC_a}{s+1/R_aC_a} \tag{8.39}$$

Using the configuration of Figure 8.31, the resulting leapfrog realization is shown in Figure 8.38 with a designation of the purpose of each of the networks used. A similar realization of the network of Figure 8.33 using the configuration of Figure 8.32 is possible. The resulting network is shown in Figure 8.39.

To illustrate the design of a doubly terminated LC low-pass filter of odd order, we consider the ladder structure of Figure 8.40, which is redrawn as in Figure 8.41. The branch relationships are found to be

$$H_1(s)=Y_1(s)=\frac{1/L_1}{s+R_1/L_1} \tag{8.40a}$$

$$H_2(s)=-Z_2(s)=-\frac{1}{sC_2} \tag{8.40b}$$

$$H_3(s)=Y_3(s)=\frac{1}{sL_3} \tag{8.40c}$$

$$H_4(s)=-Z_4(s)=-\frac{1}{sC_4} \tag{8.40d}$$

$$H_5(s)=Y_5(s)=\frac{1/L_5}{s+R_2/L_5} \tag{8.40e}$$

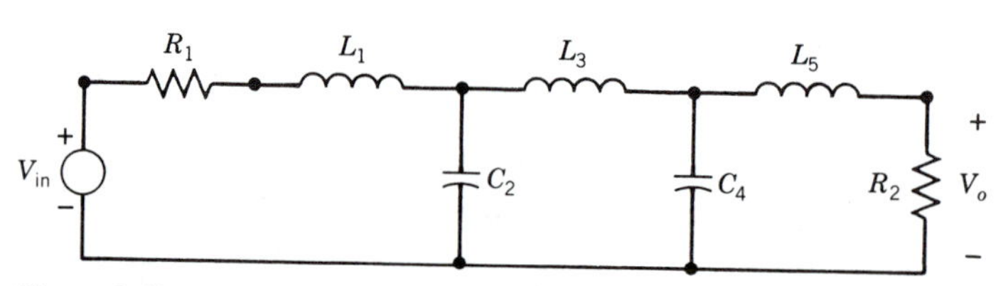

Figure 8.40

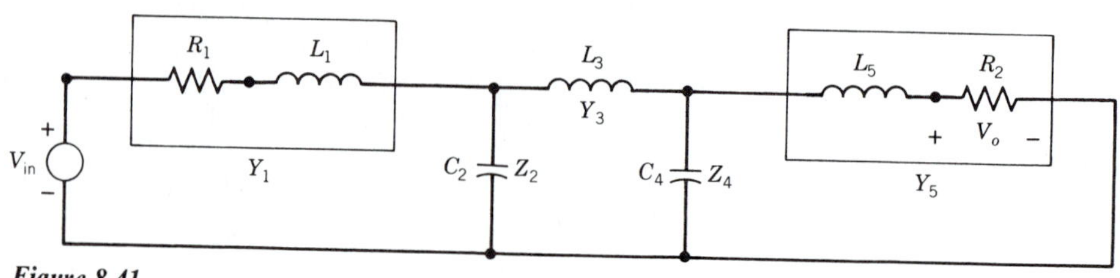

Figure 8.41

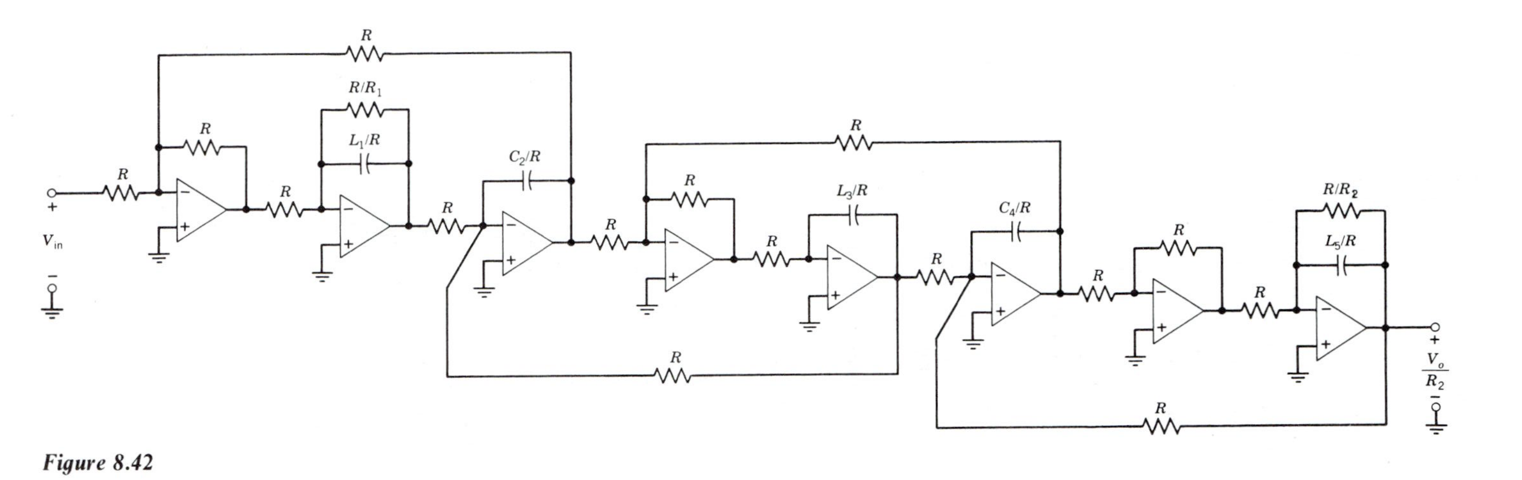

Figure 8.42

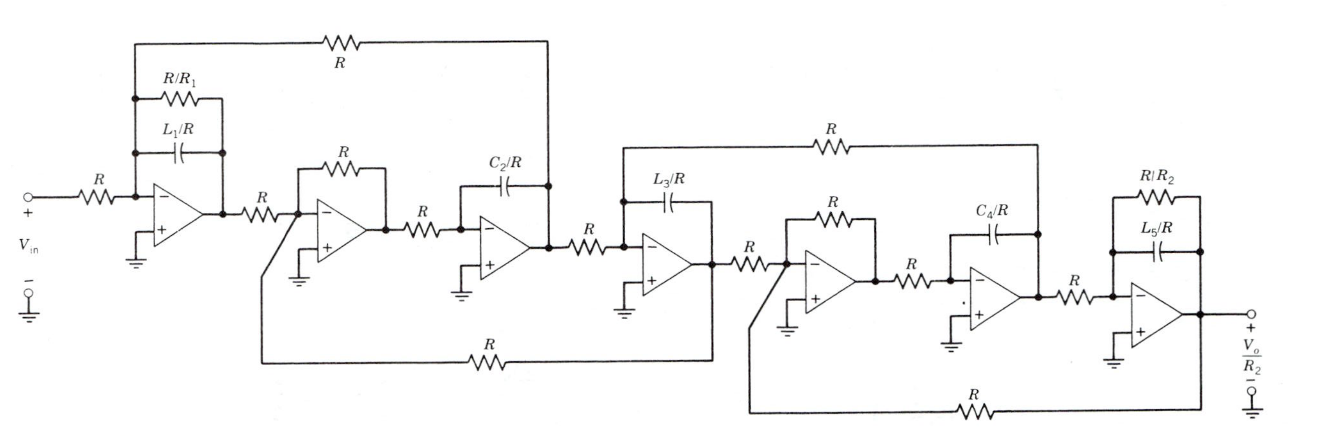

Figure 8.43

where H_i are the transfer functions for the individual blocks of Figure 8.31. For the configuration of Figure 8.32, the transfer functions for the individual blocks are $-H_i$. Using the form of Figure 8.31, the resulting leapfrog realization is shown in Figure 8.42. Similarly, if the configuration of Figure 8.32 is used, the corresponding realization is given in Figure 8.43. It is interesting to observe that the latter requires only two unity-gain inverters, resulting in one less amplifier. Also, note that in all the realizations the element values are specified as functions of an arbitrary resistance R. This resistance may be chosen to achieve a desired impedance normalization or scaling. Once R is selected, all other element values are determined.

In summary, the design procedure for leapfrog filters may be accomplished in the following steps:

1 From the specifications for a low-pass filter, design a suitable normalized low-pass prototype, using explicit design formulas as given in Section 4.5 or the design tables.
2 Select the general configuration of either Figure 8.31 or Figure 8.32 and identify the various immittances Y_i and Z_i from Figure 8.44.
3 Find an active RC network to realize each of the individual blocks. These may be taken from a catalog of realizations.
4 Connect the individual blocks with the necessary summers and inverters to yield the leapfrog filter.
5 Perform the necessary frequency-scaling to meet the specifications, and magnitude-scaling to give convenient practical element values.

Example 8.5 It is desired to obtain a low-pass leapfrog filter that realizes the fifth-order Butterworth response with a cutoff frequency $\omega_c = 10^4$ rad/s. The filter is to be operated between a resistive generator and a resistive load.

We follow the steps outlined above for designing the leapfrog filter:

Step 1. From the specifications for a low-pass filter, we use the explicit design formulas of Section 4.5 to obtain the desired fifth-order Butterworth LC ladder network. This was previously done in Example 4.12 and the resulting prototype is given in Figure 4.22 and is redrawn in Figure 8.45.

Step 2. Suppose that we select the general configuration of Figure 8.31. Comparing the networks of Figures 8.40 and 8.45, we can make the following identifications:

$$R_1 = 100\ \Omega, \qquad R_2 = 200\ \Omega, \qquad L_1 = 31.33\text{ mH}, \qquad L_3 = 30.51\text{ mH} \tag{8.41a}$$

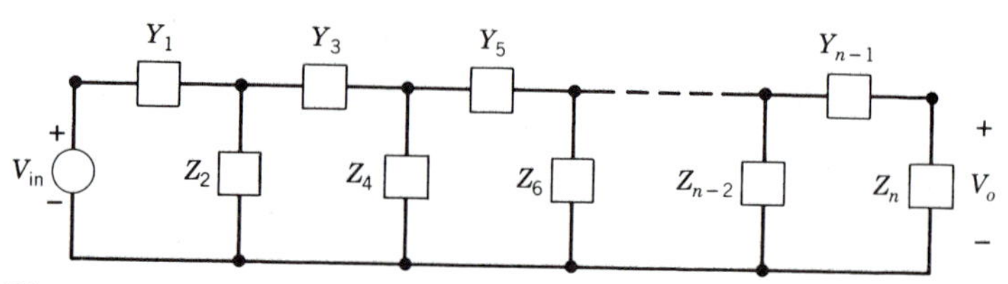

Figure 8.44

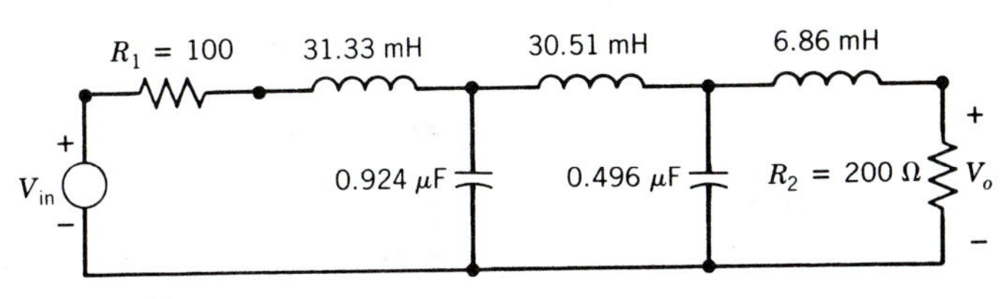

Figure 8.45

$$L_5 = 6.86 \text{ mH}, \qquad C_2 = 0.924\ \mu\text{F}, \qquad C_4 = 0.496\ \mu\text{F} \tag{8.41b}$$

Step 3. The transfer functions H_i $(i = 1, 2, 3, 4, 5)$ for individual blocks are shown in (8.40) and can be realized by the active RC networks of Figures 8.35, 8.36, and 8.37.

Step 4. The individual block realizations are connected together with the necessary summers and inverters to yield the leapfrog filter of Figure 8.42.

Step 5. Using the element values of (8.41) and choosing $R = 20$ kΩ, the final design becomes that of Figure 8.46.

8.4.1 Band-Pass Leapfrog Filters

In this section, we demonstrate that the leapfrog technique described above for low-pass filters is also applicable to band-pass filters. As in the foregoing, we start with a low-pass prototype filter and apply the low-pass to band-pass transformation as developed in Section 2.5 of Chapter 2 to each element of the low-pass prototype. The general form of the resulting series and shunt branches is shown in Figure 8.47, where B is the bandwidth and ω_0 the mid-band frequency. Consider, for example, the low-pass prototype network of Figure 8.33. Applying the low-pass to band-pass frequency transformation (2.146), the resulting network is shown in Figure 8.48 in which the terminating resistors R_1 and R_2 have been combined with the resonant networks of Figure 8.47 to give the general RLC networks of Figure 8.49. We use the admittance to characterize the network of Figure 8.49(a) and the impedance for the network of Figure 8.49(b). This gives

$$Y(s) = \frac{(1/L)s}{s^2 + (R/L)s + 1/LC} \tag{8.42}$$

for Figure 8.49(a) and

$$Z(s) = \frac{(1/C)s}{s^2 + (1/RC)s + 1/LC} \tag{8.43}$$

for Figure 8.49(b). Setting $R = 0$ in (8.42) yields the admittance of the lossless series resonant circuit, and $R = \infty$ in (8.43) yields the impedance of the lossless parallel resonant circuit. Similarly, for the low-pass structure of Figure 8.50 the source resistor R_1 can be incorporated in the impedance Z_1 by a source transformation from a voltage source to a current source after applying the low-pass to band-pass frequency transformation. This is permissible because our objective is to find the analog

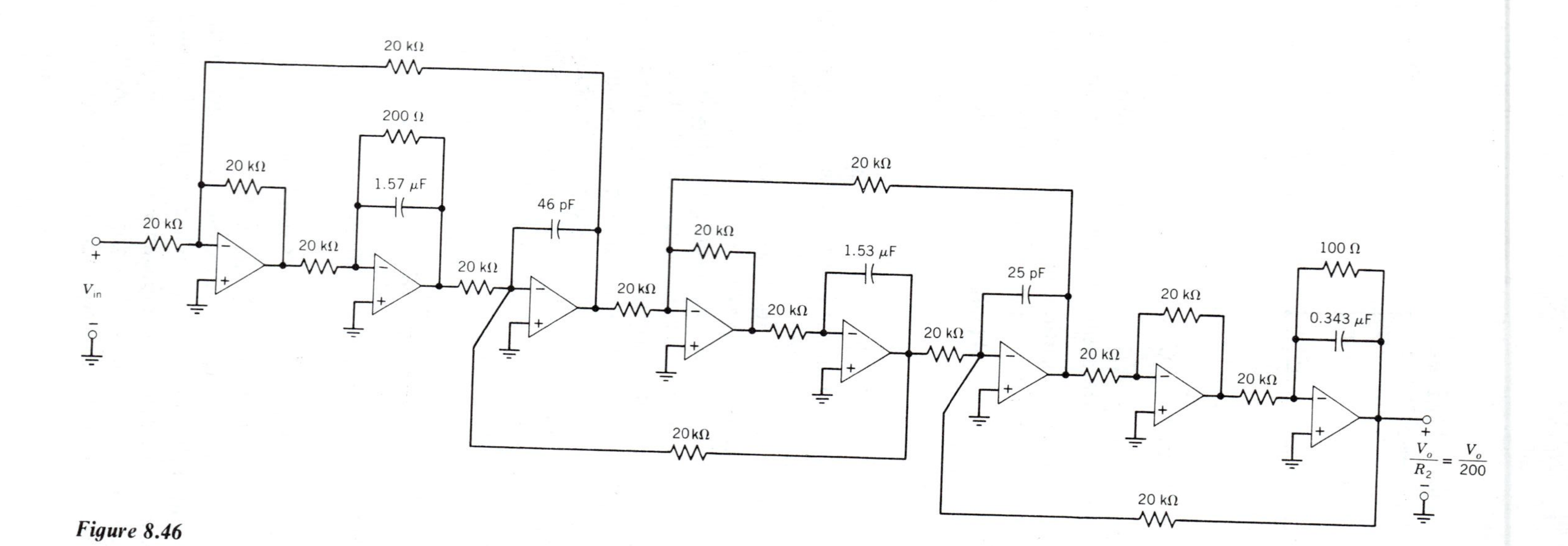

20 kΩ
200 Ω
20 kΩ
1.57 μF
46 pF
20 kΩ
+
V_{in}
−
20 kΩ
20 kΩ
20 kΩ
20 kΩ
1.53 μF
25 pF
100 Ω
20 kΩ
0.343 μF
20 kΩ
20 kΩ
20 kΩ
20 kΩ
20kΩ
20 kΩ
+
$\frac{V_o}{R_2} = \frac{V_o}{200}$
−

Figure 8.46

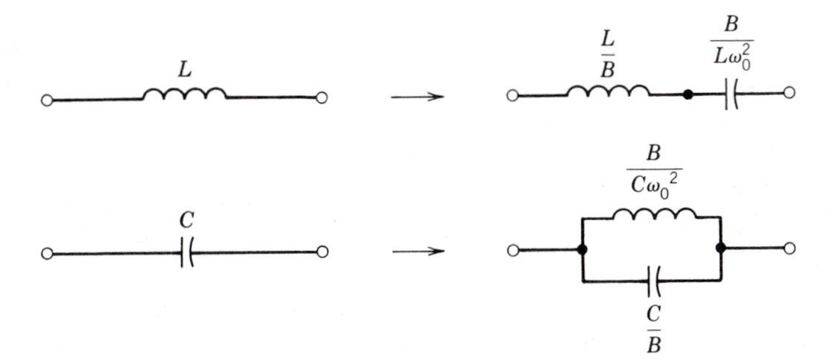

Figure 8.47

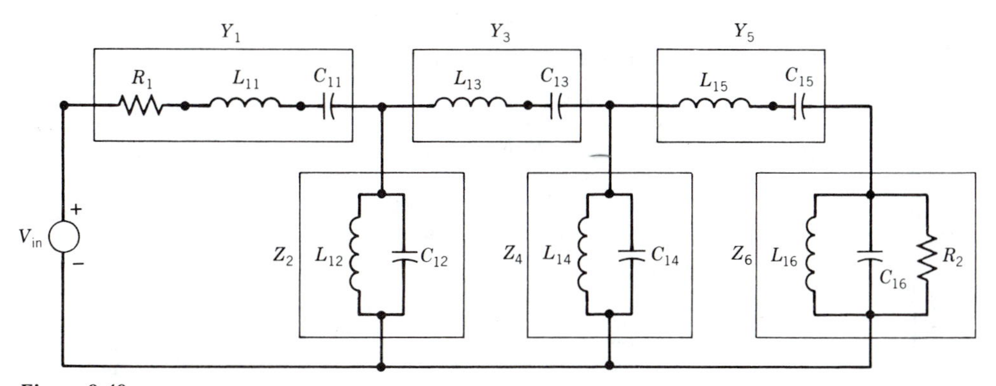

Figure 8.48

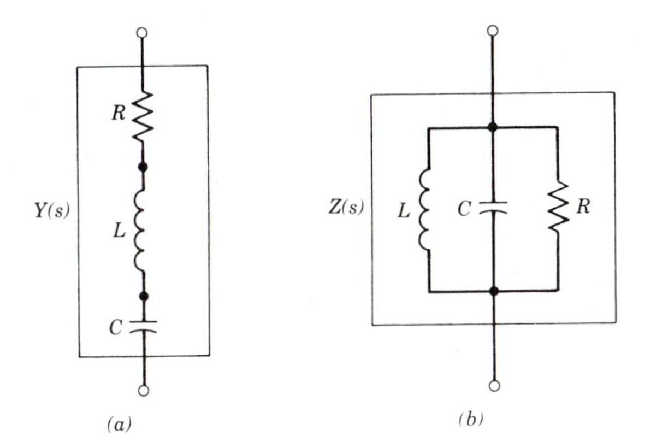

Figure 8.49

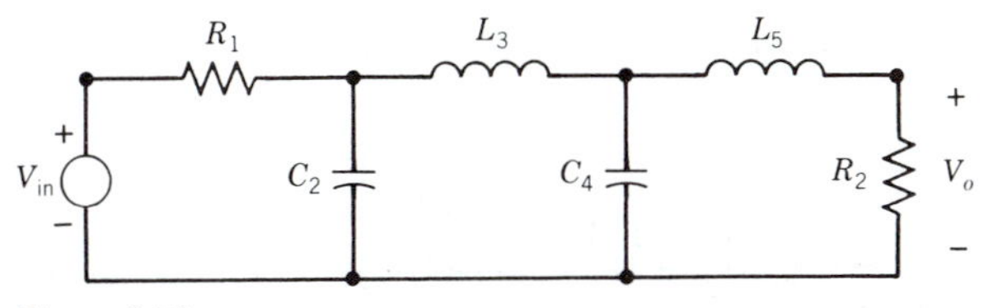

Figure 8.50

of the passive band-pass network. It does not matter whether it is driven by a voltage source or by a current source. The resulting band-pass filter is shown in Figure 8.51 in which Z_1, Y_2, Z_3, and Y_4 are identified.

The next step is to find active RC networks that will simulate the two kinds of RLC networks of Figure 8.49 used in a leapfrog realization having the transfer functions (8.42) and (8.43). Of the many band-pass filters we have studied so far, we single out three for this application.

The first network to be considered is the band-pass filter of Figure 5.27, which is redrawn in Figure 8.52, for which the transfer voltage-ratio function is from (5.51)

$$H(s)=\frac{V_o}{V_{in}}=-\frac{s/R_1C_2}{s^2+(1/R_2C_1+1/R_2C_2)s+1/R_1R_2C_1C_2} \tag{8.44}$$

The pole frequency and pole Q are identified as

$$\omega_p=\frac{1}{\sqrt{R_1R_2C_1C_2}} \tag{8.45a}$$

$$Q_p=\frac{1/\sqrt{R_1R_2C_1C_2}}{1/R_2C_1+1/R_2C_2} \tag{8.45b}$$

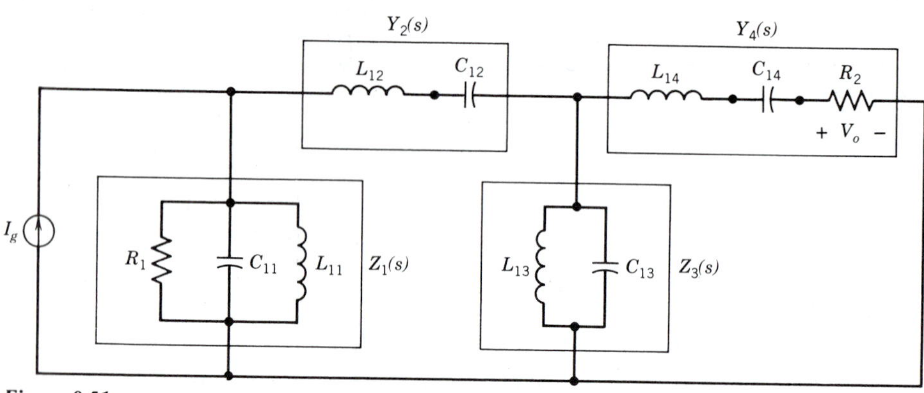

Figure 8.51

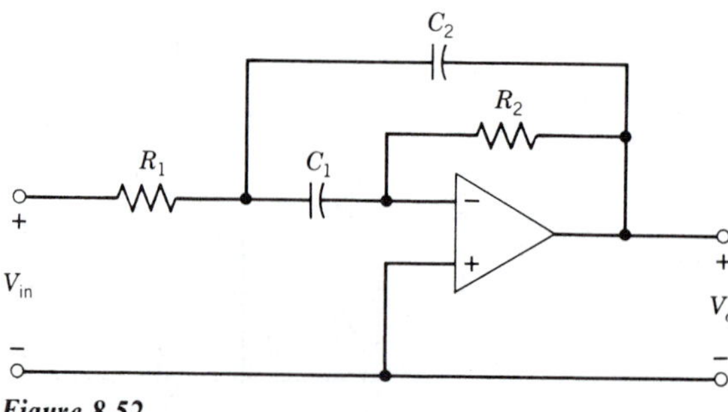

Figure 8.52

For the equal-capacitance design with

$$C_1 = C_2 = C \tag{8.46}$$

the above equations can be solved for R_1 and R_2 in terms of ω_p, Q_p, and C. The results are given by

$$R_1 = \frac{1}{2C\omega_p Q_p}, \qquad R_2 = \frac{2Q_p}{C\omega_p} \tag{8.47}$$

The second network is the band-pass SAB of Figure 7.9, which is redrawn in Figure 8.53. The general transfer voltage-ratio function is from (7.87)

$$H(s) = -\frac{\dfrac{s}{(1-1/k)R_1C_2}}{s^2 + [1/R_2C_1 + 1/R_2C_2 - 1/(k-1)R_1C_2]s + 1/R_1R_2C_1C_2} \tag{8.48}$$

giving

$$\omega_p = \frac{1}{\sqrt{R_1R_2C_1C_2}} \tag{8.49a}$$

$$Q_p = \frac{\sqrt{\dfrac{R_2}{R_1}}}{\sqrt{\dfrac{C_2}{C_1}} + \sqrt{\dfrac{C_1}{C_2}} - \dfrac{1}{k-1}\cdot\dfrac{R_2}{R_1}\sqrt{\dfrac{C_1}{C_2}}} \tag{8.49b}$$

For the equal-capacitance design with

$$C_1 = C_2 = 1\text{ F}, \qquad \gamma = r_1/r_2 \tag{8.50}$$

we can solve for R_1 and R_2, and from (7.89) we have

$$R_1 = \frac{2\gamma Q_p}{\omega_p(-1+\sqrt{1+8\gamma Q_p^2})} \tag{8.51a}$$

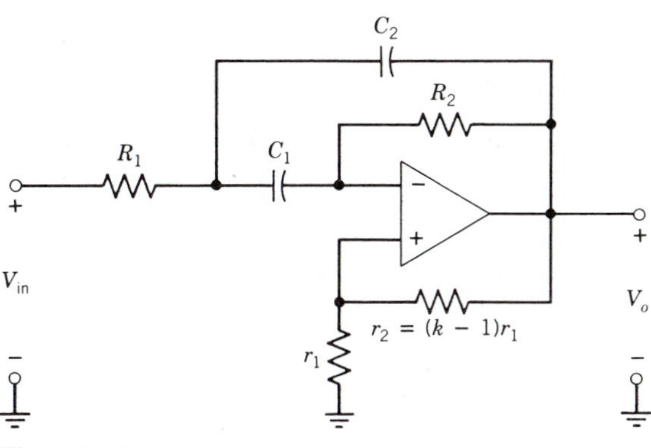

Figure 8.53

$$R_2 = \frac{1}{R_1 \omega_p^2} \tag{8.51b}$$

The parameter γ can be chosen to minimize the sensitivity of the network. The sensitivities of the biquadratic parameters ω_p and Q_p to the passive and active elements were computed earlier in Chapter 7 and are given in (7.90), (7.100) and (7.111). If we let $Q_p = \infty$ and choose $\gamma = 1/2Q_0^2$, R_1 and R_2 become

$$R_1 = \frac{1}{2\omega_p Q_0}, \qquad R_2 = \frac{2Q_0}{\omega_p} \tag{8.52}$$

The transfer function (8.48) reduces to

$$H(s) = -\frac{(2Q_0 + 1/Q_0)\omega_p s}{s^2 + \omega_p^2} \tag{8.53}$$

which represents the immittance of a lossless resonant circuit. The resulting network is shown in Figure 8.54. Here we may select Q_0 to obtain the required gain. In the case where the input voltage is larger than desired and more than one input to the networks of Figures 8.52 and 8.54 is required, we may employ the input attenuation technique of Section 5.4. The networks that result are shown in Figures 8.55 and 8.56.

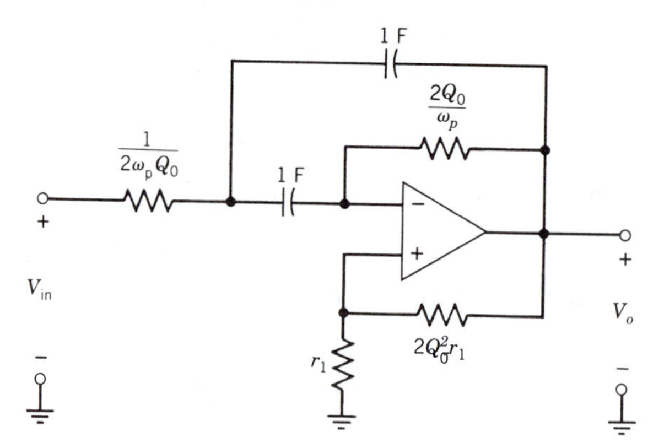

Figure 8.54

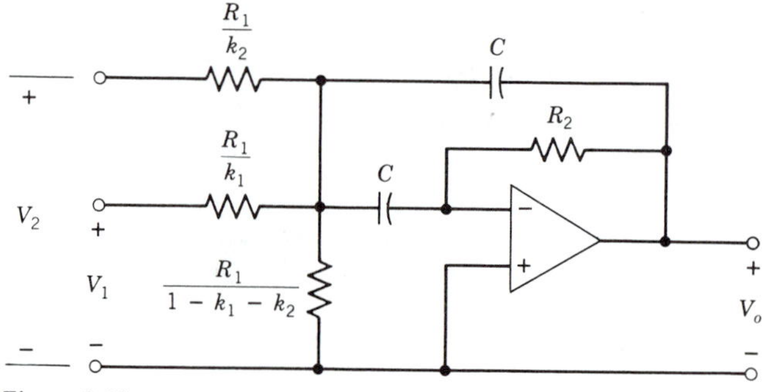

Figure 8.55

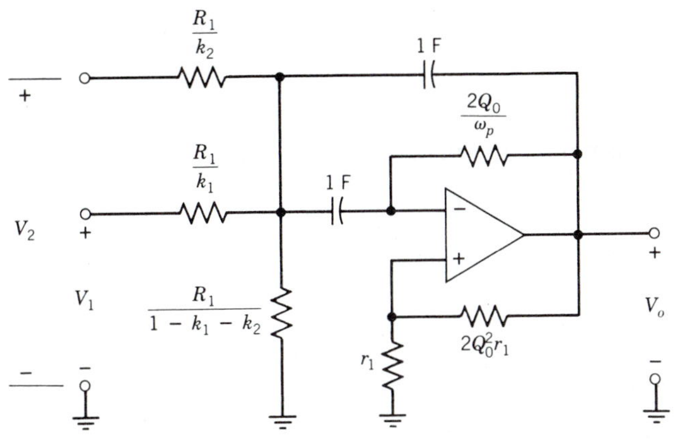

Figure 8.56

The output voltages of these networks to the inputs V_1 and V_2 are found to be

$$V_o(s) = -\frac{s/R_1C}{s^2 + (2/R_2C)s + 1/R_1R_2C^2}(k_1V_1 + k_2V_2) \tag{8.54}$$

for Figure 8.55, and

$$V_o(s) = -\frac{(2Q_0 + 1/Q_0)\omega_p s}{s^2 + \omega_p^2}(k_1V_1 + k_2V_2) \tag{8.55}$$

for Figure 8.56. By choosing k_1 and k_2 appropriately, the networks provide the necessary gain and summing requirements for the band-pass leapfrog design.

The third network to be considered is the multiple-amplifier biquad of Figure 7.18 by interchanging the second and third stages as illustrated in Figure 8.57 with $R_5 = R_6 = R_0$. This does not alter the transfer function of the filter, but it does provide us with an inverting output shown as V_{2a} and a noninverting output V_{2b}. The network

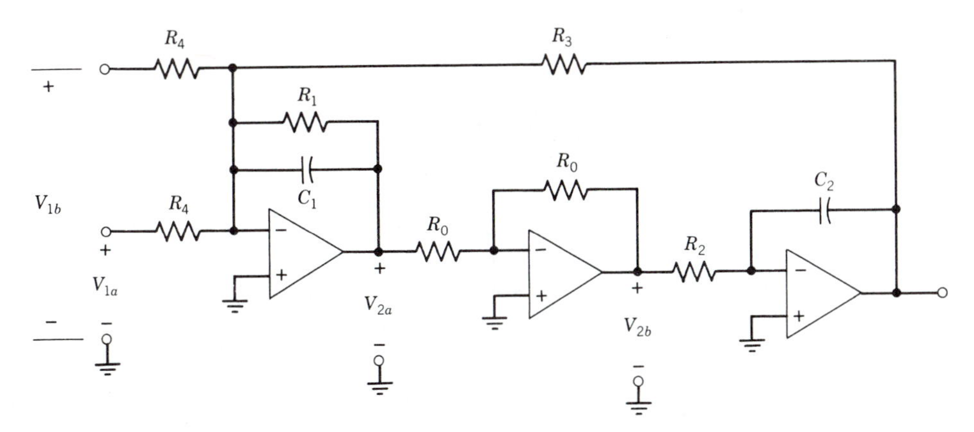

Figure 8.57

also provides a summing capability. The outputs are found from (7.128) to be

$$V_{2a} = -V_{2b} = -\frac{s/R_4C_1}{s^2 + (1/R_1C_1)s + 1/R_2R_3C_1C_2}(V_{1a} + V_{1b}) \tag{8.56}$$

Equating this transfer function to (8.42) and choosing

$$C_1 = C_2 = C_0, \qquad R_2 = R_3 = R^* \tag{8.57}$$

the other element values are obtained as

$$R_1 = \frac{R^*}{R}\sqrt{\frac{L}{C}}, \qquad R_4 = R^*\sqrt{\frac{L}{C}}, \qquad C_0 = \frac{\sqrt{LC}}{R^*} \tag{8.58}$$

Likewise, equating (8.56) to (8.43) and using (8.57) yields

$$R_1 = R^*R\sqrt{\frac{C}{L}}, \qquad R_4 = R^*\sqrt{\frac{C}{L}}, \qquad C_0 = \frac{\sqrt{LC}}{R^*} \tag{8.59}$$

Example 8.6 It is required to design a band-pass fourth-order Chebyshev filter using the leapfrog method. The passband extends from $\omega = 1000$ rad/s to $\omega = 4000$ rad/s with peak-to-peak ripple not to exceed 1 dB. The filter is to be operated between a resistive generator and a resistive load, and is of the ladder type.

From specifications, the normalized low-pass prototype is obtained in Figure 8.58, which is the same as that given in Figure 2.23. The required bandwidth and mid-band frequency are

$$B = \omega_2 - \omega_1 = 4000 - 1000 = 3000 \text{ rad/s} \tag{8.60a}$$

$$\omega_0 = \sqrt{\omega_1\omega_2} = \sqrt{1000 \times 4000} = 2000 \text{ rad/s} \tag{8.60b}$$

Applying the low-pass to band-pass transformation of Figure 8.47, we obtain the band-pass network of Figure 8.59, which we wish to simulate using the leapfrog method. To simplify the computations, we perform frequency-scaling by a factor of 10^{-4} and magnitude-scaling by a factor of 10^{-2}. The resulting network is shown in Figure 8.60 with

$$Z_1(s) = \frac{0.2632s}{s^2 + 0.0658s + 0.04} \tag{8.61a}$$

$$Y_2(s) = \frac{0.1389s}{s^2 + 0.04} \tag{8.61b}$$

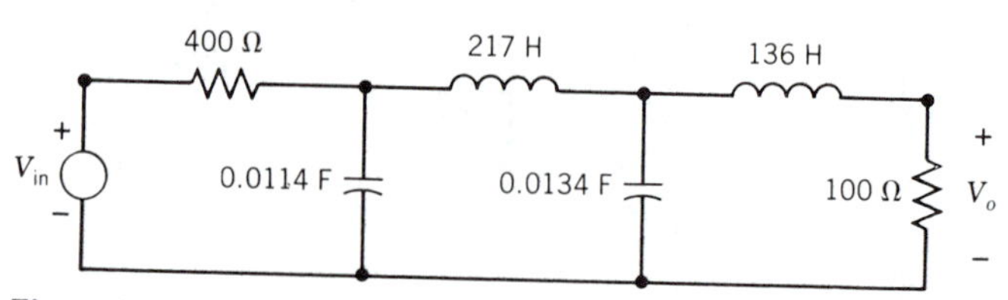

Figure 8.58

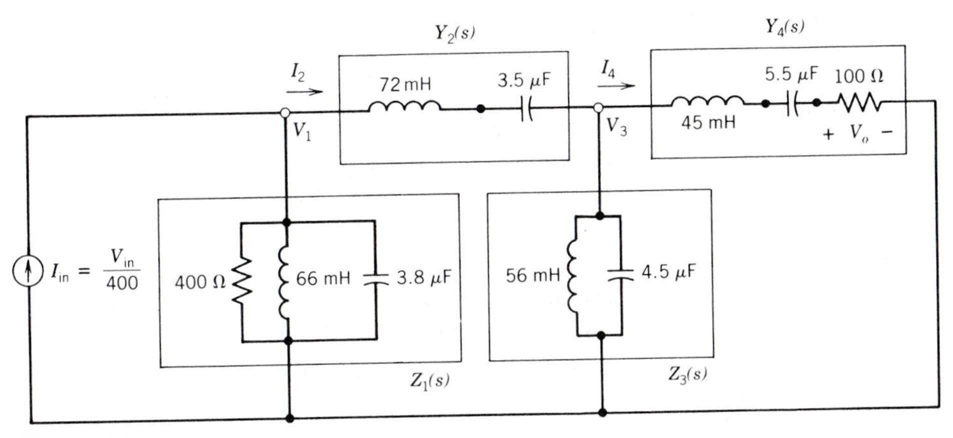

Figure 8.59

$$Z_3(s) = \frac{0.2222s}{s^2 + 0.04} \tag{8.61c}$$

$$Y_4(s) = \frac{0.2222s}{s^2 + 0.222s + 0.04} \tag{8.61d}$$

Using the references as indicated in Figure 8.60, the branch v-i relationships of the ladder are described by the equations

$$V_1 = Z_1(I_{in} - I_2) \tag{8.62a}$$

$$I_2 = Y_2(V_1 - V_3) \tag{8.62b}$$

$$V_3 = Z_3(I_2 - I_4) \tag{8.62c}$$

$$I_4 = Y_4 V_3 \tag{8.62d}$$

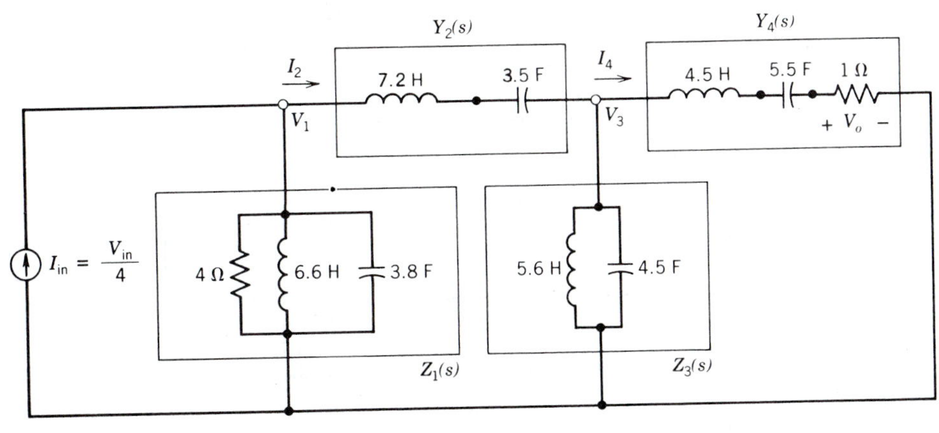

Figure 8.60

These equations may be modified somewhat to yield

$$-V_1 = -Z_1(I_{in} - I_2) \tag{8.63a}$$

$$-I_2 = Y_2(-V_1 + V_3) \tag{8.63b}$$

$$V_3 = -Z_3(-I_2 + I_4) \tag{8.63c}$$

$$I_4 = Y_4 V_3 \tag{8.63d}$$

Equations (8.63) can be interpreted by the block diagram of Figure 8.61(*a*). Observe that by eliminating the intermediate variables V_1, I_2, and V_3 among the equations (8.63), we obtain the transfer current-ratio function I_4/I_{in}. This function can also be obtained as the ratio E_4/E_{in} by eliminating the intermediate variables V_1, E_2, and V_3 among the equations

$$-V_1 = -Z_1(E_{in} - E_2) \tag{8.64a}$$

$$-E_2 = Y_2(-V_1 + V_3) \tag{8.64b}$$

$$V_3 = -Z_3(-E_2 + E_4) \tag{8.64c}$$

$$E_4 = Y_4 V_3 \tag{8.64d}$$

where E_{in}, E_2, and E_4 are considered as voltage variables. Thus, we can simulate equations (8.63) using the block diagram of Figure 8.61(*b*), where Z_1, Y_2, Z_3, and Y_4 are the transfer voltage-ratio functions of the individual blocks, even though we still use the immittance symbol Y or Z. Since $V_{in} = 4I_{in}$ and $V_o = 1 \times I_4$, the original transfer voltage-ratio function V_o/V_{in} is related to the ratio E_4/E_{in} by the equation

$$\frac{V_o}{V_{in}} = \frac{1I_4}{4I_{in}} = 0.25\,\frac{E_4}{E_{in}} \tag{8.65}$$

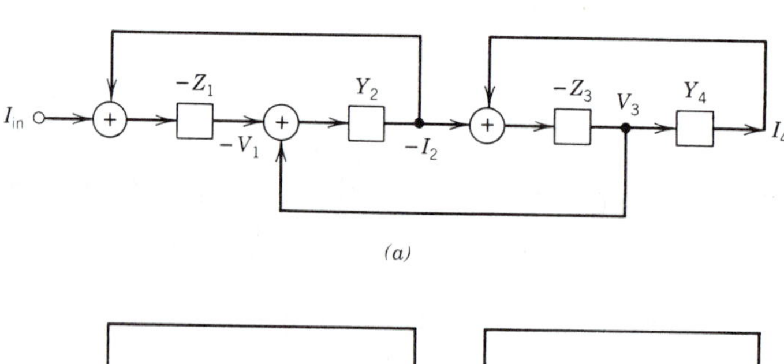

(*a*)

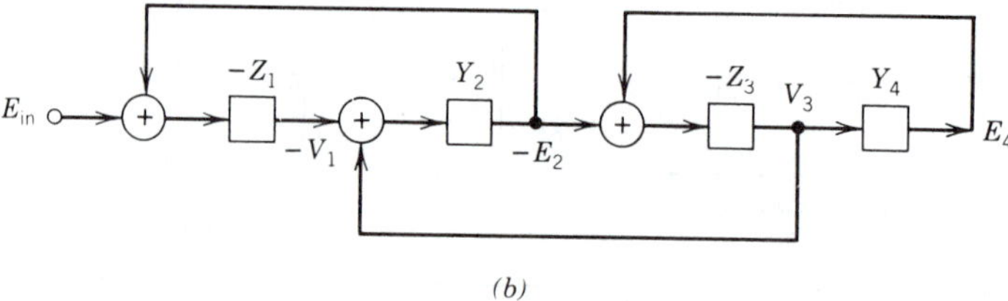

(*b*)

Figure 8.61

We first consider the design of active RC networks that will yield the transfer voltage-ratio functions $H_1(s)=Z_1(s)$ and $H_4(s)=Y_4(s)$. For this we use the network of Figure 8.55. Choosing $C=1$ F and $k_1=k_2=k$, we obtain from (8.47) and (8.61a)

$$R_1=0.822\ \Omega, \qquad R_2=30.4\ \Omega \tag{8.66}$$

and set $k/R_1C=0.2632$, giving

$$k_1=k_2=k=0.2164 \tag{8.67}$$

The network realizing (8.61a) is shown in Figure 8.62. Likewise, for $H_4(s)=Y_4(s)$ the parameter values are

$$C=1\text{ F}, \qquad R_1=2.75\ \Omega, \qquad R_2=9.01\ \Omega \tag{8.68a}$$

$$k_1=0.6108, \qquad k_2=0 \tag{8.68b}$$

Turning to the design of networks, the transfer voltage-ratio functions of which are $H_2(s)=Y_2(s)$ and $H_3(s)=Z_3(s)$, we use the network of Figure 8.56. In this case, we choose

$$Q_0=1, \qquad C=1\text{ F} \tag{8.69}$$

and obtain from (8.52)

$$R_1=2.5\ \Omega, \qquad R_2=10\ \Omega, \qquad k_1=0.2315, \qquad k_2=0 \tag{8.70}$$

for the transfer function (8.61b), and

$$R_1=2.5\ \Omega, \qquad R_2=10\ \Omega, \qquad k_1=k_2=0.3703 \tag{8.71}$$

for (8.61c). The network realizing (8.61b) is shown in Figure 8.63.

The next task is to fit the individual realizations together to form a leapfrog network in accordance with the configuration of Figure 8.61(*b*). In doing so we must pay particular attention to inverting or noninverting networks and make provision for summing at the three specified points in the network. The manner in which this is accomplished is shown in Figure 8.64. The resulting normalized parameter values are also indicated in the figure.

The final step in the design is scaling. To attain the original values required in Figure 8.60, we frequency-scale the network of Figure 8.64 by a factor of 10^4. However,

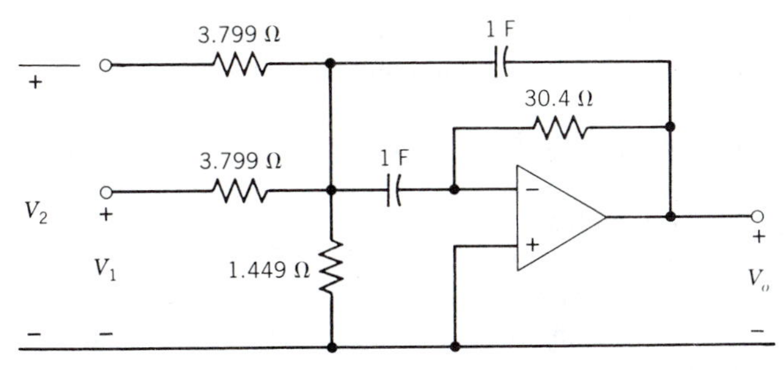

Figure 8.62

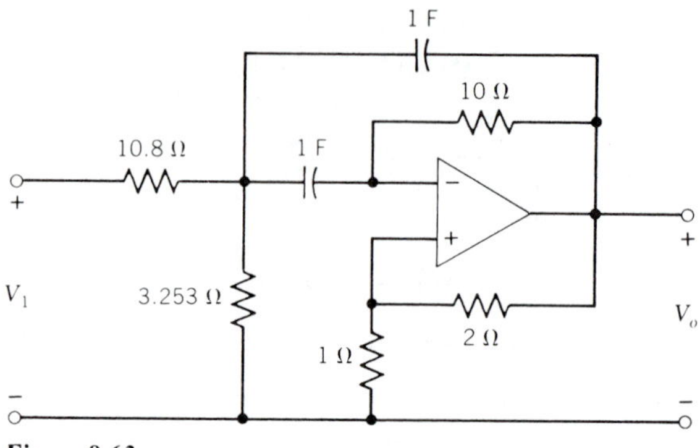

Figure 8.63

we may also magnitude-scale by a factor of 10^4 to obtain convenient element values. The final realization is shown in Figure 8.65. The example shows the ease with which realizations are found using the leapfrog method.

Doubly terminated lossless *LC* ladder networks always appear in two alternative forms. The first form is shown in Figure 8.33 with inductor L_1 appearing as the first reactive element in the ladder, and is known as the *minimum capacitance form.* The second form is shown in Figure 8.58 with capacitor C_1 being the first reactive element, and is known as the *minimum inductance form.* The minimum capacitance form was used to illustrate the design of low-pass filters using the leapfrog method. As a change, we used the minimum inductance form to demonstrate the realization of a band-pass leapfrog filter. The leapfrog method is probably one of the best methods available for the realizations of band-pass active filters. The reasons are that it makes use of the band-pass biquads developed in Chapter 7 and yet it provides sensitivities very close to those *LC* ladder networks. In many situations, the individual biquads can be detuned several percent before a noticeable change occurs in the overall frequency response. Furthermore, the leapfrog technique makes repeated use of nearly identical component networks as building blocks of the overall realizations, a great saving in manufacturing costs.

8.4.2 **Alternative Form of the Leapfrog Realization**

In this section, we introduce an alternative form of the leapfrog realization. The form is completely equivalent to the original configuration except that it is redrawn and put in an entirely different form.

Consider, for example, the leapfrog structure of Figure 8.66. This structure can be redrawn as shown in Figure 8.67. For illustrative purposes, specific points on the two block diagrams are indicated by the letters $A, B, \ldots, N$. It is straightforward to verify that these two block diagrams are completely equivalent.

A specific case of a leapfrog realization of a band-pass filter is shown in Figure 8.65. As shown in Example 8.6, this network is a realization of the leapfrog structure of Figure 8.66, which can be redrawn in the form of Figure 8.67. The final result is

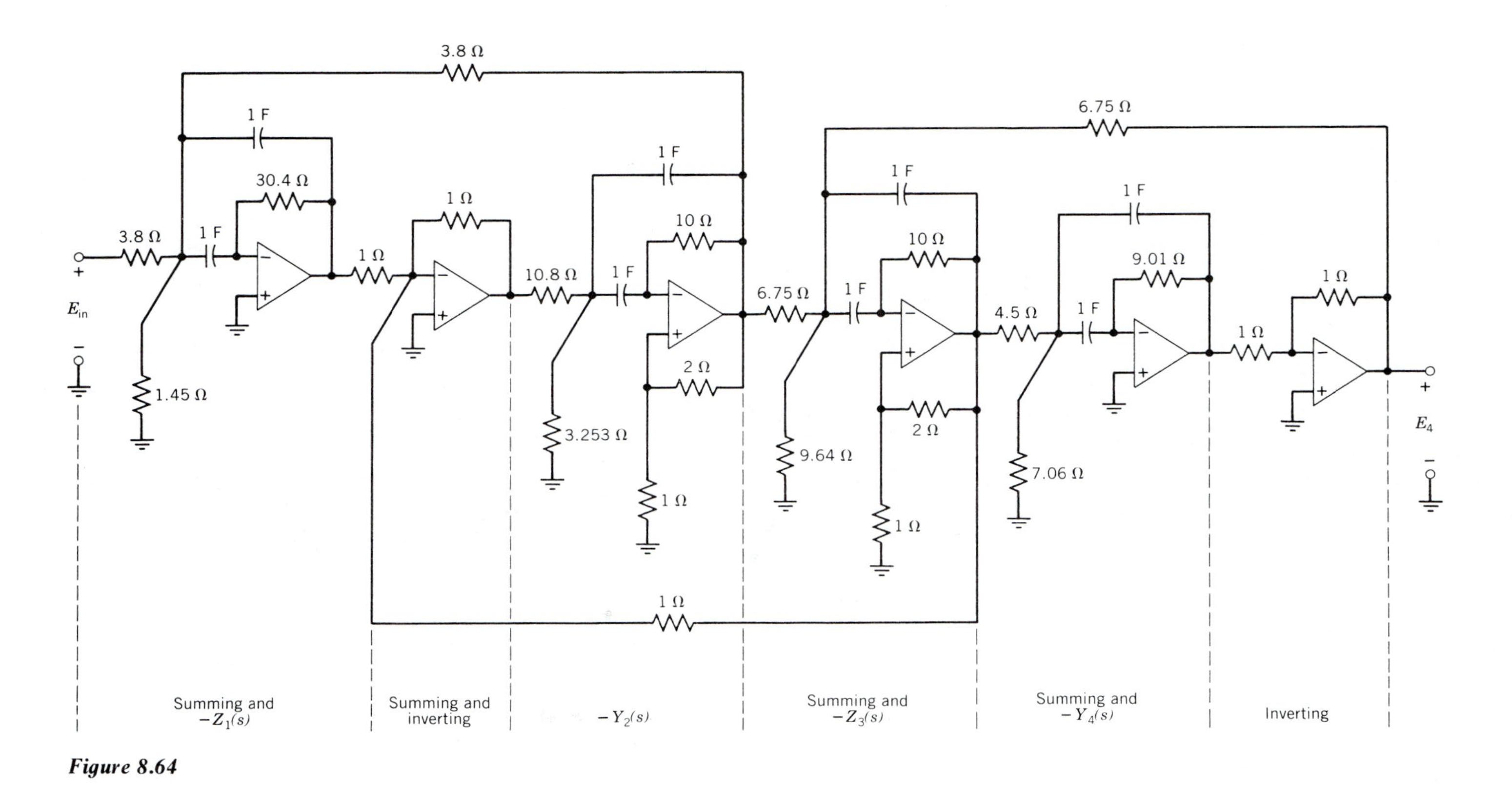

3.8 Ω
1 F
30.4 Ω
1 F
3.8 Ω
1.45 Ω
E_{in}
+
−
1 Ω
1 Ω
10.8 Ω
3.253 Ω
1 F
1 F
10 Ω
2 Ω
1 Ω
6.75 Ω
9.64 Ω
1 F
1 F
10 Ω
2 Ω
1 Ω
6.75 Ω
4.5 Ω
7.06 Ω
1 F
1 F
9.01 Ω
1 Ω
1 Ω
E_4
1 Ω
Summing and $-Z_1(s)$
Summing and inverting
$-Y_2(s)$
Summing and $-Z_3(s)$
Summing and $-Y_4(s)$
Inverting

Figure 8.64

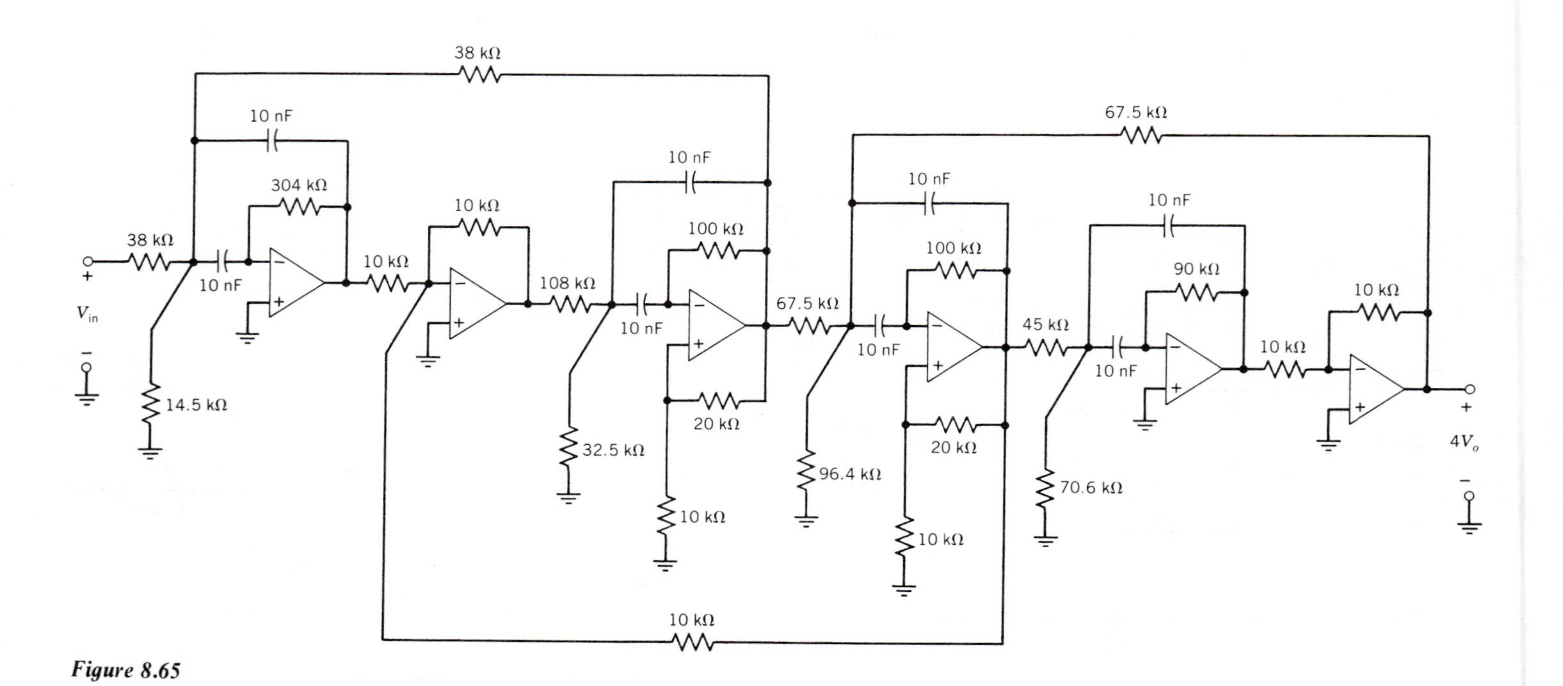

38 kΩ
10 nF
304 kΩ
38 kΩ
10 nF
+
V_{in}
−
14.5 kΩ
10 kΩ
10 kΩ
108 kΩ
10 nF
100 kΩ
10 nF
32.5 kΩ
20 kΩ
10 kΩ
67.5 kΩ
67.5 kΩ
10 nF
100 kΩ
10 nF
96.4 kΩ
20 kΩ
10 kΩ
10 kΩ
45 kΩ
10 nF
90 kΩ
10 nF
70.6 kΩ
10 kΩ
10 kΩ
+
$4V_o$
−

Figure 8.65

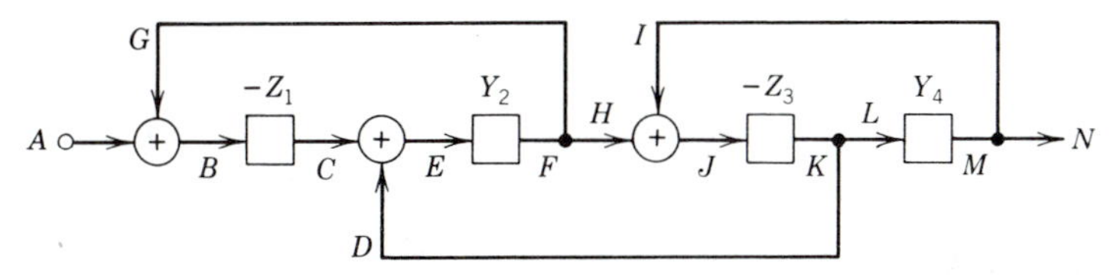

Figure 8.66

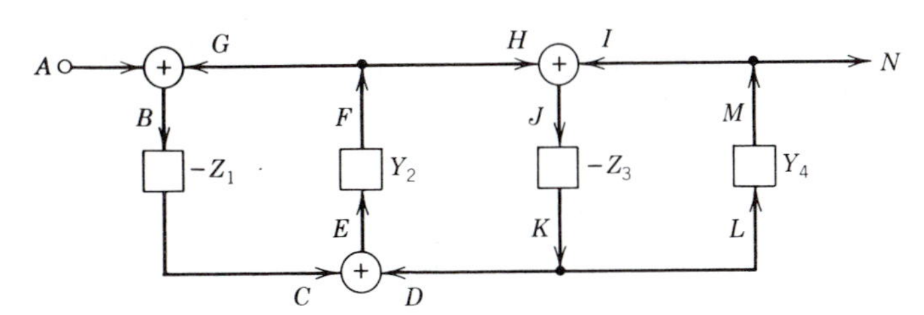

Figure 8.67

shown in Figure 8.68. The reader is urged to check all points in the figures to understand the equivalence of the connections.

We next consider the leapfrog realization of the low-pass filter of Figure 8.69, which can be simulated by the block diagram of Figure 8.70. In practice, the terminating resistors are usually incorporated in other blocks as shown in Figure 8.71, for which the leapfrog block diagram is presented in Figure 8.72. We now modify the block diagram of Figure 8.70 to match the form of Figure 8.72. One possible solution is shown in Figure 8.73 with the alternative form given in Figure 8.74. The block diagram of Figure 8.74 can be realized by the *RC* amplifier network of Figure 8.75, in which, excluding the provisions for the terminating resistors, all blocks are realized by integrators, some inverting and some noninverting. By equating gains of corresponding loops in the network of Figure 8.75 and the block diagram of Figure 8.74, we obtain the following set of design formulas:

$$R_1C'_1 = R_1C_1 \tag{8.72a}$$

$$R'_2C'_1R_3C'_2 = C_1L_2 \tag{8.72b}$$

$$R_5C'_2R_4C'_3 = L_2C_3 \tag{8.72c}$$

$$R_6C'_3R_7C'_4 = C_3L_4 \tag{8.72d}$$

$$R_8C'_4 = L_4/R_2 \tag{8.72e}$$

There are five equations in the eleven unknown variables. The additional degrees of freedom can be used to satisfy other practical requirements. For the present, we select all integrator capacitors to be equal to a practically convenient value C and require the time constants of each two-integrator loop to be equal, or

$$C'_1 = C'_2 = C'_3 = C'_4 = C \tag{8.73a}$$

$$R'_2C'_1 = R_3C'_2, \qquad R_5C'_2 = R_4C'_3, \qquad R_6C'_3 = R_7C'_4 \tag{8.73b}$$

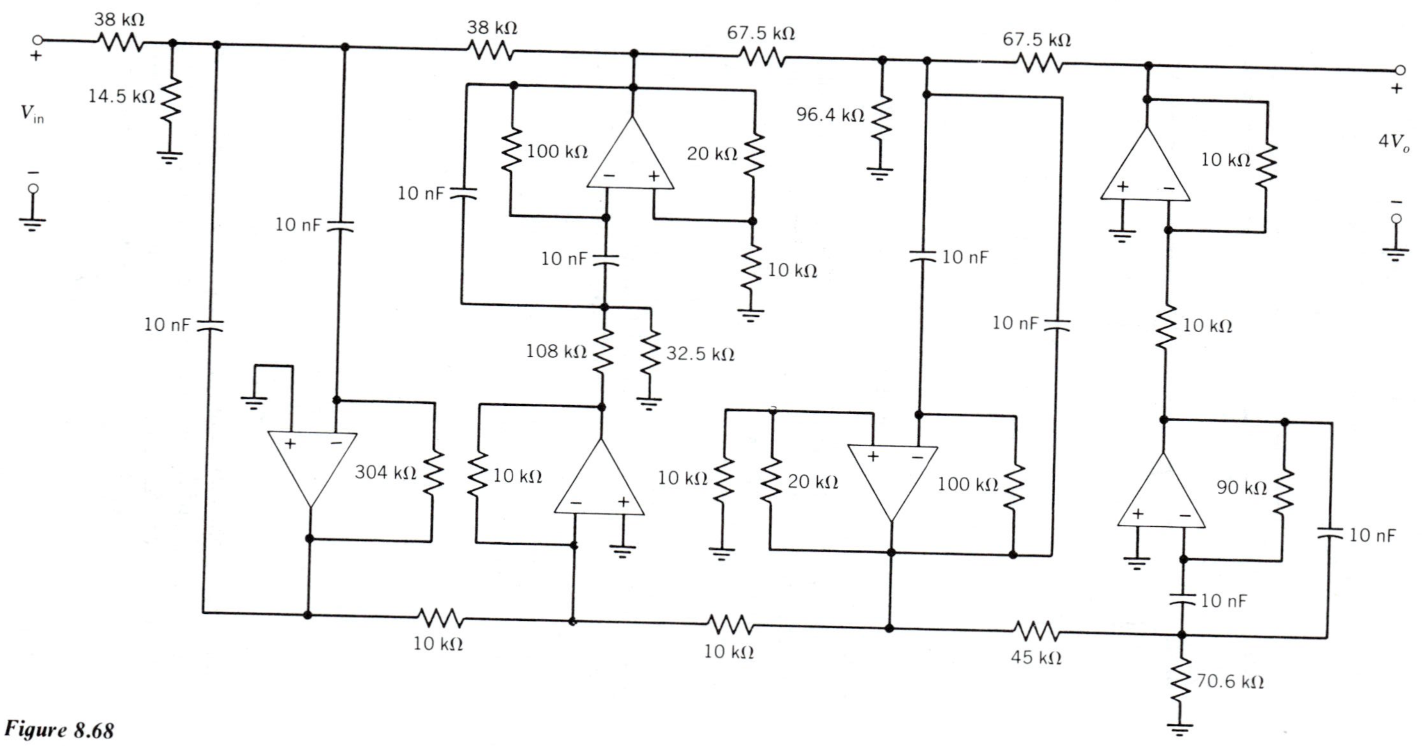

38 kΩ
14.5 kΩ
V_{in}
+
−
10 nF
10 nF
38 kΩ
10 nF
100 kΩ
20 kΩ
10 nF
10 kΩ
108 kΩ
32.5 kΩ
304 kΩ
10 kΩ
10 kΩ
10 kΩ
67.5 kΩ
96.4 kΩ
10 nF
10 nF
10 kΩ
20 kΩ
100 kΩ
10 kΩ
45 kΩ
67.5 kΩ
10 kΩ
10 kΩ
90 kΩ
10 nF
10 nF
70.6 kΩ
$4V_o$

Figure 8.68

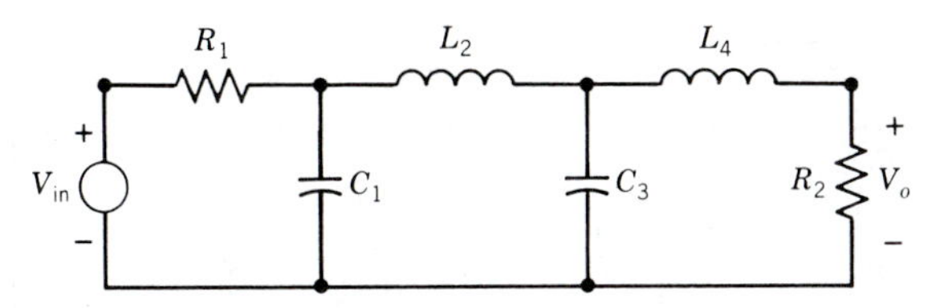

Figure 8.69

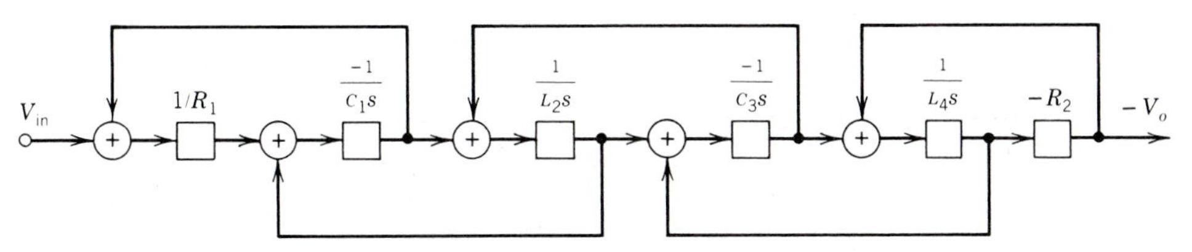

Figure 8.70

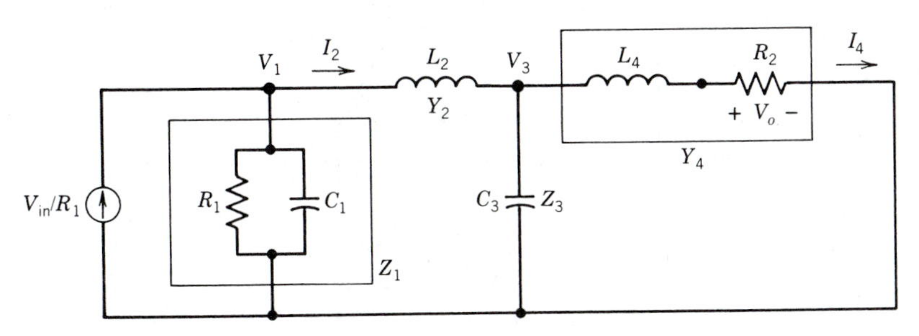

Figure 8.71

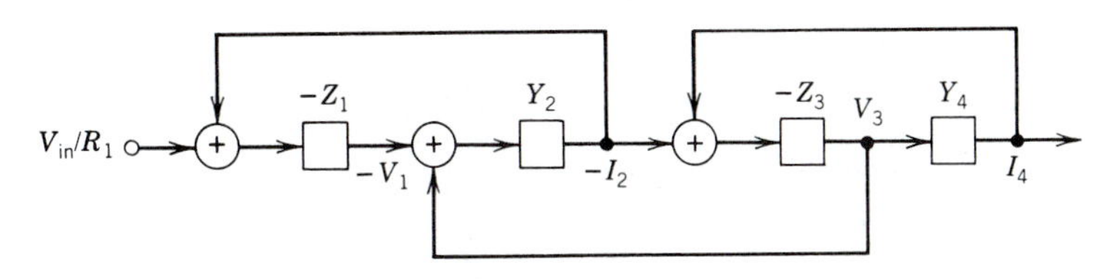

Figure 8.72

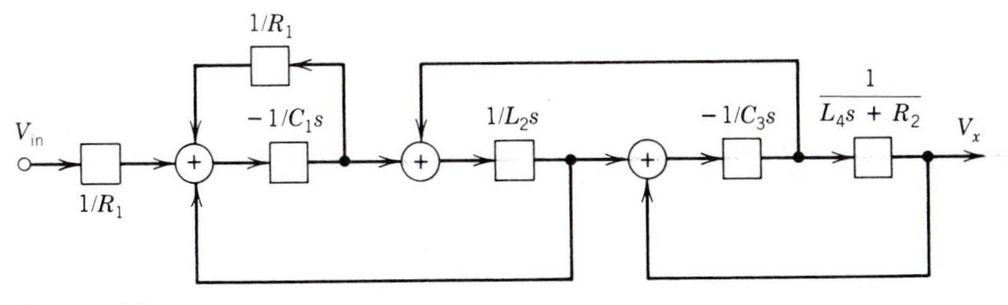

Figure 8.73

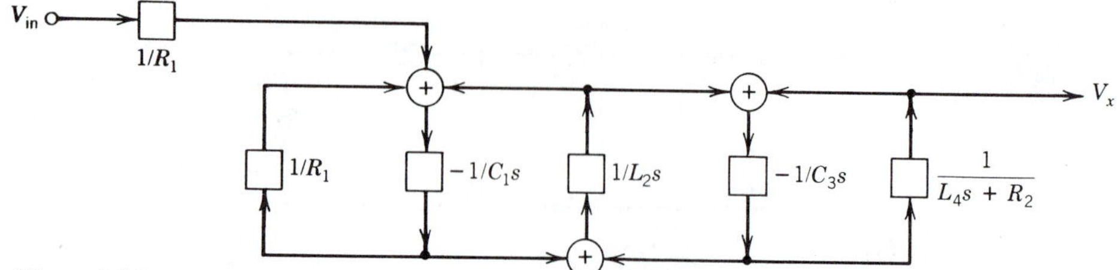

Figure 8.74

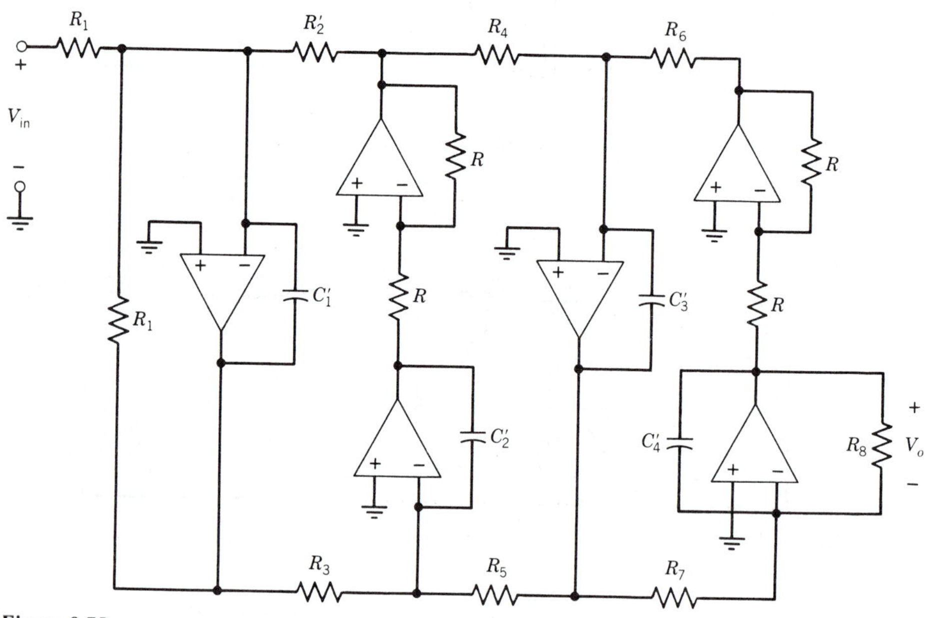

Figure 8.75

With these the above design formulas become

$$C_1' = C_2' = C_3' = C_4' = C = C_1 \tag{8.74a}$$

$$R_2' = R_3 = \sqrt{\frac{L_2}{C_1}} \tag{8.74b}$$

$$R_4 = R_5 = \frac{\sqrt{L_2 C_3}}{C_1} \tag{8.74c}$$

$$R_6 = R_7 = \frac{\sqrt{C_3 L_4}}{C_1} \tag{8.74d}$$

$$R_8 = \frac{L_4}{C_1 R_2} \tag{8.74e}$$

For added flexibility in controlling the dc gain, we replace the source resistor R_1 by R_s as shown in Figure 8.76. Then the dc gain of the resulting active filter will be

$$K = R_1/R_s \tag{8.75}$$

times that of the network of Figure 8.75. Instead of the equal-capacitance and equal-loop-time-constant realization of Figure 8.75, suppose that we select

$$C_2' = L_2, \qquad C_3' = C_3, \qquad C_4' = L_4 \tag{8.76}$$

for the element values in Figure 8.75, the resulting network of which is shown in Figure 8.76 with

$$R_2'R_3 = R_4R_5 = R_6R_7 = 1, \qquad R_8 = 1/R_2 \tag{8.77}$$

We illustrate the above procedure by the following example.

Example 8.7 It is desired to design a leapfrog active filter realizing a fourth-order Chebyshev low-pass response. The peak-to-peak ripple in the passband must not exceed 1 dB and the ripple bandwidth is 10^4 rad/s. The filter is to be operated between a resistive generator and a resistive load.

The desired LC prototype network may be obtained by means of the explicit formulas (4.169). We have already found an LC ladder realization for this response as shown in Figure 4.27, which is redrawn in Figure 8.77. Comparing this network with that of Figure 8.69, we can make the following identifications:

$$R_1 = 100\ \Omega, \qquad C_1 = 1.68\ \mu\text{F}, \qquad L_2 = 11.9\ \text{mH} \tag{8.78a}$$

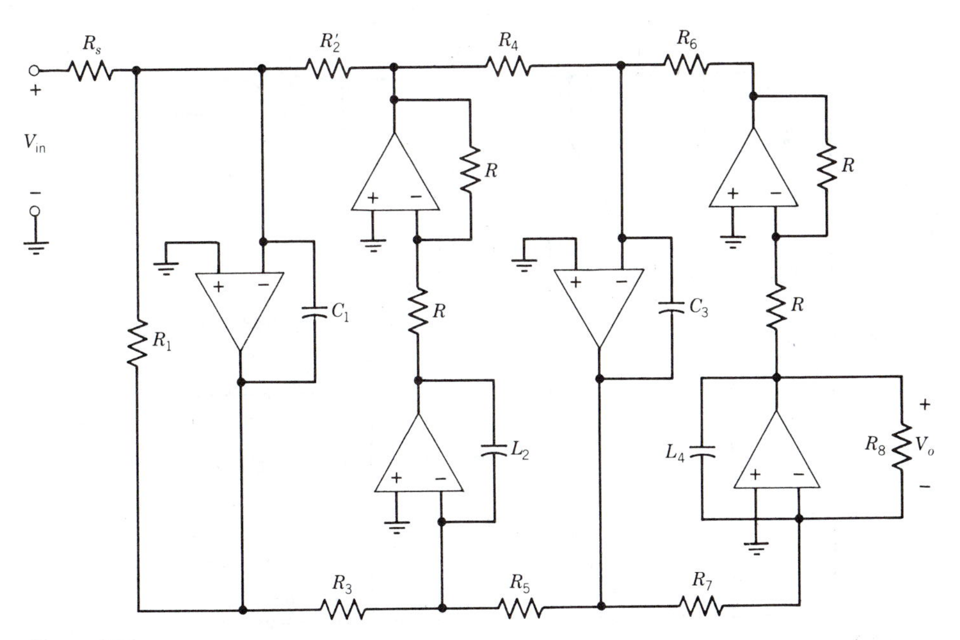

Figure 8.76

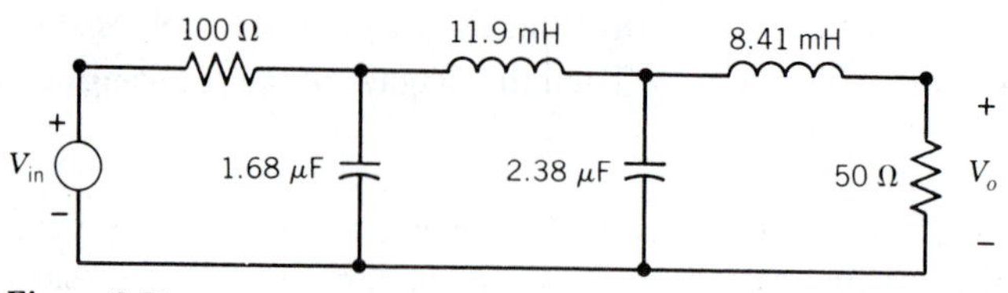

Figure 8.77

$$C_3 = 2.38\ \mu\text{F}, \qquad L_4 = 8.41\ \text{mH}, \qquad R_2 = 50\ \Omega \tag{8.78b}$$

Substituting these in (8.74) yields the element values

$$R_2' = R_3 = 84\ \Omega, \qquad R_4 = R_5 = 100\ \Omega \tag{8.79a}$$

$$R_6 = R_7 = 84\ \Omega, \qquad R_8 = 100\ \Omega \tag{8.79b}$$

We next consider the problem of adjusting the overall network gain so that the minimum attenuation is 0 dB in the passband. For the low-pass filter of Figure 8.77, the dc gain is

$$\frac{V_o}{V_{\text{in}}} = \frac{50}{100+50} = 0.333 \tag{8.80}$$

From (2.36) and (2.38b), the maximum value in the passband is greater than 0.333

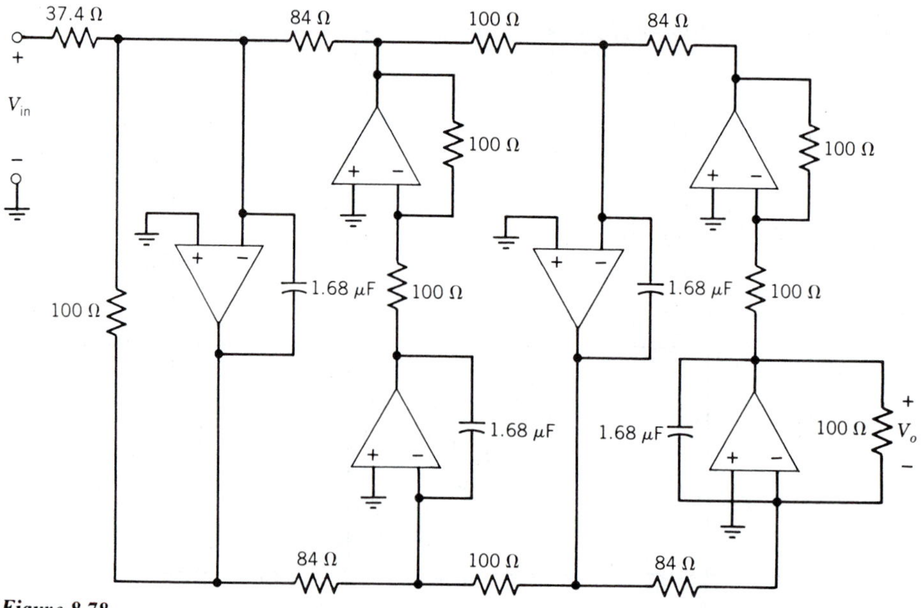

Figure 8.78

by the factor $\sqrt{1+\varepsilon^2}$, where $\varepsilon = 0.509$ is the ripple factor corresponding to 1-dB ripple, giving the passband maxima

$$\left.\frac{V_o}{V_{\text{in}}}\right|_{\max} = 0.333\sqrt{1+\varepsilon^2} = 0.374 \tag{8.81}$$

We require that this value be 1 corresponding to the 0-dB requirement. This implies that the overall dc gain must be increased by $1/0.374 = 2.674$. To accomplish this, we replace the source resistor R_1 by R_s and obtain from (8.75)

$$R_s = \frac{R_1}{K} = \frac{100}{2.674} = 37.4\ \Omega \tag{8.82}$$

leaving all other element values in the network of Figure 8.75 unaltered. Inserting these values in Figure 8.75 gives a realization shown in Figure 8.78. Finally, to obtain convenient element values, we magnitude-scale the network by a factor of 100, yielding the desired leapfrog design shown in Figure 8.79.

This example illustrates the ease with which the low-pass realizations are found using the design method for leapfrog networks, which simulate the doubly terminated LC prototype ladder networks. Other multiple-loop methods such as the follow-the-leader-feedback (FLF) -type filters, the primary-resonator-block (PRB) -type filters, and the generalized FLF (GFLF) -type filters are also available. The reader is referred to references [5] and [8] for details.

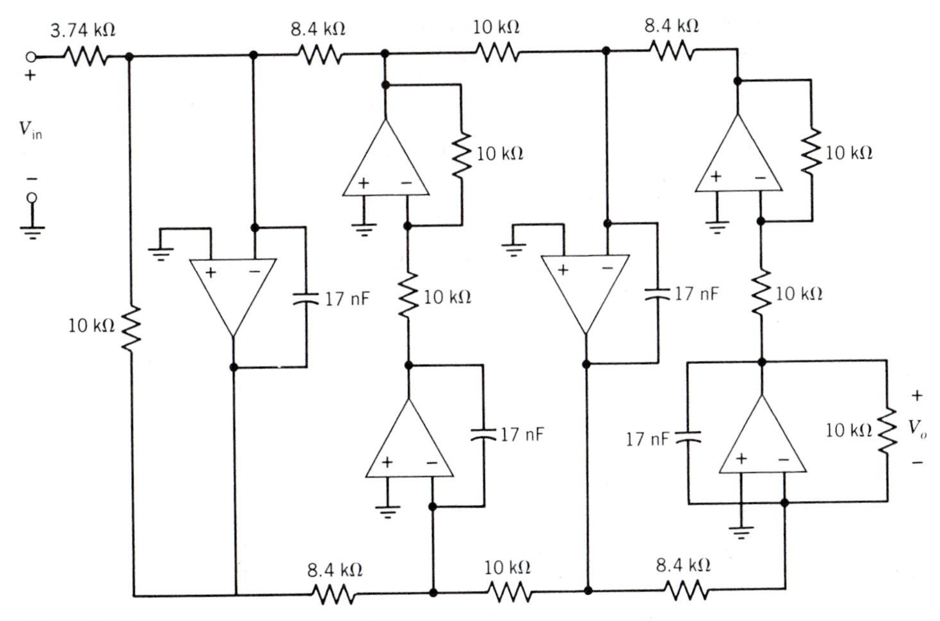

Figure 8.79

8.5 SENSITIVITY ANALYSIS

In the foregoing, we have introduced techniques for simulating passive networks by using simulated inductance or frequency-dependent negative resistors, or by simulating equations describing the passive ladder structure with analog computer simulation. In the present section, we discuss various aspects of the sensitivity problem as it relates to these techniques.

The network function of the filter is, in general, a rational function of the complex-frequency variable s, being the ratio of two polynomials of the form

$$H(s)=\frac{P(s)}{Q(s)}=\frac{b_m s^m+b_{m-1}s^{m-1}+\cdots+b_1 s+b_0}{a_n s^n+a_{n-1}s^{n-1}+\cdots+a_1 s+a_0} \tag{8.83}$$

where $n \geqq m$. Since some of the passive elements in the network are replaced by their active RC equivalents, the sensitivity of the network function to the changes of the element values in these active RC equivalents is important in that active elements such as op amps are much more susceptible to environmental changes than passive elements. To compute this sensitivity, let a replaced impedance be designated as

$$Z(s)=\beta s^k \tag{8.84}$$

where k is a positive or negative integer, zero included. For example, if $k=0$, then Z is the impedance of a resistor and $\beta=R$. Likewise, if $k=1$ then $\beta=L$, the inductance of an inductor; and if $k=-1$ then $1/\beta=C$, the capacitance of a capacitor. Finally, for $k=-2$, $Z=\beta/s^2$ denotes the impedance of a frequency-dependent negative resistor.

To proceed we next compute the sensitivities of the coefficients a_i and b_i to any arbitrary network element x. From (6.1) we define the *coefficient sensitivities* as

$$S_x^{a_i}=\frac{x}{a_i}\cdot\frac{\partial a_i}{\partial x}, \qquad S_x^{b_i}=\frac{x}{b_i}\cdot\frac{\partial b_i}{\partial x} \tag{8.85}$$

Using (6.1) in conjunction with (8.83), the network function sensitivity can be written as

$$\begin{aligned} S_x^H&=\frac{x}{H}\cdot\frac{\partial H}{\partial x}=\frac{x}{P}\cdot\frac{\partial P}{\partial x}-\frac{x}{Q}\cdot\frac{\partial Q}{\partial x} \\ &=\frac{1}{P}\sum_{i=0}^{m} x s^i\frac{\partial b_i}{\partial x}-\frac{1}{Q}\sum_{i=0}^{n} x s^i\frac{\partial a_i}{\partial x} \end{aligned} \tag{8.86}$$

Combining this with (8.85) yields

$$S_x^H=\frac{1}{P}\sum_{i=0}^{m} S_x^{b_i} b_i s^i-\frac{1}{Q}\sum_{i=0}^{n} S_x^{a_i} a_i s^i \tag{8.87}$$

Since the coefficients a_i and b_i are functions of the parameter β, which in turn is a function of some parameter x, the sensitivities of the coefficients a_i and b_i to x can be expressed as

$$S_x^{a_i}=S_\beta^{a_i}S_x^\beta, \qquad S_x^{b_i}=S_\beta^{b_i}S_x^\beta \tag{8.88}$$

Substituting these in (8.87) gives

$$S_x^H = \frac{1}{P}\sum_{i=0}^{m} S_\beta^{b_i} S_x^\beta b_i s^i - \frac{1}{Q}\sum_{i=0}^{n} S_\beta^{a_i} S_x^\beta a_i s^i \tag{8.89}$$

Since S_x^β is independent of the summations, (8.89) can be factored as

$$S_x^H = S_x^\beta \left(\frac{1}{P}\sum_{i=0}^{m} S_\beta^{b_i} b_i s^i - \frac{1}{Q}\sum_{i=0}^{n} S_\beta^{a_i} a_i s^i \right) \tag{8.90}$$

This equation expresses the transfer-function sensitivity in terms of the coefficient sensitivities to β and the sensitivity of β to x. We illustrate the use of this formula for three active RC networks simulating three passive ladder networks.

Example 8.8 A doubly terminated third-order high-pass filter is shown in Figure 8.80, the transfer voltage-ratio function of which is found to be

$$H(s) = \frac{V_o}{V_{in}}$$

$$= \frac{C_1L_2C_3R_4s^3}{(C_1L_2C_3R_4 + C_1R_1L_2C_3)s^3 + (C_1R_1C_3R_4 + L_2C_3 + C_1L_2)s^2 + (C_3R_4 + C_1R_1)s + 1}$$

$$= \frac{b_3s^3}{a_3s^3 + a_2s^2 + a_1s + a_0} = \frac{4s^3}{6s^3 + 6s^2 + 3s + 1} \tag{8.91}$$

We wish to compute the sensitivity of the transfer function with respect to each of the passive elements of the synthetic inductor used to provide a realization. The simulated inductor of Figure 8.4 is used in the active RC realization by choosing

$$Z_1 = R_{12}, \qquad Z_2 = 1/sC_{22}, \qquad Z_3 = R_{32}, \qquad Z_4 = R_{42}, \qquad Z_5 = R_{52} \tag{8.92}$$

the input impedance of which is from (8.14)

$$Z_{in}(s) = \frac{Z_1Z_3Z_5}{Z_2Z_4} = \frac{R_{12}R_{32}R_{52}C_{22}}{R_{42}}\, s \tag{8.93}$$

giving the value of the synthetic inductance as

$$L_2 = \frac{R_{12}R_{32}R_{52}C_{22}}{R_{42}} \tag{8.94}$$

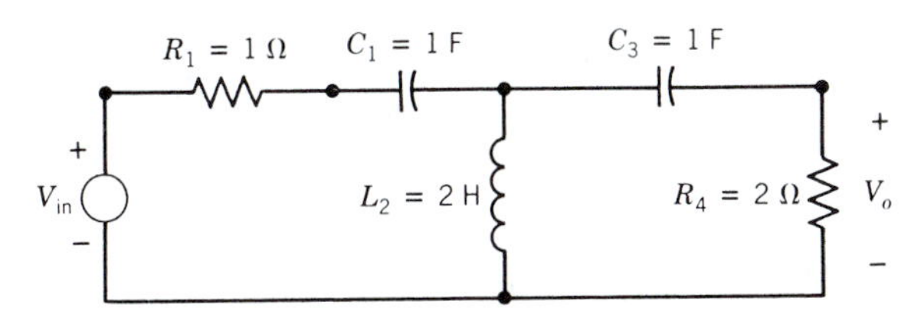

Figure 8.80

From (8.91) the coefficient sensitivities with respect to L_2 are found from (8.85) to be

$$S_{L_2}^{b_0}=S_{L_2}^{b_1}=S_{L_2}^{b_2}=0, \qquad S_{L_2}^{b_3}=1 \tag{8.95a}$$

$$S_{L_2}^{a_0}=S_{L_2}^{a_1}=0, \qquad S_{L_2}^{a_3}=1 \tag{8.95b}$$

$$S_{L_2}^{a_2}=\frac{L_2(C_1+C_3)}{C_1R_1C_3R_4+L_2C_3+C_1L_2}=\frac{2}{3} \tag{8.95c}$$

Substituting these in (8.90) in conjunction with (8.91), the transfer-function sensitivity becomes

$$\begin{aligned} S_x^H &= S_x^{L_2}\left(1-\frac{a_3s^3+S_{L_2}^{a_2}a_2s^2}{a_3s^3+a_2s^2+a_1s+a_0}\right) \\ &= S_x^{L_2}\frac{2s^2+3s+1}{6s^3+6s^2+3s+1} \end{aligned} \tag{8.96}$$

where x may be any of the quantities R_{12}, R_{32}, R_{42}, R_{52}, and C_{22}. From (8.94) the sensitivities of L_2 to x are obtained as

$$S_{R_{12},R_{32},C_{22},R_{52}}^{L_2}=1, \qquad S_{R_{42}}^{L_2}=-1 \tag{8.97}$$

where the meaning of $S_{x_1,x_2,x_3x_4}^H$ is defined the same as in (6.18). Thus, we have

$$S_{R_{12},R_{32},C_{22},R_{52}}^{H(s)}=-S_{R_{42}}^{H(s)}=\frac{2s^2+3s+1}{6s^3+6s^2+3s+1} \tag{8.98}$$

On the $j\omega$-axis, the function (8.98) can be written as

$$S_{R_{12},R_{32},C_{22},R_{52}}^{H(j\omega)}=-S_{R_{42}}^{H(j\omega)}=\frac{(1+\omega^2-6\omega^4)-j(6\omega^3+12\omega^5)}{1-3\omega^2+36\omega^6} \tag{8.99}$$

Finally, from (6.57) the fractional change in $H(j\omega)$ due to the variations in R_{12}, R_{32}, R_{42}, R_{52}, and C_{22} is approximated by the expression

$$\frac{\Delta H(j\omega)}{H(j\omega)}\approx S_{R_{12}}^H\frac{\Delta R_{12}}{R_{12}}+S_{R_{32}}^H\frac{\Delta R_{32}}{R_{32}}+S_{R_{42}}^H\frac{\Delta R_{42}}{R_{42}}+S_{R_{52}}^H\frac{\Delta R_{52}}{R_{52}}+S_{C_{22}}^H\frac{\Delta C_{22}}{C_{22}} \tag{8.100}$$

Our objective is to find the fractional change in the magnitude of $H(j\omega)$ and the increment of its phase. To this end, we appeal to (6.34) and (6.35), which can be rewritten as

$$\frac{x}{|H(j\omega)|}\cdot\frac{|\Delta H(j\omega)|}{\Delta x}=\operatorname{Re}\left(\frac{x}{H(j\omega)}\cdot\frac{\Delta H(j\omega)}{\Delta x}\right) \tag{8.101a}$$

$$\frac{x}{\phi(j\omega)}\cdot\frac{\Delta\phi(j\omega)}{\Delta x}=\frac{1}{\phi(j\omega)}\operatorname{Im}\left(\frac{x}{H(j\omega)}\cdot\frac{\Delta H(j\omega)}{\Delta x}\right) \tag{8.101b}$$

or

$$\frac{|\Delta H(j\omega)|}{|H(j\omega)|}=\operatorname{Re}\frac{\Delta H(j\omega)}{H(j\omega)} \tag{8.102a}$$

$$\Delta\phi(j\omega) = \Delta \arg H(j\omega) = \operatorname{Im} \frac{\Delta H(j\omega)}{H(j\omega)} \tag{8.102b}$$

Substituting (8.100) in (8.102) gives

$$\frac{|\Delta H(j\omega)|}{|H(j\omega)|} \approx \frac{\Delta R_{12}}{R_{12}} \operatorname{Re} S_{R_{12}}^{H(j\omega)} + \frac{\Delta R_{32}}{R_{32}} \operatorname{Re} S_{R_{32}}^{H(j\omega)} + \frac{\Delta R_{42}}{R_{42}} \operatorname{Re} S_{R_{42}}^{H(j\omega)} + \frac{\Delta R_{52}}{R_{52}} \operatorname{Re} S_{R_{42}}^{H(j\omega)}$$

$$+ \frac{\Delta C_{22}}{C_{22}} \operatorname{Re} S_{C_{22}}^{H(j\omega)} \tag{8.103}$$

$$\Delta \arg H(j\omega) \approx \frac{\Delta R_{12}}{R_{12}} \operatorname{Im} S_{R_{12}}^{H(j\omega)} + \frac{\Delta R_{32}}{R_{32}} \operatorname{Im} S_{R_{32}}^{H(j\omega)} + \frac{\Delta R_{42}}{R_{42}} \operatorname{Im} S_{R_{42}}^{H(j\omega)}$$

$$+ \frac{\Delta R_{52}}{R_{52}} \operatorname{Im} S_{R_{52}}^{H(j\omega)} + \frac{\Delta C_{22}}{C_{22}} \operatorname{Im} S_{C_{22}}^{H(j\omega)} \tag{8.104}$$

Using (8.99) in (8.103) and (8.104) yields

$$\frac{|\Delta H(j\omega)|}{|H(j\omega)|} \approx \frac{1+\omega^2-6\omega^4}{1-3\omega^2+36\omega^6}\left(\frac{\Delta R_{12}}{R_{12}} + \frac{\Delta R_{32}}{R_{32}} + \frac{\Delta R_{52}}{R_{52}} + \frac{\Delta C_{22}}{C_{22}} - \frac{\Delta R_{42}}{R_{42}}\right) \tag{8.105}$$

$$\Delta \arg H(j\omega) \approx -\frac{6\omega^3(1+2\omega^2)}{1-3\omega^2+36\omega^6}\left(\frac{\Delta R_{12}}{R_{12}} + \frac{\Delta R_{32}}{R_{32}} + \frac{\Delta R_{52}}{R_{52}} + \frac{\Delta C_{22}}{C_{22}} - \frac{\Delta R_{42}}{R_{42}}\right) \tag{8.106}$$

Clearly, by an appropriate choice of the variation of the individual passive elements of the simulated inductor of Figure 8.4, the above sensitivities can be minimized.

Example 8.9 Consider the low-pass third-order filter of Figure 8.81. Applying the magnitude-scaling by the factor of $1/s$, we obtain the FDNR network of Figure

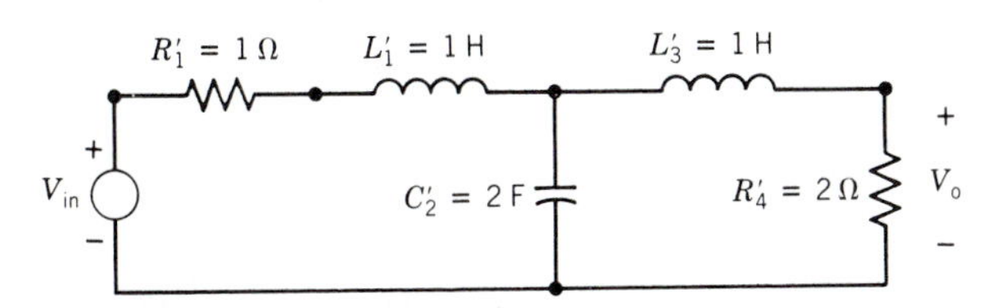

Figure 8.81

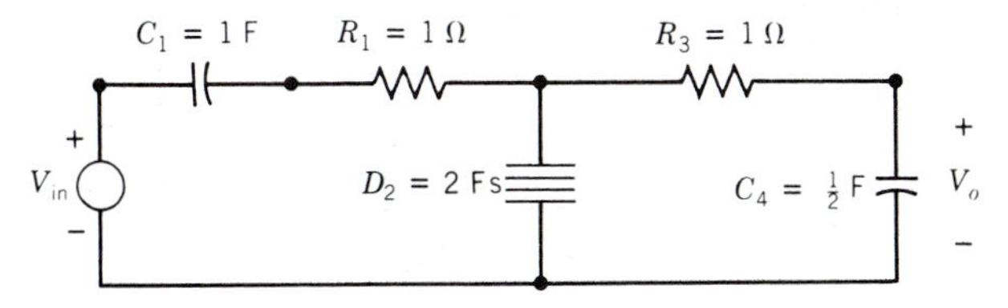

Figure 8.82

8.82, the transfer voltage-ratio function of which is given by

$$H(s)=\frac{V_o}{V_{in}}=\frac{b_0}{a_3s^3+a_2s^2+a_1s+a_0} \tag{8.107}$$

where $b_0=C_1=1$ and

$$a_3=C_1R_1D_2R_3C_4=1, \qquad a_2=C_1R_1D_2+D_2R_3C_4=3 \tag{8.108a}$$

$$a_1=C_1R_3C_4+C_1R_1C_4+D_2=3, \qquad a_0=C_1+C_4=1.5 \tag{8.108b}$$

We now investigate the sensitivities of these coefficients to the FDNR constant D_2. From (8.85) we have

$$S_{D_2}^{b_0}=S_{D_2}^{a_0}=0, \qquad S_{D_2}^{a_2}=S_{D_2}^{a_3}=1 \tag{8.109a}$$

$$S_{D_2}^{a_1}=\frac{D_2}{C_1C_4(R_1+R_3)+D_2}=\frac{2}{3} \tag{8.109b}$$

Using the network of Figure 8.12 with the following choice of elements to realize the FDNR,

$$Z_1=1/sC_{12}, \qquad Z_2=R_{22}, \qquad Z_3=1/sC_{32}, \qquad Z_4=R_{42}, \qquad Z_5=R_{52} \tag{8.110}$$

the resulting network is shown in Figure 8.83, the input impedance of which is from (8.29)

$$Z_{in}(s)=\frac{Z_1Z_3Z_5}{Z_2Z_4}=\frac{R_{52}}{R_{22}R_{42}C_{12}C_{32}s^2} \tag{8.111}$$

giving the D_2 value of the FDNR as

$$D_2=\frac{R_{22}R_{42}C_{12}C_{32}}{R_{52}} \tag{8.112}$$

The sensitivities of D_2 with respect to the FDNR elements are now found to be

$$S_{R_{22},R_{42},C_{12},C_{32}}^{D_2}=-S_{R_{52}}^{D_2}=1 \tag{8.113}$$

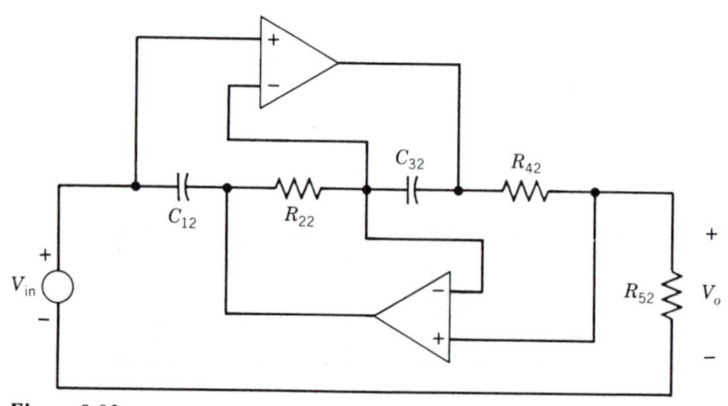

Figure 8.83

Substituting the above results in (8.90) gives the transfer-function sensitivity with respect to x as

$$S_x^H = S_x^{D_2}\left(S_{D_2}^{b_0} - \frac{S_{D_2}^{a_3}a_3s^3 + S_{D_2}^{a_2}a_2s^2 + S_{D_2}^{a_1}a_1s + S_{D_2}^{a_0}a_0}{a_3s^3 + a_2s^2 + a_1s + a_0}\right)$$

$$= -S_x^{D_2}\frac{s(s^2+3s+2)}{s^3+3s^2+3s+1.5} \tag{8.114}$$

where x may represent R_{22}, R_{42}, R_{52}, C_{12}, or C_{32}. Combining this with (8.113) gives

$$S_{R_{22},R_{42},C_{12},C_{32}}^{H(s)} = -S_{R_{52}}^{H(s)} = -\frac{s(s^2+3s+2)}{s^3+3s^2+3s+1.5} \tag{8.115}$$

From (6.57) and (8.102), the fractional change in the magnitude of $H(j\omega)$ and the increment in its phase are obtained as

$$\frac{|\Delta H(j\omega)|}{|H(j\omega)|} \approx -\frac{\omega^2(\omega^4+4\omega^2+1.5)}{\omega^6+3\omega^4+2.25}\left(\frac{\Delta R_{22}}{R_{22}} + \frac{\Delta R_{42}}{R_{42}} + \frac{\Delta C_{12}}{C_{12}} + \frac{\Delta C_{32}}{C_{32}} - \frac{\Delta R_{52}}{R_{52}}\right) \tag{8.116}$$

$$\Delta \arg H(j\omega) \approx -\frac{1.5\omega^3+3\omega}{\omega^6+3\omega^4+2.25}\left(\frac{\Delta R_{22}}{R_{22}} + \frac{\Delta R_{42}}{R_{42}} + \frac{\Delta C_{12}}{C_{12}} + \frac{\Delta C_{32}}{C_{32}} - \frac{\Delta R_{52}}{R_{52}}\right) \tag{8.117}$$

where

$$S_{R_{52}}^{H(j\omega)} = \frac{\omega^2(\omega^4+4\omega^2+1.5) + j\omega(1.5\omega^2+3)}{\omega^6+3\omega^4+2.25} \tag{8.118}$$

If the magnitude sensitivity of $H(j\omega)$ is required, then from (6.34) and (8.115) we obtain

$$S_{R_{22},R_{42},C_{12},C_{32}}^{|H(j\omega)|} = -S_{R_{52}}^{|H(j\omega)|} = \operatorname{Re} S_{R_{22},R_{42},C_{12},C_{32}}^{H(j\omega)} = -\operatorname{Re} S_{R_{52}}^{H(j\omega)}$$

$$= -\frac{\omega^2(\omega^4+4\omega^2+1.5)}{\omega^6+3\omega^4+2.25} \tag{8.119}$$

Example 8.10 Consider again the low-pass third-order filter of Figure 8.81, which is redrawn in Figure 8.84. The transfer voltage-ratio function is

$$H(s) = \frac{R_4}{L_1C_2L_3s^3 + (L_1C_2R_4 + R_1C_2L_3)s^2 + (R_1C_2R_4 + L_1 + L_3)s + R_1 + R_4}$$

$$= \frac{2}{2s^3+6s^2+6s+3} \tag{8.120}$$

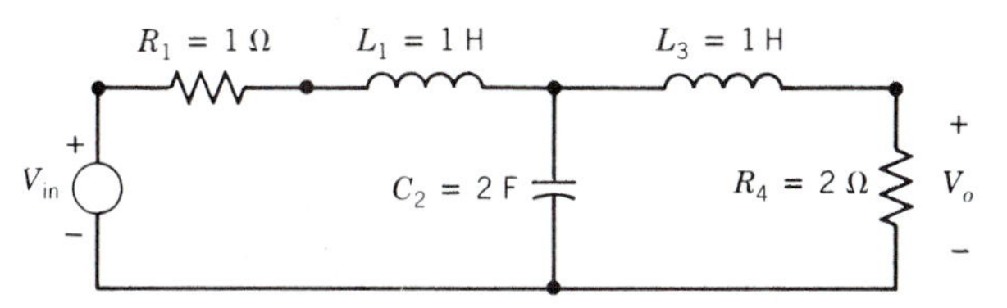

Figure 8.84

We wish to use the leapfrog technique to simulate the low-pass ladder. For the low-pass ladder of Figure 8.44, the general form of the series and shunt immittances is shown in Figure 8.85 with

$$Y_k(s) = \frac{1/L_k}{s + R_k/L_k}, \qquad Z_k(s) = \frac{1/C_k}{s + 1/R_k C_k} \tag{8.121}$$

where R_k may be zero in Y_k or infinite in Z_k. As indicated in (8.39), the active network of Figure 8.37, which is redrawn in Figure 8.86, can be used to simulate these functions. Its transfer voltage-ratio function is

$$H_\beta(s) = \frac{V_\beta}{V_\alpha} = -\frac{1/R_{1k}C_{3k}}{s + 1/R_{2k}C_{3k}} \tag{8.122}$$

For admittance Y_k, the necessary choice of the elements is given by

$$R_k = R_{1k}/R_{2k}, \qquad L_k = R_{1k}C_{3k} \tag{8.123}$$

and for Z_k we choose

$$R_k = R_{2k}/R_{1k}, \qquad C_k = R_{1k}C_{3k} \tag{8.124}$$

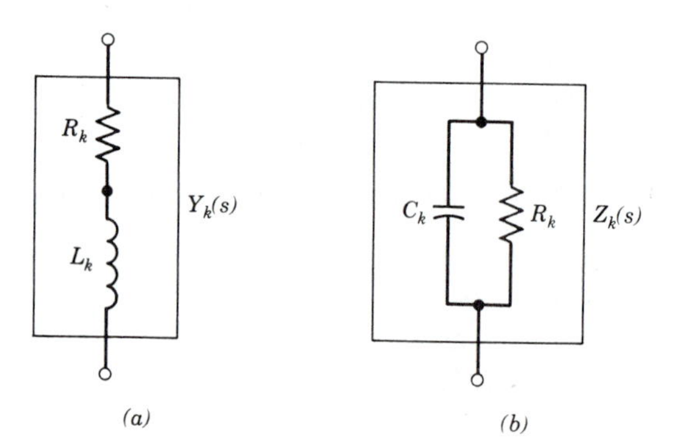

Figure 8.85

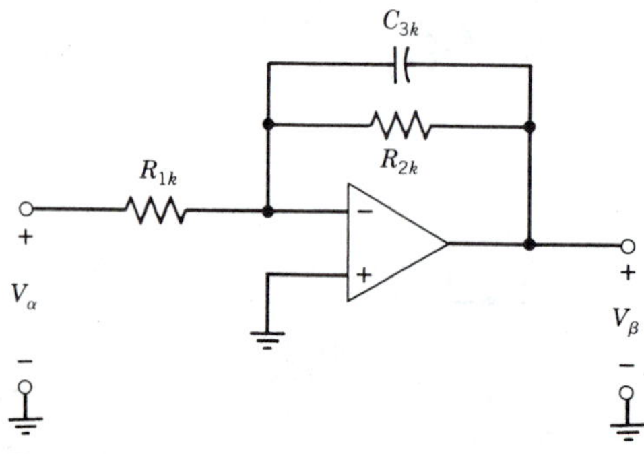

Figure 8.86

Our objective is to calculate the transfer-function sensitivity with respect to the elements R_{11}, R_{21}, and C_{31} of the network of Figure 8.86 with $k=1$, which realizes the first series branch R_1+sL_1 of Figure 8.84. From (8.123) we choose

$$R_1=R_{11}/R_{21}, \qquad L_1=R_{11}C_{31} \tag{8.125}$$

Setting $R_{11}=1\ \Omega$, we obtain from (8.125) and Figure 8.84 the other element values

$$R_{21}=1\ \Omega, \qquad C_{31}=1\ \text{F} \tag{8.126}$$

Substituting (8.125) in (8.120) yields

$$H(s)=\frac{b_0}{a_3s^3+a_2s^2+a_1s+a_0}=\frac{2}{2s^3+6s^2+6s+3} \tag{8.127}$$

where $b_0=R_{21}R_4=2$ and

$$a_3=R_{11}C_{31}R_{21}L_3C_2=2, \qquad a_2=R_{11}C_{31}R_{21}C_2R_4+R_{11}C_2L_3=6 \tag{8.128a}$$

$$a_1=R_{11}C_2R_4+R_{11}C_{31}R_{21}+R_{21}L_3=6, \qquad a_0=R_{11}+R_{21}R_4=3 \tag{8.128b}$$

The coefficient sensitivities with respect to R_{11}, R_{21}, and C_{31} are found from (8.128), as follows:

$$S^{a_0}_{R_{11}}=\frac{R_{11}}{a_0}=\frac{1}{3}, \qquad S^{a_1}_{R_{11}}=\frac{R_{11}}{a_1}(C_2R_4+C_{31}R_{21})=\frac{5}{6} \tag{8.129a}$$

$$S^{a_2}_{R_{11}}=\frac{R_{11}}{a_2}(R_{21}C_{31}C_2R_4+C_2L_3)=1, \qquad S^{a_3}_{R_{11}}=\frac{R_{11}}{a_3}R_{21}C_{31}C_2L_3=1 \tag{8.129b}$$

$$S^{a_0}_{R_{21}}=\frac{R_{21}}{a_0}R_4=\frac{2}{3}, \qquad S^{a_1}_{R_{21}}=\frac{R_{21}}{a_1}(L_3+R_{11}C_{31})=\frac{1}{3} \tag{8.129c}$$

$$S^{a_2}_{R_{21}}=\frac{R_{21}}{a_2}R_{11}C_{31}C_2R_4=\frac{2}{3}, \qquad S^{a_3}_{R_{21}}=\frac{R_{21}}{a_3}R_{11}C_{31}C_2L_3=1 \tag{8.129d}$$

$$S^{a_0}_{C_{31}}=0, \qquad S^{a_1}_{C_{31}}=\frac{C_{31}}{a_1}R_{11}R_{21}=\frac{1}{6} \tag{8.129e}$$

$$S^{a_2}_{C_{31}}=\frac{C_{31}}{a_2}R_{11}R_{21}C_2R_4=\frac{2}{3}, \qquad S^{a_3}_{C_{31}}=\frac{C_{31}}{a_3}R_{11}R_{21}C_2L_3=1 \tag{8.129f}$$

$$S^{b_0}_{R_{11}}=S^{b_0}_{C_{31}}=0, \qquad S^{b_0}_{R_{21}}=\frac{R_{21}}{b_0}R_4=1 \tag{8.129g}$$

To calculate the transfer-function sensitivity, we use formula (8.87) instead of (8.89), and obtain

$$S^H_x=S^{b_0}_x-\frac{S^{a_3}_x a_3s^3+S^{a_2}_x a_2s^2+S^{a_1}_x a_1s+S^{a_0}_x a_0}{a_3s^3+a_2s^2+a_1s+a_0} \tag{8.130}$$

where x represents R_{11}, R_{21}, or C_{31}. Substituting (8.128) and (8.129) in (8.130) gives the transfer-function sensitivities

$$S_{R_{11}}^{H(s)} = -\frac{2s^3+6s^2+5s+1}{2s^3+6s^2+6s+3} \tag{8.131a}$$

$$S_{R_{21}}^{H(s)} = \frac{2s^2+4s+1}{2s^3+6s^2+6s+3} \tag{8.131b}$$

$$S_{C_{31}}^{H(s)} = -\frac{s(2s^2+4s+1)}{2s^3+6s^2+6s+3} \tag{8.131c}$$

On the $j\omega$-axis, they become

$$S_{R_{11}}^{H(j\omega)} = -\frac{(4\omega^6+14\omega^4+6\omega^2+3)-j(2\omega^3+9\omega)}{4\omega^6+12\omega^4+9} \tag{8.132a}$$

$$S_{R_{21}}^{H(j\omega)} = \frac{(4\omega^4+12\omega^2+3)-j(4\omega^5+10\omega^3-6\omega)}{4\omega^6+12\omega^4+9} \tag{8.132b}$$

$$S_{C_{31}}^{H(j\omega)} = -\frac{(4\omega^6+10\omega^4-6\omega^2)+j(4\omega^5+12\omega^3+3\omega)}{4\omega^6+12\omega^4+9} \tag{8.132c}$$

Appealing to (6.34), the magnitude sensitivities of $H(j\omega)$ are found to be

$$S_{R_{11}}^{|H(j\omega)|} = \text{Re}\, S_{R_{11}}^{H(j\omega)} = -\frac{4\omega^6+14\omega^4+6\omega^2+3}{4\omega^6+12\omega^4+9} \tag{8.133a}$$

$$S_{R_{21}}^{|H(j\omega)|} = \text{Re}\, S_{R_{21}}^{H(j\omega)} = \frac{4\omega^4+12\omega^2+3}{4\omega^6+12\omega^4+9} \tag{8.133b}$$

$$S_{C_{31}}^{|H(j\omega)|} = \text{Re}\, S_{C_{31}}^{H(j\omega)} = -\frac{2\omega^2(2\omega^4+5\omega^2-3)}{4\omega^6+12\omega^4+9} \tag{8.133c}$$

8.6 SUMMARY AND SUGGESTED READINGS

In this chapter, we demonstrated that based on passive prototype networks, certain portions of the passive realizations could be simulated by using active *RC* networks. Such a technique is referred to as a *passive network simulation method*. One of the important advantages of this approach is that it emulates the low-sensitivity property of the passive prototype network. In addition, there is a large body of knowledge and experience with the passive networks, especially with the passive ladders, where tables of element values and explicit formulas are readily available.

A useful and practical structure from an engineering viewpoint is the doubly terminated passive ladder network. It is attractive in that it is unbalanced so that all the shunt branches can be grounded. Furthermore, the structure has very low sensitivity, which can be emulated by using active *RC* networks. This emulation is accomplished in at least three different ways.

The first approach is known as the *inductance simulation*, where the realization of the active RC filters from its passive prototype structure is achieved by direct replacement of the inductors with equivalent active RC networks. The procedure consists of first obtaining the resistively terminated LC network and then replacing each inductor by an active RC equivalent. The resulting filter is expected to have the same low sensitivity as its passive counterpart except for the imperfections in the realization of the inductor using active RC network. One way to simulate an inductor makes use of the gyrator. When a gyrator is terminated in a capacitor, the input port behaves like an inductor. Another network realization of an inductor that we introduced in this chapter uses two op amps. These networks can only realize grounded inductors. To simulate a floating inductor, we connect two gyrators in cascade or connect two identical op-amp networks back-to-back.

The second approach makes use of the *frequency-dependent negative resistor* (FDNR), a device whose impedance is inversely proportional to s^2. An FDNR can be realized by using the terminated generalized immittance converter (GIC). To apply the FDNR to the synthesis of active filters, we first realize a given approximation function by an LC doubly terminated ladder or other suitable LC prototypes. The equivalent active RC network is then obtained by replacing each resistor of resistance R by a capacitor of capacitance $1/Rb$, each inductor of inductance L by a resistor of resistance Lb, and each capacitor of capacitance C by an FDNR whose D value is C/b, where b is any appropriate positive-real constant. This transformation does not affect the topology of its prototype, nor does it affect the transfer voltage-ratio function realized. As a result, the transformed network possesses the same low sensitivity as its passive prototype except for the increase in the number of elements used and the imperfections of the op amps employed to realize the FDNR.

A common feature of the above two approaches is that only one particular element type was realized actively using RC and op-amp networks. The third approach that we discussed is by simulating equations describing the passive ladder structure with analog operations such as summing, differencing, or integration. The use of analog computer simulation for the filter appears to be impractical. However, such is not the case at all, and the filters that result are very competitive with those discussed in the foregoing. The basic idea is that we represent the equations describing a ladder network by a flow graph or block diagram. In the block diagram, we recognize that we cannot realize currents. Therefore, we have to simulate the currents as voltages. In addition, the blocks with specified immittances are implemented with active RC networks having these immittances as transfer functions. The resulting active RC network realizes the prescribed transfer voltage-ratio function. The method is known as the *leapfrog technique*. One of the disadvantages of the leapfrog realizations is that since individual blocks are not isolated from one another, any change in one block affects the response in all other blocks. This will make the tuning of the complete network more difficult.

Finally, we discussed the sensitivity problem as it relates to the above three approaches. Specifically, we derived formulas expressing the transfer-function sensitivity in terms of the coefficient sensitivities and the sensitivities of the elements realized by the active RC networks to the changes of the element values in these active RC equivalents.

An excellent introductory exposition on topics of this chapter is given by Daryanani [4] and Van Valkenburg [9]. For network structures other than those discussed in this chapter, the reader is referred to Huelsman and Allen [6] and Ghausi and Laker [5]. For explicit formulas on doubly terminated *LC* ladder networks, see Chen [3], Weinberg [10], Williams [11], and Zverev [12].

REFERENCES

1. P. Bowron and F. W. Stephenson 1979, *Active Filters for Communications and Instrumentation*, Chapter 8. New York: McGraw–Hill.
2. L. T. Bruton 1980, *RC-Active Circuits: Theory and Design*, Chapters 9 and 10. Englewood Cliffs, NJ: Prentice–Hall.
3. W. K. Chen 1976, *Theory and Design of Broadband Matching Networks*, Chapter 3. New York: Pergamon Press.
4. G. Daryanani 1976, *Principles of Active Network Synthesis and Design*, Chapter 11. New York: John Wiley.
5. M. S. Ghausi and K. R. Laker 1981, *Modern Filter Design: Active RC and Switched Capacitor*, Chapter 5. Englewood Cliffs, NJ: Prentice–Hall.
6. L. P. Huelsman and P. E. Allen 1980, *Introduction to the Theory and Design of Active Filters*, Chapter 6. New York: McGraw–Hill.
7. R. H. S. Riordan, "Simulated inductors using differential amplifiers," *Electronic Letters*, vol. 3, 1967, pp. 50–51.
8. A. S. Sedra and P. O. Brackett 1978, *Filter Theory and Design: Active and Passive*, Chapters 11 and 12. Portland, OR: Matrix Publishers.
9. M. E. Van Valkenburg 1982, *Analog Filter Design*, Chapters 14–16. New York: Holt, Rinehart & Winston.
10. L. Weinberg 1962, *Network Analysis and Synthesis*. New York: McGraw–Hill. Reissued by R. E. Krieger Publishing Co., Melbourne, FL, 1975, Chapters 12 and 13.
11. A. B. Williams 1981, *Electronic Filter Design Handbook*. New York: McGraw–Hill.
12. A. I. Zverev 1967, *Handbook of Filter Synthesis*. New York: John Wiley.

PROBLEMS

8.1 Verify that the active *RC* network of Figure 8.10 realizes a floating inductor of Figure 8.7.

8.2 The *LC* filter of Figure 4.22 realizes a fifth-order Butterworth response. Find an active *RC* realization of this filter using the simulated inductor technique.

8.3 Repeat Problem 8.2 for the *LC* filter of Figure 4.23.

8.4 The *LC* filter of Figure 4.26 realizes a fourth-order Chebyshev response with peak-to-peak ripple in the passband not exceeding 1 dB. Find an active *RC* realization of this network using the simulated inductor technique.

8.5 Repeat Problem 8.4 for the *LC* filter of Figure 4.27.

8.6 Show that the single op-amp network of Figure 8.87 realizes a grounded inductor with inductance $4C$.

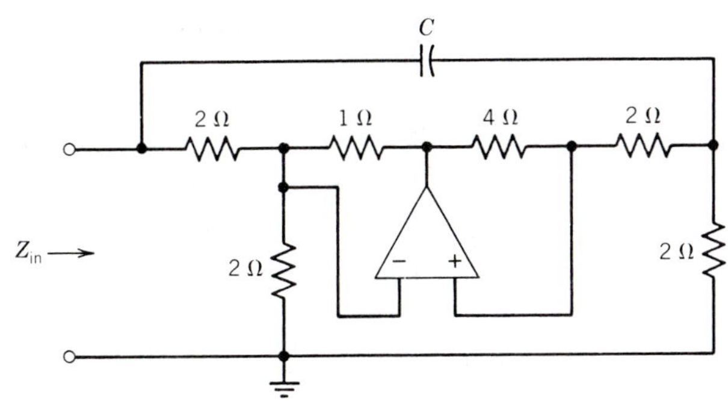

Figure 8.87

8.7 The network of Figure 8.5 is a realization of an inductor. If the two op amps are ideal, the realized inductance is R^2C. Suppose that the op amps are not ideal with both gains being approximated by the expression

$$A(s) = A_0 s_0 / s \tag{8.134}$$

as in (7.105). Show that the inductance realized is frequency dependent, and at the resonant frequency $\omega = 1/RC$, the inductance realized is

$$L \approx R^2C(1 + 4\omega/A_0 s_0) \tag{8.135}$$

8.8 The network of Figure 2.32 is a fourth-order Chebyshev band-elimination filter with peak-to-peak ripple in the passband not to exceed 1 dB. Suppose that we multiply the element values of the inductors and capacitors of the network by a factor of 100, so that the rejection band extends from $\omega = 1000$ rad/s to $\omega = 4000$ rad/s. Find an active *RC* realization of the resulting *LC* prototype.

8.9 Find an FDNR–*RC* realization for the fifth-order Butterworth low-pass filter of Figure 4.22.

8.10 Find an FDNR–*RC* realization for the fifth-order Butterworth low-pass filter of Figure 4.23.

8.11 Find an FDNR–*RC* realization for the fourth-order Chebyshev low-pass filter of Figure 4.26.

8.12 Repeat Problem 8.11 for the *LC* filter of Figure 4.27.

8.13 Determine the transfer voltage-ratio function V_o/V_{in} for each of the networks of Figure 8.88. Are these transfer functions stable?

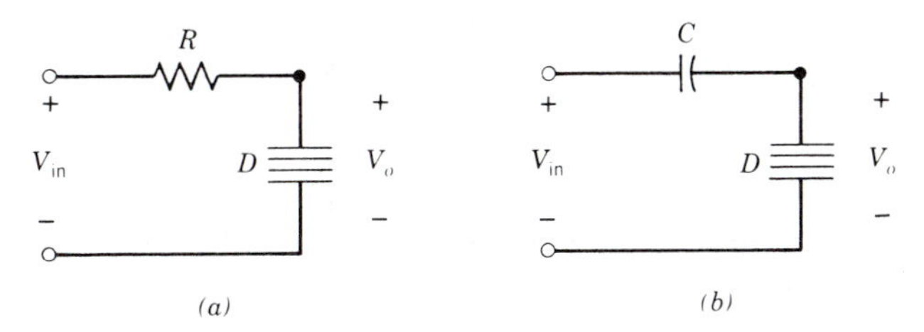

Figure 8.88

8.14 Find the input impedance of the active *RC* network of Figure 8.89. Compare the result with (8.14).

8.15 Find the input impedance of the active *RC* network of Figure 8.90. Compare the result with (8.14).

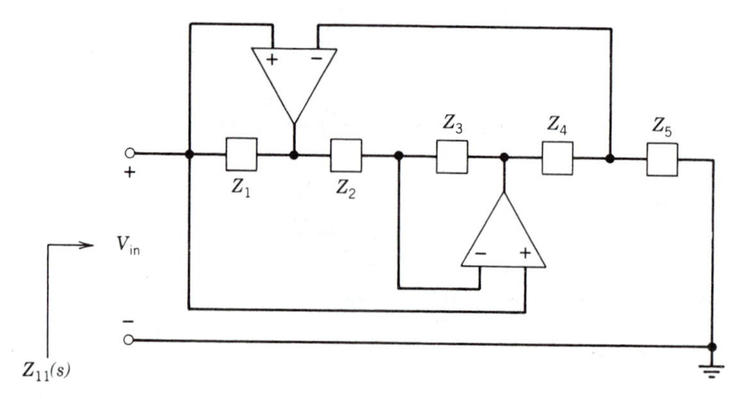

Figure 8.89

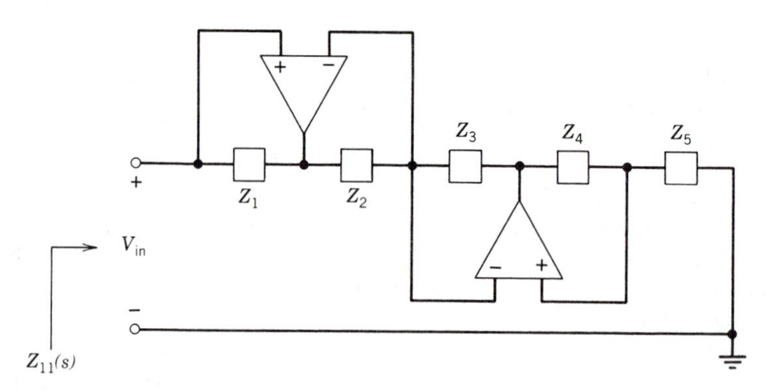

Figure 8.90

8.16 An equivalent network for a typical tunnel diode is shown in Figure 8.91. Simulate this network without using any inductors or negative resistors.

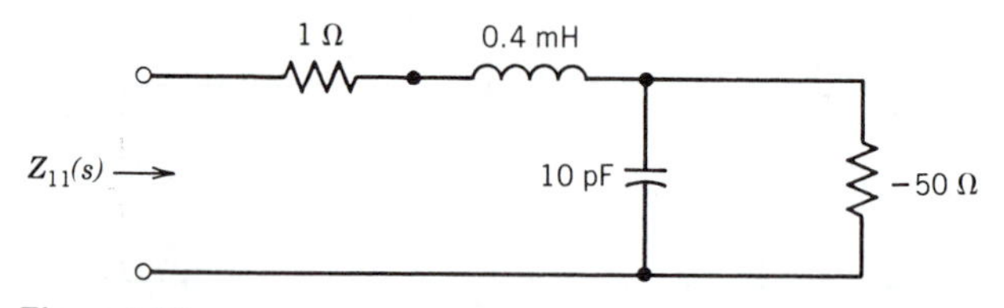

Figure 8.91

8.17 Verify that the equivalent network for the gyrator network of Figure 8.8 is shown in Figure 8.9 with element values given by (8.17).

8.18 Find an FDNR–*RC* realization for the fourth-order Chebyshev high-pass filter of Figure 8.92. The filter is of equal-ripple type with peak-to-peak ripple in the passband not exceeding 1 dB.

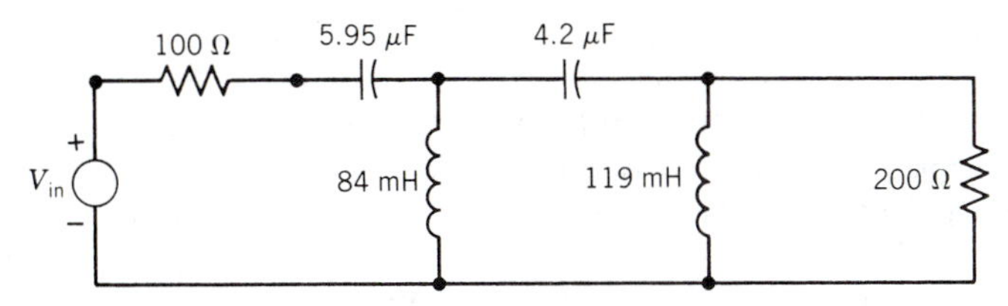

Figure 8.92

8.19 Repeat Problem 8.18 using inductance simulation.

8.20 Find an FDNR–RC realization for the fourth-order Chebyshev band-pass filter of Figure 8.93. The filter is of equal-ripple type with peak-to-peak ripple in the passband not exceeding 1 dB.

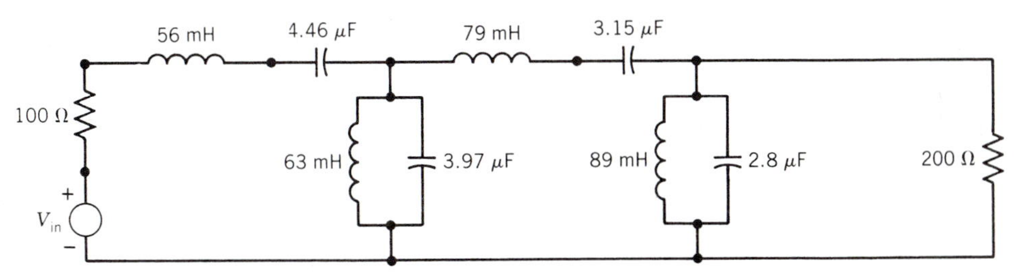

Figure 8.93

8.21 Repeat Problem 8.20 using inductance simulation.

8.22 Use the leapfrog method to design an active RC low-pass filter having a fifth-order Butterworth transfer voltage-ratio function. The cutoff frequency is required to be 5000 rad/s. The filter should have equal-resistance termination of 10 kΩ.

8.23 Use the leapfrog method to design an active RC filter, the LC prototype of which is shown in Figure 4.22.

8.24 Repeat Problem 8.23 for the prototype network of Figure 4.23.

8.25 Design a fifth-order Chebyshev band-pass active RC filter with a transfer voltage-ratio function that has a mid-band frequency of 1000 rad/s and an octave bandwidth in which the ripple is 1 dB. The filter is required to have equal-resistance termination of 1 kΩ.

8.26 The LC filter of Figure 4.26 realizes a fourth-order Chebyshev response with peak-to-peak ripple in the passband not exceeding 1 dB. Using the leapfrog method, find an active RC realization of this network.

8.27 Repeat Problem 8.26 for the LC prototype of Figure 4.27.

8.28 Design a fifth-order Butterworth band-pass active RC filter with a transfer voltage-ratio function that has a mid-band frequency of 1000 rad/s and an octave bandwidth. The filter is required to have equal-resistance termination of 1 kΩ.

8.29 Repeat Problem 8.25 if fifth order is changed to sixth order.

8.30 Repeat Problem 8.28 if fifth order is changed to sixth order.

8.31 Repeat Example 8.8 for the doubly terminated fourth-order high-pass filter of Figure 8.14 by computing the transfer-function sensitivities with respect to the elements of an active RC network used to simulate the first shunt inductor.

8.32 Repeat Example 8.9 for the doubly terminated fourth-order low-pass filter of Figure 8.23 by computing the transfer-function sensitivities with respect to the elements of an active *RC* network used to simulate the first FDNR in Figure 8.24.

8.33 Repeat Example 8.10 for the doubly terminated fourth-order low-pass filter of Figure 8.23 by computing the transfer-function sensitivities with respect to the elements of the active *RC* network of Figure 8.86 used to simulate the admittance composed of the series connection of the 136-H inductor and the 100-Ω resistor.

chapter 9

DESIGN OF BROADBAND MATCHING NETWORKS

In Section 4.4 of Chapter 4, we discussed the match between a resistive generator and a resistive load having a preassigned transducer power-gain characteristic. In the present chapter, we consider the more general problem of matching a resistive generator to a frequency-dependent load. We discuss conditions under which a lossless matching network exists that when operated between the resistive generator and the given load will yield the preassigned transducer power-gain characteristic. Explicit formulas for the design of ladder networks with a general *RLC* load having a prescribed Butterworth or Chebyshev transducer power-gain characteristic of arbitrary order will be presented.

9.1 THE BROADBAND MATCHING PROBLEM

In most practical situations, the source can usually be represented as an ideal voltage generator in series with a pure resistor, which may be the Thévenin equivalent of some other network. The load impedance is assumed to be strictly passive over a frequency band of interest. The reason for this is that the matching problem cannot be meaningfully defined if the load is purely reactive. Our problem is to design a lossless equalizer to match out a frequency-dependent load and a resistive generator and to achieve a preassigned transducer power-gain characteristic over the entire sinusoidal frequency spectrum. The arrangement is depicted schematically in Figure 9.1. As is well known, the maximum power transfer between the source and the load is achieved when the impedance presented to the generator is equal to the source

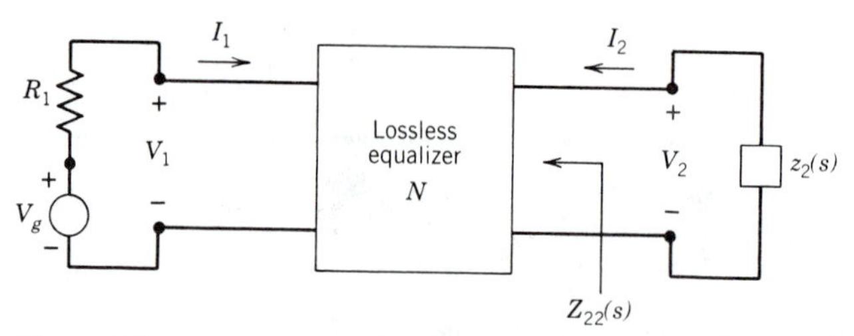

Figure 9.1

resistance. A different version of our problem is then to synthesize a coupling network that will transform a given frequency-dependent load into a constant resistance over the frequency band of interest. With the exception that the load impedance is purely resistive, it will be shown that it is not always possible to do so. Thus, to achieve the match, we must include the maximum tolerance on the match as well as the minimum bandwidth within which the match is to be obtained.

Refer to Figure 9.1. Let the output reflection coefficient be

$$\rho_2(s)=\frac{Z_{22}(s)-z_2(-s)}{Z_{22}(s)+z_2(s)} \tag{9.1}$$

where Z_{22} is the impedance looking into the output port when the input port is terminated in the source resistance R_1. As shown in (4.114), the transducer power gain $G(\omega^2)$ is related to the transmission coefficient S_{21} by the equation

$$G(\omega^2)=|S_{21}(j\omega)|^2=1-|\rho_2(j\omega)|^2 \tag{9.2}$$

Recall that in computing $\rho_1(s)$ from $\rho_1(s)\rho_1(-s)$ in (4.120) or (4.145), we assign all of the LHS poles of $\rho_1(s)\rho_1(-s)$ to $\rho_1(s)$ because with resistive load $z_2=R_2$, $\rho_1(s)$ is devoid of poles in the RHS. For the present situation of complex load, the poles of $\rho_2(s)$ include those of $z_2(-s)$, which may lie in the open RHS. As a result, the assignment of poles of $\rho_2(s)\rho_2(-s)$ to $\rho_2(s)$ is not unique. Furthermore, the nonanalyticity of $\rho_2(s)$ in the open RHS leaves much to be desired in terms of our ability to manipulate. For these reasons, we consider the normalized reflection coefficient defined by

$$\rho(s)=A(s)\rho_2(s)=A(s)\,\frac{Z_{22}(s)-z_2(-s)}{Z_{22}(s)+z_2(s)} \tag{9.3}$$

where

$$A(s)=\prod_{i=1}^{q}\frac{s-s_i}{s+s_i}, \qquad \operatorname{Re} s_i>0 \tag{9.4}$$

is the real all-pass *function* defined by the open RHS poles $s_i(i=1,2,\ldots,q)$ of $z_2(-s)$. The all-pass function is analytic in the closed RHS and such that

$$A(s)A(-s)=1 \tag{9.5}$$

On the $j\omega$-axis, the magnitude of $A(j\omega)$ is unity, meaning that an all-pass function has a flat magnitude for all sinusoidal frequencies. As a result, we can write

$$|\rho(j\omega)|=|A(j\omega)\rho_2(j\omega)|=|\rho_2(j\omega)| \tag{9.6}$$

and (9.2) becomes

$$G(\omega^2)=1-|\rho(j\omega)|^2 \tag{9.7}$$

This equation together with the normalized reflection coefficient $\rho(s)$ of (9.3), which is analytic in the open RHS by construction, forms the cornerstone of Youla's theory† on broadband matching.

9.2 ZEROS OF TRANSMISSION FOR A ONE-PORT IMPEDANCE

In Chapter 3, Section 3.3, we define the zeros of transmission for a terminated two-port network as the frequencies at which a zero output results for a finite input. In the present section, we extend this concept by defining the zeros of transmission for a one-port impedance.

Definition 9.1 : ***Zero of transmission.*** For a given impedance $z_2(s)$, a closed RHS zero of multiplicity k of the function

$$w(s)\triangleq\frac{r_2(s)}{z_2(s)} \tag{9.8}$$

where

$$r_2(s)=\text{Ev}\; z_2(s)=\tfrac{1}{2}[z_2(s)+z_2(-s)] \tag{9.9}$$

is the even part of $z_2(s)$, is said to be a zero of transmission of order k of $z_2(s)$.

This extension has its origin in Darlington synthesis discussed in Chapter 4; for if we realize impedance $z_2(s)$ as the driving-point impedance of a lossless two-port network terminated in a 1-Ω resistor, the magnitude squared of the transfer impedance function $Z_{12}(j\omega)$ between the 1-Ω resistor and the input equals the real part of the input impedance,

$$|Z_{12}(j\omega)|^2=\text{Re}\; z_2(j\omega)=r_2(j\omega) \tag{9.10}$$

as previously derived in (4.71). After substituting ω by $-js$, it is easy to see that the zeros of $r_2(s)$ would be the zeros of transmission of the lossless two-port network. Our definition (9.8) is slightly different in that we consider only the closed RHS zeros of $r_2(s)/z_2(s)$ as the zeros of transmission of z_2. This is not really a restriction since the zeros of an even function $r_2(s)$ must occur in quadrantal symmetry as depicted in Figure 3.1 if they are complex, and $z_2(s)$ is devoid of zeros and poles in the open

†D. C. Youla, "A new theory of broad-band matching," *IRE Trans. Circuit Theory*, vol. CT-11, 1964, pp. 30–50.

RHS. The physical realizability conditions imposed by the load impedance z_2 at its zeros of transmission will be automatically instituted at the LHS zeros of r_2/z_2.

As an example, consider the load impedance of Figure 9.2,

$$z_2(s)=\hat{R}_1+\frac{R_2}{R_2Cs+1} \tag{9.11}$$

the even part of which is found to be

$$r_2(s)=\tfrac{1}{2}[z_2(s)+z_2(-s)]=\frac{\hat{R}_1+R_2-\hat{R}_1R_2^2C^2s^2}{1-R_2^2C^2s^2} \tag{9.12}$$

giving

$$w(s)=\frac{r_2(s)}{z_2(s)}=\frac{R_2Cs^2-(\hat{R}_1+R_2)/\hat{R}_1R_2C}{(R_2Cs-1)[s+(\hat{R}_1+R_2)/\hat{R}_1R_2C]} \tag{9.13}$$

The impedance z_2 has a zero of transmission of order 1 at

$$s=\sigma_0=\frac{1}{R_2C}\sqrt{1+\frac{R_2}{\hat{R}_1}} \tag{9.14}$$

For convenience, the zeros of transmission of z_2 are divided into four mutually exclusive classes.

Definition 9.2: ***Classification of zeros of transmission.*** Let $s_0=\sigma_0+j\omega_0$ be a zero of transmission of order k of an impedance $z_2(s)$. Then s_0 belongs to one of the following four mutually exclusive classes depending on σ_0 and $z_2(s_0)$, as follows:

Class I: $\sigma_0>0$, which includes all the open RHS zeros of transmission.

Class II: $\sigma_0=0$ and $z_2(j\omega_0)=0$.

Class III: $\sigma_0=0$ and $0<|z_2(j\omega_0)|<\infty$.

Class IV: $\sigma_0=0$ and $|z_2(j\omega_0)|=\infty$.

For the impedance of Figure 9.2, the zero of transmission is located at $s=\sigma_0$ of (9.14). This zero belongs to the Class I zero of transmission. On the other hand, for the RC impedance of Figure 9.3,

$$z_2(s)=\frac{R}{RCs+1} \tag{9.15}$$

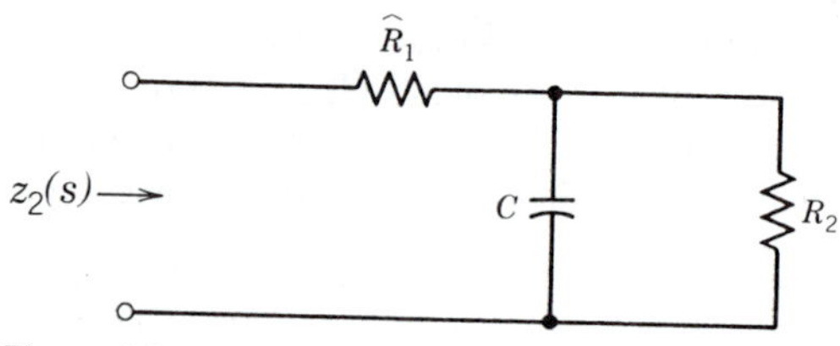

Figure 9.2

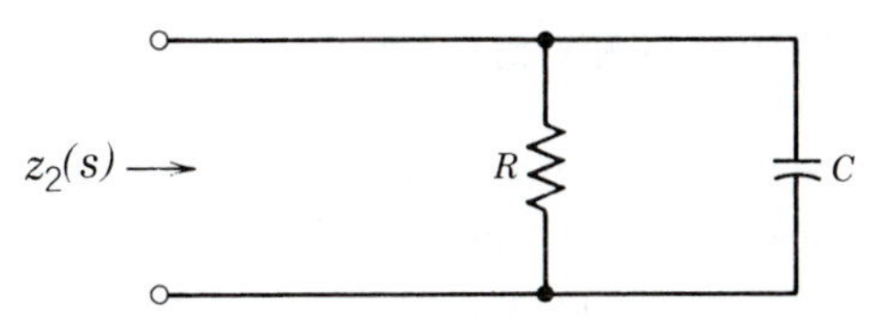

Figure 9.3

its even part is found to be

$$r_2(s) = \text{Ev}\ z_2(s) = \frac{-R}{R^2C^2s^2 - 1} \tag{9.16}$$

Since in network theory it is convenient to consider infinity as part of the $j\omega$-axis, the function

$$w(s) = \frac{r_2(s)}{z_2(s)} = -\frac{1}{RCs - 1} \tag{9.17}$$

has a zero of order 1 at infinity. Thus, z_2 has a zero of transmission of order 1 at $s = \infty$. To determine its classification, we compute

$$z_2(s_0) = z_2(j\infty) = 0 \tag{9.18}$$

According to Definition 9.2, $s = \infty$ is a Class II zero of transmission of order 1 of the impedance z_2 of (9.15).

Finally, for the *RLC* load of Figure 9.4, the input impedance is

$$z_2(s) = Ls + \frac{R}{RCs + 1} \tag{9.19}$$

the even part of which is the same as that given in (9.16), giving

$$w(s) = \frac{r_2(s)}{z_2(s)} = \frac{-R}{(RCs - 1)(RLCs^2 + Ls + R)} \tag{9.20}$$

This indicates that $w(s)$ has a zero of order 3 at infinity. Since

$$|z_2(j\infty)| = \infty \tag{9.21}$$

we conclude that the impedance of Figure 9.4 possesses a Class IV zero of transmission of order 3.

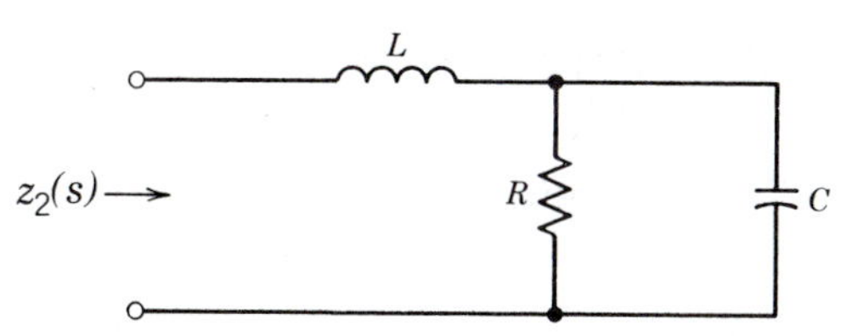

Figure 9.4

9.3 BASIC CONSTRAINTS ON $\rho(s)$

The basic constraints imposed on the normalized reflection coefficient $\rho(s)$ by the load impedance $z_2(s)$ of Figure 9.1 will be presented in this section. These constraints are important in that they are necessary and sufficient for the physical realizability of $\rho(s)$.

The restrictions imposed on $\rho(s)$ are most conveniently formulated in terms of the coefficients of the Laurent series expansions of the following quantities about a zero of transmission $s_0 = \sigma_0 + j\omega_0$ of order k of z_2:

$$\rho(s) = \rho_0 + \rho_1(s-s_0) + \rho_2(s-s_0)^2 + \cdots = \sum_{m=0}^{\infty} \rho_m(s-s_0)^m \tag{9.22a}$$

$$A(s) = A_0 + A_1(s-s_0) + A_2(s-s_0)^2 + \cdots = \sum_{m=0}^{\infty} A_m(s-s_0)^m \tag{9.22b}$$

$$F(s) \triangleq 2r_2(s)A(s) = F_0 + F_1(s-s_0) + F_2(s-s_0)^2 + \cdots = \sum_{m=0}^{\infty} F_m(s-s_0)^m \tag{9.22c}$$

Note that for a given load z_2 and a preassigned transducer power-gain characteristic $G(\omega^2)$, the functions A and F are determined by z_2 through (9.4) and (9.22c) and $\rho(s)$ from (9.7). The expansions (9.22) of the Laurent type can be found by any process. This follows from the fact that the Laurent series expansion of an analytic function over a given annulus is unique. In other words, if an expansion of the Laurent type is found by any method, it must be *the* Laurent series expansion. For a zero of transmission at infinity, the expansions take the forms

$$\rho(s) = \rho_0 + \frac{\rho_1}{s} + \frac{\rho_2}{s^2} + \cdots = \sum_{m=0}^{\infty} \frac{\rho_m}{s^m} \tag{9.23a}$$

$$A(s) = A_0 + \frac{A_1}{s} + \frac{A_2}{s^2} + \cdots = \sum_{m=0}^{\infty} \frac{A_m}{s^m} \tag{9.23b}$$

$$F(s) = 2r_2(s)A(s) = F_0 + \frac{F_1}{s} + \frac{F_2}{s^2} + \cdots = \sum_{m=0}^{\infty} \frac{F_m}{s^m} \tag{9.23c}$$

This follows from the observation that if we wish to expand, say, $F(s)$ about infinity, we first expand the function $F(1/s)$ about the origin known as the Maclaurin series:

$$F(1/s) = F_0 + F_1 s + F_2 s^2 + \cdots = \sum_{m=0}^{\infty} F_m s^m \tag{9.24}$$

Replacing s by $1/s$ in the above equation yields the expansion of $F(s)$ about infinity as shown in (9.23c).

Example 9.1 Consider the load impedance of Figure 9.2, as given by (9.11). To compute the all-pass function A associated with this load through (9.4), we determine the open RHS pole of $z_2(-s)$,

$$z_2(-s)=\hat{R}_1+\frac{R_2}{-R_2Cs+1} \tag{9.25}$$

which occurs at $s=1/R_2C$. This gives

$$A(s)=\frac{s-1/R_2C}{s+1/R_2C}=\frac{R_2Cs-1}{R_2Cs+1} \tag{9.26}$$

Using this in conjunction with (9.12), we obtain

$$F(s)=2r_2(s)A(s)=\frac{2\hat{R}_1R_2^2C^2s^2-2(\hat{R}_1+R_2)}{(1+R_2Cs)^2} \tag{9.27}$$

For illustrative purposes, let $\hat{R}_1=1\,\Omega$, $R_2=3\,\Omega$, and $C=1/3$ F. The Class I zero of transmission of order 1 is found from (9.14) to be

$$s_0=\sigma_0=2 \tag{9.28}$$

To obtain the Laurent series expansion about this zero of transmission we write

$$A(s)=\frac{s-1}{s+1}=A_0+A_1(s-2)+A_2(s-2)^2+\cdots \tag{9.29}$$

To ascertain A_0, we set $s=2$ in (9.29) and obtain

$$A_0=A(2)=\frac{2-1}{2+1}=\frac{1}{3} \tag{9.30}$$

For A_1 we take derivatives on both sides of (9.29) and then set $s=2$, giving

$$A_1=\left.\frac{dA(s)}{ds}\right|_{s=2}=\left.\frac{2}{(s+1)^2}\right|_{s=2}=\frac{2}{9} \tag{9.31}$$

Likewise, for A_2 we take second derivatives on both sides of (9.29) and then set $s=2$. The result is

$$A_2=\left.\frac{1}{2}\frac{d^2A(s)}{ds^2}\right|_{s=2}=\left.\frac{-2}{(s+1)^3}\right|_{s=2}=-\frac{2}{27} \tag{9.32}$$

The Laurent series expansion of the all-pass function A about the zero of transmission of z_2 at $s=2$ becomes

$$A(s)=\frac{s-1}{s+1}=\frac{1}{3}+\frac{2}{9}(s-2)-\frac{2}{27}(s-2)^2+\cdots \tag{9.33}$$

The above example demonstrates a method for obtaining the Laurent series expansion of an analytic function about a finite zero of transmission. For the zero of transmission at infinity, we may use the binomial expansion formula

$$(s+c)^n=s^n+ns^{n-1}c+\frac{n(n-1)}{2!}s^{n-2}c^2+\cdots \tag{9.34}$$

which is valid for all values of n if $|s|>|c|$, and is valid only for nonnegative integers

n if $|s| \leqslant |c|$. That we can apply such procedures to obtain the Laurent series expansion again follows from the uniqueness of the expansion, as mentioned previously.

Example 9.2 For the load impedance of Figure 9.3, we consider

$$z_2(-s) = \frac{R}{1 - RCs} \tag{9.35}$$

The function has an open RHS pole at $s = 1/RC$, giving

$$A(s) = \frac{s - 1/RC}{s + 1/RC} = \frac{RCs - 1}{RCs + 1} \tag{9.36}$$

This equation together with (9.16) leads to

$$F(s) = 2r_2(s)A(s) = \frac{-2R}{(RCs + 1)^2} \tag{9.37}$$

As indicated in (9.17) and (9.18), the load has a Class II zero of transmission of order 1 located at $s = \infty$. To obtain the Laurent series expansions of A and F about this zero of transmission, we apply (9.34) as follows:

$$\begin{aligned} A(s) &= \frac{RCs - 1}{RCs + 1} = (RCs - 1)(RCs + 1)^{-1} \\ &= (RCs - 1)[(RCs)^{-1} - (RCs)^{-2} + (RCs)^{-3} + \cdots] \\ &= 1 - \frac{2}{RCs} + \frac{2}{R^2C^2s^2} + \cdots \end{aligned} \tag{9.38}$$

$$\begin{aligned} F(s) &= \frac{-2R}{(RCs + 1)^2} = (-2R)(RCs + 1)^{-2} \\ &= (-2R)[(RCs)^{-2} - 2(RCs)^{-3} + 3(RCs)^{-4} + \cdots] \\ &= -\frac{2}{RC^2s^2} + \frac{4}{R^2C^3s^3} + \cdots \end{aligned} \tag{9.39}$$

from which we can make the following identifications:

$$A_0 = 1, \quad A_1 = -2/RC, \quad A_2 = 2/R^2C^2, \quad F_0 = F_1 = 0, \quad F_2 = -2/RC^2 \tag{9.40}$$

Basic constraints on $\rho(s)$

The basic constraints imposed on the normalized reflection coefficient $\rho(s)$ by a load impedance $z_2(s)$ are most succinctly expressed in terms of the coefficients of the Laurent series expansions of the functions $\rho(s)$, $A(s)$, and $F(s)$ as shown in (9.23) about each zero of transmission $s_0 = \sigma_0 + j\omega_0$ of order k of $z_2(s)$. Depending on the classification of the zero of transmission, one of the following four sets of coefficient conditions must be satisfied:

$$\textit{Class I:} \quad A_x = \rho_x \text{ for } x = 0, 1, 2, \ldots, k-1 \tag{9.41a}$$

Class II: $A_x = \rho_x$ for $x = 0, 1, 2, \ldots, k-1$, and

$$\frac{A_k - \rho_k}{F_{k+1}} \geqq 0 \tag{9.41b}$$

Class III: $A_x = \rho_x$ for $x = 0, 1, 2, \ldots, k-2$, $k \geqq 2$, and

$$\frac{A_{k-1} - \rho_{k-1}}{F_k} \geqq 0 \tag{9.41c}$$

Class IV: $A_x = \rho_x$ for $x = 0, 1, 2, \ldots, k-1$, and

$$\frac{F_{k-1}}{A_k - \rho_k} \geqq a_{-1}, \quad \text{the residue of } z_2(s) \text{ at the pole } s = j\omega_0. \tag{9.41d}$$

We next consider the determination of the normalized reflection coefficient from a preassigned transducer power-gain characteristic $G(\omega^2)$. In (9.7) we replace ω by $-js$ and obtain

$$\rho(s)\rho(-s) = 1 - G(-s^2) \tag{9.42}$$

Our task, then, is to determine the function $\rho(s)$, knowing $\rho(s)\rho(-s)$, which must be the ratio of two even polynomials. Thus, the zeros and poles of $\rho(s)\rho(-s)$ must appear in quadrantal symmetry, being symmetric with respect to both the real and the imaginary axes of the s-plane. The question now is how to pick the zeros and poles of $\rho(s)$ from among those of $\rho(s)\rho(-s)$. For the poles, the answer is simple. Since $\rho(s)$ is analytic in the closed RHS and since the poles of $\rho(-s)$ are the negatives of the poles of $\rho(s)$, the poles of $\rho(s)\rho(-s)$ can be uniquely distributed: the open LHS poles of $\rho(s)\rho(-s)$ belong to $\rho(s)$, whereas those in the open RHS belong to $\rho(-s)$. Note that, for the lumped system considered in this book, $\rho(s)$ is devoid of poles on the $j\omega$-axis.

As for the zeros, there are no unique ways to assign them. Since $\rho(s)$ may have closed RHS zeros, we need not assign all the LHS zeros of $\rho(s)\rho(-s)$ to $\rho(s)$. The only requirement is that the complex-conjugate pair of zeros must be assigned together. However, if it is specified that $\rho(s)$ be made a minimum-phase function, then all the open LHS zeros of $\rho(s)\rho(-s)$ are assigned to $\rho(s)$. The $j\omega$-axis zeros of $\rho(s)\rho(-s)$ are of even multiplicity, and thus they are divided equally between $\rho(s)$ and $\rho(-s)$. In other words, $\rho(s)$ is uniquely determined by the zeros and poles of $\rho(s)\rho(-s)$ only if $\rho(s)$ is required to be minimum-phase.

Let $\hat{\rho}(s)$ be the minimum-phase factorization of $\rho(s)\rho(-s)$ by the procedure outlined above. Then any solution of the form

$$\rho(s) = \pm\eta(s)\hat{\rho}(s) \tag{9.43}$$

is admissible, where $\eta(s)$ is an arbitrary real all-pass function possessing the property that

$$\eta(s)\eta(-s) = 1 \tag{9.44}$$

Having obtained the normalized reflection coefficient $\rho(s)$, we now require that it satisfy the basic coefficient constraints (9.41). The significance of these constraints is that they are both necessary and sufficient for the physical realizability of $\rho(s)$, and is summarized in the following theorem. The proof of this theorem is beyond the scope of this book and can be found in Chen.†

Theorem 9.1 Given a strictly passive impedance $z_2(s)$, the function defined by the equation

$$Z_{22}(s) \triangleq \frac{F(s)}{A(s)-\rho(s)} - z_2(s) \tag{9.45}$$

is positive real if and only if $|\rho(j\omega)| \leqslant 1$ for all ω and the coefficient conditions (9.41) are satisfied, where $A(s)$, $F(s)$, and $\rho(s)$ are as constructed above,

The function defined in (9.45) is actually the back-end impedance of a desired equalizer. To see this, we refer to Figure 9.1 and equation (9.3). If we solve for Z_{22} in (9.3), the result is

$$Z_{22}(s) = \frac{A(s)z_2(-s)+\rho(s)z_2(s)}{A(s)-\rho(s)} = \frac{A(s)[z_2(s)+z_2(-s)]}{A(s)-\rho(s)} - z_2(s)$$

$$= \frac{F(s)}{A(s)-\rho(s)} - z_2(s) \tag{9.46}$$

Theorem 9.1 guarantees that the equalizer back-end impedance Z_{22} constructed in this way is positive real. As a result, it can be realized as the input impedance of a lossless two-port network terminated in a resistor, using the Darlington method if necessary. The removal of this resistor gives the desired matching network. An ideal transformer may be needed at the input port to compensate for the actual level of the generator resistance R_1.

Example 9.3 Design a lossless matching network to equalize the load as shown in Figure 9.5 to a resistive generator of internal resistance of 100 Ω and to achieve the fourth-order Butterworth transducer power-gain characteristic with a maximal dc gain. The cutoff frequency is 10^8 rad/s.

To simplify the computation, we magnitude-scale the network by a factor of 10^{-2} and frequency-scale it by a factor of 10^{-8}. Thus, s denotes the normalized

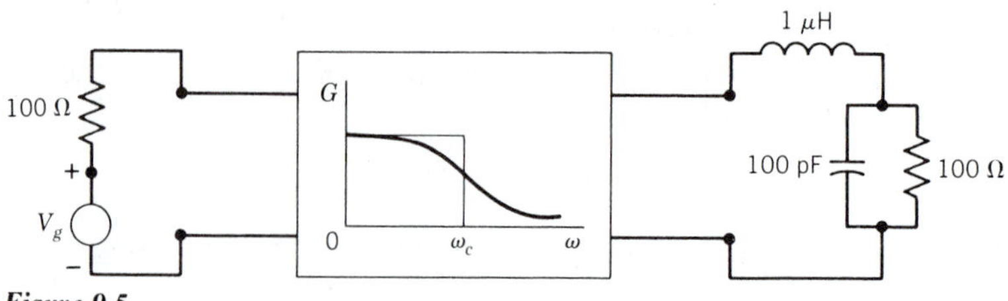

Figure 9.5

†W. K. Chen 1976, *Theory and Design of Broadband Matching Networks*, Chapter 4. New York: Pergamon Press.

complex frequency and ω the normalized real frequency. The load impedance is found to be

$$z_2(s)=\frac{s^2+s+1}{s+1} \tag{9.47}$$

the even part of which is

$$r_2(s)=\tfrac{1}{2}[z_2(s)+z_2(-s)]=\frac{1}{1-s^2} \tag{9.48}$$

Since $z_2(-s)$ has an open RHS pole at $s=1$, the all-pass function defined by this pole is given by

$$A(s)=\frac{s-1}{s+1} \tag{9.49}$$

This leads to the function

$$F(s)=2r_2(s)A(s)=\frac{-2}{(s+1)^2} \tag{9.50}$$

To ascertain the zero of transmission, we compute

$$w(s)=\frac{r_2(s)}{z_2(s)}=\frac{1}{(s^2+s+1)(1-s)} \tag{9.51}$$

indicating that $s=\infty$ is a Class IV zero of transmission of order 3.

For fourth-order Butterworth transducer power-gain characteristic,

$$G(\omega^2)=\frac{K_4}{1+\omega^8}, \qquad 0\leqslant K_4\leqslant 1 \tag{9.52}$$

Appealing to (9.42) gives

$$\rho(s)\rho(-s)=1-G(-s^2)=\frac{1-K_4+s^8}{1+s^8}=\alpha^8\,\frac{1+x^8}{1+s^8} \tag{9.53}$$

where $x=s/\alpha$ and

$$\alpha=(1-K_4)^{1/8} \tag{9.54}$$

Using formula (2.16) or from Table 2.2 in Chapter 2, the minimum-phase solution of (9.53) is found to be

$$\hat{\rho}(s)=\alpha^4\,\frac{x^4+2.6131x^3+3.4142x^2+2.6131x+1}{s^4+2.6131s^3+3.4142s^2+2.6131s+1} \tag{9.55}$$

The dc gain K_4 will be determined later from the coefficient conditions.

The physical realizability imposed by the load z_2 is given in terms of coefficient constraints (9.41). To this end, we expand A, F, and ρ in Laurent series about the zero of transmission, which is at infinity. The results are given by

$$A(s)=\frac{s-1}{s+1}=1-\frac{2}{s}+\frac{2}{s^2}-\frac{2}{s^3}+\cdots \tag{9.56a}$$

$$F(s)=\frac{-2}{(s+1)^2}=0+0-\frac{2}{s^2}+\frac{4}{s^3}+\cdots \tag{9.56b}$$

$$\hat{\rho}(s)=1+\frac{2.6131(\alpha-1)}{s}+\frac{3.4142(\alpha^2-2\alpha+1)}{s^2}$$

$$+\frac{2.6131(\alpha^3-3.4142\alpha^2+3.4142\alpha-1)}{s^3}+\cdots \tag{9.56c}$$

For a Class IV zero of transmission of order 3, the coefficient conditions are, from (9.41d) with $k=3$,

$$A_m=\rho_m, \qquad m=0,\,1,\,2 \tag{9.57a}$$

$$\frac{F_2}{A_3-\rho_3}\geqq a_{-1}(\infty)=1 \tag{9.57b}$$

where $a_{-1}(\infty)=1$ is the residue of z_2 at the pole $s=\infty$, which is also the zero of transmission of z_2. Substituting the coefficients of the Laurent series (9.56) in (9.57) yields the constraints imposed on K_4:

$$A_0=1=\rho_0 \tag{9.58a}$$

$$A_1=-2=\rho_1=2.6131(\alpha-1) \tag{9.58b}$$

$$A_2=2=\rho_2=3.4142(\alpha-1)^2 \tag{9.58c}$$

$$\frac{F_2}{A_3-\rho_3}=\frac{-2}{-2-2.6131(\alpha-1)(\alpha^2-2.4142\alpha+1)}\geqq 1 \tag{9.58d}$$

giving

$$\alpha=0.23463 \tag{9.59}$$

and from (9.54), the dc gain is determined to be

$$K_4=1-\alpha^8=0.99999 \tag{9.60}$$

Thus, with this choice of K_4, all the coefficient constraints are satisfied without inserting any open RHS zeros in $\rho(s)$, as indicated in (9.43). This corresponds to setting $\pm\eta=1$ and $\rho(s)=\hat{\rho}(s)$.

Finally, to realize the matching network we compute the equalizer back-end impedance from (9.46). The result is given by

$$Z_{22}(s)=\frac{F(s)}{A(s)-\rho(s)}-z_2(s)$$

$$=\frac{\dfrac{-2}{(s+1)^2}}{\dfrac{s-1}{s+1}-\dfrac{s^4+0.6131s^3+0.1880s^2+0.0338s+0.0030}{s^4+2.6131s^3+3.4142s^2+2.6131s+1}}-\frac{s^2+s+1}{s+1}$$

$$= \frac{0.98s^4 + 2.556s^3 + 3.158s^2 + 2.576s + 1}{(s+1)(1.02s^2 + 1.65s + 1)}$$

$$= \frac{0.98s^3 + 1.576s^2 + 1.582s + 1}{1.02s^2 + 1.65s + 1} \tag{9.61}$$

which is guaranteed to be positive real by Theorem 9.1. This impedance can be realized by an LC ladder terminated in a resistor. For this we expand Z_{22} in a continued fraction and obtain

$$Z_{22}(s) = 0.96s + \cfrac{1}{1.64s + \cfrac{1}{0.622s + \cfrac{1}{1}}} \tag{9.62}$$

which is identified as an LC ladder terminated in a 1-Ω resistor as shown in Figure 9.6. Denormalizing the element values with magnitude-scaling by a factor of 100 and frequency-scaling by a factor of 10^8 yields a final realization as shown in Figure 9.7.

To compute the transducer power gain from the realized network, we again consider the normalized network of Figure 9.6, and compute the input impedance when the output port is terminated in the load z_2, as depicted in Figure 9.8. The result is given by

$$Z_{11}(s) = \frac{2s^4 + 2s^3 + 3.6s^2 + 2.58s + 1}{3.21s^3 + 3.21s^2 + 2.64s + 1} \tag{9.63}$$

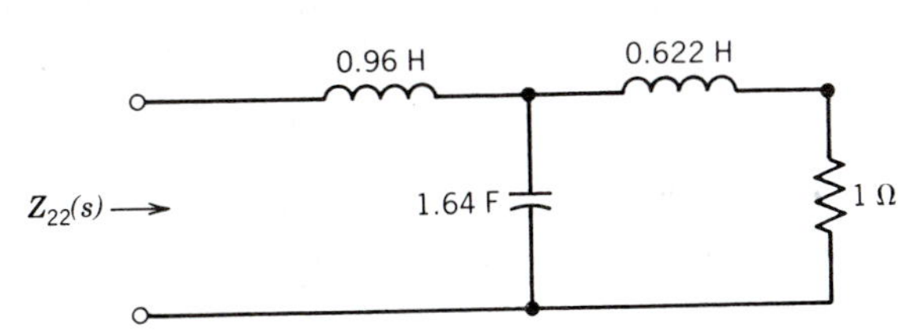

Figure 9.6

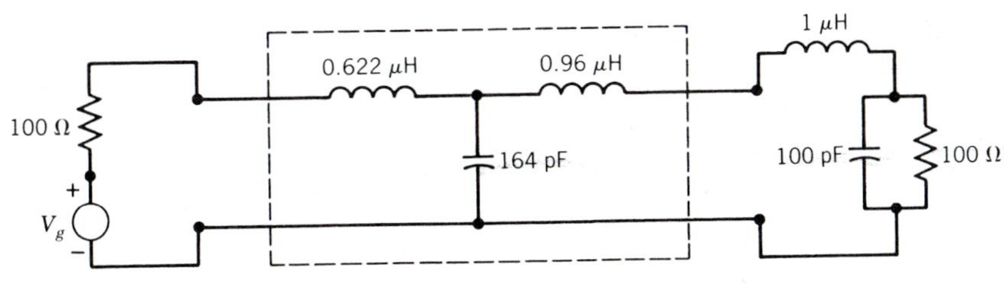

Figure 9.7

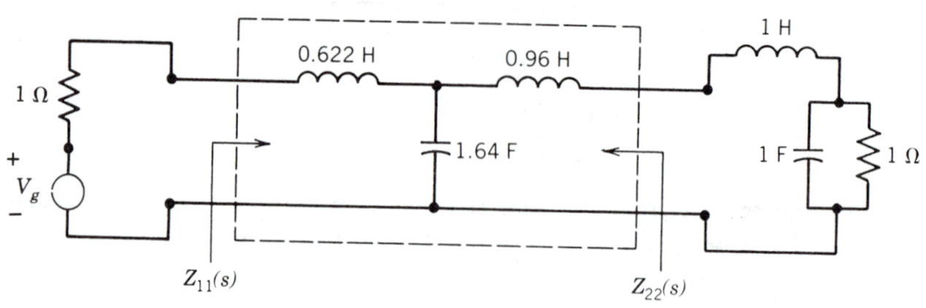

Figure 9.8

Using this in conjunction with (4.99), the transducer power gain is found to be

$$G(\omega^2)=\frac{4\,\mathrm{Re}\,Z_{11}(j\omega)}{|Z_{11}(j\omega)+1|^2}=\frac{1}{1+\omega^8} \tag{9.64}$$

confirming our design.

9.4 DESIGN PROCEDURE FOR THE EQUALIZERS

From the preceding discussions, we now outline a simple procedure for the design of an optimum lossless matching network of Figure 9.1 that equalizes a frequency-dependent load impedance z_2 to a resistive generator of internal resistance R_1 and achieves a preassigned transducer power-gain characteristic $G(\omega^2)$ over the entire sinusoidal frequency spectrum. The procedure is stated in eight steps, as follows:

Step 1. From a preassigned transducer power-gain characteristic $G(\omega^2)$ verify that $G(\omega^2)$ is an even rational real function and satisfies the inequality

$$0 \leqslant G(\omega^2) \leqslant 1 \qquad \text{for all } \omega \tag{9.65}$$

The gain level is usually not specified to allow desired flexibility.

Step 2. From a prescribed strictly passive load impedance z_2, compute

$$r_2(s)=\mathrm{Ev}\,z_2(s)=\tfrac{1}{2}[z_2(s)+z_2(-s)] \tag{9.66}$$

$$A(s)=\prod_{i=1}^{q}\frac{s-s_i}{s+s_i}, \qquad \mathrm{Re}\,s_i>0 \tag{9.67}$$

where s_i $(i=1, 2, \ldots, q)$ are the open RHS poles of $z_2(-s)$, and

$$F(s)=2r_2(s)A(s) \tag{9.68}$$

Step 3. Determine the locations and the orders of the zeros of transmission of z_2, which are defined as the closed RHS zeros of the function

$$w(s)=r_2(s)/z_2(s) \tag{9.69}$$

and divide them into respective classes according to Definition 9.2.

Step 4. Perform the unique factorization of the function

$$\hat{\rho}(s)\hat{\rho}(-s)=1-G(-s^2) \tag{9.70}$$

in which the numerator of $\hat{\rho}(s)$ is a Hurwitz polynomial and the denominator of $\hat{\rho}(s)$ is a strictly Hurwitz polynomial. In other words, $\hat{\rho}(s)$ is a minimum-phase solution of (9.70).

Step 5. Obtain the Laurent series expansions of the functions A, F, and $\hat{\rho}$ about each zero of transmission s_0 of z_2:

$$A(s)=\sum_{m=0}^{\infty} A_m(s-s_0)^m \tag{9.71a}$$

$$F(s)=\sum_{m=0}^{\infty} F_m(s-s_0)^m \tag{9.71b}$$

$$\hat{\rho}(s)=\sum_{m=0}^{\infty} \rho_m(s-s_0)^m \tag{9.71c}$$

These expansions may be obtained by any available techniques.

Step 6. According to the classes of zeros of transmission, list the basic constraints (9.41) imposed on the coefficients of (9.71). The gain level is determined from these constraints. If the constraints cannot all be satisfied, we consider the general solution

$$\rho(s)=\pm\eta(s)\hat{\rho}(s) \tag{9.72}$$

where η is an arbitrary real all-pass function. Then repeat Step 5 for $\rho(s)$. Of course, we should start with lower-order η. If the constraints still cannot all be satisfied, we must modify the preassigned transducer power-gain characteristic $G(\omega^2)$. Otherwise, no match exists.

Step 7. Having successfully carried out Step 6, the equalizer back-end driving-point impedance is determined by the equation

$$Z_{22}(s)=\frac{F(s)}{A(s)-\rho(s)}-z_2(s) \tag{9.73}$$

where $\rho(s)$ may be $\hat{\rho}(s)$. Z_{22} is guaranteed to be positive real.

Step 8. Using Darlington's procedure if necessary, realize the positive-real impedance Z_{22} as the driving-point impedance of a lossless two-port network terminated in a 1-Ω resistor. An ideal transformer may be required at the input port to compensate for the actual level of the generator resistance R_1. This completes the design of an equalizer.

Before we proceed with illustrations, we mention that it is sometimes convenient to use magnitude- and frequency-scalings to simplify the numerical computation, as we did in Example 9.3.

Example 9.4 It is desired to design a lossless matching network N to equalize the RC load of Figure 9.9 to a generator of internal resistance of 100 Ω and to achieve

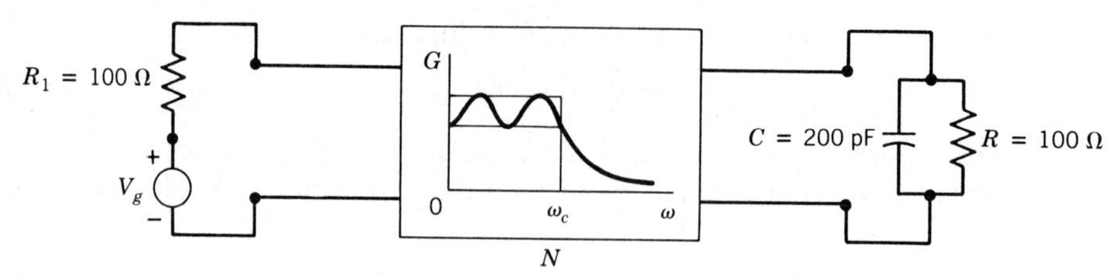

Figure 9.9

the fifth-order Chebyshev transducer power gain. The passband tolerance is 1 dB and the cutoff frequency is 10^8 rad/s.

For computational purposes, the network of Figure 9.9 is first magnitude-scaled down by a factor 10^{-2}, and frequency-scaled down by 10^{-8}. This results in the following normalized quantities:

$$R_1=1\ \Omega,\qquad R=1\ \Omega,\qquad C=2\ \text{F},\qquad \omega_c=1\ \text{rad/s}$$

For a 1-dB peak-to-peak ripple in the passband, the corresponding ripple factor is from (2.41)

$$\varepsilon=\sqrt{10^{0.1}-1}=0.50885 \tag{9.74}$$

We now follow the eight steps outlined above to obtain an equalizer that meets the desired specifications.

Step 1. The fifth-order Chebyshev transducer power gain is from (2.23)

$$G(\omega^2)=\frac{K_5}{1+\varepsilon^2 C_5^2(\omega)},\qquad 0\leqslant K_5\leqslant 1 \tag{9.75}$$

We wish to maximize K_5.

Step 2. From the load impedance

$$z_2(s)=\frac{1}{2s+1} \tag{9.76}$$

we next compute the functions

$$r_2(s)=\tfrac{1}{2}[z_2(s)+z_2(-s)]=\frac{-1}{4s^2-1} \tag{9.77}$$

$$A(s)=\frac{s-\frac{1}{2}}{s+\frac{1}{2}}=\frac{2s-1}{2s+1} \tag{9.78}$$

where $s=\frac{1}{2}$ is the open RHS pole of $z_2(-s)$, and

$$F(s)=2r_2(s)A(s)=\frac{-2}{(2s+1)^2} \tag{9.79}$$

Step 3. The zero of transmission of z_2 is defined by the closed RHS zero of the function

$$w(s)=\frac{r_2(s)}{z_2(s)}=\frac{-1}{2s-1} \tag{9.80}$$

indicating that $s=\infty$ is a Class II zero of transmission of order 1 of z_2.

Step 4. Substituting (9.75) in (9.70) with $-js$ replacing ω gives

$$\hat{\rho}(s)\hat{\rho}(-s)=1-G(-s^2)=(1-K_5)\frac{1+\hat{\varepsilon}^2C_5^2(-js)}{1+\varepsilon^2C_5^2(-js)} \tag{9.81}$$

where

$$\hat{\varepsilon}=\frac{\varepsilon}{\sqrt{1-K_5}} \tag{9.82}$$

In the case $K_5=1$, (9.81) reduces to

$$\hat{\rho}(s)\hat{\rho}(-s)=\frac{\varepsilon^2C_5^2(-js)}{1+\varepsilon^2C_5^2(-js)} \tag{9.83}$$

The numerator and denominator polynomials of (9.81) can be factored in terms of the roots of the equation

$$1+v^2C_5^2(-js)=0 \tag{9.84}$$

where $v=\varepsilon$ or $\hat{\varepsilon}$. For $v=\varepsilon$ the denominator polynomial is determined either by formula (2.62) or directly from Table 2.5 in Chapter 2. As a result, the minimum-phase solution of (9.81) can be written in the form

$$\hat{\rho}(s)=\frac{s^5+\hat{b}_4s^4+\hat{b}_3s^3+\hat{b}_2s^2+\hat{b}_1s+\hat{b}_0}{s^5+0.937s^4+1.689s^3+0.974s^2+0.580s+0.123} \tag{9.85}$$

The constant K_5 has not yet been determined at this point. It will be ascertained when the coefficient conditions are imposed.

Step 5. The Laurent series expansions of A, F, and $\hat{\rho}$ about the zero of transmission, which is at infinity, are obtained as follows:

$$\begin{aligned}A(s)&=\frac{2s-1}{2s+1}=(2s-1)(2s+1)^{-1}\\&=(2s-1)[(2s)^{-1}-(2s)^{-2}+(2s)^{-3}+\cdots]\\&=1-\frac{1}{s}+\frac{1}{2s^2}+\cdots\end{aligned} \tag{9.86a}$$

$$\begin{aligned}F(s)&=\frac{-2}{(2s+1)^2}=-2(2s+1)^{-2}=-2[(2s)^{-2}-2(2s)^{-3}+\cdots]\\&=0+0-\frac{1}{2s^2}+\frac{1}{2s^3}+\cdots\end{aligned} \tag{9.86b}$$

$$\hat{\rho}(s)=(s^5+\hat{b}_4s^4+\cdots+\hat{b}_0)(s^5+0.937s^4+\cdots+0.123)^{-1}$$
$$=(s^5+\hat{b}_4s^4+\cdots+\hat{b}_0)[s^{-5}-s^{-10}(0.937s^4+\cdots+0.123)+\cdots]$$
$$=1+\frac{\hat{b}_4-0.937}{s}+\cdots \tag{9.86c}$$

Step 6. For a Class II zero of transmission of order 1, the coefficient conditions (9.41b) become

$$A_0=\rho_0 \tag{9.87a}$$

$$\frac{A_1-\rho_1}{F_2}\geqq 0 \tag{9.87b}$$

Substituting the coefficients of the Laurent series (9.86) in (9.87) yields the constraints on the constant K_5:

$$A_0=1=\rho_0 \tag{9.88a}$$

$$\frac{A_1-\rho_1}{F_2}=\frac{-1-\hat{b}_4+0.937}{-\frac{1}{2}}\geqq 0 \tag{9.88b}$$

giving

$$\hat{b}_4\geqq -0.063 \tag{9.89}$$

But for a Hurwitz polynomial, $\hat{b}_4$ is nonnegative, implying that the constraints (9.87) are satisfied for all values of K_5 between 0 and 1. To maximize K_5, we choose

$$K_5=1 \tag{9.90}$$

the maximum permissible value. With this value of K_5, (9.83) applies and the minimum-phase solution becomes

$$\hat{\rho}(s)=\frac{s^5+1.25s^3+0.312s}{s^5+0.937s^4+1.689s^3+0.974s^2+0.580s+0.123} \tag{9.91}$$

The numerator is obtained from (2.31) after replacing ω by $-js$ and making the leading coefficient unity.

Step 7. The equalizer back-end impedance is determined by the equation

$$Z_{22}(s)=\frac{F(s)}{A(s)-\hat{\rho}(s)}-z_2(s)$$

$$=\frac{\dfrac{-2}{(2s+1)^2}}{\dfrac{2s-1}{2s+1}-\dfrac{s^5+1.25s^3+0.312s}{s^5+0.937s^4+1.689s^3+0.974s^2+0.580s+0.123}}-\frac{1}{2s+1}$$

$$=\frac{0.937s^4+0.439s^3+0.974s^2+0.268s+0.123}{0.126s^5+0.059s^4+0.991s^3+0.439s^2+0.647s+0.123} \tag{9.92}$$

This impedance is guaranteed to be positive real, and therefore is physically realizable.

Step 8. Expanding Z_{22} in a continued fraction results in

$$Z_{22}(s)=\cfrac{1}{0.135s+\cfrac{1}{1.090s+\cfrac{1}{2.991s+\cfrac{1}{1.100s+\cfrac{1}{2.134s+\cfrac{1}{1}}}}}} \tag{9.93}$$

which can be identified as an LC ladder network terminated in a 1-Ω resistor, as shown in Figure 9.10. Denormalizing the element values with regard to magnitude-scaling by a factor of 100 and frequency-scaling by a factor of 10^8 gives the final design of the equalizer as indicated in Figure 9.11.

For illustrative purposes, we compute the transducer power gain in the normalized network of Figure 9.11. To this end, we compute the output reflection coefficient $\rho_2(s)$ by means of (9.92). The result is

$$\rho_2(s)=\frac{Z_{22}(s)-z_2(-s)}{Z_{22}(s)+z_2(s)}$$

$$=\frac{-(1+2s)(2s^5+2.50s^3+0.625s)}{(1-2s)(2s^5+1.87s^4+3.378s^3+1.948s^2+1.16s+0.246)} \tag{9.94}$$

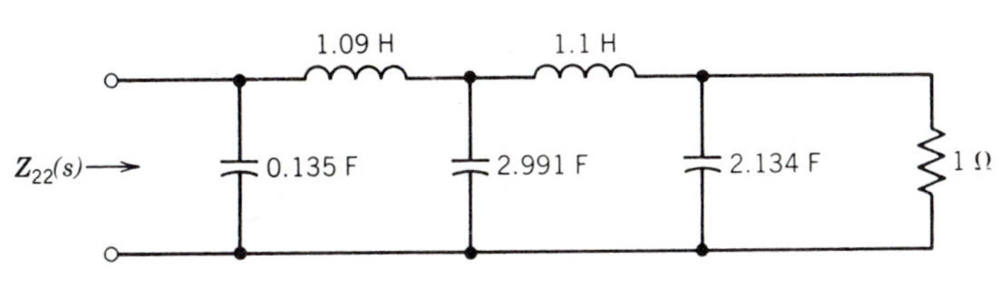

Figure 9.10

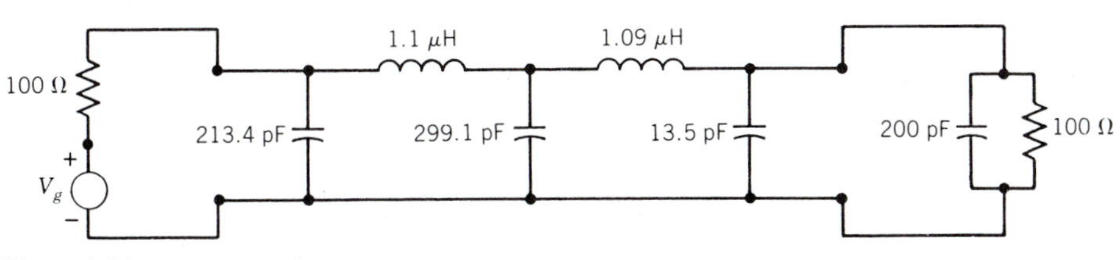

Figure 9.11

The transducer power gain is found from (9.2) or (9.7) as

$$G(\omega^2)=1-|\rho(j\omega)|^2=1-|\rho_2(j\omega)|^2$$
$$=\frac{0.0151}{\omega^{10}-2.5\omega^8+2.188\omega^6-0.781\omega^4+0.0975\omega^2+0.0151} \tag{9.95}$$

A plot of $G(\omega^2)$ as a function of the normalized frequency ω is presented in Figure 9.12.

Example 9.5 In Example 9.4, we showed that a match is possible for any value of K_5 between 0 and 1. To maximize K_5 we chose $K_5=1$. In this example, we shall arbitrarily choose

$$K_5=0.795 \tag{9.96}$$

and proceed to the design of the equalizer.

From (9.82) we first calculate

$$\hat{\varepsilon}=\frac{\varepsilon}{\sqrt{1-K_5}}=1.124 \tag{9.97}$$

With $v=\hat{\varepsilon}$ in (9.84), the LHS roots of the equation are obtained by means of formula (2.62), as follows:

$$\hat{a}=\frac{1}{5}\sinh^{-1}\frac{1}{1.124}=0.160 \tag{9.98}$$

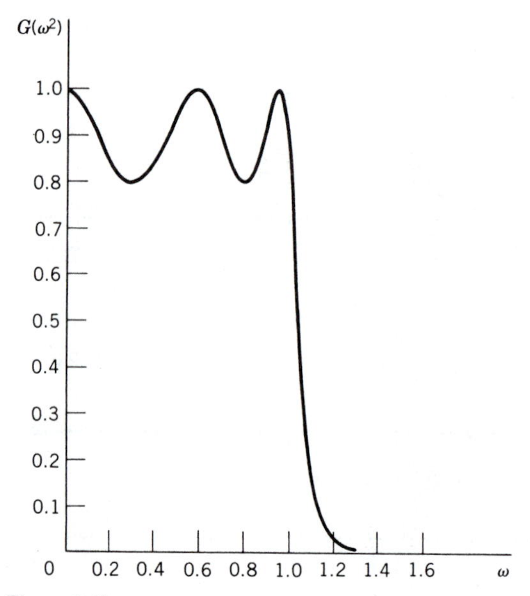

Figure 9.12

$$s_k = -\sinh \hat{a} \sin \frac{(2k-1)\pi}{10} + j \cosh \hat{a} \cos \frac{(2k-1)\pi}{10} \tag{9.99}$$

giving

$$s_1, s_5 = -0.04969 \pm j0.96327 \tag{9.100a}$$

$$s_2, s_4 = -0.13008 \pm j0.59533 \tag{9.100b}$$

$$s_3 = -0.16078 \tag{9.100c}$$

The minimum-phase solution of (9.81) becomes

$$\hat{\rho}(s) = \frac{s^5 + 0.520s^4 + 1.385s^3 + 0.492s^2 + 0.390s + 0.055}{s^5 + 0.937s^4 + 1.689s^3 + 0.974s^2 + 0.580s + 0.123} \tag{9.101}$$

Substituting (9.76), (9.78), (9.79), and (9.101) in (9.73), the equalizer back-end impedance is obtained as

$$\begin{aligned} Z_{22}(s) &= \frac{0.834s^5 + 1.025s^4 + 1.268s^3 + 0.862s^2 + 0.325s + 0.067}{(2s+1)(1.166s^5 + 0.849s^4 + 2.110s^3 + 1.086s^2 + 0.835s + 0.178)} \\ &= 0.357 \frac{s^4 + 0.729s^3 + 1.156s^2 + 0.455s + 0.162}{s^5 + 0.728s^4 + 1.809s^3 + 0.931s^2 + 0.716s + 0.153} \\ &= \cfrac{1}{2.796s + \cfrac{1}{0.547s + \cfrac{1}{5.930s + \cfrac{1}{0.523s + \cfrac{1}{3.641s + \cfrac{1}{0.378}}}}}} \end{aligned} \tag{9.102}$$

which can be identified as an LC ladder terminated in a resistor, as shown in Figure 9.13. After denormalization, the required equalizer together with its terminations is presented in Figure 9.14.

As a check, we compute the transducer power gain from the realized network

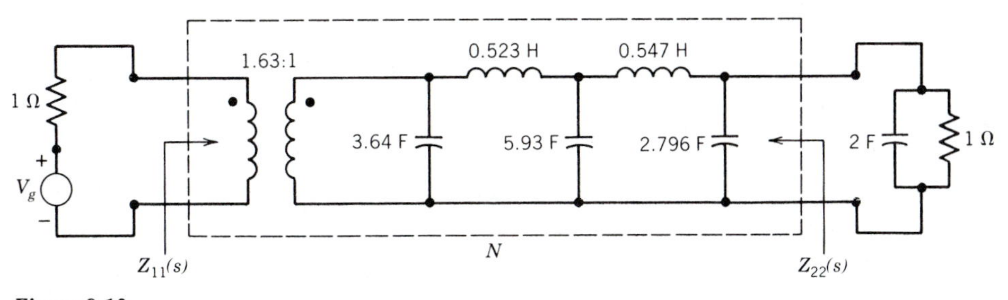

Figure 9.13

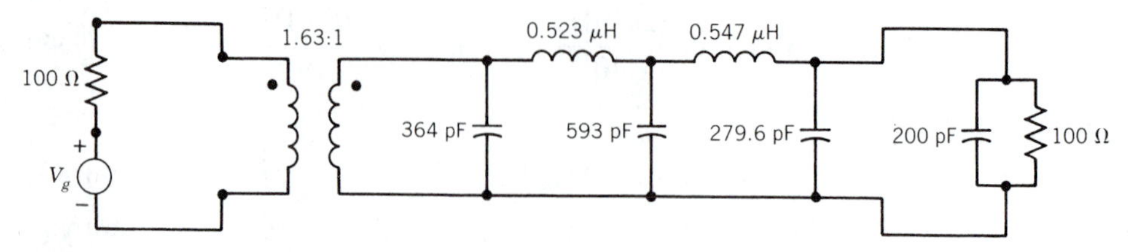

Figure 9.14

of Figure 9.13. We first compute the impedance looking into the input port with output port terminating in z_2. The result is

$$Z_{11}(s)=0.726\frac{s^4+0.208s^3+1.012s^2+0.131s+0.123}{s^5+0.208s^4+1.537s^3+0.241s^2+0.485s+0.034} \tag{9.103}$$

The transducer power gain is found from (4.99) to be

$$G(\omega^2)=\frac{4\ \mathrm{Re}\ Z_{11}(j\omega)}{|Z_{11}(j\omega)+1|^2}$$

$$=\frac{0.012}{\omega^{10}-2.503\omega^8+2.187\omega^6-0.778\omega^4+0.097\omega^2+0.015} \tag{9.104}$$

A plot of $G(\omega^2)$ as a function of ω is shown in Figure 9.15.

Example 9.6 Design a lossless equalizer that matches the load impedance of Figure 9.16 to a generator of internal resistance R_1 to achieve a maximum truly-flat transducer power gain over the entire sinusoidal frequency spectrum.

Step 1. For a truly-flat transducer power gain, let

$$G(\omega^2)=K, \qquad 0\leqslant K\leqslant 1 \tag{9.105}$$

Step 2. For the load impedance

$$z_2(s)=\frac{2s+1}{s+2} \tag{9.106}$$

we compute the functions

$$r_2(s)=\frac{2(s^2-1)}{s^2-4} \tag{9.107a}$$

$$A(s)=\frac{s-2}{s+2} \tag{9.107b}$$

$$F(s)=\frac{4(s^2-1)}{(s+2)^2} \tag{9.107c}$$

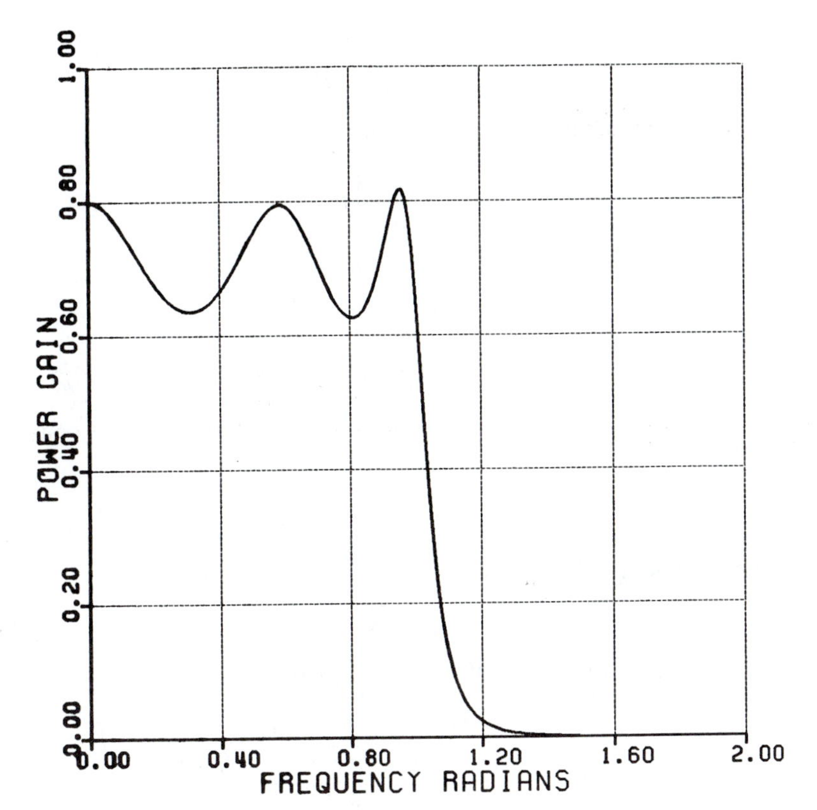

Figure 9.15

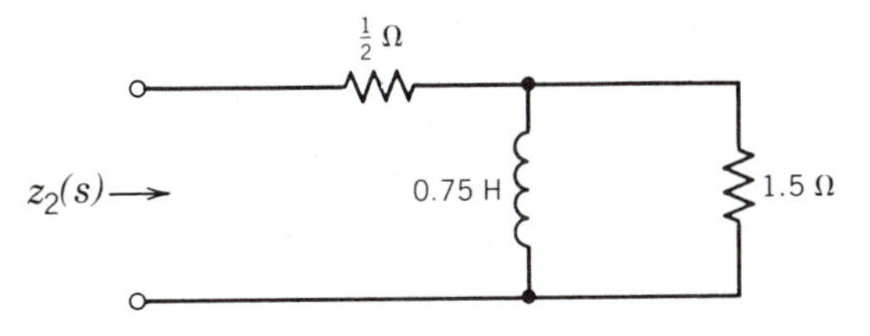

Figure 9.16

Step 3. Since

$$w(s) = \frac{r_2(s)}{z_2(s)} = \frac{2(s^2-1)}{(s-2)(2s+1)} \tag{9.108}$$

the load impedance z_2 has a Class I zero of transmission of order 1 at $s=1$.

Step 4. Substituting (9.105) in (9.70) gives

$$\hat{\rho}(s)\hat{\rho}(-s) = 1 - K \tag{9.109}$$

or

$$\hat{\rho}(s) = \pm\sqrt{1-K} \tag{9.110}$$

Step 5. The Laurent series expansions about the zero of transmission are given by

$$A(s) = \frac{s-2}{s+2} = -\frac{1}{3} + \frac{4}{9}(s-1) + \cdots \tag{9.111a}$$

$$F(s) = \frac{4(s^2-1)}{(s+2)^2} = 0 + \frac{8}{9}(s-1) + \cdots \tag{9.111b}$$

$$\hat{\rho}(s) = \pm\sqrt{1-K} + 0(s-1) + 0(s-1)^2 + \cdots \tag{9.111c}$$

Step 6. For a Class I zero of transmission of order 1, the coefficient constraint is from (9.41a)

$$A_0 = -\tfrac{1}{3} = \rho_0 = \pm\sqrt{1-K} \tag{9.112}$$

or $K = 8/9$ by choosing the minus sign for ρ_0. From (9.110) we obtain

$$\hat{\rho} = -\sqrt{1-K} = -\tfrac{1}{3} \tag{9.113}$$

Step 7. The equalizer back-end impedance is found to be

$$Z_{22}(s) = \frac{F(s)}{A(s) - \hat{\rho}(s)} - z_2(s) = \frac{\dfrac{4(s^2-1)}{(s+2)^2}}{\dfrac{s-2}{s+2} + \dfrac{1}{3}} - \frac{2s+1}{s+2}$$

$$= \frac{3(s+1)}{s+2} - \frac{2s+1}{s+2} = 1 \tag{9.114}$$

Step 8. Since the internal resistance of the generator is R_1, the matching network is simply an ideal transformer with turns ratio $\sqrt{R_1}:1$, as shown in Figure 9.17.

As a check, we compute the transducer power gain of the network of Figure

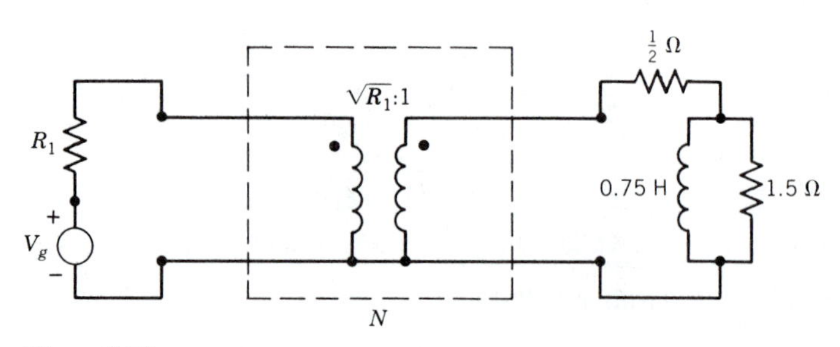

Figure 9.17

9.17, obtaining from (4.99)

$$G(\omega^2)=\frac{4r_2(j\omega)}{|1+z_2(j\omega)|^2}=\frac{8(1+\omega^2)}{4+\omega^2}\cdot\frac{4+\omega^2}{9(1+\omega^2)}$$
$$=\frac{8}{9}=K \tag{9.115}$$

as designed.

The above examples illustrated the procedure for the design of an optimum lossless equalizer. However, none of them requires the use of an all-pass function in Step 6. The following two examples illustrate the use of the all-pass function $\eta(s)$.

Example 9.7 Design a lossless equalizer to match the load of Figure 9.2 with $\hat{R}_1=100\,\Omega$, $R_2=200\,\Omega$, and $C=50$ pF to a resistive generator and to achieve the second-order Butterworth transducer power-gain characteristic having a maximum attainable dc gain. The cutoff frequency is 10^8 rad/s.

To simplify the computation, we first magnitude-scale the network by a factor of 10^{-2} and frequency-scale it by a factor of 10^{-8}, resulting in the normalized values

$$\hat{R}_1=1\,\Omega, \qquad R_2=2\,\Omega, \qquad C=0.5\text{ F}, \qquad \omega_c=1\text{ rad/s}, \qquad n=2 \tag{9.116}$$

We now follow the eight steps to complete the design.

Step 1. The transducer power-gain characteristic is required to be

$$G(\omega^2)=\frac{K_2}{1+\omega^4}, \qquad 0\leqslant K_2\leqslant 1 \tag{9.117}$$

Step 2. For the load

$$z_2(s)=1+\frac{2}{s+1}=\frac{s+3}{s+1} \tag{9.118}$$

we compute the functions

$$r_2(s)=\frac{s^2-3}{s^2-1} \tag{9.119a}$$

$$A(s)=\frac{s-1}{s+1} \tag{9.119b}$$

$$F(s)=\frac{2(s^2-3)}{(s+1)^2} \tag{9.119c}$$

Step 3. Since

$$w(s)=\frac{r_2(s)}{z_2(s)}=\frac{s^2-3}{(s+3)(s-1)} \tag{9.120}$$

the load impedance z_2 possesses a Class I zero of transmission of order 1 at $s=\sqrt{3}$.

Step 4. Substituting (9.117) in (9.42) gives

$$\rho(s)\rho(-s)=1-G(-s^2)=1-\frac{K_2}{1+s^4}$$

$$=\frac{1-K_2+s^4}{1+s^4}=\alpha^4\frac{1+x^4}{1+s^4} \tag{9.121}$$

where $x=s/\alpha$ and

$$\alpha=(1-K_2)^{1/4} \tag{9.122}$$

The minimum-phase solution of (9.121) is found to be

$$\hat{\rho}(s)=\frac{s^2+\sqrt{2}\alpha s+\alpha^2}{s^2+\sqrt{2}s+1} \tag{9.123}$$

Step 5. Since for a Class I zero of transmission of order 1 the coefficient condition is $A_0=\rho_0$, which is equivalent to

$$A_0=A(\sqrt{3})=\rho_0=\rho(\sqrt{3}) \tag{9.124}$$

the Laurent series expansions of the functions A, F, and $\hat{\rho}$ about the zero of transmission at $s=\sqrt{3}$ are not required.

Step 6. Substituting (9.119b) and (9.123) in (9.124) yields the constraint on α or K_2, as follows:

$$\frac{\sqrt{3}-1}{\sqrt{3}+1}=\frac{3+\sqrt{2}\alpha\sqrt{3}+\alpha^2}{3+\sqrt{2}\sqrt{3}+1} \tag{9.125}$$

or

$$\alpha^2+2.45\alpha+1.272=0 \tag{9.126}$$

giving $\alpha=-1.703$, -0.747. This shows that (9.124) cannot be satisfied without the insertion of an all-pass function $\eta(s)$. Consider the more general non-minimum-phase solution of (9.121), given as

$$\rho(s)=\eta(s)\hat{\rho}(s)=\frac{s-\sigma_1}{s+\sigma_1}\cdot\frac{s^2+\sqrt{2}\alpha s+\alpha^2}{s^2+\sqrt{2}s+1} \tag{9.127}$$

Substituting this in (9.124) in conjunction with (9.119b) results in

$$\frac{\sqrt{3}-1}{\sqrt{3}+1}=\frac{\sqrt{3}-\sigma_1}{\sqrt{3}+\sigma_1}\cdot\frac{3+\sqrt{2}\alpha\sqrt{3}+\alpha^2}{3+\sqrt{2}\sqrt{3}+1} \tag{9.128}$$

Equation (9.128) is a relation involving α and σ_1. To maximize K_2, let us choose $\alpha=0$, giving from (9.122) and (9.128)

$$K_2=1 \tag{9.129}$$

$$\sigma_1=0.466 \tag{9.130}$$

$$\rho(s)=\frac{(s-0.466)s^2}{(s+0.466)(s^2+1.414s+1)} \tag{9.131}$$

***Step* 7.** The equalizer back-end impedance is computed as

$$Z_{22}(s)=\frac{F(s)}{A(s)-\rho(s)}-z_2(s)=\frac{(1+\rho)s-3(1-\rho)}{(1-\rho)s-(1+\rho)}$$

$$=\frac{2s^4+1.414s^3-5.379s^2-4.511s-1.398}{0.346s^3+0.245s^2-1.193s-0.466}$$

$$=\frac{2s^3+4.878s^2+3.070s+0.807}{0.346s^2+0.844s+0.269} \tag{9.132}$$

which is guaranteed to be positive real.

***Step* 8.** The positive-real impedance Z_{22} is realized as a lossless two-port network terminated in a resistor, using Darlington's procedure if necessary. The details are omitted.

Example 9.8 It is desired to equalize the load impedance

$$z_2(s)=\frac{5s^2+3s+4}{s^2+2s+2} \tag{9.133}$$

to a resistive generator and to achieve a truly-flat transducer power gain over the entire $j\omega$-axis. Obtain the maximum attainable constant transducer power gain.

***Step* 1.** For a truly-flat transducer power gain, let

$$G(\omega^2)=K, \qquad 0\leqslant K\leqslant 1 \tag{9.134}$$

***Step* 2.** The following functions are computed from z_2:

$$r_2(s)=\frac{5s^4+8s^2+8}{(s^2+2s+2)(s^2-2s+2)} \tag{9.135a}$$

$$A(s)=\frac{s^2-2s+2}{s^2+2s+2} \tag{9.135b}$$

$$F(s)=\frac{2(5s^4+8s^2+8)}{(s^2+2s+2)^2} \tag{9.135c}$$

***Step* 3.** Since

$$w(s)=\frac{r_2(s)}{z_2(s)}=\frac{5s^4+8s^2+8}{(s^2-2s+2)(5s^2+3s+4)} \tag{9.136}$$

the load impedance z_2 possesses two Class I zeros of transmission of order 1 located at

$$s_1, s_2=0.482\pm j1.016 \tag{9.137}$$

***Step* 4.** Substituting (9.134) in (9.42) gives

$$\rho(s)\rho(-s)=1-K \tag{9.138}$$

or the minimum-phase solution is found to be

$$\hat{\rho}(s) = \pm\sqrt{1-K} \tag{9.139}$$

Step 5. For Class I zeros of transmission of order 1, the coefficient conditions are

$$A(s_i) = \hat{\rho}(s_i), \qquad i=1, 2 \tag{9.140}$$

As a result, the Laurent series expansions of the functions A, F, and $\hat{\rho}$ about the zeros of transmission at s_1 and s_2 are not required.

Step 6. Substituting (9.135b) and (9.139) in (9.140) yields

$$A(s_i) = -0.193 \mp j0.217 \neq \hat{\rho}(s_i) = \pm\sqrt{1-K} \tag{9.141}$$

Since the coefficient constraints are not satisfied, we insert the first-order all-pass function $\eta(s)$ in $\hat{\rho}(s)$ and obtain

$$\rho(s) = \eta(s)\hat{\rho}(s) = \frac{s-\sigma_1}{s+\sigma_1}\hat{\rho}(s) \tag{9.142}$$

Using this $\rho(s)$ in (9.140) results in the equation

$$(s_i+\sigma_1)A(s_i) = \pm(s_i-\sigma_1)\sqrt{1-K} \tag{9.143}$$

which after substituting s_i and $A(s_i)$ from (9.137) and (9.141) leads to the constraint on σ_1:

$$\sigma_1^2 + 1.808\sigma_1 - 1.266 = 0 \tag{9.144}$$

This equation is solved for σ_1 to give $\sigma_1 = 0.541$ or -2.341. For our purposes, we must choose $\sigma_1 = 0.541$. Using this σ_1 in (9.143), the maximum permissible flat transducer power gain is found to be

$$K = 0.831 \tag{9.145}$$

Step 7. The equalizer back-end impedance is determined as

$$\begin{aligned} Z_{22}(s) &= \frac{F(s)}{A(s)-\rho(s)} - z_2(s) = \frac{A(s)z_2(-s)+\rho(s)z_2(s)}{A(s)-\rho(s)} \\ &= \frac{2.938s^3 - 0.417s^2 + 1.396s + 3.057}{1.413s^3 - 0.857s^2 + 1.297s + 0.636} \\ &= 2.08\,\frac{s+0.822}{s+0.355} \end{aligned} \tag{9.146}$$

Step 8. The positive-real impedance Z_{22} can be realized as the input impedance of a lossless two-port network terminated in a resistor. The details are omitted.

9.5 EXPLICIT FORMULAS FOR THE *RLC* LOAD

In most practical cases, the source can usually be represented as an ideal voltage source in series with a pure resistor, which may be the Thévenin equivalent of some other network. The load impedance is composed of the parallel combination of a resistor and a capacitor and then in series with an inductor, as shown in Figure 9.18, which may include the parasitic effects of a practical device. The problem is to match out this load and source over a given frequency band to within a given tolerance, which recurs constantly in broadband amplifier design. In this section, we shall present explicit formulas for the design of optimum Butterworth and Chebyshev matching networks for any *RLC* load of the type shown in Figure 9.18, thus avoiding the necessity of applying the coefficient constraints and solving the nonlinear equations for selecting the optimum design parameters. As a consequence, we reduce the design of these matching networks to simple arithmetic.

9.5.1 **Butterworth Networks**

Refer to Figure 9.18. Our objective is to match out the load impedance

$$z_2(s) = Ls + \frac{R}{RCs+1} \tag{9.147}$$

to a resistive generator and to achieve the nth-order Butterworth transducer power-gain characteristic

$$G(\omega^2) = \frac{K_n}{1+(\omega/\omega_c)^{2n}}, \qquad 0 \leqslant K_n \leqslant 1 \tag{9.148}$$

with maximum attainable dc gain K_n, where ω_c is the 3-dB bandwidth or the radian cutoff frequency.

As shown in (9.20), the load impedance z_2 possesses a Class IV zero of transmission of order 3. The even part of z_2 is found to be

$$r_2(s) = \text{Ev}\, z_2(s) = \frac{-R}{R^2C^2s^2-1}. \tag{9.149}$$

Since $z_2(-s)$ has an open RHS pole at $s = 1/RC$, the all-pass function defined by this pole is given by

$$A(s) = \frac{s-1/RC}{s+1/RC} = \frac{RCs-1}{RCs+1} \tag{9.150}$$

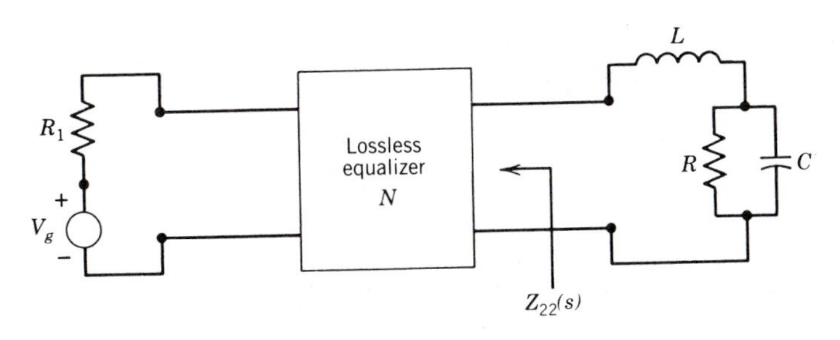

Figure 9.18

giving

$$F(s)=2r_2(s)A(s)=\frac{-2R}{(RCs+1)^2} \tag{9.151}$$

We next replace ω by $-js$ in (9.148) and substitute the resulting equation in (9.42), obtaining

$$\rho(s)\rho(-s)=\alpha^{2n}\frac{1+(-1)^n x^{2n}}{1+(-1)^n y^{2n}} \tag{9.152}$$

where as in (4.121)

$$y=s/\omega_c \tag{9.153a}$$

$$\alpha=(1-K_n)^{1/2n} \tag{9.153b}$$

$$x=y/\alpha \tag{9.153c}$$

The minimum-phase solution of (9.152) is found from (2.14) and (2.15) to be

$$\hat{\rho}(s)=\alpha^n\frac{q(x)}{q(y)} \tag{9.154}$$

as previously given in (4.122). For our purposes, we shall consider the more general solution

$$\rho(s)=\pm\eta(s)\hat{\rho}(s) \tag{9.155}$$

where $\eta(s)$ is an arbitrary first-order all-pass function of the form

$$\eta(s)=\frac{s-\sigma_1}{s+\sigma_1} \tag{9.156}$$

The above results follow from the application of the first four steps outlined in Section 9.4. To continue we next expand the functions A, F, and ρ in Laurent series about the zero of transmission at infinity and apply the coefficient conditions

$$A_m=\rho_m, \qquad m=0, 1, 2 \tag{9.157a}$$

$$L_a \geqq L \tag{9.157b}$$

where

$$L_a \triangleq \frac{F_2}{A_3-\rho_3} \tag{9.158}$$

To satisfy the first three constraints (9.157a), we require†

$$K_n=1-\left[1-\frac{2(1-RC\sigma_1)\sin\gamma_1}{RC\omega_c}\right]^{2n} \tag{9.159}$$

†For a more detailed derivation, the reader is referred to the original paper by W. K. Chen, "Explicit formulas for the synthesis of optimum broad-band impedance-matching networks," *IEEE Trans. Circuits Syst.*, vol. CAS-24, 1977, pp. 157–169.

where as in (4.130)

$$\gamma_m = m\pi/2n, \qquad m=1, 2, \ldots \tag{9.160}$$

For constraint (9.157b), we compute L_a by substituting the coefficients F_2, A_3, and ρ_3 in (9.158). After considerable mathematical manipulation, the inductance L_a can be simplified. The result is surprisingly simple and is given by the expression‡

$$L_a = \frac{4R \sin \gamma_1 \sin \gamma_3}{(1 - RC\sigma_1)[RC\omega_c^2(\alpha^2 - 2\alpha \cos \gamma_2 + 1) + 4\sigma_1 \sin \gamma_1 \sin \gamma_3]} \tag{9.161}$$

Thus, with K_n as specified in (9.159), the matching is possible if and only if the series inductance L does not exceed L_a. To show that any *RLC* load of Figure 9.18 can be matched, we demonstrate that there exists a nonnegative real σ_1 such that L_a can be made at least as large as the given inductance L and satisfies the constraint (9.159) with $0 \leqslant K_n \leqslant 1$. To this end, four cases are distinguished, as follows: Let

$$L_{a1} = \frac{R^2 C\omega_c \sin \gamma_3}{[(RC\omega_c - \sin \gamma_1)^2 + \cos^2 \gamma_1]\omega_c \sin \gamma_1} > 0 \tag{9.162}$$

$$L_{a2} = \frac{8R \sin^2 \gamma_1 \sin \gamma_3}{[(RC\omega_c - \sin \gamma_3)^2 + (1 + 4 \sin^2 \gamma_1) \sin \gamma_1 \sin \gamma_3]\omega_c} > 0 \tag{9.163}$$

Case 1: *$RC\omega_c \geqq 2 \sin \gamma_1$ and $L_{a1} \geqq L$*. Under this situation, $\sigma_1 = 0$ and the maximum attainable K_n is given by (9.159). The equalizer back-end impedance Z_{22} can be expanded in a continued fraction as§

$$Z_{22}(s) = (L_{a1} - L)s + \cfrac{1}{C_2 s + \cfrac{1}{L_3 s + \cfrac{1}{\ddots + \cfrac{1}{W}}}} \tag{9.164}$$

where W is a constant representing either a resistance or a conductance, and

$$L_1 = L_{a1} \tag{9.165a}$$

$$C_{2m} L_{2m-1} = \frac{4 \sin \gamma_{4m-1} \sin \gamma_{4m+1}}{\omega_c^2(1 - 2\alpha \cos \gamma_{4m} + \alpha^2)}, \qquad m \leqslant \tfrac{1}{2}(n-1) \tag{9.165b}$$

$$C_{2m} L_{2m+1} = \frac{4 \sin \gamma_{4m+1} \sin \gamma_{4m+3}}{\omega_c^2(1 - 2\alpha \cos \gamma_{4m+2} + \alpha^2)}, \qquad m < \tfrac{1}{2}(n-1) \tag{9.165c}$$

where $m = 1, 2, \ldots, [\tfrac{1}{2}(n-1)]$, $n > 1$. In addition, the final reactive element can be

‡*Loc. cit.*
§*Loc. cit.*

computed directly by the formulas

$$C_{n-1}=\frac{2(1+\alpha^n)\sin\gamma_1}{R(1-\alpha^n)(1+\alpha)\omega_c},\qquad n\text{ odd} \tag{9.166a}$$

$$L_{n-1}=\frac{2R(1-\alpha^n)\sin\gamma_1}{(1+\alpha^n)(1+\alpha)\omega_c},\qquad n\text{ even} \tag{9.166b}$$

Equation (9.164) can be identified as an LC ladder network terminated in a resistor, as depicted in Figure 9.19. The terminating resistance of the ladder is found to be

$$R_{22}=R\frac{1-\alpha^n}{1+\alpha^n} \tag{9.167}$$

Case 2: *$RC\omega_c \geqq 2\sin\gamma_1$, and $L_{a1}<L$.* Under this condition, σ_1 is nonzero and can be determined by the formula†

$$\sigma_1=\frac{1}{RC}\left[1+2\sqrt{p}\sinh\frac{\phi}{3}-\frac{2RC\omega_c\sin^2\gamma_1+\sin\gamma_3}{3\sin\gamma_1}\right] \tag{9.168}$$

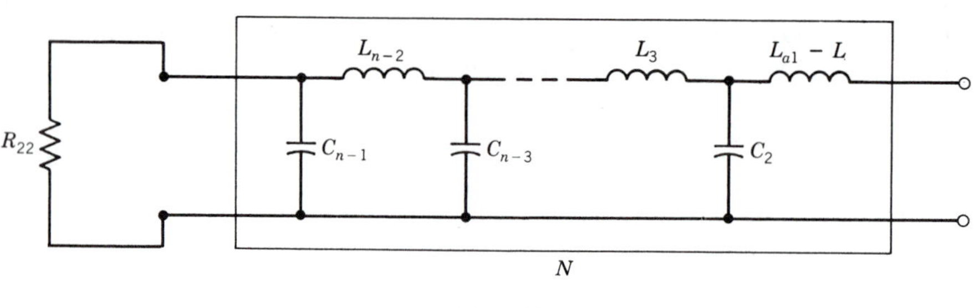

(a) n odd

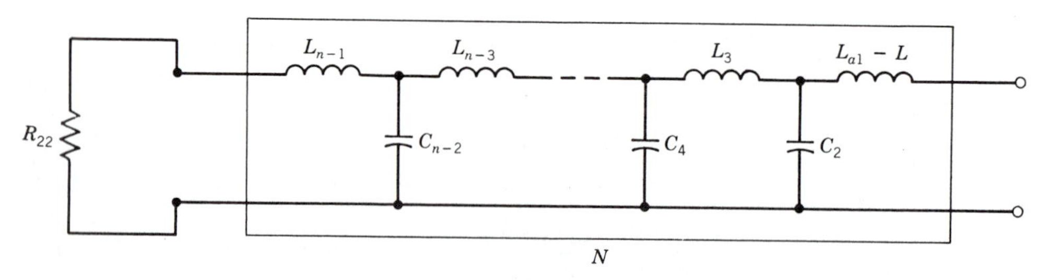

(b) n even

Figure 9.19

†*Loc. cit.*

where

$$p=\frac{(RC\omega_c-2\sin\gamma_1)^2\sin\gamma_3}{9\sin\gamma_1}>0 \tag{9.169a}$$

$$w=\frac{(2RC\omega_c\sin^2\gamma_1+\sin\gamma_3)}{54\sin^3\gamma_1}[3(RC\omega_c-2\sin\gamma_1)^2\sin\gamma_1\sin\gamma_3$$

$$+(2RC\omega_c\sin^2\gamma_1+\sin\gamma_3)^2]-\frac{R^2C\omega_c\sin\gamma_3}{2\omega_cL\sin\gamma_1} \tag{9.169b}$$

$$\phi=\sinh^{-1}\frac{w}{(\sqrt{p})^3} \tag{9.169c}$$

Using this value of σ_1, the dc gain K_n is determined by (9.159).

Case 3: $RC\omega_c<2\sin\gamma_1$ *and* $L_{a2}\geqq L$. Then we have

$$K_n=1 \tag{9.170a}$$

$$\sigma_1=\frac{1}{RC}\left(1-\frac{RC\omega_c}{2\sin\gamma_1}\right)>0 \tag{9.170b}$$

Case 4: $RC\omega_c<2\sin\gamma_1$ *and* $L_{a2}<L$. Under this situation, the desired σ_1 can be computed by formula (9.168). Using this σ_1, the dc gain K_n is determined by (9.159).

We illustrate the above results by the following examples.

Example 9.9 Let

$$R=100\ \Omega,\qquad C=100\text{ pF},\qquad L=0.5\ \mu\text{H}$$
$$n=5,\qquad \omega_c=10^8\text{ rad/s}$$

From (9.162) we first compute

$$L_{a1}=\frac{100\sin 54^\circ}{[(1-\sin 18^\circ)^2+\cos^2 18^\circ]\times 10^8\sin 18^\circ}=1.89443\ \mu\text{H} \tag{9.171}$$

where $RC\omega_c=1$. Since

$$RC\omega_c=1>2\sin 18^\circ=0.61803 \tag{9.172}$$

and $L_{a1}\geqq L$, Case 1 applies and the matching network N can be realized as an LC ladder terminated in a resistor as shown in Figure 9.19(a). With $\sigma_1=0$, the maximum attainable dc gain K_5 is from (9.159)

$$K_5=1-\left(1-\frac{2\sin 18^\circ}{RC\omega_c}\right)^{10}=1-(1-2\sin 18^\circ)^{10}=0.99993 \tag{9.173}$$

giving from (9.153b)

$$\alpha=(1-K_5)^{1/10}=0.38197 \tag{9.174}$$

Applying the formulas (9.165) gives the element values of the LC ladder network, as follows:

$$L_1 = L_{a1} = 1.89443\ \mu\text{H}, \tag{9.175a}$$

$$C_2 = \frac{4 \sin 54^\circ \sin 90^\circ}{1.89443 \times 10^{-6} \times 10^{16}(1 - 2 \times 0.38197 \cos 72^\circ + 0.14590)}$$

$$= 187.74978\ \text{pF} \tag{9.175b}$$

$$L_3 = \frac{4 \sin 90^\circ \sin 126^\circ}{187.74978 \times 10^{-12} \times 10^{16}(1 - 2 \times 0.38197 \cos 108^\circ + 0.14590)}$$

$$= 1.24721\ \mu\text{H} \tag{9.175c}$$

$$C_4 = \frac{4 \sin 126^\circ \sin 162^\circ}{1.24721 \times 10^{-6} \times 10^{16}(1 - 2 \times 0.38197 \cos 144^\circ + 0.14590)}$$

$$= 45.45455\ \text{pF} \tag{9.175d}$$

The last reactive element C_4 can also be ascertained directly from (9.166a), yielding

$$C_4 = \frac{2(1 + 0.38197^5) \sin 18^\circ}{100 \times (1 - 0.38197^5)(1 + 0.38197) \times 10^8} = 45.45455\ \text{pF} \tag{9.176}$$

Finally, the terminating resistance is determined from (9.167) as

$$R_{22} = 100\, \frac{1 - 0.38197^5}{1 + 0.38197^5} = 98.38699 \tag{9.177}$$

The matching network together with its termination is shown in Figure 9.20. Note that for computational accuracy we retain five significant figures in all the computations. In practice, one or two significant digits are sufficient, as indicated in Figure 9.20.

Example 9.10 We consider the same problem as in Example 9.9 except that now we raise the series inductance L from 0.5 to 3 μH, everything else being the same.

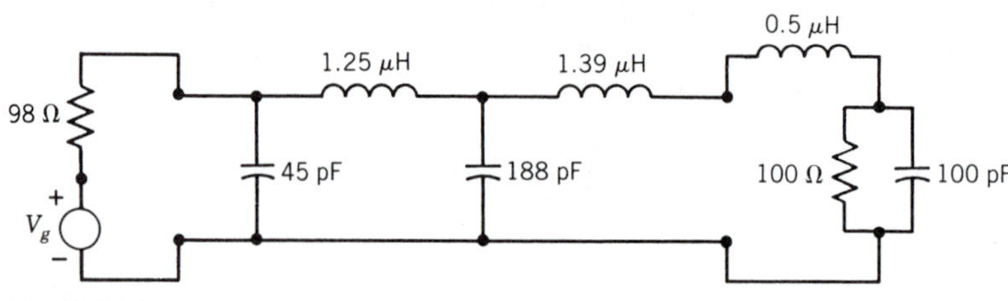

Figure 9.20

Thus, Case 2 applies and an extra all-pass function is needed. From (9.169) we obtain

$$p = \frac{(1-2\sin 18^\circ)^2 \sin 54^\circ}{9 \sin 18^\circ} = 0.04244 \tag{9.178a}$$

$$\phi = \sinh^{-1} \frac{0.25990}{(\sqrt{0.04244})^3} = 4.08543 \tag{9.178b}$$

where $w = 0.25990$. Substituting these in (9.168) yields the desired value for σ_1, as follows:

$$\begin{aligned}\sigma_1 &= 10^8 \left[1 + 2\sqrt{0.04244}\sinh 1.36181 - \frac{2\sin^2 18^\circ + \sin 54^\circ}{3 \sin 18^\circ}\right] \\ &= 0.67265 \times 10^8 \end{aligned} \tag{9.179}$$

giving from (9.159) the maximum attainable dc gain

$$K_5 = 1 - \left[1 - \frac{2(1-0.67265)\sin 18^\circ}{1}\right]^{10} = 0.89570 \tag{9.180}$$

and from (9.154)–(9.156) in conjunction with (2.15) and Table 2.2

$$\begin{aligned}\rho(s) &= \alpha^5 \frac{(y-0.67265)(x^5 + 3.23607x^4 + 5.23607x^3 + 5.23607x^2 + 3.23607x + 1)}{(y+0.67265)(y^5 + 3.23607y^4 + 5.23607y^3 + 5.23607y^2 + 3.23607y + 1)} \\ &= \frac{(y-0.67265)(y^5 + 2.58136y^4 + 3.33171y^3 + 2.65765y^2 + 1.31021y + 0.32297)}{(y+0.67265)(y^5 + 3.23607y^4 + 5.23607y^3 + 5.23607y^2 + 3.23607y + 1)}\end{aligned} \tag{9.181}$$

where from (9.153), $x = y/\alpha$ and

$$\alpha = (1 - K_5)^{1/10} = 0.79768 \tag{9.182}$$

Finally, from (9.147), (9.150), (9.151), and (9.181) the matching network back-end impedance is determined as

$$\begin{aligned}Z_{22}(s) &= \frac{F(s)}{A(s) - \rho(s)} - z_2(s) \\ &= 100 \frac{1.03554y^3 + 3.01209y^2 + 3.48051y + 1.30749}{0.97952y^4 + 2.84914y^3 + 3.74014y^2 + 2.53964y + 0.66911} \\ &= \cfrac{100}{0.94590y + \cfrac{1}{2.31186y + \cfrac{1}{0.23165y + \cfrac{1}{1.93361 \cfrac{y+0.67619}{y+0.66912}}}}}\end{aligned} \tag{9.183}$$

This impedance can be realized as the input impedance of a lossless two-port network terminated in a 195.41-Ω resistor as shown in Figure 9.21. Observe that since an all-pass function is used in (9.181), (9.183) cannot be realized as a simple LC ladder terminating in a resistor. The all-pass cycle of operations in the continued-fraction expansion (9.183) corresponds to a Darlington type-C section as shown in Figure 9.21.

9.5.2 Chebyshev Networks

Refer again to Figure 9.18. Our objective is to match the RLC load to a resistive generator and to achieve the nth-order Chebyshev transducer power-gain characteristic

$$G(\omega^2)=\frac{K_n}{1+\varepsilon^2 C_n^2(\omega/\omega_c)}, \qquad 0 \leqslant K_n \leqslant 1 \tag{9.184}$$

having maximum attainable constant K_n, as in (2.23). Following (9.152), we obtain

$$\rho(s)\rho(-s)=(1-K_n)\frac{1+\hat{\varepsilon}^2 C_n^2(-jy)}{1+\varepsilon^2 C_n^2(-jy)} \tag{9.185}$$

where as before $y=s/\omega_c$ and

$$\hat{\varepsilon}=\frac{\varepsilon}{\sqrt{1-K_n}} \tag{9.186}$$

As in (9.154), let $\hat{\rho}(s)$ be the minimum-phase solution of (9.185). For our purposes, we shall consider the more general solution

$$\rho(s)=\frac{s-\sigma_1}{s+\sigma_1}\hat{\rho}(s) \tag{9.187}$$

We next expand the functions A, F, and ρ from (9.150), (9.151), and (9.187) in Laurent series about the zero of transmission at infinity and apply the coefficient conditions (9.157). The first three conditions (9.157a) lead to the constraint on K_n, requiring†

$$K_n=1-\varepsilon^2 \sinh^2\left(n \sinh^{-1}\left[\sinh a-\frac{2(1-RC\sigma_1)\sin\gamma_1}{RC\omega_c}\right]\right) \tag{9.188}$$

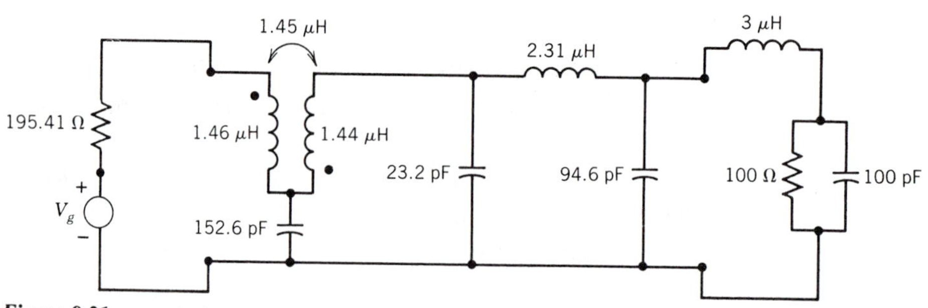

Figure 9.21

†*Loc. cit.*

where γ_1 is defined in (9.160) and

$$a = \frac{1}{n} \sinh^{-1} \frac{1}{\varepsilon} \tag{9.189}$$

To apply the fourth constraint, we rewrite (9.157b) as

$$L_b \triangleq \frac{F_2}{A_3 - \rho_3} \geqq L \tag{9.190}$$

After substituting the coefficients F_2, A_3, and ρ_3 from the Laurent series expansions of F, A, and ρ in (9.190) and after considerable mathematical manipulation, the left-hand side of the inequality (9.190) becomes‡

$$L_b = \frac{4R \sin \gamma_1 \sin \gamma_3}{(1 - RC\sigma_1)[RC\omega_c^2 f_1(\sinh a, \sinh \hat{a}) + 4\sigma_1 \sin \gamma_1 \sin \gamma_3]} \tag{9.191}$$

where§

$$\hat{a} = \frac{1}{n} \sinh^{-1} \frac{1}{\hat{\varepsilon}} = \frac{1}{n} \sinh^{-1} \frac{\sqrt{1 - K_n}}{\varepsilon} \tag{9.192}$$

$$f_m(\sinh a, \sinh \hat{a}) = \sinh^2 a + \sinh^2 \hat{a} + \sin^2 \gamma_{2m} - 2 \sinh a \sinh \hat{a} \cos \gamma_{2m}, \qquad m = 1, 2, \ldots, [\tfrac{1}{2}n] \tag{9.193}$$

Thus, with K_n as specified in (9.188), the matching is possible if and only if the series inductance L does not exceed L_b. Now we demonstrate that there exists a nonnegative real σ_1 such that L_b can be made at least as large as the given L and satisfies (9.188) with $0 \leqq K_n \leqq 1$. To this end, four cases are distinguished, as follows: Let

$$L_{b1} = \frac{R^2 C\omega_c \sin \gamma_3}{[(1 - RC\omega_c \sinh a \sin \gamma_1)^2 + R^2 C^2 \omega_c^2 \cosh^2 a \cos^2 \gamma_1]\omega_c \sin \gamma_1} > 0 \tag{9.194}$$

$$L_{b2} = \frac{8R \sin^2 \gamma_1 \sin \gamma_3}{[(RC\omega_c \sinh a - \sin \gamma_3)^2 + (1 + 4\sin^2 \gamma_1) \sin \gamma_1 \sin \gamma_3 + R^2 C^2 \omega_c^2 \sin^2 \gamma_2]\omega_c \sinh a} > 0 \tag{9.195}$$

Observe that both L_{b1} and L_{b2} are positive.

Case 1: *$RC\omega_c$* sinh *a* ≧ 2 sin *γ_1* and *$L_{b1} \geqq L$*. Under this situation, $\sigma_1 = 0$ and the maximum attainable constant K_n is given by

$$K_n = 1 - \varepsilon^2 \sinh^2 \left(n \sinh^{-1} \left[\sinh a - \frac{2 \sin \gamma_1}{RC\omega_c} \right] \right) \tag{9.196}$$

The equalizer back-end impedance Z_{22} can be expanded in a continued fraction as in (9.164) with L_{b1} replacing L_{a1} and realized by the LC ladders of Figure 9.22 with

‡*Loc. cit.*

§The general form for $f_m(\sinh a, \sinh \hat{a})$ is required later.

$$L_1 = L_{b1} \tag{9.197a}$$

$$C_{2m}L_{2m-1} = \frac{4 \sin \gamma_{4m-1} \sin \gamma_{4m+1}}{\omega_c^2 f_{2m}(\sinh a, \sinh \hat{a})}, \qquad m \leqq \tfrac{1}{2}(n-1) \tag{9.197b}$$

$$C_{2m}L_{2m+1} = \frac{4 \sin \gamma_{4m+1} \sin \gamma_{4m+3}}{\omega_c^2 f_{2m+1}(\sinh a, \sinh \hat{a})}, \qquad m < \tfrac{1}{2}(n-1) \tag{9.197c}$$

where $m = 1, 2, \ldots, [\frac{1}{2}(n-1)]$, $n > 1$. In addition, the final reactive element can be computed directly by the formulas

$$C_{n-1} = \frac{2(\sinh na + \sinh n\hat{a}) \sin \gamma_1}{\omega_c R(\sinh a + \sinh \hat{a})(\sinh na - \sinh n\hat{a})}, \qquad n \text{ odd} \tag{9.198a}$$

$$L_{n-1} = \frac{2R(\cosh na - \cosh n\hat{a}) \sin \gamma_1}{\omega_c(\sinh a + \sinh \hat{a})(\cosh na + \cosh n\hat{a})}, \qquad n \text{ even} \tag{9.198b}$$

The terminating resistance of the LC ladder networks is given by

$$R_{22} = R\,\frac{\sinh na - \sinh n\hat{a}}{\sinh na + \sinh n\hat{a}}, \qquad n \text{ odd} \tag{9.199a}$$

$$R_{22} = R\,\frac{\cosh na - \cosh n\hat{a}}{\cosh na + \cosh n\hat{a}}, \qquad n \text{ even} \tag{9.199b}$$

Case 2: ***$RC\omega_c \sinh a \geqq 2 \sin \gamma_1$*, and *$L_{b1} < L$*.** Under this condition, σ_1 is nonzero and can be determined by the formula†

$$\sigma_1 = \frac{1}{RC}\left(1 + 2\sqrt{q}\sinh\frac{\phi}{3} - \frac{2RC\omega_c \sin^2 \gamma_1 \sinh a + \sin \gamma_3}{3 \sin \gamma_1}\right) \tag{9.200}$$

where

$$q = \frac{(RC\omega_c \sinh a - 2 \sin \gamma_1)^2 \sin \gamma_3 + 3R^2C^2\omega_c^2 \sin \gamma_1 \cos^2 \gamma_1}{9 \sin \gamma_1} > 0 \tag{9.201a}$$

$$\begin{aligned}\zeta = {} & \frac{(2RC\omega_c \sin^2 \gamma_1 \sinh a + \sin \gamma_3)}{54 \sin^3 \gamma_1}[3(RC\omega_c \sinh a - 2 \sin \gamma_1)^2 \sin \gamma_1 \sin \gamma_3 \\ & + 2.25R^2C^2\omega_c^2 \sin^2 \gamma_2 + (2RC\omega_c \sin^2 \gamma_1 \sinh a + \sin \gamma_3)^2] \\ & - \frac{R^2C \sin \gamma_3}{2L \sin \gamma_1}\end{aligned} \tag{9.201b}$$

$$\phi = \sinh^{-1}\frac{\zeta}{(\sqrt{q})^3} \tag{9.201c}$$

Using this value of σ_1, the maximum constant K_n is determined by (9.188).

†*Loc. cit.*

Case 3: ***$RC\omega_c \sinh a < 2 \sin \gamma_1$ and $L_{b2} \geqq L$.*** Then we have

$$K_n = 1 \tag{9.202a}$$

$$\sigma_1 = \frac{1}{RC}\left(1 - \frac{RC\omega_c \sinh a}{2 \sin \gamma_1}\right) > 0 \tag{9.202b}$$

Case 4: ***$RC\omega_c \sinh a < 2 \sin \gamma_1$ and $L_{b2} < L$.*** Under this condition, the desired value of σ_1 can be computed by formula (9.200). Using this σ_1, the constant K_n is determined by (9.188).

We illustrate the use of the above formulas by the following examples.

Example 9.11 Given

$$R = 100\ \Omega, \qquad C = 500\text{ pF}, \qquad L = 0.5\ \mu\text{H}$$
$$n = 5, \qquad \varepsilon = 0.50885 \text{ (1-dB ripple)}, \qquad \omega_c = 10^8 \text{ rad/s}$$

obtain a lossless Chebyshev network meeting these specifications.

We first compute

$$RC\omega_c \sinh a = 5 \sinh 0.28560 = 1.44747 > 2 \sin 18^\circ = 0.61803 \tag{9.203}$$

where $a = 0.28560$. We next compute the inductance L_{b1} from (9.194):

$$L_{b1} = \frac{5 \times 100 \sin 54^\circ}{[(1 - 5 \sinh 0.28560 \sin 18^\circ)^2 + 25 \cosh^2 0.28560 \cos^2 18^\circ]10^8 \sin 18^\circ}$$
$$= 0.52755\ \mu\text{H} \tag{9.204}$$

Since $L_{b1} > L$, Case 1 applies and the matching network can be realized as an LC ladder as shown in Figure 9.22(*a*). From (9.196), the maximum attainable K_5 is found to be

$$K_5 = 1 - 0.50885^2 \sinh^2 [5 \sinh^{-1} (\sinh 0.28560 - 0.4 \sin 18^\circ)] = 0.77954 \tag{9.205}$$

giving from (9.186) and (9.192)

$$\hat{\varepsilon} = \frac{0.50885}{\sqrt{1 - 0.77954}} = 1.08373 \tag{9.206a}$$

$$\hat{a} = \frac{1}{5} \sinh^{-1} \frac{1}{1.08373} = 0.16514 \tag{9.206b}$$

Applying formulas (9.197) gives the element values of the desired LC ladder network, as follows: $L_1 = L_{b1} = 0.52755\ \mu\text{H}$

$$C_2 = \frac{4 \sin 54^\circ \sin 90^\circ}{0.52755 \times 10^{-6} \times 10^{16} f_2(\sinh 0.28560, \sinh 0.16514)} = 622.03127\text{ pF} \tag{9.207a}$$

$$L_3 = \frac{4 \sin 90^\circ \sin 126^\circ}{6.22031 \times 10^{-10} \times 10^{16} f_3(\sinh 0.28560, \sinh 0.16514)} = 0.49760\ \mu\text{H} \tag{9.207b}$$

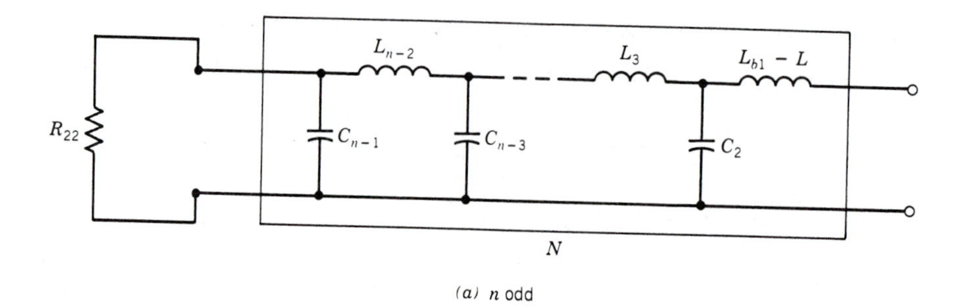

(a) n odd

(b) n even

Figure 9.22

$$C_4 = \frac{4 \sin 126° \sin 162°}{0.49760 \times 10^{-6} \times 10^{16} f_4(\sinh 0.28560, \sinh 0.16514)} = 375.97653 \text{ pF} \qquad (9.207c)$$

As a check, the last reactive element C_4 can also be determined from (9.198a), giving

$$C_4 = \frac{2(\sinh 5 \times 0.28560 + \sinh 5 \times 0.16514) \sin 18°}{10^8 \times 100(\sinh 0.28560 + \sinh 0.16514)(\sinh 5 \times 0.28560 - \sinh 5 \times 0.16514)}$$
$$= 375.97653 \text{ pF} \qquad (9.208)$$

Finally, the terminating resistance of the LC ladder network is obtained from (9.199a) as

$$R_{22} = 100 \frac{\sinh 5 \times 0.28560 - \sinh 5 \times 0.16514}{\sinh 5 \times 0.28560 + \sinh 5 \times 0.16514} = 36.09755 \ \Omega \qquad (9.209)$$

The matching network together with its terminations is shown in Figure 9.23.

Example 9.12 Given

$$R = 100 \ \Omega, \qquad C = 100 \text{ pF}, \qquad L = 1.5 \ \mu\text{H}$$
$$n = 4, \qquad \varepsilon = 0.50885 \text{ (1-dB ripple)}, \qquad \omega_c = 10^8 \text{ rad/s}$$

obtain a lossless Chebyshev network to meet these specifications.

As before, we first compute

$$RC\omega_c \sinh a = 0.36463 < 2 \sin 22.5° = 0.76537 \qquad (9.210)$$

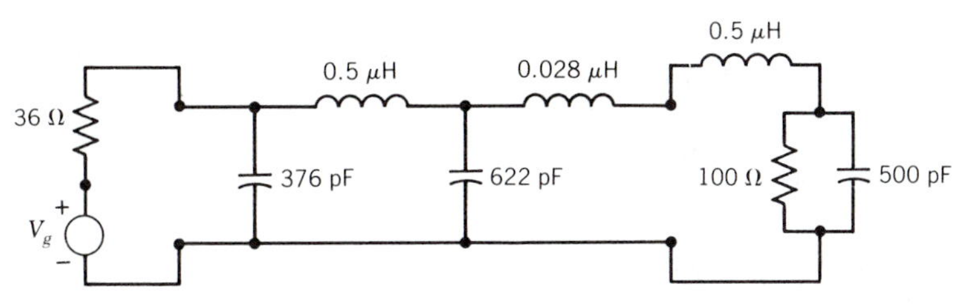

Figure 9.23

where $a=0.35699$ and $RC\omega_c=1$. We next compute the inductance L_{b2} from (9.195) and the result is given by

$$L_{b2}=\frac{(8\times 100\sin^2 22.5^\circ\sin 67.5^\circ)10^{-8}}{\xi}=2.16139\ \mu\text{H} \tag{9.211}$$

where

$$\xi=[(\sinh 0.35699-\sin 67.5^\circ)^2+\sin^2 45^\circ+(1+4\sin^2 22.5^\circ)\sin 22.5^\circ\sin 67.5^\circ]\sinh a=0.50079 \tag{9.212}$$

Since $L_{b2}>L$, Case 3 applies and

$$K_4=1 \tag{9.213}$$

can be achieved. The desired value of σ_1 is determined from (9.202b) as

$$\sigma_1=\frac{1}{100\times 10^{-10}}\left(1-\frac{100\times 10^{-10}\times 10^8\sinh 0.35699}{2\sin 22.5^\circ}\right)$$
$$=0.52359\times 10^8 \tag{9.214}$$

For $K_4=1$, (9.185) degenerates into

$$\rho(s)\rho(-s)=\frac{\varepsilon^2C_4^2(-jy)}{1+\varepsilon^2C_4^2(-jy)} \tag{9.215}$$

Applying formula (2.62) or Table 2.5, the minimum-phase solution of (9.215) is found to be

$$\hat{\rho}(s)=\frac{y^4+y^2+0.125}{y^4+0.95281y^3+1.45392y^2+0.74262y+0.27563} \tag{9.216}$$

The numerator of (9.216) is actually the fourth-order Chebyshev polynomial as given in (2.31) after making the leading coefficient unity and replacing ω by $-jy$. From (9.187) we obtain

$$\rho(s)=\frac{(y-0.52359)(y^4+y^2+0.125)}{(y+0.52359)(y^4+0.95281y^3+1.45392y^2+0.74262y+0.27563)} \tag{9.217}$$

Finally, from (9.147), (9.150), (9.151), and (9.217) the equalizer back-end impedance Z_{22} is found from (9.73)

$$Z_{22}(s)=\frac{F(s)}{A(s)-\rho(s)}-z_2(s)$$

$$=0.66139\times10^{-6}s+\cfrac{1}{1.19469\times10^{-10}s+\cfrac{1}{2.35354\times10^{-6}s+\cfrac{1}{Z_b(s)}}} \tag{9.218}$$

where

$$Z_b(s)=55.72136\,\frac{s+1.14394\times10^8}{s+0.23966\times10^8} \tag{9.219}$$

The impedance Z_{22} can be realized as the input impedance of a lossless two-port network terminating in a resistor as shown in Figure 9.24. The all-pass cycle of operations corresponding to Z_b in (9.218) is realized by a Darlington type-C section as indicated in Figure 9.24 with unit coupling coefficient. To avoid the use of a perfectly coupled transformer, we merge the series inductances 2.35 μH and 0.73 μH to yield a transformer whose coefficient of coupling is less than unity, as shown in Figure 9.25.

Example 9.13 Consider the same problem as in Example 9.12 except that now we raise the series inductance L from 1.5 to 3 μH, everything else being the same. Thus Case 4 applies and an extra all-pass function is required for the match. From (9.201) we obtain

$$q=0.57236 \tag{9.220a}$$

$$\zeta=0.40057 \tag{9.220b}$$

$$\phi=1.50301 \tag{9.220c}$$

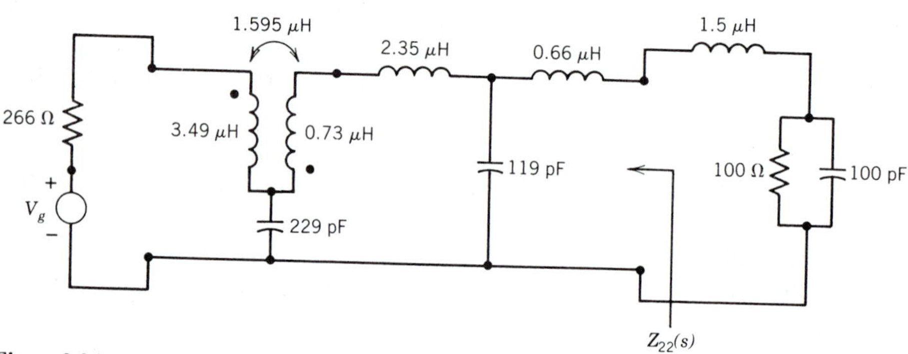

Figure 9.24

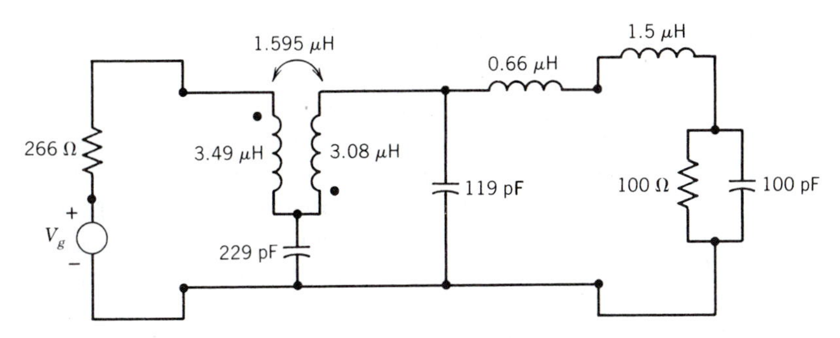

Figure 9.25

Substituting these in (9.200) yields the desired value for σ_1:

$$\sigma_1 = 0.70004 \times 10^8 \tag{9.221}$$

giving from (9.188) and (9.186)

$$K_4 = 0.91735 \tag{9.222}$$

$$\hat{\varepsilon} = 1.77001 \tag{9.223}$$

With these we can compute the equalizer back-end impedance Z_{22}, which can then be expanded in a continued fraction and realized as an LC ladder and a Darlington type-C section, as demonstrated in the previous two examples. However, the details are omitted.

9.6 SUMMARY AND SUGGESTED READINGS

In this chapter, we discussed the broadband matching between a frequency-dependent load and a resistive generator. Conditions were presented under which a lossless matching network exists that when operated between the resistive generator and the given load will yield the preassigned transducer power-gain characteristic. These conditions are most conveniently and succinctly expressed in terms of the coefficients of the Laurent series expansions of the functions defined by the given load and the preassigned transducer power-gain characteristic. For this we extended the concept of a zero of transmission for a two-port network to that for a one-port impedance. We then divided these zeros of transmission into four mutually exclusive classes. Basic coefficient constraints imposed on the normalized reflection coefficient by a given load impedance were stated. They depend on the classification of the zero of transmission. The significance of these constraints is that they are both necessary and sufficient for the physical realizability of the matching networks. To assist the designers, we outlined a simple procedure, consisting of eight steps, for the realization of an optimum lossless matching network that equalizes a frequency-dependent load to a resistive generator and achieves a preassigned transducer power-gain characteristic over the entire sinusoidal frequency spectrum.

In many practical situations, the load can usually be represented as a parallel combination of a resistor and a capacitor and then in series with an inductor. The problem to match out this load to a resistive generator over a given frequency band to within a given tolerance recurs constantly in broadband amplifier design. For this we presented explicit formulas for the design of optimum Butterworth and Chebyshev matching networks of arbitrary order, thus avoiding the necessity of applying the coefficient conditions and solving the related nonlinear equations for selecting the optimum design parameters. As a consequence, we reduce the design of these matching networks to simple arithmetic.

For additional study on topics of this chapter, the reader is referred to Chen [1]. For explicit formulas on *RLC* load, see Chen [5] and Chen and Kourounis [7] for low-pass networks, and Chen and Chaisrakeo [6] for band-pass networks having Butterworth and Chebyshev responses of arbitrary order. To study the relation between compatible impedances and the broadband matching problem, see Satyanarayana and Chen [13]. For a general broadband matching between a frequency-dependent source and a frequency-dependent load, see Chien [10], Chen [3], and Chen and Satyanarayana [8].

REFERENCES

1. W. K. Chen 1976, *Theory and Design of Broadband Matching Networks*, Chapter 4. New York: Pergamon Press.
2. W. K. Chen, "On the design of broadband elliptic impedance-matching networks," *J. Franklin Inst.*, vol. 301, 1976, pp. 451–463.
3. W. K. Chen, "Unified theory of broadband matching," *J. Franklin Inst.*, vol. 310, 1980, pp. 287–301.
4. W. K. Chen, "Explicit formulas for the synthesis of Chebyshev impedance-matching networks," *Electronics Letters*, vol. 12, 1976, pp. 412–413.
5. W. K. Chen, "Explicit formulas for the synthesis of optimum broadband impedance-matching networks," *IEEE Trans. Circuits Syst.*, vol. CAS-24, 1977, pp. 157–169.
6. W. K. Chen and T. Chaisrakeo, "Explicit formulas for the synthesis of optimum bandpass Butterworth and Chebyshev impedance-matching networks," *IEEE Trans. Circuits Syst.*, vol. CAS-27, 1980, pp. 928–942.
7. W. K. Chen and K. G. Kourounis, "Explicit formulas for the synthesis of optimum broadband impedance-matching networks II," *IEEE Trans. Circuits Syst.*, vol. CAS-25, 1978, pp. 609–620.
8. W. K. Chen and C. Satyanarayana, "General theory of broadband matching," *Proc. IEE*, vol. 129, Pt. G, 1982, pp. 96–102.
9. W. K. Chen and C. K. Tsai, "A general theory of broadband matching of an active load," *Circuits, Systems and Signal Processing*, vol. 1, 1982, pp. 105–122.
10. T. M. Chien, "A theory of broadband matching of a frequency-dependent generator and load," *J. Franklin Inst.*, vol. 298, 1974, pp. 181–221.
11. R. M. Fano, "Theoretical limitations on the broadband matching of arbitrary impedances," *J. Franklin Inst.*, vol. 249, 1950, pp. 57–83 and 139–154.

12. C. W. Ho and N. Balabanian, "Synthesis of active and passive compatible impedances," *IEEE Trans. Circuit Theory*, vol. CT-14, 1967, pp. 118–128.
13. C. Satyanarayana and W. K. Chen, "Theory of broadband matching and the problem of compatible impedances," *J. Franklin Inst.*, vol. 309, 1980, pp. 267–280.
14. D. C. Youla, "A new theory of cascade synthesis," *IRE Trans. Circuit Theory*, vol. CT-8, 1961, pp. 244–260.
15. D. C. Youla, "A new theory of broad-band matching," *IEEE Trans. Circuit Theory*, vol. CT-11, 1964, pp. 30–50.

PROBLEMS

9.1 It is required to equalize the load impedance

$$z_2(s)=\frac{s^2+9s+8}{s^2+2s+2} \tag{9.224}$$

to a resistive generator of internal resistance 100 Ω and to achieve the largest flat transducer power gain over the entire sinusoidal frequency spectrum. Verify that the largest flat transducer power gain that can be obtained for this load is $G_{\max}=0.82684$, and obtain the desired lossless equalizer.

9.2 It is desired to design a lossless matching network to equalize the load impedance

$$z_2(s)=\frac{s}{s^2+2s+1} \tag{9.225}$$

to a resistive generator and to achieve the transducer power-gain characteristic

$$G(\omega^2)=\frac{K\omega^2}{\omega^4-\omega^2+1} \tag{9.226}$$

having a maximum constant K. Verify that $K=1$ and that the equalizer back-end impedance is required to be

$$Z_{22}(s)=\frac{s}{3s^2+2s+3} \tag{9.227}$$

9.3 Repeat Example 9.3 for $n=5$, everything else being the same.

9.4 Repeat Example 9.4 for $n=4$, everything else being the same.

9.5 Repeat Example 9.7 for $n=3$, everything else being the same.

9.6 It is desired to design a lossless matching network to equalize the load of Figure 9.18 with

$$R=100\ \Omega, \qquad C=50\ \text{pF}, \qquad L=0.5\ \mu\text{H} \tag{9.228}$$

to a resistive generator and to achieve the fourth-order Butterworth transducer power gain having a maximum dc gain K_4. The radian cutoff frequency is $\omega_c=10^8$ rad/s.

9.7 Repeat Problem 9.6 for $n=5$, everything else being the same.

9.8 Repeat Problem 9.6 if L is increased from 0.5 to 3 μH, everything else being the same.

9.9 Repeat Example 9.9 for $n=6$, everything else being the same.

9.10 Repeat Example 9.11 for $n=4$, everything else being the same.

9.11 Repeat Example 9.11 for $n=6$, everything else being the same.

9.12 It is desired to design a lossless matching network to equalize the load of Figure 9.18 with

$$R=100\ \Omega, \qquad C=100\ \text{pF}, \qquad L=0.5\ \mu\text{H} \tag{9.229}$$

to a resistive generator and to achieve the fourth-order Chebyshev transducer power-gain characteristic having maximum constant K_4. The peak-to-peak ripple in the passband should not exceed 2 dB and the radian cutoff frequency is $\omega_c=10^8$ rad/s.

9.13 Repeat Problem 9.12 for $n=5$, everything else being the same.

9.14 Repeat Problem 9.12 if L is increased from 0.5 to 3 μH, everything else being the same.

9.15 Repeat Example 9.11 if L is increased from 0.5 to 1 μH, everything else being the same.

9.16 Design a lossless matching network to equalize the load composed of the series connection of a 100-Ω resistor and a 20-mH inductor to a resistive generator and to achieve the third-order Butterworth transducer power-gain characteristic having a maximum dc gain K_3. The radian cutoff frequency is $\omega_c=10^3$ rad/s. Verify that the equalizer back-end impedance is required to be

$$Z_{22}(s)=\frac{16s^3+16s^2+18s+10}{2s^2+2s+1} \tag{9.230}$$

9.17 Show that a resistive load R_2 can always be matched to a resistive generator to achievc nth-order Butterworth transducer power gain having unity dc gain. Derive the equalizer back-end impedance Z_{22}.

9.18 Determine if it is possible to design a lossless equalizer to match the load of Figure 9.26 to a resistive generator and to achieve the first-order Butterworth transducer power-gain characteristic having a nonzero dc gain.

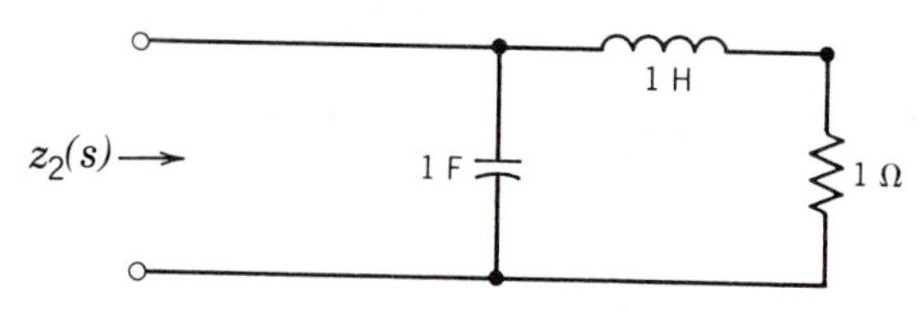

Figure 9.26

9.19 Design a lossless matching network to equalize the load impedance

$$z_2(s)=\frac{2s+3}{2s+1} \tag{9.231}$$

to a resistive generator and to achieve the second-order Butterworth transducer power-gain characteristic having a maximum dc gain. The 3-dB bandwidth is 1 rad/s. Verify that the equalizer back-end impedance can be expanded in a continued fraction as

$$Z_{22}(s)=5.53322s+\cfrac{1}{0.20898s+\cfrac{1}{2.99166}} \tag{9.232}$$

Choose appropriate denormalization factors to yield practical element values.

9.20 Design a matching network to equalize the load impedance

$$z_2(s) = \frac{s}{s+1} \tag{9.233}$$

to a resistive generator and to achieve the transducer power-gain characteristic

$$G(\omega^2) = \frac{\omega^4}{1+\omega^4} \tag{9.234}$$

Realize the desired equalizer by showing that its back-end impedance is required to be

$$Z_{22}(s) = \frac{s(s+1.41421)}{s^2+0.41421s+0.58579} \tag{9.235}$$

9.21 Design a lossless equalizer that matches the load impedance

$$z_2(s) = \frac{s+2}{s+1} \tag{9.236}$$

to a resistive generator to achieve a maximum transducer power gain that is truly flat over the entire sinusoidal frequency spectrum. Verify that the required equalizer back-end impedance is

$$Z_{22}(s) = \sqrt{2} \tag{9.237}$$

9.22 A lossless matching network is required to equalize the load composed of the parallel combination of a 1-Ω resistor and a 2-F capacitor to a resistive generator and to achieve the transducer power-gain characteristic

$$G(\omega^2) = \frac{K\omega^2}{1+\omega^4} \tag{9.238}$$

Show that the constant K is bounded by

$$0 \leqslant K \leqslant 2\sqrt{2} - 1 \tag{9.239}$$

9.23 It is desired to equalize the load impedance

$$z_2(s) = \frac{s+2}{s+1} \tag{9.240}$$

to a resistive generator to achieve the transducer power-gain characteristic

$$G(\omega^2) = \frac{K\omega^2}{1+\omega^2} \tag{9.241}$$

Show that $K=1$ can always be attained with the equalizer back-end impedance given by

$$Z_{22}(s) = \frac{s^2+2s+4(\sqrt{2}-1)}{s(s+1)} \tag{9.242}$$

9.24 Design an equalizer to match the load of Figure 9.2 with $\hat{R}_1 = 100\ \Omega$, $R_2 = 200\ \Omega$, and $C = 50$ pF to a resistive generator and to achieve the second-order Butterworth transducer power-gain characteristic having a maximum dc gain K_2. The radian cutoff frequency is $\omega_c = 10^8$ rad/s. Show that the equalizer back-end impedance is required to be

$$Z_{22}(s) = 100\,\frac{2y^3+4.878y^2+3.070y+0.807}{0.346y^2+0.844y+0.269} \tag{9.243}$$

where $y = s/10^8$.

9.25 Design a lossless matching network to equalize the load of Figure 9.27 to a resistive generator and to achieve the transducer power-gain characteristic

$$G(\omega^2)=\frac{K(\omega^2-1)^2}{\omega^4-\omega^2+1} \tag{9.244}$$

having a maximum constant K. Verify that the desired equalizer back-end impedance is required to be

$$Z_{22}(s)=1+\frac{s}{s^2+1} \tag{9.245}$$

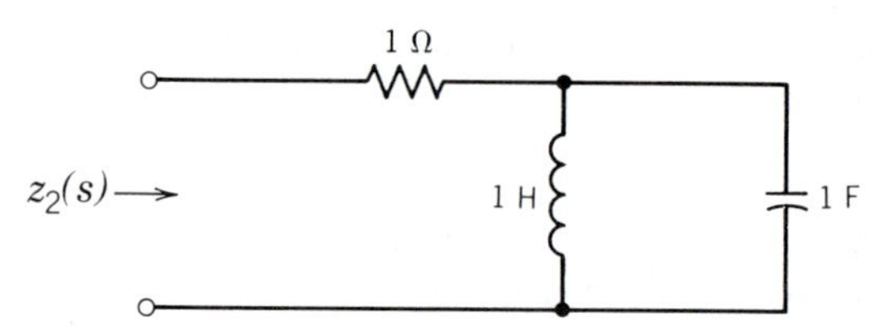

Figure 9.27

9.26 It is required to equalize the parallel combination of a 60-Ω resistor and a 150-pF capacitor to a resistive generator and to achieve the third-order Chebyshev transducer power-gain characteristic having a maximum constant K_3. The passband tolerance is 1.5 dB and the radian cutoff frequency is $\omega_c=10^8$ rad/s. Verify that the equalizer back-end impedance is, using $y=s/10^8$,

$$Z_{22}(s)=100\,\frac{5.93004y^2+2.93046y+2.94786}{11.10494y^3+5.48771y^2+15.46231y+4.91310} \tag{9.246}$$

9.27 Design a lossless matching network to equalize the load impedance

$$z_2(s)=\frac{s^2+1}{s^2+s+1} \tag{9.247}$$

to a resistive generator and to achieve the transducer power-gain characteristic

$$G(\omega^2)=K\,\frac{\omega^4-2\omega^2+1}{\omega^4-\omega^2+1} \tag{9.248}$$

having a maximum attainable K. Verify that the desired equalizer back-end impedance is required to be

$$Z_{22}(s)=z_2(s) \tag{9.249}$$

9.28 Repeat Problem 9.24 for third-order Butterworth transducer power-gain characteristic. Realize the equalizer.

9.29 Repeat Problem 9.26 for fourth-order Chebyshev transducer power-gain characteristic. Obtain the lossless equalizer.

chapter 10 THEORY OF PASSIVE CASCADE SYNTHESIS

In Chapter 4, we have shown that the immittance (impedance or admittance) parameters obtained by Darlington's procedure are always compatible in representing a lossless two-port network. In fact, we demonstrated that they can be realized by either a series or parallel connection of component lossless two-port networks, each corresponding to a pair of poles on the real-frequency ($j\omega-$) axis, or the pole at the origin or infinity. At best, the realization requires as many physical transformers as there are pairs of poles. At worst, these transformers must be perfectly coupled, except the one at the output, which must be ideal.

As an alternative, in the present chapter we shall apply the method of cascade synthesis to the realization of a rational positive-real impedance according to Darlington theory. Our approach is based primarily on the work of Hazony [5] and Youla [10].

The idea underlying the method is the following. Consider the even part

$$\text{Ev } Z(s) = r(s) = \tfrac{1}{2}[Z(s) + Z(-s)] \tag{10.1}$$

of a given rational positive-real impedance $Z(s)$. As before, we first separate the numerator and denominator polynomials of $Z(s)$ into even and odd parts, and write

$$Z(s) = \frac{m_1 + n_1}{m_2 + n_2} \tag{10.2}$$

Then we have

$$r(s) = \frac{m_1 m_2 - n_1 n_2}{m_2^2 - n_2^2} \tag{10.3}$$

showing that if s_0 is a zero or pole of $r(s)$, so is $-s_0$. Thus, the zeros and poles of $r(s)$ possess quadrantal symmetry, being symmetric with respect to both the real and imaginary axes. More specifically, they may appear in pairs on the real axis, in pairs on the $j\omega$-axis, or in the form of sets of quadruplets in the complex-frequency plane. Furthermore, for positive-real impedance $Z(s)$, the $j\omega$-axis zeros are required to be of even multiplicity in order that $\operatorname{Re} Z(j\omega)=r(j\omega)$ never be negative.

Suppose that we can extract from the given impedance $Z(s)$ a set of open-circuit impedance parameters $z_{ij}(s)$ characterizing a component two-port network, as depicted in Figure 10.1, which produces one pair of real axis zeros, one pair of $j\omega$-axis zeros, or one set of quadruplet of zeros of $r(s)$, and leaves a rational positive-real impedance $Z_1(s)$ of lower degree, whose even part $r_1(s)$ is devoid of these zeros but contains all other zeros of $r(s)$. After a finite q steps, we arrive at a rational positive-real impedance $Z_q(s)$ whose even part $r_q(s)$ is devoid of zeros in the entire complex-frequency plane, meaning that the even part of $Z_q(s)$ must be a nonnegative constant, i.e.,

$$\begin{aligned} r_q(s) &= \tfrac{1}{2}[Z_q(s)+Z_q(-s)] \\ &= c \end{aligned} \tag{10.4}$$

c being a nonnegative constant. It is straightforward to show that $Z_q(s)$ must be of the form

$$Z_q(s)=Z_{LC}(s)+c \tag{10.5}$$

$Z_{LC}(s)$ being a reactance function. $Z_q(s)$ can then be realized as a lossless two-port network terminated in a c-ohm resistor, as shown in Figure 10.2.

To present the material in an orderly fashion, we divide the subject into the following sections.

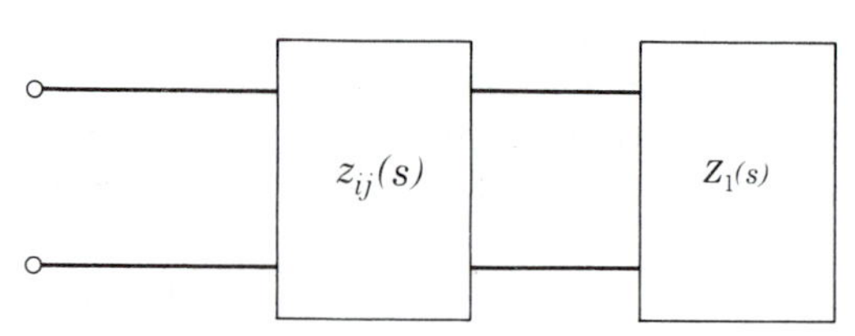

Figure 10.1

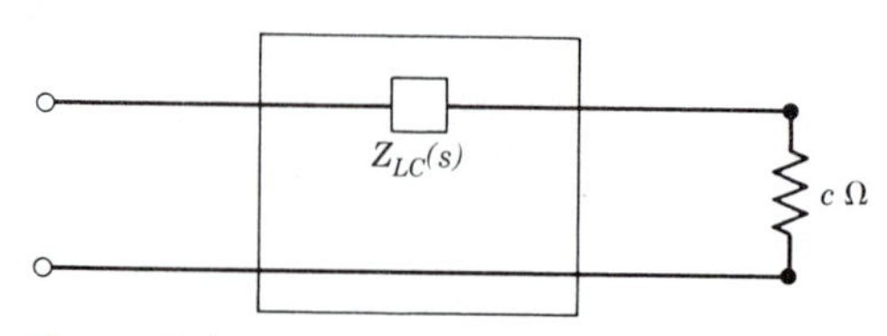

Figure 10.2

10.1 THE MAIN THEOREM

To motivate our discussion, we first present a theorem due to Richards [6], which is intimately tied up with the famous Bott–Duffin realization technique.†

Lemma 10.1 Let $Z(s)$ be a positive-real function that is neither of the form Ls nor $1/Cs$. Let k be an arbitrary positive-real constant. Then the *Richards' function*

$$W(s)=\frac{kZ(s)-sZ(k)}{kZ(k)-sZ(s)} \tag{10.6}$$

is also positive real.

Proof Consider the bilinear transformations

$$p(s)=\frac{s-k}{s+k}=\frac{s/k-1}{s/k+1} \qquad \text{or} \qquad s(p)=k\,\frac{1+p}{1-p} \tag{10.7}$$

$$f(p)=\frac{Z(s)-Z(k)}{Z(s)+Z(k)}=\frac{Z(s)/Z(k)-1}{Z(s)/Z(k)+1} \tag{10.8}$$

where $f(p)$ is considered as a function of p by replacing s on the right-hand side of (10.8) by its value in terms of p, as expressed in (10.7). It is easy to verify that, since s/k is positive real, (10.7) maps the open right-half of s-plane onto the interior of the unit circle in the p-plane. Similarly, (10.8) maps the open right-half of Z-plane onto the interior of the unit circle in the f-plane. By simple algebra, we can write

$$-\frac{f(p)}{p}=\frac{W(s)-1}{W(s)+1} \tag{10.9}$$

Now we need *Schwarz's lemma*‡ from the theory of functions of a complex variable, which states that if $f(p)$ is analytic on and within a circle of radius R and if its value vanishes at $p=0$, then at any point p inside the circle, except at the point $p=0$, the inequality

$$|f(p)|<|p|\,\frac{M}{R} \tag{10.10}$$

holds, M being the maximum value of $|f(p)|$ on the circle. At $p=0$, $f(p)=0$ so that the equality will hold. In fact, the equality can hold only if it holds identically, in which case $|f(p)|=|p|(M/R)$.

In the case of (10.8), $f(p)$ vanishes at $p=0$ or $s=k$. Thus, it satisfies the conditions of Schwarz's lemma. The radius of the circle in the p-plane is unity, i.e., $R=1$. On the circle, $|p|=1$, which corresponds to the $j\omega$-axis of the s-plane. Since $Z(s)$ is positive real, $\operatorname{Re} Z(j\omega)\geqq 0$, which corresponds to points on and within the unit circle in the

†R. Bott and R. J. Duffin, "Impedance synthesis without the use of transformers," *J. Appl. Phys.*, vol. 20, 1949, p. 816.

‡E. A. Guillemin 1949, *The Mathematics of Circuit Analysis*, pp. 327–330. Cambridge, MA: The M.I.T. Press.

f-plane. Thus, the maximum value of $|f(p)|$ on the unit circle in the p-plane is also unity, i.e., $M=1$. The inequality (10.10) becomes

$$|f(p)| \leqslant |p| \qquad \text{for } |p|<1 \tag{10.11}$$

Since (10.9) is again a bilinear transformation, the two variables being $W(s)$ and $-f(p)/p$, Re $W(s) \geqq 0$ if and only if $|f(p)/p| \leqslant 1$. Note that $p=0$ or $s=k$ is not a pole of $f(p)/p$ since $f(p)$ also vanishes there. It follows that $|p|<1$ implies Re $W(s) \geqq 0$. But $|p|<1$ is the map of Re $s>0$, showing that Re $s>0$ implies Re $W(s) \geqq 0$. In other words, Re $W(s)$ is a nonnegative and continuous function of σ and ω for all $s=\sigma+j\omega$ in the open RHS. As a limit of a continuous nonnegative function, Re $W(j\omega)$ must be nonnegative whenever it exists. Thus, $W(s)$ is positive real and the proof is completed.

The *degree* of a rational function is defined as the sum of the degrees of its relatively prime numerator and denominator polynomials. Using this definition, we see that if $Z(s)$ is rational, Richards' function $W(s)$ is also rational whose degree is not greater than that of $Z(s)$. Our approach to Darlington theory is to apply Richards' theorem twice, as will be demonstrated below. It was pointed out by Richards that if k can be chosen so that $r(k)=0$, $r(s)$ being the even part of $Z(s)$, then the degree of $W(s)$ is at least two less than that of $Z(s)$.

Recall that Richards' function contains one arbitrary positive-real constant k. In an attempt to extend his result to contain complex k, we can apply his theorem twice at the complex-conjugate points of an arbitrary constant as follows: Let

$$s_0=\sigma_0+j\omega_0 \tag{10.12}$$

be a point in the closed RHS. Then, according to Lemma 10.1, the function

$$\hat{W}_1(s)=\frac{s_0 Z(s)-sZ(s_0)}{s_0 Z(s_0)-sZ(s)} \tag{10.13}$$

is positive real if s_0 is positive real; and the function

$$W_1(s)=Z(s_0)\,\hat{W}_1(\bar{s}_0)\,\frac{\bar{s}_0\,\hat{W}_1(s)-s\,\hat{W}_1(\bar{s}_0)}{\bar{s}_0\,\hat{W}_1(\bar{s}_0)-s\,\hat{W}_1(s)} \tag{10.14}$$

is positive real if s_0 is a positive-real constant and $\hat{W}_1(s)$ is a positive-real function. Substituting (10.13) in (10.14) yields (Problem 10.2)

$$W_1(s)=\frac{D_1(s)Z(s)-B_1(s)}{-C_1(s)Z(s)+A_1(s)} \tag{10.15}$$

where

$$A_1(s)=q_4 s^2+|s_0|^2 \tag{10.16a}$$

$$B_1(s)=q_2 s \tag{10.16b}$$

$$C_1(s)=q_3 s \tag{10.16c}$$

$$D_1(s)=q_1 s^2+|s_0|^2 \tag{10.16d}$$

$$q_1 = \frac{R_0/\sigma_0 - X_0/\omega_0}{R_0/\sigma_0 + X_0/\omega_0} \tag{10.17a}$$

$$q_2 = \frac{2|Z_0|^2}{R_0/\sigma_0 + X_0/\omega_0} \tag{10.17b}$$

$$q_3 = \frac{2}{R_0/\sigma_0 - X_0/\omega_0} \tag{10.17c}$$

$$q_4 = 1/q_1 \tag{10.17d}$$

in which

$$Z(s_0) = R_0 + jX_0 \equiv Z_0 \tag{10.18}$$

In the case $\omega_0 = 0$, then X_0/ω_0 must be replaced by $Z'(\sigma_0)$:

$$X_0/\omega_0 \to Z'(\sigma_0) = \left.\frac{dZ(s)}{ds}\right|_{s=\sigma_0} \tag{10.19a}$$

For $\sigma_0 = 0$ and $R_0 = 0$, R_0/σ_0 must be replaced by $X'(\omega_0)$:

$$R_0/\sigma_0 \to X'(\omega_0) = \left.\frac{dZ(s)}{ds}\right|_{s=j\omega_0} \tag{10.19b}$$

Definition 10.1 ***Index set.*** For a given positive-real function $Z(s)$, let s_0 be any point in the open RHS or any finite nonzero point on the $j\omega$-axis where $Z(s)$ is analytic. Then the set of four real numbers q_1, q_2, q_3, and q_4, as defined in (10.17)–(10.19), is called the *index set* assigned to the point s_0 by the positive-real function $Z(s)$.

We shall illustrate this concept by the following examples.

Example 10.1 Determine the index set assigned to the point $s_0 = j1$ by the positive-real function

$$Z(s) = \frac{4s^2 + s + 2}{2s^2 + 4s + 4} \tag{10.20}$$

From definition and (10.20), we have

$$s_0 = 0 + j1 = \sigma_0 + j\omega_0 \tag{10.21a}$$

$$Z(s_0) = 0 + j0.5 = R_0 + jX_0 \equiv Z_0 \tag{10.21b}$$

Since $\sigma_0 = 0$ and $R_0 = 0$, (10.19b) applies, giving

$$X'(1) = 1.5 \tag{10.22}$$

Substituting these in (10.17) with $X'(1)$ replacing R_0/σ_0 yields

$$q_1 = (1.5 - 0.5)/(1.5 + 0.5) = 0.5 \tag{10.23a}$$

$$q_2 = 0.5/(1.5 + 0.5) = 0.25 \tag{10.23b}$$

$$q_3 = 2/(1.5 - 0.5) = 2 \tag{10.23c}$$

$$q_4 = 1/q_1 = 2 \tag{10.23d}$$

Thus, the set of four numbers 0.5, 0.25, 2, and 2 is the index set assigned to the point $s_0 = j1$ by the positive-real function $Z(s)$.

Example 10.2 Determine the index set assigned to the point

$$s_0 = 0.61139 + j1.02005 \tag{10.24}$$

by the positive-real function

$$Z(s) = \frac{6s^2 + 5s + 6}{2s^2 + 4s + 4} \tag{10.25}$$

as given in (4.21), where s_0 is an open RHS zero of (4.22).

First we compute the value of the function $Z(s)$ at s_0, which is given by

$$Z(s_0) = Z_0 = Z(0.61139 + j1.02005) = 1.56553 + j0.44814 \tag{10.26}$$

whose squared magnitude is obtained as

$$|Z_0|^2 = 2.65171 \tag{10.27}$$

Substituting these in (10.17) results in

$$q_1 = (2.56061 - 0.43933)/(2.56061 + 0.43933) = 0.70711 \tag{10.28a}$$

$$q_2 = 2 \times 2.65171/(2.56061 + 0.43933) = 1.76784 \tag{10.28b}$$

$$q_3 = 2/(2.56061 - 0.43933) = 0.94283 \tag{10.28c}$$

$$q_4 = 1/q_1 = 1.41421 \tag{10.28d}$$

These four numbers form the index set assigned to the point s_0 by the positive-real function $Z(s)$.

With these preliminaries, we now proceed to state the main theorem of this section, which forms the cornerstone of the method of cascade synthesis of a rational positive-real impedance according to Darlington theory. To exhibit the technique succinctly, we relegate the proof of the main theorem to Section 10.4 so that the realization procedure can be brought to a quick conclusion.

Theorem 10.1 Let $Z(s)$ be a positive-real function, which is neither of the form Ls nor $1/Cs$, L and C being real nonnegative constants. Let $s_0 = \sigma_0 + j\omega_0$ be a finite nonzero point in the closed RHS where $Z(s)$ is analytic, then the function

$$W_1(s) = \frac{D_1(s)Z(s) - B_1(s)}{-C_1(s)Z(s) + A_1(s)} \tag{10.29}$$

is positive real, where A_1, B_1, C_1, and D_1 are defined in (10.16) with $\{q_1, q_2, q_3, q_4\}$ being the index set assigned to the point s_0 by $Z(s)$. Furthermore, $W_1(s)$ possesses the following properties:

(i) If $Z(s)$ is rational, $W_1(s)$ is rational whose degree is not greater than that of

$Z(s)$, i.e.,

$$\text{degree } W_1(s) \leqslant \text{degree } Z(s) \tag{10.30}$$

(ii) If $Z(s)$ is rational and if s_0 is a zero of its even part $r(s)$ then

$$\text{degree } W_1(s) \leqslant \text{degree } Z(s) - 4, \qquad \omega_0 \neq 0 \tag{10.31a}$$

$$\text{degree } W_1(s) \leqslant \text{degree } Z(s) - 2, \qquad \omega_0 = 0 \tag{10.31b}$$

(iii) If $s_0 = \sigma_0 > 0$ is a real zero of $r(s)$ of at least multiplicity 2 and if $Z(s)$ is rational, then

$$\text{degree } W_1(s) \leqslant \text{degree } Z(s) - 4 \tag{10.32}$$

We remark that since $Z(s)$ is positive real, all the points in the open RHS are admissible. Any point on the real-frequency axis, exclusive of the origin and infinity, where $Z(s)$ is analytic is admissible as s_0. As before, we express the rational positive-real impedance

$$Z(s) = \frac{m_1 + n_1}{m_2 + n_2} \tag{10.33}$$

in terms of the even and odd parts of its numerator and denominator polynomials. Then the even part of $Z(s)$ is given by

$$r(s) = \frac{m_1 m_2 - n_1 n_2}{m_2^2 - n_2^2} \tag{10.34}$$

We note that the even part $r(s)$ is not the same as the real part of $Z(s)$, which is defined as

$$Z(s) = R(\sigma, \omega) + jX(\sigma, \omega) \tag{10.35}$$

However, they become identical on the $j\omega$-axis. Consider the impedance $Z(s) = s + 1/s$, whose real part is given by

$$R(\sigma, \omega) = \sigma + \sigma/(\sigma^2 + \omega^2) \tag{10.36}$$

and whose even part is $r(s) = 0$. On the $j\omega$-axis, they become identical, both being zero.

Example 10.3 Consider the impedance $Z(s)$ of (10.20), whose even part is given by

$$r(s) = \frac{2(s^2+1)^2}{s^4+4} \tag{10.37}$$

Thus, $s_0 = j1$ is a zero of $r(s)$. The index set assigned to s_0 by $Z(s)$ was computed earlier and its elements are given in (10.23). Substituting these in (10.29) in conjunction with (10.16) yields

$$\begin{aligned} W_1(s) &= \frac{(0.5s^2+1)(4s^2+s+2) - 0.25s(2s^2+4s+4)}{-2s(4s^2+s+2) + (2s^2+1)(2s^2+4s+4)} \\ &= \tfrac{1}{2} = Z(0) \end{aligned} \tag{10.38}$$

which is of degree zero, as expected from (10.31a).

Example 10.4 The zeros of the even part $r(s)$ of the impedance $Z(s)$ of (10.25) are the roots of the equation

$$m_1 m_2 - n_1 n_2 = 4(3s^4 + 4s^2 + 6) = 0 \tag{10.39}$$

Using quadratic formula, it is straightforward to show that $s_0 = 0.61139 + j1.02005$, as given in (10.24), is an open RHS zero of $r(s)$. The index set assigned to this point s_0 by $Z(s)$ was computed in Example 10.2, whose elements are listed in (10.28). Substituting these element values in (10.29) in conjunction with (10.16) gives

$$\begin{aligned} W_1(s) &= \frac{(0.70711s^2 + 1.41430)(6s^2 + 5s + 6) - 1.76784s(2s^2 + 4s + 4)}{-0.94283s(6s^2 + 5s + 6) + (1.41421s^2 + 1.41430)(2s^2 + 4s + 4)} \\ &= 1.5 = Z(0) \end{aligned} \tag{10.40}$$

which is of degree 0, confirming (10.31a).

Example 10.5 In the design of a broadband equalizer, it is required to realize the positive-real impedance (Chen [2], p. 302)

$$Z_{22}(s) = \frac{100.927s + 58.388}{41.613s + 35.403} \tag{10.41}$$

as the input impedance of a reciprocal lossless two-port network terminated in a 1-Ω resistor. To simplify our notation, let

$$Z(s) = 0.41231 Z_{22}(s) = \frac{s + 0.57852}{s + 0.85077} \tag{10.42}$$

whose even part is given by

$$r(s) = \frac{s^2 - 0.49218}{s^2 - 0.72381} \tag{10.43}$$

Thus, $\sigma_0 = 0.70156$ is an open RHS zero of $r(s)$. Substituting this value in (10.42) and its derivative yields

$$Z(\sigma_0) = Z(0.70156) = Z_0 = R_0 = 0.82462 \tag{10.44a}$$

$$Z'(\sigma_0) = Z'(0.70156) = 0.11298 \tag{10.44b}$$

Substituting these in (10.17) with $Z'(\sigma_0)$ replacing X_0/ω_0 gives

$$q_1 = (1.17541 - 0.11298)/(1.17541 + 0.11298) = 0.82462 \tag{10.45a}$$

$$q_2 = 2 \times 0.68/(1.17541 + 0.11298) = 1.05557 \tag{10.45b}$$

$$q_3 = 2/(1.17541 - 0.11298) = 1.88248 \tag{10.45c}$$

$$q_4 = 1/q_1 = 1.21268 \tag{10.45d}$$

They are the elements of the index set assigned to the point $s_0 = \sigma_0 = 0.70156$ by $Z(s)$.

As before, using (10.29) we can compute the associated function

$$W_1(s)=\frac{(0.82462s^2+0.49218)(s+0.57852)-1.05557s(s+0.85077)}{-1.88248s(s+0.57852)+(1.21268s^2+0.49218)(s+0.85077)}$$

$$=0.68=Z(0) \tag{10.46}$$

whose degree is again zero, a reduction of two from that of $Z(s)$. This confirms the result stated in (10.31b). A network realization of $Z(s)$ will be presented in Example 10.8.

In the preceding examples, we have selected s_0 to be a closed RHS zero of the even part $r(s)$ of $Z(s)$. As indicated in the theorem, this is not necessary for the associated function $W_1(s)$ to be positive real. The following example illustrates this by two different choices of s_0.

Example 10.6 Let $s_0=1+j1$ be an open RHS point in the s-plane. Determine the associated function $W_1(s)$ of the positive-real impedance

$$Z(s)=\frac{s+1}{s+2} \tag{10.47}$$

First, we compute the index set assigned to the point $s_0=1+j1$ by $Z(s)$, giving

$$q_1=(0.7-0.1)/(0.7+0.1)=3/4 \tag{10.48a}$$

$$q_2=1/(0.7+0.1)=5/4 \tag{10.48b}$$

$$q_3=2/(0.7-0.1)=10/3 \tag{10.48c}$$

$$q_4=1/q_1=4/3 \tag{10.48d}$$

where

$$Z_0=Z(1+j1)=0.7+j0.1=2^{-1/2}e^{j8°8'} \tag{10.49}$$

Substituting these in (10.29) in conjunction with (10.16) yields

$$W_1(s)=\frac{(3s^2/4+2)(s+1)-(5s/4)(s+2)}{-(10s/3)(s+1)+(4s^2/3+2)(s+2)}$$

$$=\frac{3(3s^3-2s^2-2s+8)}{8(2s^3-s^2-2s+6)}=\frac{9s+12}{16s+24} \tag{10.50}$$

which is clearly positive real, confirming Theorem 10.1. However, its degree is the same as that of $Z(s)$, since $s_0=1+j1$ is not a zero of $r(s)$.

Now suppose we choose $s_0=\sigma_0=\sqrt{2}$, an open RHS zero of $r(s)$. Since $\omega_0=0$, to compute the index set assigned to σ_0 by $Z(s)$ by (10.17), we must replace X_0/ω_0 by $Z'(\sigma_0)$, giving

$$q_1=(3+2\sqrt{2})/(4+3\sqrt{2}) \tag{10.51a}$$

$$q_2=(41+29\sqrt{2})/(24+17\sqrt{2}) \tag{10.51b}$$

$$q_3=(14+10\sqrt{2})/(3+2\sqrt{2}) \tag{10.51c}$$

$$q_4=(4+3\sqrt{2})/(3+2\sqrt{2}) \tag{10.51d}$$

where

$$Z(\sigma_0)=Z(\sqrt{2})=(1+\sqrt{2})/(2+\sqrt{2})=R_0 \tag{10.52a}$$

$$Z'(\sigma_0)=1/(2+\sqrt{2})^2=1/(6+4\sqrt{2}) \tag{10.52b}$$

Substituting these in (10.29) in conjunction with (10.16) yields

$$\begin{aligned} W_1(s)&=\frac{(q_1s^2+2)(s+1)-q_2s(s+2)}{-q_3s(s+1)+(q_4s^2+2)(s+2)}\\ &=\frac{(3+2\sqrt{2})[(140+99\sqrt{2})s+198+140\sqrt{2}]}{(198+140\sqrt{2})[(4+3\sqrt{2})s+6+4\sqrt{2}]}\\ &=0.5=Z(0) \end{aligned} \tag{10.53}$$

As expected from (10.31b), the degree of $W_1(s)$, being zero, is two less than that of $Z(s)$. We remark that the irrational terms have been retained in order to produce exact cancellations in (10.53). This is in contrast to the two preceding examples where only five significant digits are retained.

10.2 NONRECIPROCAL REALIZATION

After a detailed discussion of the main theorem, we are now in a position to realize the desired lossless two-port network. First we show that there exists a nonreciprocal realization, and then in the following section we demonstrate that by applying the main theorem twice, a lossless reciprocal two-port network results.

Our starting point is (10.29), which after solving $Z(s)$ in terms of $W_1(s)$ yields

$$Z(s)=\frac{A_1(s)W_1(s)+B_1(s)}{C_1(s)W_1(s)+D_1(s)} \tag{10.54}$$

Physically, the impedance $Z(s)$ can be interpreted as the input impedance of a two-port network N_1, which is characterized by its transmission matrix

$$\mathbf{T}_1(s)=\begin{bmatrix} A_1(s) & B_1(s) \\ C_1(s) & D_1(s) \end{bmatrix} \tag{10.55}$$

terminated in $W_1(s)$, as depicted in Figure 10.3. To see this, we first compute the corresponding impedance matrix $\mathbf{Z}_1(s)$ of N_1 from $\mathbf{T}_1(s)$, giving

$$\mathbf{Z}_1(s)=\begin{bmatrix} z_{11} & z_{12} \\ z_{21} & z_{22} \end{bmatrix}=\begin{bmatrix} A_1/C_1 & (A_1D_1-B_1C_1)/C_1 \\ 1/C_1 & D_1/C_1 \end{bmatrix} \tag{10.56}$$

The input impedance $Z_{11}(s)$ of N_1 with the output port terminating in $W_1(s)$ is ob-

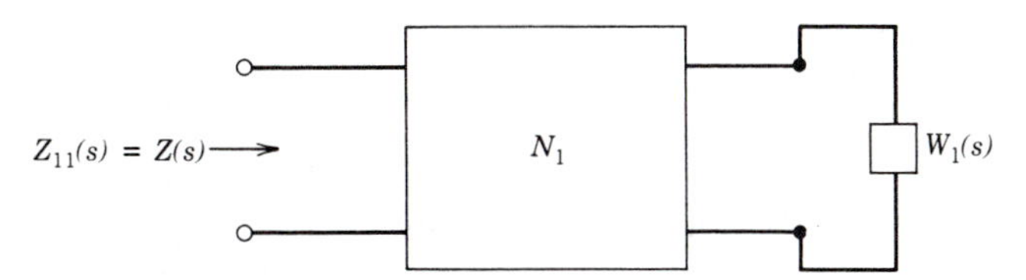

Figure 10.3

tained as

$$Z_{11}(s)=z_{11}-\frac{z_{12}z_{21}}{z_{22}+W_1}=\frac{A_1W_1+B_1}{C_1W_1+D_1}=Z(s) \tag{10.57}$$

Substituting (10.16) in (10.55), the determinant of the transmission matrix is given by

$$\det \mathbf{T}_1(s)=A_1D_1-B_1C_1=s^4+2(\omega_0^2-\sigma_0^2)s^2+|s_0|^4 \tag{10.58}$$

It is significant to note that det $\mathbf{T}_1(s)$ depends only upon the point s_0 and not on $Z(s)$. Observe that the input impedance remains unaltered if each element of $\mathbf{T}_1(s)$ is multiplied or divided by a nonzero finite quantity. To complete the realization, we must now demonstrate that the two-port network N_1 is physically realizable.

10.2.1 Type-*E* Section

Consider the lossless nonreciprocal two-port network of Figure 10.4, which is referred to as the *type-E section*. Our objective is to show that this realizes N_1.

It is straightforward to compute the impedance matrix $\mathbf{Z}_E(s)$ of the type-*E*

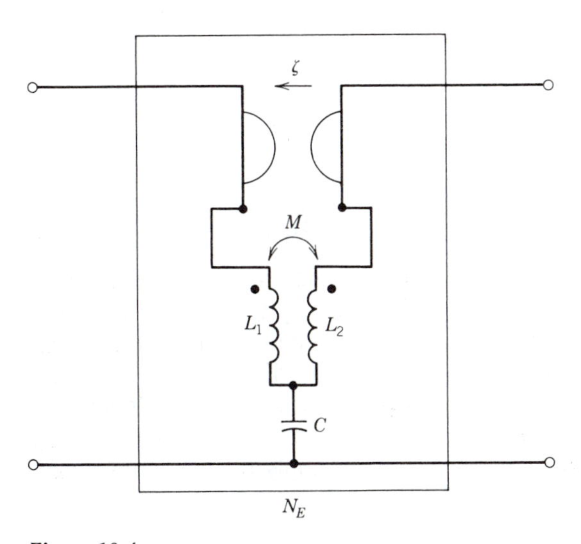

Figure 10.4

section of Figure 10.4, yielding (Problem 10.3)

$$\mathbf{Z}_E(s)=\begin{bmatrix} L_1 s+1/Cs & Ms+1/Cs+\zeta \\ Ms+1/Cs-\zeta & L_2 s+1/Cs \end{bmatrix} \tag{10.59}$$

where $M^2=L_1L_2$, whose determinant is given by

$$\det \mathbf{Z}_E(s)=(L_1+L_2-2M+\zeta^2 C)/C \tag{10.60}$$

a constant independent of s due to perfect coupling. From the impedance matrix $\mathbf{Z}_E(s)$, the transmission matrix $\mathbf{T}_E(s)$ of the type-E section is obtained as

$$\mathbf{T}_E(s)=\frac{1}{P(s)}\begin{bmatrix} L_1 Cs^2+1 & (L_1+L_2-2M+\zeta^2 C)s \\ Cs & L_2 Cs^2+1 \end{bmatrix} \tag{10.61a}$$

where

$$P(s)=MCs^2-\zeta Cs+1 \tag{10.61b}$$

To show that the type-E section realizes N_1, we divide each element of $\mathbf{T}_1(s)$ (10.55) by $|s_0|^2 P(s)$. This manipulation will not affect the input impedance $Z(s)$ but it will result in a transmission matrix having the form of $\mathbf{T}_E(s)$. Comparing this new matrix with (10.61a) in conjunction with (10.16) yields the following identifications:

$$L_1 C=q_4/|s_0|^2 \tag{10.62a}$$

$$L_1+L_2-2M+\zeta^2 C=q_2/|s_0|^2 \tag{10.62b}$$

$$C=q_3/|s_0|^2 \tag{10.62c}$$

$$L_2 C=q_1/|s_0|^2 \tag{10.62d}$$

Solving these equations for the element values of the type-E section gives

$$L_1=q_4/q_3=1/q_1q_3 \tag{10.63a}$$

$$L_2=q_1/q_3 \tag{10.63b}$$

$$C=q_3/|s_0|^2 \tag{10.63c}$$

$$M=1/q_3 \tag{10.63d}$$

$$\zeta=\pm 2\sigma_0/q_3 \tag{10.63e}$$

To show that all these elements are physical, it suffices to prove that the elements of the index set assigned to the point s_0 by the rational positive-real function $Z(s)$ are positive and finite for all admissible points s_0 in the closed RHS except at those admissible points $s_0=j\omega_0$ on the $j\omega$-axis where $R_0\neq 0$. Clearly, at any such point, R_0/σ_0 becomes infinity and $q_1=q_4=1$ and $q_2=q_3=0$. Under this situation, $W_1(s)=Z(s)$ and the corresponding two-port network N_1 degenerates into a pair of wires.

To justify the above assertion, we apply the well-known positive-real criterion (Chen [2], p. 37)

$$|\arg Z(s)|<|\arg s|, \qquad \operatorname{Re} s>0 \tag{10.64}$$

At $s=s_0=\sigma_0+j\omega_0$, $Z(s_0)=R_0+jX_0$ and the inequality (10.64) becomes

$$\frac{R_0}{\sigma_0}>\left|\frac{X_0}{\omega_0}\right|, \qquad \omega_0\neq 0 \text{ and } \sigma_0>0 \tag{10.65a}$$

$$\frac{R_0}{\sigma_0}>|Z'(\sigma_0)|, \qquad \omega_0=0 \text{ and } \sigma_0>0 \tag{10.65b}$$

$$X'(\omega_0)>\left|\frac{X_0}{\omega_0}\right|, \qquad s_0=j\omega_0 \text{ and } R_0=0 \tag{10.65c}$$

Thus, all the elements of the index set are finite and positive.

Appealing to Theorem 10.1 shows that if s_0 is chosen to be a complex open RHS zero of the even part $r(s)$ of $Z(s)$, the type-E section is capable of extracting a set of quadrantal zeros of $r(s)$ and leads to at least a four-degree reduction. For zeros of $r(s)$ on the $j\omega$-axis, the type-E section degenerates into other types of sections, to be presented shortly.

We illustrate the above procedure by the following example.

Example 10.7 We wish to realize the impedance

$$Z(s)=\frac{6s^2+5s+6}{2s^2+4s+4} \tag{10.66}$$

The elements of the index set assigned to the open RHS zero $s_0=0.61139+j1.02005$, as in (10.24), of $r(s)$ by $Z(s)$ were computed in Example 10.2, and are repeated below:

$$q_1=0.70711, \qquad q_2=1.76784 \tag{10.67a}$$

$$q_3=0.94283, \qquad q_4=1.41421 \tag{10.67b}$$

Substituting these in (10.63) yields the element values of the type-E section as

$$L_1=1/q_1q_3=1.500 \text{ H} \tag{10.68a}$$

$$L_2=q_1/q_3=0.750 \text{ H} \tag{10.68b}$$

$$C=q_3/|s_0|^2=0.667 \text{ F} \tag{10.68c}$$

$$M=1/q_3=1.061 \text{ H} \tag{10.68d}$$

$$\zeta=\pm 2\sigma_0/q_3=\pm 1.297\,\Omega \tag{10.68e}$$

where

$$|s_0|^2=1.41430 \tag{10.69}$$

The type-E section together with its terminating impedance $W_1(s)=1.5\,\Omega$, as computed in (10.40), is presented in Figure 10.5. We remark that since $W_1(s)$ is known to be a constant after a four-degree reduction from $Z(s)$, it can be determined outright from $Z(s)$ by the relation $W_1(s)=W_1(0)=Z(0)$.

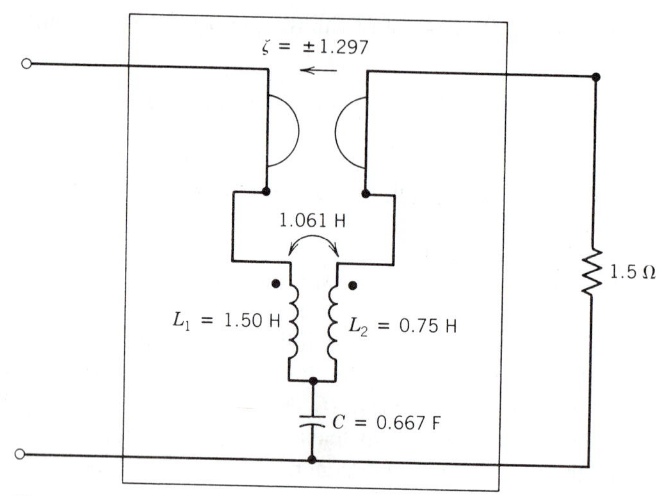

Figure 10.5

We now proceed to consider special sections to which the type-E section degenerates when $s_0 = \sigma_0$ or $j\omega_0$.

10.2.2 Degenerate Type-*E* Sections: The Type-*A*, Type-*B*, Type-*C*, and Brune Sections

For $s_0 = \sigma_0$ or $j\omega_0$, the type-E section degenerates into other familiar types of sections in network synthesis, depending upon the values of σ_0 and ω_0: the type-A section, the type-B section, the type-C section, and the Brune section, each realizing a special type of zeros of $r(s)$. To facilitate our discussion, three cases are distinguished.

Case 1: $s_0 = \sigma_0 > 0$. In solving the equations (10.62) to obtain the element values of the type-E section, we choose $M = +1/q_3$ (Problem 10.7). For $s_0 = \sigma_0 > 0$, $M = -1/q_3$ is also a permissible solution. For this choice of negative mutual inductance M, the gyrator of the type-E section can be avoided. For, under the stipulated conditions, we have $\zeta = 0$ and

$$L_1 = q_4/q_3 = 1/q_1 q_3 \tag{10.70a}$$

$$L_2 = q_1/q_3 \tag{10.70b}$$

$$C = q_3/\sigma_0^2 \tag{10.70c}$$

$$M = -1/q_3 < 0 \tag{10.70d}$$

their derivations being left as an exercise (Problem 10.8). In explicit form, they are listed in Problem 10.9. Under this situation, the type-E section degenerates to the *Darlington type-C section*, as shown in Figure 10.6. Thus, the type-C section is capable of extracting any positive σ-axis zero of $r(s)$, yielding at least a two-degree reduction.

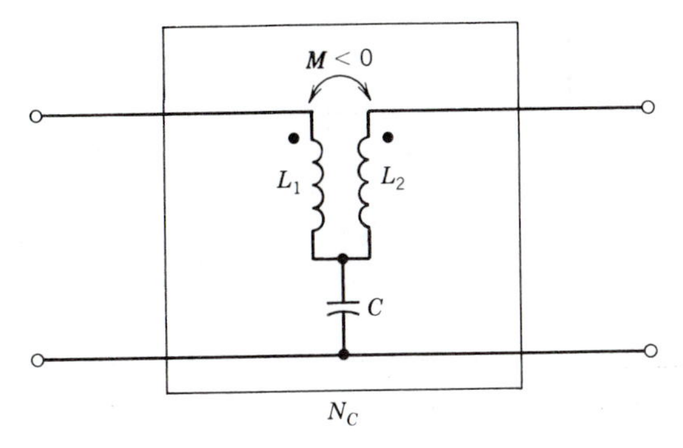

Figure 10.6

According to Theorem 10.1, if the zero is of multiplicity two or more, the type-C section is capable of diminishing the degree by four.

We demonstrate this by considering the following examples.

Example 10.8 As computed in Example 10.5, the four numbers

$$q_1 = 0.82462, \qquad q_2 = 1.05557 \tag{10.71a}$$

$$q_3 = 1.88248, \qquad q_4 = 1.21268 \tag{10.71b}$$

are the elements of the index set assigned to the open RHS zero $s_0 = \sigma_0 = 0.70156$ of $r(s)$ by the positive-real impedance

$$Z(s) = 0.41231 Z_{22}(s) = \frac{s + 0.57852}{s + 0.85077} \tag{10.72}$$

as given in (10.42). Substituting these in (10.70) yields the element values of the type-C section:

$$L_1 = 1/q_1 q_3 = 0.64419 \text{ H} \tag{10.73a}$$

$$L_2 = q_1/q_3 = 0.43805 \text{ H} \tag{10.73b}$$

$$C = q_3/\sigma_0^2 = 3.82475 \text{ F} \tag{10.73c}$$

$$M = -1/q_3 = -0.53121 \text{ H} \tag{10.73d}$$

As indicated in (10.46), the terminating impedance $W_1(s)$ of the type-C section, being equal to $Z(0)$, is 0.68 Ω. The complete network realizing the impedance $Z(s)$ is shown in Figure 10.7. Raising the impedance level by a factor of 2.42536 results in the network realization of the equalizer back-end impedance $Z_{22}(s)$, as stated in (10.41).

Example 10.9 The elements of the index set assigned to the open RHS zero $s_0 = \sigma_0 = \sqrt{2}$ of $r(s)$ by the positive-real impedance

$$Z(s) = \frac{s+1}{s+2} \tag{10.74}$$

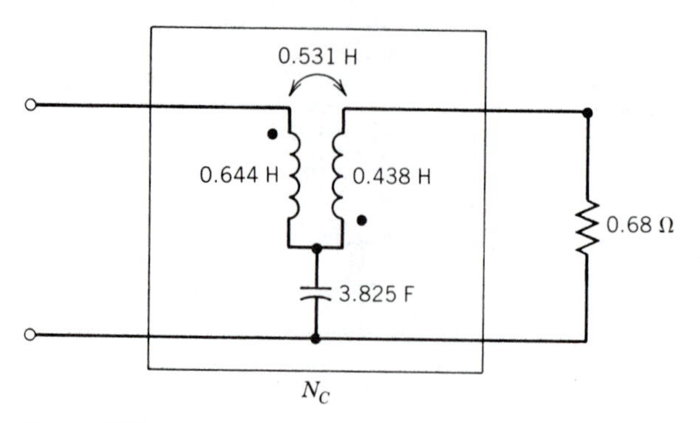

Figure 10.7

are listed in (10.51). Substituting them in (10.70) gives the desired values of the elements of the type-C section:

$$L_1=(4+3\sqrt{2})/(14+10\sqrt{2})=0.29290 \text{ H} \tag{10.75a}$$

$$L_2=(17+12\sqrt{2})/(116+82\sqrt{2})=0.14645 \text{ H} \tag{10.75b}$$

$$C=(7+5\sqrt{2})/(3+2\sqrt{2})=2.41421 \text{ F} \tag{10.75c}$$

$$M=-(3+2\sqrt{2})/(14+10\sqrt{2})=-0.20711 \text{ H} \tag{10.75d}$$

The type-C section together with its termination is presented in Figure 10.8.

Example 10.10 Consider the positive-real impedance

$$Z(s)=\frac{s^2+9s+8}{s^2+2s+2} \tag{10.76}$$

whose even part is given by

$$r(s)=\frac{(s^2-4)^2}{s^4+4} \tag{10.77}$$

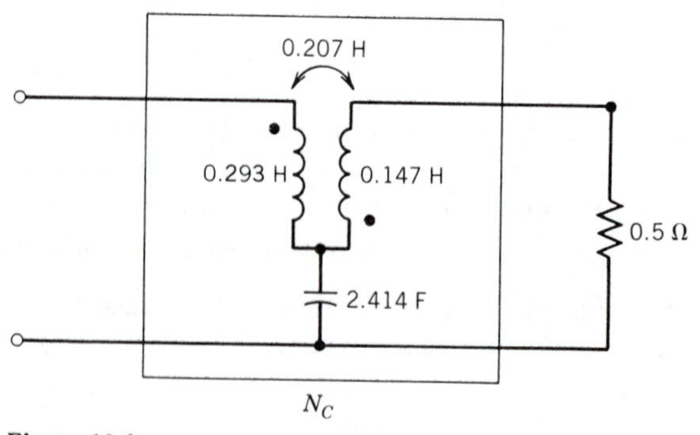

Figure 10.8

Thus, $s_0=\sigma_0=2$ is an open RHS zero of $r(s)$ of multiplicity 2. The elements of the index set assigned to the point $\sigma_0=2$ by $Z(s)$ are obtained as (Problem 10.10)

$$q_1=2, \qquad q_2=18, \qquad q_3=1, \qquad q_4=0.5 \tag{10.78}$$

yielding

$$L_1=0.5\text{ H}, \qquad L_2=2\text{ H} \tag{10.79a}$$

$$C=0.25\text{ F}, \qquad M=-1\text{ H} \tag{10.79b}$$

According to (10.32) of Theorem 10.1, the associated function $W_1(s)$ is at least four degrees less than $Z(s)$, meaning that $W_1(s)$ is a constant. Referring to (10.29), we have (Problem 10.11)

$$W_1(s)=W_1(0)=Z(0)=4\,\Omega \tag{10.80}$$

The complete network realization of the impedance $Z(s)$ is presented in Figure 10.9.

Case 2: $s_0=j\omega_0$ and $R_0=0$. In this case, R_0/σ_0 must be replaced by $X'(\omega_0)$, as indicated in (10.19b), and the gyrator in the type-*E* section can be avoided since from (10.63e), $\zeta=0$. The type-*E* section degenerates into the network of Figure 10.10, which is known as the *Brune section*, first given by Brune.† In explicit form, the element values of the Brune section are given by (Problem 10.13)

$$L_1=\frac{q_4}{q_3}=\frac{1}{q_1q_3}=\frac{\omega_0X'(\omega_0)+X_0}{2\omega_0} \tag{10.81a}$$

$$L_2=\frac{q_1}{q_3}=\frac{[\omega_0X'(\omega_0)-X_0]^2}{2\omega_0[\omega_0X'(\omega_0)+X_0]} \tag{10.81b}$$

$$C=\frac{q_3}{|s_0|^2}=\frac{2}{\omega_0[\omega_0X'(\omega_0)-X_0]} \tag{10.81c}$$

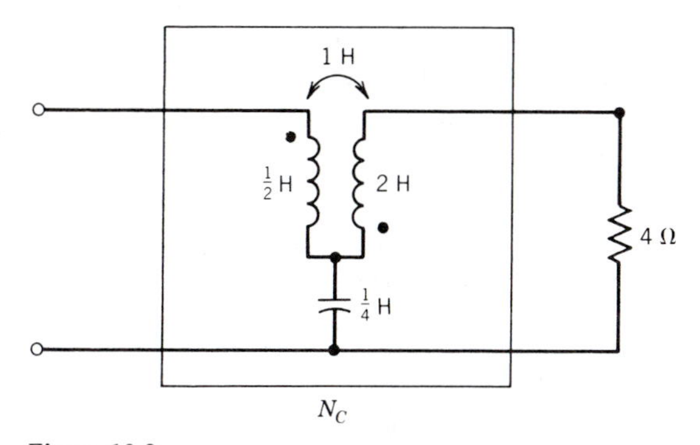

Figure 10.9

†O. Brune, "Synthesis of a finite two-terminal network whose driving-point impedance is a prescribed function of frequency," *J. Math. and Phys.*, vol. 10, 1931, pp. 191–236.

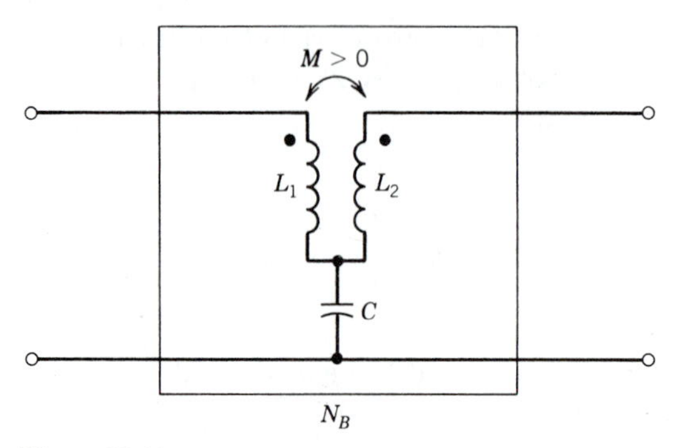

Figure 10.10

$$M = \frac{1}{q_3} = \frac{\omega_0 X'(\omega_0) - X_0}{2\omega_0} > 0 \tag{10.81d}$$

These formulas permit the determination of the element values of the Brune section directly from the given impedance $Z(s)$ without even going through the intermediate step of first computing the index set.

In particular, if $X_0 = 0$ or $Z(j\omega_0) = 0$, the Brune section of Figure 10.10 degenerates into the two-port network of Figure 10.11, whose element values are given by

$$L = \tfrac{1}{2} X'(\omega_0) \tag{10.82a}$$

$$C = \frac{2}{\omega_0^2 X'(\omega_0)} \tag{10.82b}$$

In the limit as ω_0 approaches zero, the degenerate Brune section of Figure 10.11 goes into the network of Figure 10.12, which is called the *type-A section*. The inductance is determined by the formula

$$L = \tfrac{1}{2} Z'(0) \tag{10.83}$$

As ω_0 approaches infinity, Figure 10.11 collapses into the network of Figure 10.13, which is known as the *type-B section*. Its capacitance is given by (Problem 10.14)

$$2/C = \lim_{s \to \infty} sZ(s) \tag{10.84}$$

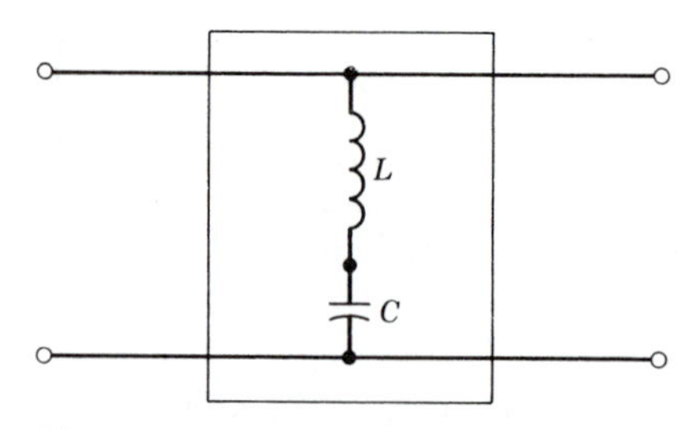

Figure 10.11

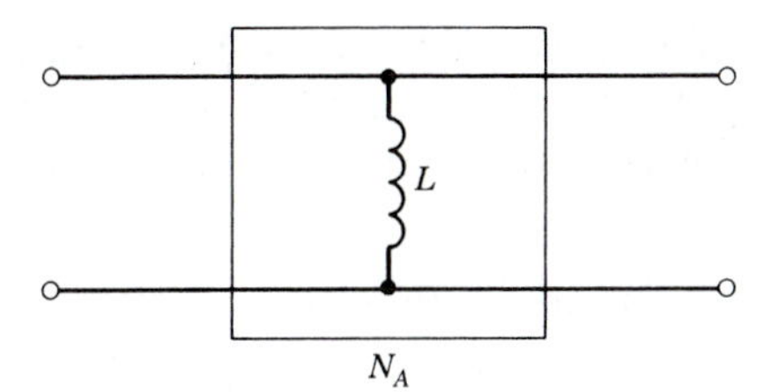

Figure 10.12

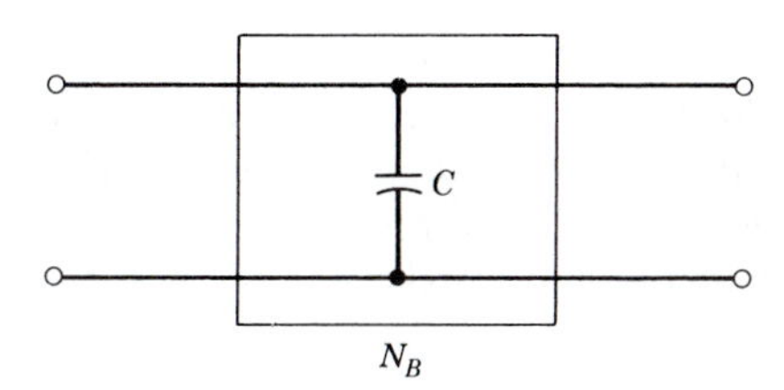

Figure 10.13

Thus, we conclude that the Brune section is capable of extracting any real-frequency zero $s_0 = j\omega_0$ of $r(s)$, and leads to at least a four-degree reduction if ω_0 is nonzero and finite, a two-degree reduction otherwise. The latter corresponds to the type-A or type-B section. (Also see Problem 10.16.)

Case 3: $s_0 = j\omega_0$ and $R_0 \neq 0$. As mentioned earlier, under this situation, R_0/σ_0 is infinity and

$$q_1 = q_4 = 1, \qquad q_2 = q_3 = 0 \tag{10.85}$$

The two-port network degenerates into a pair of wires.

We shall illustrate the Brune section by the following example.

Example 10.11 Consider the positive-real impedance

$$Z(s) = \frac{4s^2 + s + 2}{2s^2 + 4s + 4} \tag{10.86}$$

The elements of the index set assigned to the real-frequency zero $s_0 = j1$ of $r(s)$ by $Z(s)$ were computed in Example 10.1 and are given by

$$q_1 = 0.5, \qquad q_2 = 0.25, \qquad q_3 = 2, \qquad q_4 = 2 \tag{10.87}$$

Substituting these in (10.81) yields the element values of the Brune section:

$$L_1 = 1/q_1 q_3 = 1\ \text{H}, \qquad L_2 = q_1/q_3 = 0.25\ \text{H} \tag{10.88a}$$

$$C = q_3/\omega_0^2 = 2\ \text{F}, \qquad M = 1/q_3 = 0.5\ \text{H} \tag{10.88b}$$

Since ω_0 is nonzero and finite, according to (10.31a) a four-degree reduction is expected, meaning that the terminating impedance $W_1(s)$ must be of degree zero, a nonnegative constant. This resistance can be determined directly from $Z(s)$ by the relation

$$W_1(s) = W_1(0) = Z(0) = 0.5\ \Omega \tag{10.89}$$

The complete network realization of $Z(s)$ is presented in Figure 10.14.

10.2.3 **The Richards Section**

Before we turn our attention to the reciprocal realization of the two-port network N_1, we show that any positive zero $s_0 = \sigma_0$ of $r(s)$, in addition to being realized by the reciprocal Darlington type-C section, can also be realized by a nonreciprocal section called the *Richards section*, as shown in Figure 10.15.

Let $Z(s)$ be a given rational positive-real function. Then according to Lemma 10.1, for any positive σ_0 the function

$$W_1(s) = Z(\sigma_0)\,\frac{\sigma_0 Z(s) - sZ(\sigma_0)}{\sigma_0 Z(\sigma_0) - sZ(s)} \tag{10.90}$$

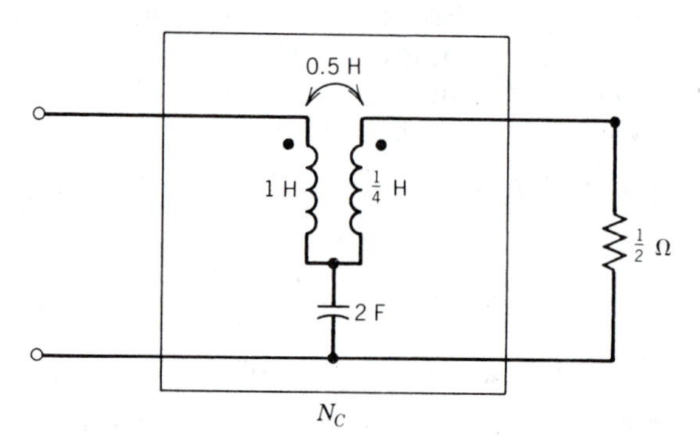

Figure 10.14

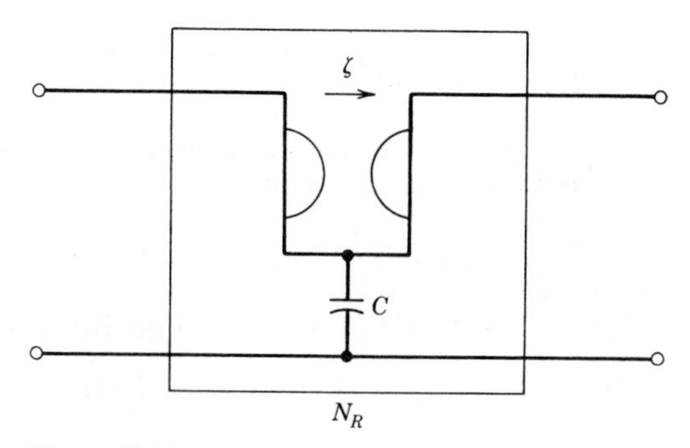

Figure 10.15

is also rational and positive real, whose degree is not greater than that of $Z(s)$. As pointed out by Richards,† if σ_0 is a zero of $r(s)$ then

$$\text{degree } W_1(s) \leqslant \text{degree } Z(s) - 2 \tag{10.91}$$

Inverting (10.90) for $Z(s)$ yields

$$Z(s) = \frac{\sigma_0 W_1(s) + sZ(\sigma_0)}{sW_1(s)/Z(\sigma_0) + \sigma_0} \tag{10.92}$$

Following (10.57), it is straightforward to show that this impedance can be realized by the Richards section terminated in $W_1(s)$, as indicated in Figure 10.16 with the element values being identified as

$$C = \frac{1}{\sigma_0 Z(\sigma_0)} \tag{10.93a}$$

$$\zeta = \pm Z(\sigma_0) \tag{10.93b}$$

The details of the derivation are left as an exercise (Problem 10.17).

We remark that the Richards section is capable of extracting any real and positive zero of $r(s)$ and leads to at least a two-degree reduction, while the type-C section achieves the same goal but reduces the degree by at least four if the zero is of multiplicity two or more, as indicated in (10.32).

Example 10.12 Suppose that we wish to realize the positive-real impedance $Z(s)$ given in (10.76) by the Richards sections. Then from (10.77) we see that $s_0 = \sigma_0 = 2$ is a real and positive zero of $r(s)$ of multiplicity 2. The element values of the Richards section are computed by (10.93), and are given by

$$C = 1/6 \text{ F}, \qquad \zeta = \pm 3\,\Omega \tag{10.94}$$

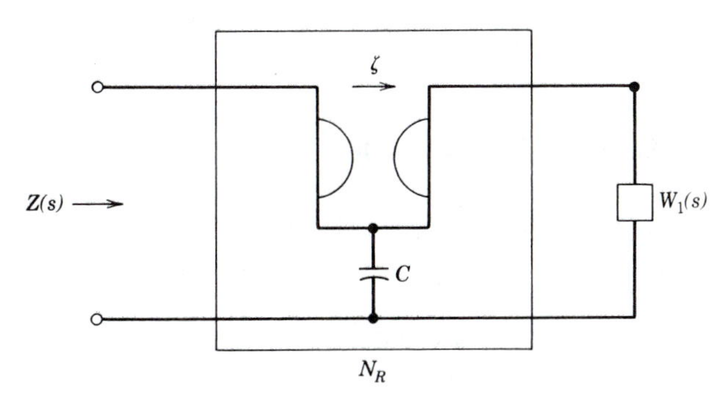

Figure 10.16

†P. I. Richards, "A special class of functions with positive real part in half-plane," *Duke Math. J.*, vol. 14, 1947, pp. 114–120.
P. I. Richards, "General impedance-function theory," *Quart. Appl. Math.*, vol. 6, 1948, pp. 21–29.

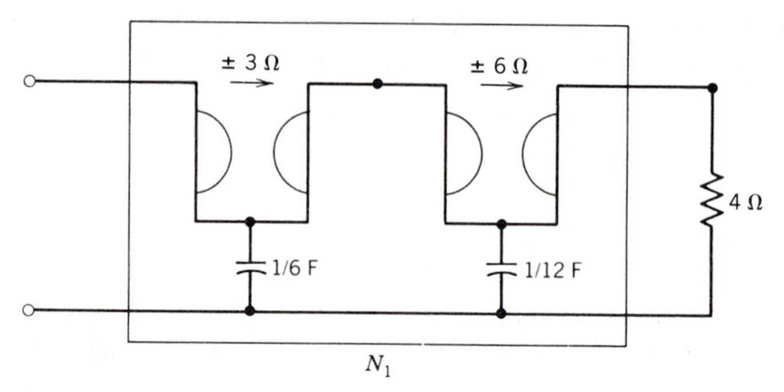

Figure 10.17

where $Z(\sigma_0)=Z(2)=3$. The terminating impedance $W_1(s)$ is determined by (10.90), yielding

$$W_1(s)=3\,\frac{2(s^2+9s+8)-3s(s^2+2s+2)}{6(s^2+2s+2)-s(s^2+9s+8)}$$

$$=\frac{3(3s+4)}{s+3} \tag{10.95}$$

a two-degree reduction from $Z(s)$. This impedance is clearly positive real and can be realized by another Richards section terminated in a resistor of 4 Ω. The complete network realization of $Z(s)$ is presented in Figure 10.17. It is of interest to observe that the even part of $W_1(s)$ possesses a real and positive zero at $s_0=\sigma_0=2$, as expected. This zero is realized by the second Richards section as shown in Figure 10.17. Comparing the network realization of Figure 10.17 with that of Figure 10.9 confirms an earlier assertion that if the positive σ-axis zero of $r(s)$ is of multiplicity two or more, the type-C section is capable of diminishing the degree by at least four, while the Richards section can only reduce it by at least two.

10.3 RECIPROCAL REALIZATION

In the foregoing, we have shown that the lossless two-port network N_1 can be realized by lossless nonreciprocal type-E sections, which degenerate into the classical type-A, type-B, type-C, and Brune sections when the zero s_0 of $r(s)$ is restricted to the $j\omega$-axis or the positive σ-axis. In the present section, we demonstrate that N_1 can also be realized by a lossless reciprocal two-port network by the application of the main theorem twice.

Our starting point is (10.29). According to Theorem 10.1, if $Z(s)$ is rational positive real and s_0 is chosen to be one of the zeros of $r(s)$, then $Z(s)$ is describable as in (10.54), $W_1(s)$ being rational positive real and at least four or two degrees less than $Z(s)$, depending on whether $\omega_0\neq 0$ or $\omega_0=0$. Now we apply Theorem 10.1 to $W_1(s)$

at the same point s_0. Then by (10.29) the function

$$W_2(s)=\frac{D_2(s)W_1(s)-B_2(s)}{-C_2(s)W_1(s)+A_2(s)} \tag{10.96}$$

is rational positive real, whose degree cannot exceed that of $W_1(s)$, being at least two or four degrees less than that of $Z(s)$, where

$$A_2(s)=p_4s^2+|s_0|^2 \tag{10.97a}$$

$$B_2(s)=p_2s \tag{10.97b}$$

$$C_2(s)=p_3s \tag{10.97c}$$

$$D_2(s)=p_1s^2+|s_0|^2 \tag{10.97d}$$

$\{p_1, p_2, p_3, p_4\}$ being the index set assigned to the point s_0 by the positive-real function $W_1(s)$. Solving $W_1(s)$ in terms of $W_2(s)$ in (10.96) gives

$$W_1(s)=\frac{A_2(s)W_2(s)+B_2(s)}{C_2(s)W_2(s)+D_2(s)} \tag{10.98}$$

As in (10.55), $W_1(s)$ can be represented physically as the input impedance of a two-port network N_2 characterized by its transmission matrix

$$\mathbf{T}_2(s)=\begin{bmatrix} A_2(s) & B_2(s) \\ C_2(s) & D_2(s) \end{bmatrix} \tag{10.99}$$

terminated in $W_2(s)$, as depicted in Figure 10.18.

Consider the cascade connection of the two-port network N_1 of Figure 10.3 and N_2 of Figure 10.18, as shown in Figure 10.19. As is well known, the transmission matrix of the overall two-port network N is simply the product of the transmission matrices of the individual two-port networks:†

$$\mathbf{T}(s)=\mathbf{T}_1(s)\mathbf{T}_2(s) \tag{10.100}$$

Evidently, when N is terminated in $W_2(s)$, its input impedance is $Z(s)$. Since from (10.58) the determinant of $\mathbf{T}_1(s)$ or $\mathbf{T}_2(s)$ is independent of $Z(s)$, it follows that

$$\det \mathbf{T}(s)=[\det\mathbf{T}_1(s)][\det \mathbf{T}_2(s)]=[s^4+2(\omega_0^2-\sigma_0^2)s^2+|s_0|^4]^2 \tag{10.101}$$

As indicated in the paragraph below (10.58), the input impedance looking into

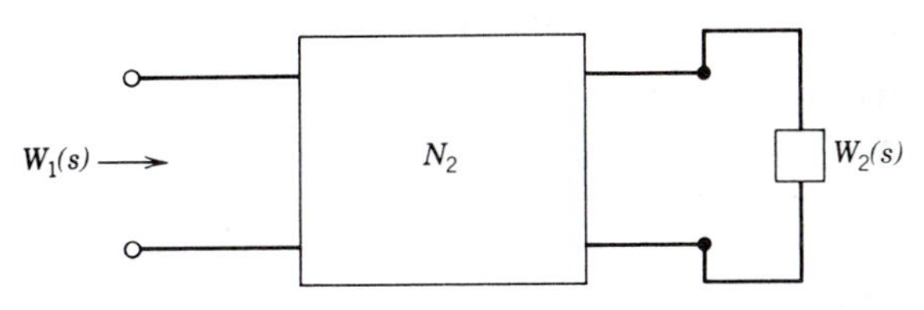

Figure 10.18

†W. K. Chen 1983, *Linear Networks and Systems*, pp. 248–249. Monterey, CA: Brooks/Cole.

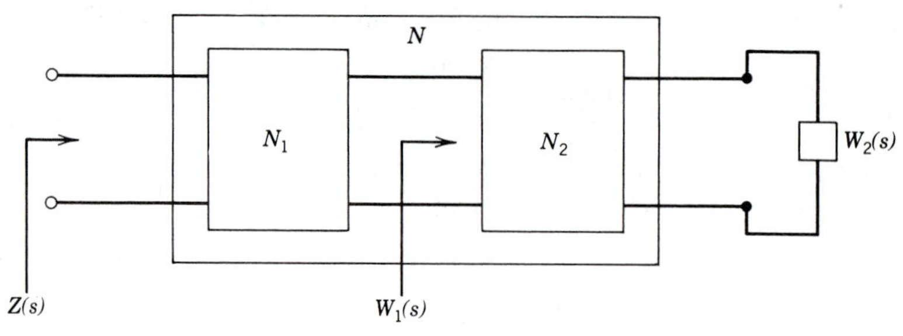

Figure 10.19

N_1 remains unaltered if each element of $\mathbf{T}_1(s)$ is divided by a nonzero finite quantity. For our purposes, we stipulate that the two-port network N_1 be characterized by the transmission matrix

$$\hat{\mathbf{T}}_1(s) = \frac{1}{\Delta(s)} \mathbf{T}_1(s) \tag{10.102}$$

where

$$\Delta(s) = s^4 + 2(\omega_0^2 - \sigma_0^2)s^2 + |s_0|^4 \tag{10.103}$$

Using this matrix in conjunction with (10.16), (10.55), (10.97), and (10.99), the transmission matrix $\hat{\mathbf{T}}(s)$ of the overall two-port network N becomes

$$\begin{aligned}\hat{\mathbf{T}}(s) &= \frac{1}{\Delta(s)} \mathbf{T}_1(s)\mathbf{T}_2(s) = \frac{1}{\Delta(s)} \mathbf{T}(s) = \frac{1}{\Delta(s)} \begin{bmatrix} A(s) & B(s) \\ C(s) & D(s) \end{bmatrix} \\ &= \frac{1}{\Delta(s)} \begin{bmatrix} p_4 q_4 s^4 + [(p_4 + q_4)|s_0|^2 & (p_2 q_4 + p_1 q_2)s^3 \\ \quad + p_3 q_2]s^2 + |s_0|^4 & \quad + (p_2 + q_2)|s_0|^2 s \\ (p_4 q_3 + p_3 q_1)s^3 & p_1 q_1 s^4 + [(p_1 + q_1)|s_0|^2 \\ \quad + (p_3 + q_3)|s_0|^2 s & \quad + p_2 q_3]s^2 + |s_0|^4 \end{bmatrix}\end{aligned} \tag{10.104}$$

Applying the standard relations among the transmission and impedance parameters, as given in (10.56), the impedance matrix $\mathbf{Z}(s)$ of the overall two-port network N is given by

$$\mathbf{Z}(s) = \frac{1}{C(s)} \begin{bmatrix} A(s) & \Delta(s) \\ \Delta(s) & D(s) \end{bmatrix} \tag{10.105}$$

showing that N is reciprocal since $\mathbf{Z}(s)$ is symmetric. Thus, the poles of the elements of $\mathbf{Z}(s)$ occur at $s = \infty$ and at the zeros of the equation

$$(p_4 q_3 + p_3 q_1)s^3 + (p_3 + q_3)|s_0|^2 s = 0 \tag{10.106}$$

whose roots are located at the points $s = 0$ and

$$s_1 = \pm j\omega_1 = \pm j|s_0| \frac{(p_3 + q_3)^{1/2}}{(p_4 q_3 + p_3 q_1)^{1/2}} \tag{10.107}$$

Expanding the elements of $\mathbf{Z}(s)$ in partial fractions yields

$$\mathbf{Z}(s)=s\mathbf{K}_{\infty}+\frac{1}{s}\mathbf{K}_{0}+\frac{2s}{s^{2}+\omega_{1}^{2}}\mathbf{K}_{1} \tag{10.108}$$

where the residue matrices $\mathbf{K}_{\infty}$, $\mathbf{K}_0$, and $\mathbf{K}_1$ are given by

$$\mathbf{K}_{\infty}=\frac{1}{p_4q_3+p_3q_1}\begin{bmatrix} p_4q_4 & 1 \\ 1 & p_1q_1 \end{bmatrix} \tag{10.109a}$$

$$\mathbf{K}_{0}=\frac{|s_0|^2}{p_3+q_3}\begin{bmatrix} 1 & 1 \\ 1 & 1 \end{bmatrix} \tag{10.109b}$$

$$\mathbf{K}_{1}=\frac{1}{2(p_4q_3+p_3q_1)}\begin{bmatrix} (p_4+q_4)|s_0|^2+p_3q_2 - p_4q_4\omega_1^2-|s_0|^4/\omega_1^2 & 2(\omega_0^2-\sigma_0^2) - \omega_1^2-|s_0|^4/\omega_1^2 \\ 2(\omega_0^2-\sigma_0^2) - \omega_1^2-|s_0|^4/\omega_1^2 & (p_1+q_1)|s_0|^2+p_2q_3 - p_1q_1\omega_1^2-|s_0|^4/\omega_1^2 \end{bmatrix} \tag{10.109c}$$

It is straightforward to show that, by direct computation, the residue matrices $\mathbf{K}_{\infty}$, $\mathbf{K}_0$, and $\mathbf{K}_1$ are singular, indicating that all the poles of the elements of the impedance matrix $\mathbf{Z}(s)$ are compact (Problem 10.19).

Consider the reciprocal lossless two-port network N_D of Figure 10.20 with two perfectly coupled transformers

$$L_1L_2=M_1^2 \tag{10.110a}$$

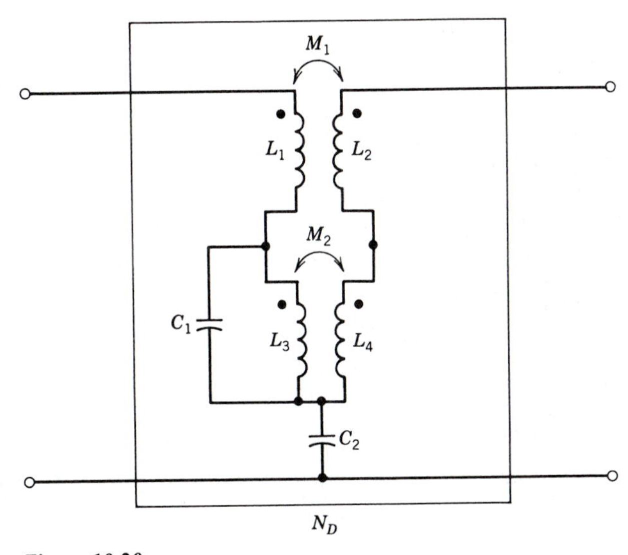

Figure 10.20

$$L_3 L_4 = M_2^2 \tag{10.110b}$$

This two-port network N_D is referred to as the *Darlington type-D section*. The impedance matrix $\mathbf{Z}_D(s)$ of N_D is obtained as (Problems 10.21 and 10.22)

$$\mathbf{Z}_D(s) = \begin{bmatrix} L_1 s + \dfrac{1}{C_2 s} + \dfrac{s/C_1}{s^2+\omega_a^2} & M_1 s + \dfrac{1}{C_2 s} + \dfrac{\omega_a^2 M_2 s}{s^2+\omega_a^2} \\ M_1 s + \dfrac{1}{C_2 s} + \dfrac{\omega_a^2 M_2 s}{s^2+\omega_a^2} & L_2 s + \dfrac{1}{C_2 s} + \dfrac{\omega_a^2 L_4 s}{s^2+\omega_a^2} \end{bmatrix} \tag{10.111}$$

where

$$\omega_a^2 = 1/C_1 L_3 \tag{10.112}$$

Performing the partial-fraction expansions of the elements of $\mathbf{Z}_D(s)$ yields

$$\mathbf{Z}_D(s) = s\begin{bmatrix} L_1 & M_1 \\ M_1 & L_2 \end{bmatrix} + \frac{1}{s}\begin{bmatrix} 1/C_2 & 1/C_2 \\ 1/C_2 & 1/C_2 \end{bmatrix} + \frac{2s}{s^2+\omega_a^2}\begin{bmatrix} 1/2C_1 & \frac{1}{2}\omega_a^2 M_2 \\ \frac{1}{2}\omega_a^2 M_2 & \frac{1}{2}\omega_a^2 L_4 \end{bmatrix} \tag{10.113}$$

A simple analysis reveals that the residue matrices are again singular, showing that all of its poles are compact, as desired.

Comparing (10.109) with (10.113) and after some algebraic manipulations, we can make the following identifications:

$$L_1 = p_4 q_4/(p_4 q_3 + p_3 q_1) \tag{10.114a}$$

$$L_2 = p_1 q_1/(p_4 q_3 + p_3 q_1) = M_1^2/L_1 \tag{10.114b}$$

$$M_1 = 1/(p_4 q_3 + p_3 q_1) = (L_1 L_2)^{1/2} \tag{10.114c}$$

$$C_2 = (p_3 + q_3)/|s_0|^2 \tag{10.114d}$$

$$\omega_a^2 = \omega_1^2 = |s_0|^2 (p_3 + q_3)/(p_4 q_3 + p_3 q_1) \tag{10.114e}$$

$$M_2 = -\frac{\omega_1^4 - 2(\omega_0^2 - \sigma_0^2)\omega_1^2 + |s_0|^4}{\omega_1^4 (p_4 q_3 + p_3 q_1)}$$

$$= -\frac{p_3^2 q_3^2 |\bar{W}_1(s_0) + Z(s_0) q_1|^2}{|s_0|^2 (p_4 q_3 + p_3 q_1)(p_3 + q_3)^2} \leqslant 0 \tag{10.114f}$$

$$L_4 = \frac{[(p_1+q_1)|s_0|^2 + p_2 q_3]\omega_1^2 - p_1 q_1 \omega_1^4 - |s_0|^4}{\omega_1^4 (p_4 q_3 + p_3 q_1)}$$

$$= \frac{p_2 q_3}{|s_0|^2 (p_3+q_3)} - \frac{(p_4 q_3 + p_3 q_1 - p_3 q_1 - q_1 q_3)(p_4 q_3 + p_3 q_1 - p_1 p_3 - p_1 q_3)}{(p_4 q_3 + p_3 q_1)(p_3+q_3)^2}$$

$$= -q_3 M_2/p_3 \tag{10.114g}$$

$$L_3 = M_2^2/L_4 = -p_3 M_2/q_3 \tag{10.114h}$$

$$C_1 = \frac{1}{\omega_1^2 L_3} = -\frac{q_3}{\omega_1^2 p_3 M_2} \tag{10.114i}$$

The details of the algebraic manipulations in obtaining these formulas are outlined in Problems 10.24–10.28. Thus, all the element values except M_2 are nonnegative and we have succeeded in proving that two successive applications of Theorem 10.1 yield the type-D section, which is lossless and reciprocal and is equivalent to the two type-E sections in cascade.

We shall illustrate this procedure by the following example.

Example 10.13 Consider the positive-real impedance

$$Z(s) = \frac{6s^2 + 5s + 6}{2s^2 + 4s + 4} \tag{10.115}$$

whose even part $r(s)$ has a zero at $s_0 = 0.61139 + j1.02005$, as given in (10.24). The elements of the index set assigned to the point s_0 by $Z(s)$ are listed in (10.28), and are repeated below:

$$q_1 = 0.70711, \qquad q_2 = 1.76784 \tag{10.116a}$$

$$q_3 = 0.94283, \qquad q_4 = 1.41421 \tag{10.116b}$$

According to (10.31a) of Theorem 10.1, the function $W_1(s)$ must be of degree zero, being a constant. Its value can be determined directly from $Z(s)$ by the relation

$$W_1(s) = W_1(0) = Z(0) = 1.5\ \Omega \tag{10.117}$$

The elements of the index set assigned to the point s_0 by $W_1(s)$ are given by

$$p_1 = 1, \qquad p_2 = 1.83417 \tag{10.118a}$$

$$p_3 = 0.81519, \qquad p_4 = 1 \tag{10.118b}$$

Substituting these in (10.114) in conjunction with (10.26) and (10.69) yields the desired element values of the type-D section:

$$L_1 = p_4 q_4/(p_4 q_3 + p_3 q_1) = 0.93086\ \text{H} \tag{10.119a}$$

$$L_2 = p_1 q_1/(p_4 q_3 + p_3 q_1) = 0.46543\ \text{H} \tag{10.119b}$$

$$M_1 = 1/(p_4 q_3 + p_3 q_1) = 0.65822\ \text{H} \tag{10.119c}$$

$$C_2 = (p_3 + q_3)/|s_0|^2 = 1.24303\ \text{F} \tag{10.119d}$$

$$\omega_1^2 = |s_0|^2 (p_3 + q_3)/(p_4 q_3 + p_3 q_1) = 1.63656 \tag{10.119e}$$

$$M_2 = -\frac{\omega_1^4 - 2(\omega_0^2 - \sigma_0^2)\omega_1^2 + |s_0|^4}{\omega_1^4 (p_4 q_3 + p_3 q_1)} = -0.61350\ \text{H} \tag{10.119f}$$

$$L_4 = -q_3 M_2/p_3 = 0.70956\ \text{H} \tag{10.119g}$$

$$L_3 = -p_3 M_2/q_3 = 0.53044\ \text{H} \tag{10.119h}$$

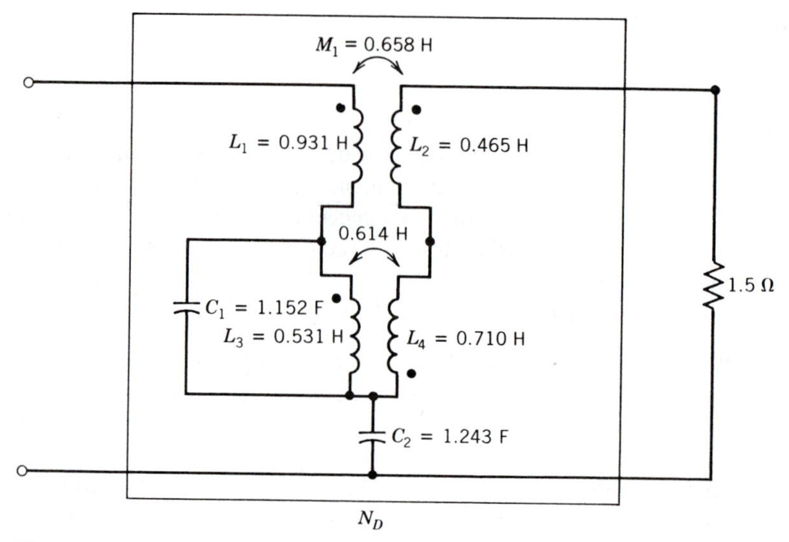

Figure 10.21

$$C_1 = -\frac{q_3}{\omega_1^2 p_3 M_2} = 1.15193 \text{ F} \tag{10.119i}$$

The complete network together with its termination is presented in Figure 10.21.

From a computational viewpoint, it is advantageous first to strip a given positive-real impedance $Z(s)$ of all of its $j\omega$-axis poles and zeros by the usual partial-fraction expansions. The process is termed the *Foster preamble* to the realization of $Z(s)$, because of its obvious analogy with the Foster procedure for reactance functions. The remaining impedance, being devoid of both poles and zeros on the $j\omega$-axis, is still positive real, and can be realized by the removal of appropriate sections, depending upon the locations of the zeros of its even part.

10.4 PROOF OF THE MAIN THEOREM

Having demonstrated the usefulness of Theorem 10.1, we now proceed to prove its validity. For this, we first show that $W_1(s)$ is positive real, and if, in addition, $Z(s)$ is rational with s_0 being a closed RHS zero of its even part, then $W_1(s)$ possesses the attributes (10.30)–(10.32).

10.4.1 The Positive-Real Function $W_1(s)$

Let $Z(s)$ be a given positive-real function and let $s_0 = \sigma_0 + j\omega_0$ be a point in the open RHS. Write

$$Z(s_0) = Z_0 = R_0 + jX_0 \tag{10.120}$$

For the sake of definiteness, assume that $\omega_0 \geqq 0$ and $X_0 \leqq 0$ to begin with. Then the impedance function

$$z(s) = r_0 + L_0 s \tag{10.121}$$

where

$$r_0 = 2\sigma_0/q_3 \tag{10.122a}$$

$$L_0 = |Z_0|^2(q_1 - 1)/q_2 \tag{10.122b}$$

is positive real such that the function $f(s) = Z(s) - z(-s)$ satisfies either

$$f(s_0) = f(\bar{s}_0) = 0, \qquad \omega_0 > 0 \tag{10.123a}$$

or

$$f(\sigma_0) = f'(\sigma_0) = 0, \qquad \omega_0 = 0 \tag{10.123b}$$

$f'(\sigma_0)$ being the derivative of $f(s)$ evaluated at σ_0. Since the even part of $z(s)$ is devoid of zeros and poles in the entire s-plane, the reflection coefficient defined by the relation

$$\rho_1(s) = \frac{Z(s) - z(-s)}{Z(s) + z(s)} \tag{10.124}$$

is analytic in the closed RHS and possesses the property that

$$|\rho_1(s)| \leqq 1, \qquad \text{Re}\, s \geqq 0 \tag{10.125}$$

Such a reflection coefficient is termed a *bounded-real reflection coefficient*. Since from (10.123) the numerator of $\rho_1(s)$ contains the factor

$$g(-s) = (s - s_0)(s - \bar{s}_0) = s^2 - 2\sigma_0 s + |s_0|^2 \tag{10.126}$$

it follows that the function

$$\rho_2(s) = \eta(-s)\rho_1(s) \tag{10.127a}$$

where

$$\eta(s) = \frac{(s - s_0)(s - \bar{s}_0)}{(s + s_0)(s + \bar{s}_0)} = \frac{g(-s)}{g(s)} \tag{10.127b}$$

is an all-pass function, is also a bounded-real reflection coefficient (Problem 10.30). Consequently, the function

$$W(s) = \frac{1 + \rho_2(s)}{1 - \rho_2(s)} \tag{10.128}$$

is positive real. Using (10.124)–(10.127), (10.128) can be expressed in terms of $g(s)$, $Z(s)$, and $z(s)$ as

$$W(s) = \frac{[g(-s) + g(s)]Z(s) + g(-s)z(s) - g(s)z(-s)}{[g(-s) - g(s)]Z(s) + g(-s)z(s) + g(s)z(-s)} \tag{10.129a}$$

$$= \frac{(s^2 + |s_0|^2)Z(s) + L_0 s^3 + (|s_0|^2 L_0 - 2\sigma_0 r_0)s}{-2\sigma_0 s Z(s) + (r_0 - 2\sigma_0 L_0)s^2 + r_0|s_0|^2} \tag{10.129b}$$

For a positive-real impedance $Z(s)$, we have

$$\lim_{s\to\infty} Z'(s) = \lim_{s\to\infty} \frac{Z(s)}{s} = L \tag{10.130}$$

L being a real nonnegative number, and such that the function

$$Z_a(s) = Z(s) - Ls \tag{10.131}$$

is still positive real. Applying this result to $W(s)$, which is known to be positive real, shows that

$$L_a = \frac{L + L_0}{r_0 - 2\sigma_0(L_0 + L)} \geqq 0 \tag{10.132}$$

and the function

$$W_a(s) = W(s) - L_a s \tag{10.133}$$

is positive real. Let

$$L_b = \frac{L_0}{r_0 - 2\sigma_0 L_0} \tag{10.134}$$

Then

$$0 \leqq L_b \leqq L_a \tag{10.135}$$

and the function

$$r_0[W_a(s) + (L_a - L_b)s] = r_0 W(s) - r_0 L_b s \tag{10.136}$$

is clearly also positive real. Substituting (10.134) and (10.129b) in the right-hand side of (10.136) gives

$$\begin{aligned} r_0[W_a(s) + (L_a - L_b)s] &= r_0 W(s) - r_0 L_b s \\ &= \frac{[r_0 s^2/(r_0 - 2\sigma_0 L_0) + |s_0|^2]Z(s) - 2\sigma_0 s[r_0 + |s_0|^2 L_0^2/(r_0 - 2\sigma_0 L_0)]}{(-2\sigma_0 s/r_0)Z(s) + (1 - 2\sigma_0 L_0/r_0)s^2 + |s_0|^2} \end{aligned} \tag{10.137}$$

Since the element values r_0 and L_0 can be expressed in terms of the real and imaginary parts of Z_0 and s_0 as

$$r_0 = 2\sigma_0/q_3 = R_0 - \sigma_0 X_0/\omega_0 \tag{10.138a}$$

$$L_0 = -X_0/\omega_0 \geqq 0 \tag{10.138b}$$

(10.137) simplifies to

$$r_0[W_a(s) + (L_a - L_b)s] = \frac{[q_1 s^2 + |s_0|^2]Z(s) - q_2 s}{-q_3 s Z(s) + q_4 s^2 + |s_0|^2} = W_1(s) \tag{10.139}$$

Thus, $W_1(s)$ is a positive-real function under the stipulation that $\omega_0 \geqq 0$ and $X_0 \leqq 0$. In a similar fashion, we can show that $W_1(s)$ is positive real for $\omega_0 < 0$ or $X_0 > 0$, the details being left as exercises (Problems 10.33 and 10.34). Thus, $W_1(s)$ is positive real for any s_0 in the open RHS.

To complete our proof, we must remove the restriction that s_0 be an open RHS point. For this let the point s_0 approach any finite point on the $j\omega$-axis, exclusive of the origin, along an arbitrary ray not parallel to the axis. At all the points where $Z(s)$ is analytic, the elements of the index sets are well defined by (10.17)–(10.19) and the function $W_1(s)$, as defined by (10.29), is positive real for any point s_0 on the $j\omega$-axis. This completes the proof of the first part of the theorem.

10.4.2 **The Degree of $W_1(s)$**

In the present section, we show that if $Z(s)$ is rational, then $W_1(s)$ as defined in (10.29) possesses the attributes (10.30)–(10.32).

Again, for the sake of definiteness, assume that $\omega_0 \gtreqless 0$ and $X_0 \lesseqgtr 0$, leaving the other two possibilities as exercises (Problems 10.35 and 10.36). Our starting point is (10.139), which after appealing to (10.136) can be expressed as

$$W_1(s) = r_0 \frac{N_W(s) - L_b D_W(s) s}{D_W(s)} \tag{10.140}$$

where

$$W(s) = \frac{N_W(s)}{D_W(s)} \tag{10.141}$$

$N_W(s)$ and $D_W(s)$ being the numerator and denominator of $W(s)$, as given in (10.129a).

Referring to (10.29), compare the degrees of $W_1(s)$ and $Z(s)$. Since $Z(s)$, by hypothesis, is rational and positive real, the degrees of its numerator and denominator polynomials can at most differ by one, as indicated below:

		$Z(s)$	
numerator	k	$k+1$	k
denominator	k	k	$k+1$

The corresponding degree sequence for $W_1(s)$ is given by

numerator	$k+2$	$k+3$	$k+2$
denominator	$k+2$	$k+2$	$k+3$

and in each case the *apparent* increase in degree for $W_1(s)$ is four. In the following, we show that a decrease of its degree by four, six, or eight is always brought about by the cancellation of the common factors from its numerator and denominator functions. To this end, we consider three situations, each being presented in a separate section.

10.4.2.1 THE COMMON QUADRATIC FACTOR $g(-s)$. We show that $g(-s)$, as defined in (10.126), is a common factor of $N_W(s)$ and $D_W(s)$. Clearly, we have $g(-s_0) = g(-\bar{s}_0) = 0$. Using this in conjunction with (10.123) and (10.129a) shows that

$$N_W(s_0) = N_W(\bar{s}_0) = 0, \qquad \omega_0 \neq 0 \tag{10.142a}$$

$$D_W(s_0) = D_W(\bar{s}_0) = 0, \qquad \omega_0 \neq 0 \tag{10.142b}$$

and

$$N_W(\sigma_0) = N'_W(\sigma_0) = 0, \qquad \omega_0 = 0 \tag{10.143a}$$

$$D_W(\sigma_0) = D'_W(\sigma_0) = 0, \qquad \omega_0 = 0 \tag{10.143b}$$

where the prime denotes the derivative of a function with respect to s. This shows that $g(-s)$ is a common factor of $N_W(s)$ and $D_W(s)$. Invoking (10.140) reveals that the degree of $W_1(s)$ is decreased at least by four from its apparent degree indicated earlier. Thus, we conclude that

$$\text{degree } W_1(s) \leqq \text{degree } Z(s) \tag{10.144}$$

This proves (10.30).

10.4.2.2 THE COMMON QUADRATIC FACTOR $g(s)$. We show that if $s_0 = \sigma_0 + j\omega_0$, $\omega_0 \neq 0$, is a zero of the even part $r(s)$ of $Z(s)$, then $N_W(s)$ and $D_W(s)$, in addition to having $g(-s)$ as a common factor, also possess the common quadratic factor $g(s)$.

From (10.129a) and (10.123a) in conjunction with the facts that $g(-s_0) = g(-\bar{s}_0) = 0$ and $Z(s_0) = -Z(-s_0)$, we obtain

$$N_W(-s_0) = D_W(-s_0) = -g(s_0)[Z(s_0) - z(-s_0)] = 0 \tag{10.145a}$$

$$N_W(-\bar{s}_0) = D_W(-\bar{s}_0) = -g(\bar{s}_0)[Z(\bar{s}_0) - z(-\bar{s}_0)] = 0 \tag{10.145b}$$

Thus, $g(s)$ as defined in (10.126) is also a common factor of $N_W(s)$ and $D_W(s)$. Appealing to (10.140) shows that the degree of $W_1(s)$ is decreased at least by eight from its apparent degree because of the cancellation of the factors $g(s)$ and $g(-s)$, implying that

$$\text{degree } W_1(s) \leqq \text{degree } Z(s) - 4, \qquad \omega_0 \neq 0 \tag{10.146}$$

This proves (10.31a).

10.4.2.3 THE COMMON FACTORS $(s+\sigma_0)$ and $(s+\sigma_0)^2$. We show that if $s_0 = \sigma_0$ is a first-order zero of the even part $r(s)$ of $Z(s)$, then $(s+\sigma_0)$ is a common factor of $N_W(s)$ and $D_W(s)$; and if it is a zero of multiplicity at least two, they possess the common quadratic factor $(s+\sigma_0)^2$.

From (10.129a) and (10.123b) in conjunction with the facts that $g(-\sigma_0) = 0$ and $Z(\sigma_0) = -Z(-\sigma_0)$, we have

$$N_W(-\sigma_0) = D_W(-\sigma_0) = -g(\sigma_0)[Z(\sigma_0) - z(-\sigma_0)] = 0 \tag{10.147}$$

implying that $(s+\sigma_0)$ is a common factor of $N_W(s)$ and $D_W(s)$. From (10.140), we conclude that

$$\text{degree } W_1(s) \leqq \text{degree } Z(s) - 2, \qquad \omega_0 = 0 \tag{10.148}$$

proving (10.31b).

Finally, suppose that $s_0 = \sigma_0$ is a zero of $r(s)$ of multiplicity at least two. Then

$$Z'(\sigma_0) = Z'(-\sigma_0) \tag{10.149}$$

Differentiating the numerator $N_W(s)$ and denominator $D_W(s)$ of (10.129a) and evaluating their values at $s=-\sigma_0$, we obtain

$$\begin{aligned} N'_W(-\sigma_0)=D'_W(-\sigma_0)&=-g'(\sigma_0)[Z(-\sigma_0)+z(-\sigma_0)]+g(\sigma_0)[Z'(-\sigma_0)+z'(-\sigma_0)] \\ &=g'(\sigma_0)[Z(\sigma_0)-z(-\sigma_0)]+g(\sigma_0)[Z'(\sigma_0)+z'(-\sigma_0)] \\ &=0+g(\sigma_0)f'(\sigma_0)=0 \end{aligned} \tag{10.150}$$

The last line follows directly from (10.147) and (10.123b). This shows that $s=-\sigma_0$ is a zero of both $N_W(s)$ and $D_W(s)$ of multiplicity at least two. From (10.150), we conclude that

$$\text{degree } W_1(s) \leqslant \text{degree } Z(s)-4 \tag{10.151}$$

provided that $s_0=\sigma_0$ is a zero of $r(s)$ of multiplicity at least two. This proves (10.32).

To complete the proof, we must remove the restrictions that $\omega_0 \geqq 0$ and $X_0 \leqslant 0$. This can easily be done by observing that the index sets assigned to the points s_0 and $\bar{s}_0$ by $Z(s)$ are identical (Problem 10.32). The details of the justifications, as mentioned at the beginning of Section 10.4.2, are left as exercises. This completes the proof of the main theorem.

10.5 SUMMARY AND SUGGESTED READINGS

According to Darlington theory, any rational positive-real impedance can be realized as the input impedance of a lossless reciprocal two-port network terminated in a 1-Ω resistor. In Chapter 4, we demonstrated that, for a given positive-real impedance, it is always possible to identify a set of immittance parameters, which can be realized by a lossless reciprocal structure composed of the series or parallel connection of component two-port networks, each corresponding to a pair of poles or the pole at the origin or infinity of the associated immittance matrix. In the present chapter, we presented a technique that realizes the desired two-port network by the cascade connection of a number of two-port networks, each of which realizes a set of zeros of the even part of the given impedance. The essence of the technique is to extract from the given impedance a lossless two-port network, which possesses a set of zeros of the even part, in such a way as to leave a physically realizable remainder impedance that does not possess these zeros in its even part and that is simpler than the original impedance. In other words, a lossless two-port network must be removed from the given impedance so as to leave a remainder impedance that is compatible with the given impedance.

To this end, we demonstrated by construction that the *type-E section* is capable of extracting any complex zero of the even part of a given rational positive-real impedance and leads to a four-degree reduction. The *type-D section* achieves the same objective but can lead to an eight-degree reduction if this zero is of multiplicity two or more. In a similar manner, the *Richards section* extracts any σ-axis zero of the even part, resulting in a two-degree reduction, while the *type-C section* performs the same operation but is capable of diminishing the degree by four if this zero is of multiplicity two or more. Finally, for any finite nonzero $j\omega$-axis zero of the even part, the *Brune section* is capable of extracting this zero, yielding a four-degree reduction. The familiar *type-A* and *type-B sections* are used to remove zeros at the origin and at infinity, respectively.

A discussion of other aspects of the Richards' function can be found in Hazony [5] and Richards [6 and 7]. A chart constructed to facilitate the computations of the design is given by Youla [10]. Application of the cascade synthesis technique to the design of broadband matching networks can be found in Chen [2].

REFERENCES

1. O. Brune, "Synthesis of a finite two-terminal network whose driving point impedance is a prescribed function of frequency," *J. Math. Phys.*, vol. 10, 1931, pp. 191–236.
2. W. K. Chen 1976, *Theory and Design of Broadband Matching Networks*, Chapters 1 and 2. Oxford, England: Pergamon Press.
3. S. Darlington, "Synthesis of reactance 4-poles which produce prescribed insertion loss characteristics," *J. Math. Phys.*, vol. 18, 1939, pp. 257–353.
4. E. A. Guillemin 1957, *Synthesis of Passive Networks*. New York: John Wiley.
5. D. Hazony, "Zero cancellation synthesis using impedance operators," *IRE Trans. Circuit Theory*, vol. CT-8, 1961, pp. 114–120.
6. P. I. Richards, "A special class of functions with positive real part in half-plane," *Duke Math. J.*, vol. 14, 1947, pp. 777–786.
7. P. I. Richards, "General impedance-function theory," *Quart. Appl. Math.*, vol. 6, 1948, pp. 21–29.
8. D. F. Tuttle, Jr. 1958, *Network Synthesis*, Vol. I. New York: John Wiley.
9. D. C. Youla, "Physical realizability criteria," *IRE Trans. Circuit Theory* (Special Supplement), vol. CT-7, 1960, pp. 50–68.
10. D. C. Youla, "A new theory of cascade synthesis," *IRE Trans. Circuit Theory*, vol. CT-8, 1961, pp. 244–260.

PROBLEMS

10.1 Given the impedance function

$$Z(s)=\frac{s^2+s+1}{s^2+s+2} \tag{10.152}$$

synthesize a one-port network for this $Z(s)$ by the cascade technique.

10.2 Substituting (10.13) in (10.14), show that (10.14) becomes (10.15).

10.3 Confirm that the impedance matrix $\mathbf{Z}_E(s)$ of the type-E section of Figure 10.4 is given by (10.59).

10.4 Using cascade synthesis technique, repeat Problem 4.14.

10.5 Using (10.64), derive (10.65). [*Hint:* Expand $Z(s)$ by Taylor series.]

10.6 Using cascade synthesis technique, repeat Problem 4.15.

10.7 From (10.62), derive formulas (10.63) for the element values of the type-E section, as shown in Figure 10.4.

10.8 Derive formulas (10.70) for the element values of the Darlington type-C section of Figure 10.6.

10.9 In explicit form, show that formulas (10.70) can be expressed as

$$L_1 = \frac{Z(\sigma_0)+\sigma_0 Z'(\sigma_0)}{2\sigma_0} \tag{10.153a}$$

$$L_2 = \frac{[Z(\sigma_0)-\sigma_0 Z'(\sigma_0)]^2}{2\sigma_0[Z(\sigma_0)+\sigma_0 Z'(\sigma_0)]} \tag{10.153b}$$

$$C = \frac{2}{\sigma_0[Z(\sigma_0)-\sigma_0 Z'(\sigma_0)]} \tag{10.153c}$$

$$M = -\frac{Z(\sigma_0)-\sigma_0 Z'(\sigma_0)}{2\sigma_0} \tag{10.153d}$$

10.10 Confirm that the elements of the index set assigned to the point $\sigma_0 = 2$ by the impedance (10.76) are given by (10.78).

10.11 In Example 10.10, using (10.29) confirm (10.80).

10.12 Show that in solving the equations (10.62) for the element values of the type-E section, if $M = -1/q_3$ and $\omega_0 \neq 0$ then ζ is imaginary.

10.13 Derive the explicit formulas (10.81) for the element values of the Brune section of Figure 10.10.

10.14 Show that (10.82a) and (10.82b) degenerate to (10.83) and (10.84), respectively, as ω_0 approaches zero and infinity.

10.15 Using cascade technique, realize the impedance given in Problem 4.17.

10.16 Quite often, the two-port networks of Figures 10.22(*a*) and (*b*) are also referred to as the *type-A* and *type-B sections*, respectively. Show that Figure 10.22(*a*) cannot alter the even part $r(s)$ of a rational positive-real impedance $Z(s)$, and that Figure 10.22(*b*) is capable of removing some finite nonzero real-frequency axis zeros of $r(s)$.

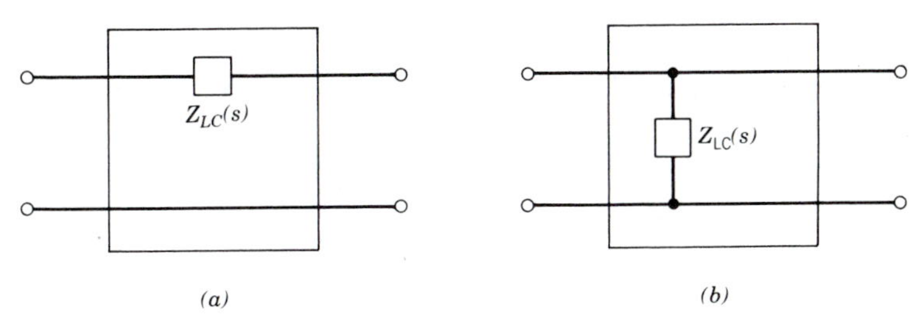

Figure 10.22

10.17 Derive the formulas (10.93) for the element values of the Richards section of Figure 10.16. [*Hint*: Proceed as outlined in (10.59)–(10.62).]

10.18 Assume that $s_0 = j\omega_0$ and $Z(j\omega_0) = R_0 + jX_0$, $R_0 \neq 0$. Show that the elements of the index set assigned to the point $j\omega_0$ by $Z(s)$ are given

$$q_1 = q_4 = 1 \qquad \text{and} \qquad q_2 = q_3 = 0 \tag{10.154}$$

10.19 Show that (10.109c) is singular.

10.20 Using the cascade technique, synthesize the impedance

$$Z(s)=\frac{6s^2+5s+3}{s^2+3s+2} \tag{10.155}$$

10.21 Confirm that the impedance matrix $\mathbf{Z}_D(s)$ of the type-D section of Figure 10.20 is given by (10.111).

10.22 Show that the two-port network of Figure 10.23 possesses the impedance matrix of (10.111), meaning that it is equivalent to the type-D section of Figure 10.20. Also determine the capacitance C in terms of other element values of the type-D section of Figure 10.20.

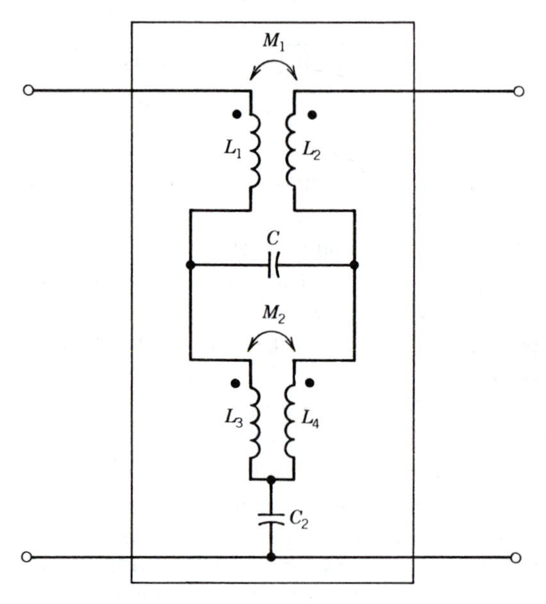

Figure 10.23

10.23 Using the cascade technique, synthesize the impedance given in Problem 4.6.

10.24 Referring to (10.114), let

$$x=p_4q_3+p_3q_1 \tag{10.156a}$$

$$y=p_3+q_3 \tag{10.156b}$$

Show that

$$x+y=p_3q_3(U_0+R_0q_1)/\sigma_0 \tag{10.157a}$$

$$x-y=p_3q_3(V_0-X_0q_1)/\omega_0 \tag{10.157b}$$

where

$$Z(s_0)=R_0+jX_0 \tag{10.158a}$$

$$W_1(s_0)=U_0+jV_0 \tag{10.158b}$$

10.25 Using (10.156), show that the first equation of (10.114f) can be put in the form

$$M_2 = -\frac{\sigma_0^2(x+y)^2 + \omega_0^2(x-y)^2}{xy^2|s_0|^2} \tag{10.159}$$

10.26 Using (10.156), show that the first equation of (10.114g) can be put in the form

$$L_4 = \frac{xyp_2q_3 - |s_0|^2(x-yq_1)(x-yp_1)}{xy^2|s_0|^2} \tag{10.160}$$

Using this in conjunction with (10.159), show that

$$L_4 = -q_3M_2/p_3 \tag{10.161}$$

10.27 Like the first equation of (10.114g), show that L_3 of the type-D section can be expressed as

$$L_3 = \frac{[(p_4+q_4)|s_0|^2 + p_3q_2]\omega_1^2 - p_4q_4\omega_1^2 - |s_0|^2}{\omega_1^4(p_4q_3+p_3q_1)} \tag{10.162}$$

10.28 Using (10.156), show that (10.162) can be put in the form

$$L_3 = \frac{xyp_3q_2 - |s_0|^2(x-yq_4)(x-yp_4)}{xy^2|s_0|^2} \tag{10.163}$$

From this, deduce (10.114h).

10.29 Using cascade technique, repeat Problem 4.7.

10.30 Show that the function $\rho_2(s)$, as given in (10.127a), is a bounded-real reflection coefficient. [*Hint:* Invoke the maximum modulus theorem.]

10.31 Let $\{q_1, q_2, q_3, q_4\}$ be the index set assigned to the point s_0 by the positive-real impedance $Z(s)$. Show that the elements of the index set can be expressed as

$$q_1 = \frac{\sin(\theta-\phi)}{\sin(\theta+\phi)} \tag{10.164a}$$

$$q_2 = \frac{|s_0Z(s_0)| \sin 2\theta}{\sin(\theta+\phi)} \tag{10.164b}$$

$$q_3 = \frac{|s_0| \sin 2\theta}{|Z(s_0)| \sin(\theta-\phi)} \tag{10.164c}$$

$$q_4 = \frac{\sin(\theta+\phi)}{\sin(\theta-\phi)} \tag{10.164d}$$

where $\theta = \arg s_0$ and $\phi = \arg Z(s_0)$.

10.32 Let

$$\Psi_Z(s_0) = \{q_1, q_2, q_3, q_4\} \tag{10.165}$$

be the index set assigned to the closed RHS point $s_0 \neq 0$ by the positive-real function $Z(s)$. Using the results in Problem 10.31, show that

$$\Psi_Z(s_0) = \Psi_Z(\bar{s}_0) \tag{10.166a}$$

$$\Psi_Y(s_0) = \{q_4, q_3, q_2, q_1\} \tag{10.166b}$$

where $Y(s) = 1/Z(s)$.

10.33 Show that the function $W_1(s)$, as given in (10.29), is positive real for $\omega_0 < 0$. [*Hint:* Apply (10.166a).]

10.34 Show that the function $W_1(s)$, as given in (10.29), is positive real for $X_0 > 0$. [*Hint:* Apply (10.166b).]

10.35 Prove that the conclusions reached in Section 10.4.2 remain valid for $\omega_0 < 0$. [*Hint:* Apply (10.166).]

10.36 Prove that the conclusions reached in Section 10.4.2 remain valid for $X_0 > 0$. [*Hint:* Apply (10.166).]

10.37 Using cascade synthesis technique, realize the impedance

$$Z(s) = \frac{4s^2 + s + 2}{2s^2 + 4s + 4} \tag{10.167}$$

chapter 11

GENERAL THEORY OF COMPATIBLE IMPEDANCES

In Chapter 4, we introduced the concept of compatible impedances. Two positive-real impedances are said to be compatible if one can be realized as the input impedance of a lossless two-port network terminated in the other. We then presented Darlington's theory, which states that any positive-real impedance is compatible with any non-negative resistance, which may be zero. We justify this by showing that for the given impedance it is always possible to identify a set of immittance parameters that are compatible in representing a lossless reciprocal two-port network. The resulting network is either a series or parallel combination of the lossless component two-port networks. The configuration is not attractive in that it contains an excessive number of perfectly coupled and/or ideal transformers. As an alternative, in Chapter 10 we presented a technique that realizes the desired two-port network by the cascade connection of a number of two-port networks, each realizing a set of zeros of the even part of the given impedance.

The general problem of compatibility of two impedances was not considered because its solution requires the introduction of the concept of the scattering matrix. Like other two-port matrices, the *scattering matrix* of a two-port network is a square matrix of order 2, whose diagonal elements are the reflection coefficients $\rho_i(s)$ at the two ports and whose off-diagonal elements are the transmission coefficients $S_{ij}(s)$, as defined in Section 4.4 of Chapter 4.

In the present chapter, we shall consider the general problem of compatibility with the aid of the complex-normalized scattering matrix, and show that it essentially reduces to the problem of the existence of a certain type of all-pass functions.

11.1 NORMALIZATION OF THE SCATTERING MATRIX

Referring to the network of Figure 11.1, our objective is to derive a set of necessary and sufficient conditions under which two arbitrary rational positive-real impedances

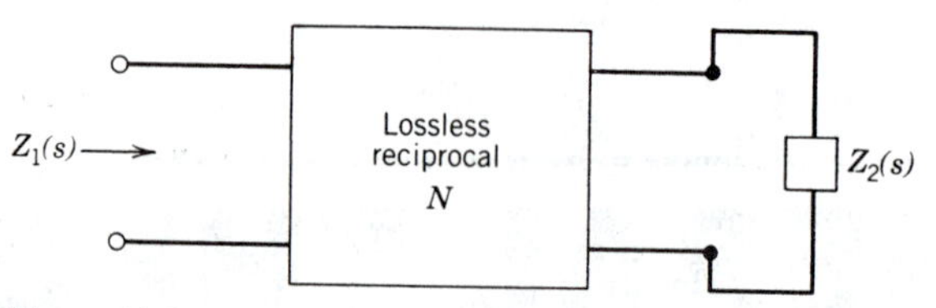

Figure 11.1

$Z_1(s)$ and $Z_2(s)$ are compatible. That is, given $Z_1(s)$ and $Z_2(s)$, determine if there is a lossless two-port network N, which, when terminated in $Z_2(s)$, yields the input impedance $Z_1(s)$. To simplify the problem, in the following we shall confine ourselves to the class of reciprocal lossless two-port networks, leaving the extension to non-reciprocal realization at a later stage.

To this end, let

$$\mathbf{S}(s)=\begin{bmatrix} S_{11}(s) & S_{12}(s) \\ S_{12}(s) & S_{22}(s) \end{bmatrix} \tag{11.1}$$

be the scattering matrix of the two-port network N of Figure 11.1, normalizing to the impedances $z_1(s)=1$ and $z_2(s)=Z_2(s)$. We note that the input port is normalized to the 1-Ω resistance rather than the impedance $Z_1(s)$. The *reference impedance matrix* of $\mathbf{S}(s)$ is defined by

$$\mathbf{z}(s)=\begin{bmatrix} 1 & 0 \\ 0 & Z_2(s) \end{bmatrix} \tag{11.2}$$

The *para-hermitian part* of $\mathbf{z}(s)$ is defined by the equation

$$\mathbf{p}(s)=\tfrac{1}{2}[\mathbf{z}(s)+\mathbf{z}(-s)] \tag{11.3a}$$

which can be factored into the form

$$\mathbf{p}(s)=\mathbf{h}(s)\mathbf{h}(-s) \tag{11.3b}$$

where

$$\mathbf{h}(s)=\begin{bmatrix} 1 & 0 \\ 0 & h_2(s) \end{bmatrix} \tag{11.4a}$$

and the even part $r_2(s)$ of $Z_2(s)$ is decomposed as

$$r_2(s)=\tfrac{1}{2}[Z_2(s)+Z_2(-s)]=h_2(s)h_2(-s) \tag{11.4b}$$

the zeros and poles of $h_2(s)$ being chosen in accordance with the following three requirements:

(i) $h_2(s)$ is analytic in the open RHS;
(ii) $h_2^{-1}(-s)$ is analytic in the open RHS;
(iii) $h_2(s)$ is the ratio of two polynomials of minimal order.

Thus, all the poles of $h_2(s)$ are in the open LHS and all of its zeros are restricted to the closed RHS. We remark that for a positive-real $Z_2(s)$, its even part $r_2(s)$ is devoid of poles on the $j\omega$-axis but may have zeros of even multiplicity on this axis.

From its definition (4.92), the input reflection coefficient $S_{11}(s)$ can be expressed in terms of the input impedance $Z_1(s)$ and the reference impedance $z_1(s)=1$ as

$$S_{11}(s)=\frac{Z_1(s)-1}{Z_1(s)+1} \tag{11.5}$$

The problem is now similar to the Darlington procedure for which the input reflection coefficient $S_{11}(s)$ is specified and we wish to determine $S_{12}(s)$ and $S_{22}(s)$ so that the three parameters $S_{11}(s)$, $S_{22}(s)$, and $S_{12}(s)=S_{21}(s)$, known as the *scattering parameters*, are compatible for the realization of a two-port network N. It can be shown that for a lossless two-port network N, its scattering matrix must be para-unitary for almost all the complex frequencies.† A scattering matrix is *para-unitary* if

$$\mathbf{S}'(-s)\mathbf{S}(s)=\mathbf{S}(s)\mathbf{S}'(-s)=\mathbf{U}_2 \tag{11.6}$$

whenever it is defined, yielding

$$S_{11}(s)S_{11}(-s)+S_{12}(s)S_{12}(-s)=1 \tag{11.7a}$$

$$S_{22}(s)=-S_{11}(-s)S_{12}(s)/S_{12}(-s) \tag{11.7b}$$

Substituting (11.5) in (11.7a) yields

$$\begin{aligned} S_{12}(s)S_{12}(-s)&=\frac{2[Z_1(s)+Z_1(-s)]}{[Z_1(s)+1][Z_1(-s)+1]} \\ &=\frac{4k_1(s)k_1(-s)}{[Z_1(s)+1][Z_1(-s)+1]} \end{aligned} \tag{11.8}$$

where the even part $r_1(s)$ of $Z_1(s)$ is decomposed as

$$r_1(s)=\tfrac{1}{2}[Z_1(s)+Z_1(-s)]=k_1(s)k_1(-s) \tag{11.9}$$

the poles and zeros of $k_1(s)$ being chosen in accordance with the following three requirements:

(i) $k_1(s)$ is analytic in the open RHS;
(ii) the reciprocal of $k_1(s)$ is analytic in the open RHS;
(iii) $k_1(s)$ is the ratio of two polynomials of minimal order.

Thus, like $h_2(s)$, all the poles of $k_1(s)$ are in the open LHS. However, the zeros of $k_1(s)$ are restricted to the closed LHS. That this decomposition is always possible follows directly from the facts that the poles and zeros of the even part $r_1(s)$ of $Z_1(s)$ possess the quadrantal symmetry, and that, as for $r_2(s)$, on the $j\omega$-axis, $r_1(s)$ is devoid of any poles but may have zeros of even multiplicity. We remark that since $Z_1(s)$ is not used

†See, for example, W. K. Chen 1976, *Theory and Design of Broadband Matching Networks*, Chapter 2. New York: Pergamon Press.

in the normalization of $\mathbf{S}(s)$, we have freedom in choosing its decomposition $k_1(s)$ as outlined above. From (11.8), the most general solution for $S_{12}(s)$ can be identified as

$$S_{12}(s)=\eta(s)\frac{2k_1(s)}{Z_1(s)+1} \tag{11.10}$$

$\eta(s)$ being an all-pass function.

Finally, to identify $S_{22}(s)$, we substitute (11.5) and (11.10) in (11.7b), giving

$$S_{22}(s)=\eta^2(s)\frac{k_1(s)}{k_1(-s)}\cdot\frac{1-Z_1(-s)}{1+Z_1(s)} \tag{11.11}$$

11.2 COMPATIBILITY THEOREM

We now proceed to determine the conditions imposed on the all-pass function $\eta(s)$ so that the three scattering parameters defined in (11.5), (11.10), and (11.11) characterize a physical two-port network N. For this we shall appeal to a result that expresses the scattering matrix of a two-port network in terms of the admittance matrix of its augmented two-port network. It can be shown that the scattering matrix $\mathbf{S}(s)$ of N can be expressed in terms of the admittance matrix $\mathbf{Y}_{a1}(s)$ of the augmented two-port network N_{a1} of Figure 11.2 by the relation†

$$\mathbf{S}(s)=\mathbf{h}(s)\mathbf{h}^{-1}(-s)-2\mathbf{h}(s)\mathbf{Y}_{a1}(s)\mathbf{h}(s) \tag{11.12}$$

where

$$\mathbf{Y}_{a1}(s)=[\mathbf{Z}(s)+\mathbf{z}(s)]^{-1} \tag{11.13}$$

$\mathbf{Z}(s)$ being the impedance matrix of N. Solving for $Y_{a1}(s)$ in (11.12) yields

$$\mathbf{Y}_{a1}(s)=\tfrac{1}{2}\mathbf{h}^{-1}(s)[\mathbf{h}(s)\mathbf{h}^{-1}(-s)-\mathbf{S}(s)]\mathbf{h}^{-1}(s) \tag{11.14}$$

Writing

$$\mathbf{Y}_{a1}(s)=\begin{bmatrix} y_{a11}(s) & y_{a12}(s) \\ y_{a21}(s) & y_{a22}(s)\end{bmatrix} \tag{11.15}$$

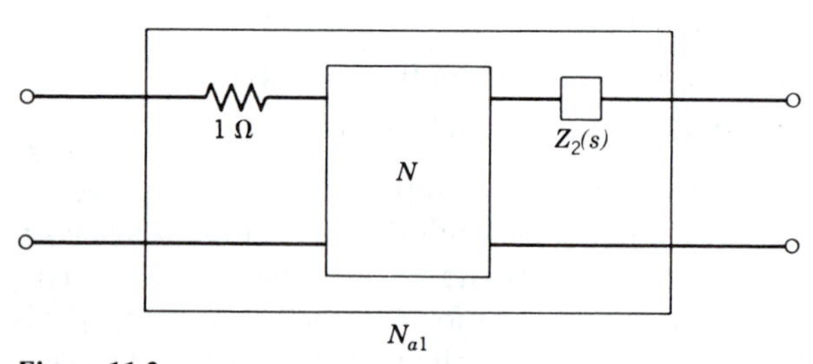

Figure 11.2

†W. K. Chen 1976, *Theory and Design of Broadband Matching Networks*, Chapter 2. New York: Pergamon Press.

and substituting the appropriate quantities in (11.14) gives the short-circuit admittance parameters of the augmented two-port network N_{a1}:

$$y_{a11}(s)=\frac{1}{Z_1(s)+1} \tag{11.16a}$$

$$y_{a12}(s)=y_{a21}(s)=-\frac{k_1(s)}{h_2(s)}\cdot\frac{\eta(s)}{Z_1(s)+1} \tag{11.16b}$$

$$y_{a22}(s)=\frac{1}{2h_2^2(s)}\left[\frac{h_2(s)}{h_2(-s)}-\eta^2(s)\frac{k_1(s)}{k_1(-s)}\cdot\frac{1-Z_1(-s)}{1+Z_1(s)}\right] \tag{11.16c}$$

According to (11.13), the impedance matrix $\mathbf{Z}(s)$ of N, if it exists, must be given by (Problem 11.1)

$$\mathbf{Z}(s)=\begin{bmatrix} Z_1(s)+2k_1^2(s)\eta^2(s)/q(s) & 2k_1(s)h_2(s)\eta(s)/q(s) \\ 2k_1(s)h_2(s)\eta(s)/q(s) & -Z_2(s)+2h_2^2(s)/q(s)\end{bmatrix} \tag{11.17}$$

where

$$q(s)=\frac{h_2(s)}{h_2(-s)}-\eta^2(s)\frac{k_1(s)}{k_1(-s)} \tag{11.18}$$

For the admittance matrix $\mathbf{Y}_{a1}(s)$ to be physically realizable, the two given impedances $Z_1(s)$ and $Z_2(s)$ must satisfy certain conditions. These conditions reduce essentially to the existence of an all-pass function $\eta(s)$ having certain properties. To exhibit the constraints succinctly, we state the results as a theorem, relegating its proof to Section 11.5. The result was first stated by Wohlers.†

Theorem 11.1 Let $Z_1(s)$ and $Z_2(s)$ be two rational positive-real impedances, which are not reactance functions. Let $Z_2(s)$ be minimum reactance, being devoid of poles on the real-frequency axis. Then the necessary and sufficient conditions that $Z_1(s)$ and $Z_2(s)$ be compatible, possessing a reciprocal realization, are that

(i) The even part $r_1(s)$ of $Z_1(s)$ must contain all the zeros of the even part $r_2(s)$ of $Z_2(s)$ to at least the same multiplicity. Furthermore, the order of the zeros of $r_1(s)$, plus the order of the zeros of $r_2(s)$, must be an even integer at the open RHS zeros of $r_2(s)$.

(ii) There exists an all-pass function $\hat{\eta}(s)$ so that the function $f(s)$ defined by the relation

$$f(s)=\frac{1}{2h_2^2(s)}\left[\frac{h_2(s)}{h_2(-s)}-\hat{\eta}^2(s)\frac{k_1(s)}{k_1(-s)}\cdot\frac{n_1(-s)}{n_1(s)}\cdot\frac{n_2(-s)}{n_2(s)}\right] \tag{11.19}$$

is analytic in the open RHS, where $r_1(s)=k_1(s)k_1(-s)$ and $r_2(s)=$

†M. R. Wohlers, "Complex normalization of scattering matrices and the problem of compatible impedances," *IEEE Trans. Circuit Theory*, vol. CT-12, 1965, pp. 528–535.

$h_2(s)h_2(-s)$, whose factorizations $k_1(s)$ and $h_2(s)$ are described in paragraphs below (11.9) and (11.4b), respectively, and

$n_1(-s)$ = the polynomial formed by all the open RHS zeros of $r_1(s)$, which coincide with the zeros of $r_2(s)$,

$n_2(-s)$ = the polynomial formed by all the open RHS zeros of $r_2(s)$.

(iii) The function

$$y_{a22}(s) = \frac{1}{2h_2^2(s)}\left[\frac{h_2(s)}{h_2(-s)} - \eta^2(s)\frac{k_1(s)}{k_1(-s)}\cdot\frac{1-Z_1(-s)}{1+Z_1(s)}\right] \tag{11.20}$$

can at most have simple poles on the $j\omega$-axis with positive-real residues, where

$$\eta^2(s) = \hat{\eta}^2(s)\frac{n_1(-s)n_2(-s)}{n_1(s)n_2(s)} \tag{11.21}$$

It appears that the compatibility problem has simply been recast into a new form that may be more inconvenient. A careful examination of (11.19) reveals that condition (ii) reduces essentially to the existence of an all-pass function $\hat{\eta}(s)$ that interpolates to prescribed values at desired points, being the zeros of $r_2(s)$, in the closed RHS. To see this, we first observe that the function defined by the ratio $h_2(s)/h_2(-s)$ is an all-pass function, being analytic in the closed RHS and containing all the RHS zeros of $h_2(s)$ to exactly the same multiplicity. Let s_0 be an open RHS zero of multiplicity m of $r_2(s)$. Write explicitly

$$h_2(s) = (s-s_0)^m\hat{h}_2(s) \tag{11.22a}$$

$$k_1(-s) = (s-s_0)^m\hat{k}_1(-s) \tag{11.22b}$$

$$n_1(-s) = (s-s_0)^m\hat{n}_1(-s) \tag{11.23a}$$

$$n_2(-s) = (s-s_0)^m\hat{n}_2(-s) \tag{11.23b}$$

Substituting these in (11.19) yields

$$f(s) = \frac{1}{2(s-s_0)^m\hat{h}_2^2(s)}\left[\frac{\hat{h}_2(s)}{h_2(-s)} - \hat{\eta}^2(s)\frac{k_1(s)\hat{n}_1(-s)\hat{n}_2(-s)}{\hat{k}_1(-s)n_1(s)n_2(s)}\right] \tag{11.24}$$

Since $\hat{h}_2(s)$ is finite nonzero at $s=s_0$, for $f(s)$ to be analytic in the open RHS, it is necessary that the function inside the brackets contain a zero at $s=s_0$ to at least the multiplicity m, or equivalently

$$\frac{d^u}{ds^u}\left[\frac{\hat{h}_2(s)}{h_2(-s)}\right]\Bigg|_{s=s_0} = \frac{d^u}{ds^u}\left[\hat{\eta}^2(s)\frac{k_1(s)\hat{n}_1(-s)\hat{n}_2(-s)}{\hat{k}_1(-s)n_1(s)n_2(s)}\right]\Bigg|_{s=s_0}, \qquad u=0,1,2,\ldots,m-1 \tag{11.25}$$

Thus, to satisfy condition (ii), an all-pass function $\hat{\eta}(s)$ must exist, so that it assumes the prescribed values, including its derivatives to the order $m-1$, at each open RHS zero s_0 of multiplicity m of $r_2(s)$.

We shall illustrate the theorem by the following examples.

Example 11.1 Consider the positive-real impedances

$$Z_1(s)=\frac{14s^2+9s+4}{14s^3+9s^2+6s+1} \tag{11.26a}$$

$$Z_2(s)=\frac{2s+4}{2s+1} \tag{11.26b}$$

as given in (4.1) and (4.2), whose even parts are obtained as

$$r_1(s)=\frac{4(1+s)(1-s)}{(14s^3+9s^2+6s+1)(-14s^3+9s^2-6s+1)} \tag{11.27a}$$

$$r_2(s)=\frac{4(1+s)(1-s)}{(1+2s)(1-2s)} \tag{11.27b}$$

Thus, $s=1$ is an open RHS zero of $r_1(s)$ and $r_2(s)$ and of multiplicity one, and condition (i) is satisfied. To test condition (ii), we perform the factorizations of $r_1(s)$ and $r_2(s)$ according to the rules outlined earlier. This gives

$$k_1(s)=\frac{2(s+1)}{14s^3+9s^2+6s+1} \tag{11.28a}$$

$$h_2(s)=\frac{2(1-s)}{1+2s} \tag{11.28b}$$

Also, we have

$$n_1(s)=s+1 \tag{11.29a}$$

$$n_2(s)=s+1 \tag{11.29b}$$

Substituting (11.28) and (11.29) in (11.19) yields

$$f(s)=\frac{(2s+1)^2}{8(s+1)(s-1)}\left[\frac{2s-1}{2s+1}-\hat{\eta}^2(s)\,\frac{14s^3-9s^2+6s-1}{14s^3+9s^2+6s+1}\right] \tag{11.30}$$

For $f(s)$ to be analytic in the open RHS, it is necessary that there be an all-pass function $\hat{\eta}(s)$ that assumes the value

$$\hat{\eta}^2(1)=\left.\frac{(2s-1)(14s^3+9s^2+6s+1)}{(2s+1)(14s^3-9s^2+6s-1)}\right|_{s=1}=1 \tag{11.31}$$

at the open RHS zero $s_0=1$ of $r_2(s)$, yielding $\hat{\eta}(s)=1$. This shows that condition (ii) is satisfied.

Finally, to test condition (iii), from (11.21) we obtain

$$\eta^2(s)=\hat{\eta}^2(s)\,\frac{n_1(-s)n_2(-s)}{n_1(s)n_2(s)}=\frac{(s-1)^2}{(s+1)^2} \tag{11.32}$$

giving

$$\eta(s)=\pm\frac{s-1}{s+1} \tag{11.33}$$

Substituting (11.26a), (11.28), and (11.32) in (11.20) yields

$$y_{a22}(s)=\frac{(2s+1)(-s^3+s^2-s-1)}{(s^2-1)(14s^3+23s^2+15s+5)}$$

$$=\frac{(2s+1)(s+1)}{14s^3+23s^2+15s+5} \tag{11.34}$$

which is analytic in the closed RHS. In fact, $y_{a22}(s)$ is a positive-real function, as it must be. Thus, condition (iii) is automatically satisfied, and the given impedances $Z_1(s)$ and $Z_2(s)$ are compatible.

Using (11.5), (11.10), and (11.11), the scattering matrix of the lossless reciprocal two-port network N, normalizing to the impedances 1 and $Z_2(s)$, can now be determined, and is given by

$$\mathbf{S}(s)=\begin{bmatrix} \dfrac{-14s^3+5s^2+3s+3}{14s^3+23s^2+15s+5} & \dfrac{\pm 4(s-1)}{14s^3+23s^2+15s+5} \\ \dfrac{\pm 4(s-1)}{14s^3+23s^2+15s+5} & \dfrac{(s-1)(14s^3+5s^2-3s+3)}{(s+1)(14s^3+23s^2+15s+5)} \end{bmatrix} \tag{11.35}$$

It is easy to confirm that this matrix, being analytic in the closed RHS, is para-unitary whenever it is defined (Problem 11.2).

Example 11.2 Determine the compatibility of the following two positive-real impedances:

$$Z_1(s)=\frac{6s^4+6s^3+12s^2+7s+3}{6s^5+6s^4+18s^3+13s^2+9s+2} \tag{11.36a}$$

$$Z_2(s)=\frac{6s^2+5s+3}{s^2+3s+2} \tag{11.36b}$$

The even parts of these two impedances are given by

$$r_1(s)=\frac{6(s^4+1)}{(6s^5+6s^4+18s^3+13s^2+9s+2)(-6s^5+6s^4-18s^3+13s^2-9s+2)}$$
$$=k_1(s)k_1(-s) \tag{11.37a}$$

$$r_2(s)=\frac{6(s^4+1)}{(s^2+3s+2)(s^2-3s+2)}=h_2(s)h_2(-s) \tag{11.37b}$$

whose unique factorizations $k_1(s)$ and $h_2(s)$, up to a sign, are obtained as

$$k_1(s)=\frac{\sqrt{6}(s^2+\sqrt{2}s+1)}{6s^5+6s^4+18s^3+13s^2+9s+2} \tag{11.38a}$$

$$h_2(s)=\frac{\sqrt{6}(s^2-\sqrt{2}s+1)}{s^2+3s+2} \tag{11.38b}$$

Also, we have

$$n_1(s) = n_2(s) = s^2 + \sqrt{2}s + 1 \tag{11.39}$$

Thus, condition (i) of the theorem is satisfied.

To test condition (ii), we substitute (11.38) and (11.39) in (11.19) yielding

$$f(s) = \frac{(s^2+3s+2)^2}{12(s^4+1)}\left[\frac{s^2-3s+2}{s^2+3s+2} - \hat{\eta}^2(s)\frac{w(-s)}{w(s)}\right] \tag{11.40}$$

where

$$w(s) = 6s^5 + 6s^4 + 18s^3 + 13s^2 + 9s + 2 \tag{11.41}$$

Since $s^4 + 1 = 0$ has two open RHS roots located at the points

$$s_0,\, \bar{s}_0 = \frac{1}{\sqrt{2}} \pm j\frac{1}{\sqrt{2}} \tag{11.42}$$

for $f(s)$ to be analytic in the open RHS, we require that there be an all-pass function $\hat{\eta}(s)$ that assumes the values

$$\hat{\eta}^2(1/\sqrt{2} \pm j1/\sqrt{2}) = \left.\frac{(s^2-3s+2)w(s)}{(s^2+3s+2)w(-s)}\right|_{s=1/\sqrt{2}\pm j1/\sqrt{2}} \tag{11.43}$$

at the open RHS points s_0 and $\bar{s}_0$. A straightforward computation gives

$$\hat{\eta}^2(1/\sqrt{2} \pm j1/\sqrt{2}) = \frac{(-4 \mp j3\sqrt{2})(-148 \pm j111\sqrt{2})}{(14+9\sqrt{2})(518-333\sqrt{2})} = 1 \tag{11.44}$$

showing that $\hat{\eta}(s) = 1$ and condition (ii) is fulfilled.

Finally, to test condition (iii), we obtain from (11.21)

$$\eta(s) = \pm\frac{s^2 - \sqrt{2}s + 1}{s^2 + \sqrt{2}s + 1} \tag{11.45}$$

Substituting (11.36a), (11.38), and (11.45) in (11.20) yields

$$y_{a22}(s) = \frac{(s^2+3s+2)(s^7+s^6+2s^5+s^4+s^3+s^2+2s+1)}{(s^4+1)(6s^5+12s^4+24s^3+25s^2+16s+5)}$$

$$= \frac{(s^2+3s+2)(s^3+s^2+2s+1)}{6s^5+12s^4+24s^3+25s^2+16s+5} \tag{11.46}$$

Appealing to the Hurwitz test shows that it is analytic in the closed RHS. Thus, condition (iii) is also satisfied, and the impedances $Z_1(s)$ and $Z_2(s)$ are compatible.

Using (11.5), (11.10), and (11.11), the scattering matrix of the lossless reciprocal two-port network N, normalizing to the reference impedances 1 and $Z_2(s)$, is obtained as

$$\mathbf{S}(s) = \frac{1}{g(s)}\begin{bmatrix} -6s^5 - 12s^3 - s^2 - 2s + 1 & \pm 2\sqrt{6}(s^2 - \sqrt{2}s + 1) \\ \pm 2\sqrt{6}(s^2 - \sqrt{2}s + 1) & \dfrac{(s^2-\sqrt{2}s+1)}{(s^2+\sqrt{2}s+1)}(-6s^5 - 12s^3 + s^2 - 2s - 1) \end{bmatrix} \tag{11.47a}$$

where

$$g(s) = 6s^5 + 12s^4 + 24s^3 + 25s^2 + 16s + 5 \tag{11.47b}$$

It is not difficult to confirm that this matrix is para-unitary whenever it is defined (Problem 11.7).

The realization of the two-port network N will be considered in the following section.

11.3 REALIZATION OF THE LOSSLESS RECIPROCAL TWO-PORT NETWORK *N*

Having established that two impedances $Z_1(s)$ and $Z_2(s)$ are compatible, we turn next to a study of the procedures for the realization of the two-port network N. To facilitate our discussion, two cases are distinguished, each being presented in a separate section.

11.3.1 $\mathbf{Y}_{a1}(s)$ Being Nonsingular

By hypothesis, the admittance matrix $\mathbf{Y}_{a1}(s)$ of the augmented two-port network N_{a1} of Figure 11.2 is nonsingular. Then, as indicated in (11.13), the impedance matrix

$$\mathbf{Z}(s) = \begin{bmatrix} z_{11}(s) & z_{12}(s) \\ z_{12}(s) & z_{22}(s) \end{bmatrix} \tag{11.48}$$

of N exists and is given by (11.17). Using the technique outlined in Section 4.2 of Chapter 4 with some modifications, the two-port network N can be realized. As before, we first expand the elements $z_{ij}(s)$ in partial fractions, and then realize each column of like terms in the expansions as a lossless two-port network. The only difference is that the poles may not all be compact. This can easily be taken care of by dividing the residues of $z_{11}(s)$ and $z_{22}(s)$ at each noncompact pole into two parts, so that using one of the two parts in the residue condition satisfies it by the equality sign. The terms corresponding to the excess residues can be treated as the private poles of $z_{11}(s)$ and $z_{22}(s)$, which lead to reactive networks connected in series at the input and the output ports of the component two-port networks, as discussed in Section 4.2 of Chapter 4. There is hardly any point in going through the same ground again in detail, and we are satisfied with this brief summary.

As an alternative, we can apply the method of cascade synthesis of Chapter 10 to the realization of the two-port network N. Refer to Figure 11.3. The input impedance of N, when terminated in a 1-Ω resistance, can be expressed in terms of the impedance parameters $z_{ij}(s)$ of (11.48) by the known relation (Problem 11.16)

$$Z_{11}(s) = z_{11}(s) - \frac{z_{12}^2(s)}{z_{22}(s) + 1} \tag{11.49}$$

Substituting the elements of (11.17) in (11.49) gives

$$Z_{11}(s) = Z_1(s) + \frac{2k_1^2(s)\eta^2(s)}{q(s) + 2h_2^2(s)/[1 - Z_2(s)]} \tag{11.50}$$

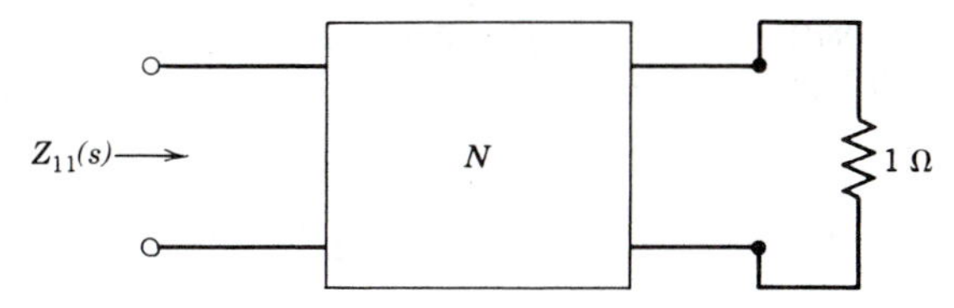

Figure 11.3

which is known to be positive real. Using the cascade synthesis technique of Chapter 10, $Z_{11}(s)$ can be realized as the input impedance of a lossless reciprocal two-port network N terminated in a 1-Ω resistor. When the resistor is removed, we obtain a lossless reciprocal two-port network, which when terminated in $Z_2(s)$ may not necessarily yield the original impedance $Z_1(s)$. However, as indicated in (11.19), the flexibility inherent in the nonuniqueness of $\hat{\eta}(s)$ can be used to guarantee the equality of the realized impedance and $Z_1(s)$.

11.3.2 $\mathbf{Y}_{a1}(s)$ Being Singular

Now suppose that the admittance matrix $\mathbf{Y}_{a1}(s)$ is singular. Then from (11.16) we find that

$$\det \mathbf{Y}_{a1}(s) = y_{a11}(s)y_{a22}(s) - y_{a12}^2(s)$$
$$= \frac{\left[\dfrac{h_2(s)}{h_2(-s)} - \eta^2(s)\dfrac{k_1(s)}{k_1(-s)}\right]}{2h_2^2(s)[Z_1(s)+1]} = 0 \tag{11.51}$$

which is equivalent to

$$\eta^2(s) = \frac{k_1(-s)h_2(s)}{k_1(s)h_2(-s)} \tag{11.52}$$

Under this situation, the admittance parameters $y_{aij}(s)$ of (11.16) become

$$y_{a11}(s) = \frac{1}{Z_1(s)+1} \tag{11.53a}$$

$$y_{a12}(s) = y_{a21}(s) = \frac{-r_1^{1/2}(s)}{r_2^{1/2}(s)[Z_1(s)+1]} \tag{11.53b}$$

$$y_{a22}(s) = \frac{r_1(s)}{r_2(s)[Z_1(s)+1]} \tag{11.53c}$$

Since $y_{a22}(s)$, being the short-circuit admittance parameter at the output port of the augmented two-port network N_{a1}, is positive real, from condition (i) of the theorem we conclude that this is possible for (11.53c) only if

$$r_1(s) = cr_2(s) \tag{11.54}$$

c being a real and positive constant (Problem 11.9). But a positive-real function is uniquely determined from its even part to within a reactance function. This implies that

$$Z_1(s) = cZ_2(s) + Z_{LC}(s) \tag{11.55}$$

$Z_{LC}(s)$ being a reactance function containing all the $j\omega$-axis poles of $Z_1(s)$ that do not occur in $Z_2(s)$. The resulting augmented two-port network N_{a1} is presented in Figure 11.4, from which we can identify the desired two-port network N as shown in Figure 11.5.

11.3.3 Illustrative Examples

Example 11.3 We wish to realize a lossless reciprocal two-port network for the two compatible impedances considered in Example 11.1.

Using (11.16) in conjunction with the results obtained in Example 11.1, we compute the short-circuit admittance parameters of N_{a1}:

$$y_{a11}(s) = \frac{14s^3 + 9s^2 + 6s + 1}{14s^3 + 23s^2 + 15s + 5} \tag{11.56a}$$

$$y_{a12}(s) = y_{a21}(s) = \pm \frac{2s + 1}{14s^3 + 23s^2 + 15s + 5} \tag{11.56b}$$

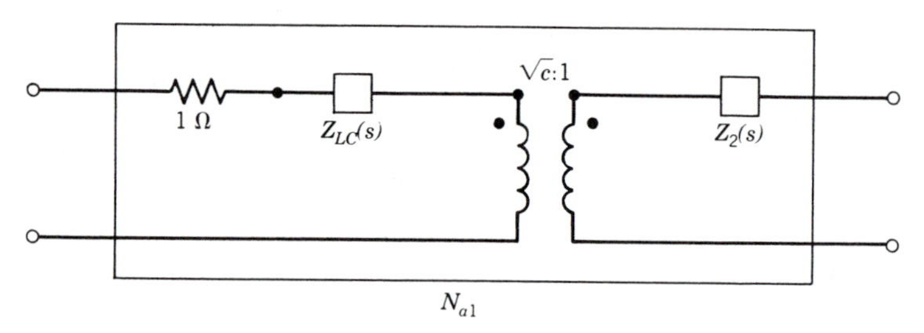

Figure 11.4

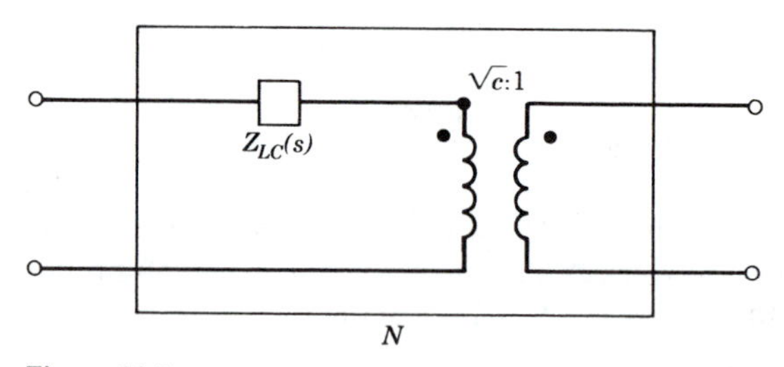

Figure 11.5

$$y_{a22}(s) = \frac{(2s+1)(s+1)}{14s^3+23s^2+15s+5} \tag{11.56c}$$

The determinant of $\mathbf{Y}_{a1}(s)$ is obtained as

$$\begin{aligned} \det \mathbf{Y}_{a1}(s) &= y_{a11}(s)y_{a22}(s) - y_{a12}^2(s) \\ &= \frac{s(2s+1)}{14s^3+23s^2+15s+5} \end{aligned} \tag{11.57}$$

Thus, the impedance matrix $\mathbf{Z}(s)$ of N exists and can be determined either directly by (11.17) or indirectly by (11.13). For illustrative purposes, we use (11.13), giving

$$\begin{aligned} \mathbf{Z}(s) &= \mathbf{Y}_{a1}^{-1}(s) - \mathbf{z}(s) \\ &= \begin{bmatrix} \dfrac{s+1}{s} & \pm\dfrac{1}{s} \\ \pm\dfrac{1}{s} & \dfrac{14s^3+9s^2+6s+1}{s(2s+1)} \end{bmatrix} - \begin{bmatrix} 1 & 0 \\ 0 & \dfrac{2s+4}{2s+1} \end{bmatrix} \\ &= \begin{bmatrix} \dfrac{1}{s} & \pm\dfrac{1}{s} \\ \pm\dfrac{1}{s} & \dfrac{7s^2+1}{s} \end{bmatrix} \end{aligned} \tag{11.58}$$

Applying (11.49) yields the input impedance of N when its output is terminated in a 1-Ω resistor:

$$Z_{11}(s) = \frac{7s+1}{7s^2+s+1} \tag{11.59}$$

This positive-real impedance can be realized as a lossless reciprocal two-port network N terminated in a 1-Ω resistor. From computational and practical viewpoints, we should first go through the usual Foster preamble by stripping $Z_{11}(s)$ of all real-frequency zeros and poles. This results in

$$Z_{11}(s) = \cfrac{1}{s + \cfrac{1}{7s+1}} \tag{11.60}$$

which can be identified as a lossless ladder terminated in a 1-Ω resistor, as depicted in Figure 11.6. Removing the 1-Ω resistor results in a lossless reciprocal two-port network N, which when terminated in $Z_2(s)$, yields the input impedance $Z_1(s)$, as shown in Figure 11.7.

Example 11.4 Consider the two compatible impedances $Z_1(s)$ and $Z_2(s)$ of (11.36). We wish to realize the lossless reciprocal two-port network N.

As before, we use (11.16) to compute the short-circuit admittance parameters

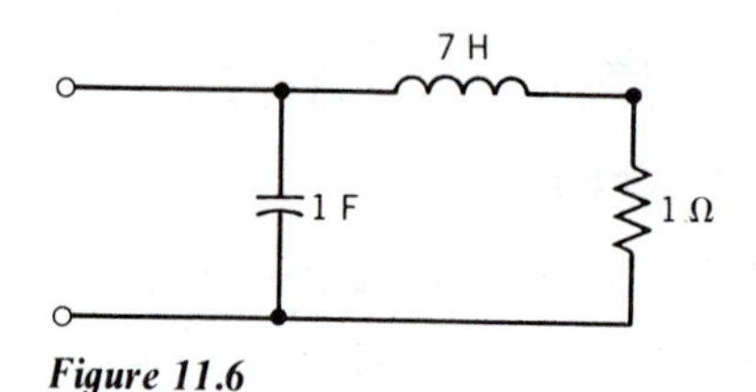

Figure 11.6

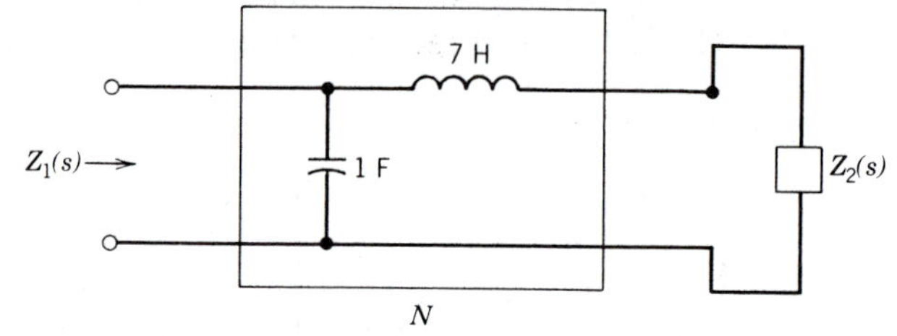

Figure 11.7

$y_{aij}(s)$ of N_{a1}, giving

$$y_{a11}(s)=\frac{6s^5+6s^4+18s^3+13s^2+9s+2}{6s^5+12s^4+24s^3+25s^2+16s+5} \tag{11.61a}$$

$$y_{a12}(s)=y_{a21}(s)=\frac{\pm(s^2+3s+2)}{6s^5+12s^4+24s^3+25s^2+16s+5} \tag{11.61b}$$

$$y_{a22}(s)=\frac{(s^2+3s+2)(s^3+s^2+2s+1)}{6s^5+12s^4+24s^3+25s^2+16s+5} \tag{11.61c}$$

The determinant of $\mathbf{Y}_{a1}(s)$ is given by

$$\det \mathbf{Y}_{a1}(s)=\frac{s(s^2+2)(s^2+3s+2)}{6s^5+12s^4+24s^3+25s^2+16s+5} \tag{11.62}$$

Finally, the impedance matrix $\mathbf{Z}(s)$ of N is obtained as

$$\begin{aligned}\mathbf{Z}(s)&=\mathbf{Y}_{a1}^{-1}(s)-\mathbf{z}(s)\\
&=\begin{bmatrix}\dfrac{s^3+s^2+2s+1}{s(s^2+2)} & \dfrac{\pm1}{s(s^2+2)}\\[2ex] \dfrac{\pm1}{s(s^2+2)} & \dfrac{6s^5+6s^4+18s^3+13s^2+9s+2}{s(s^2+2)(s^2+3s+2)}\end{bmatrix}-\begin{bmatrix}1 & 0\\ 0 & Z_2(s)\end{bmatrix}\\
&=\frac{1}{s(s^2+2)}\begin{bmatrix}s^2+1 & \pm1\\ \pm1 & s^2+1\end{bmatrix}\end{aligned} \tag{11.63}$$

The input impedance of N when terminated in a 1-Ω resistor is found to be

$$Z_{11}(s)=\frac{s^4+s^3+3s^2+2s+2}{(s^2+2)(s^3+s^2+2s+1)}=\frac{s^2+s+1}{s^3+s^2+2s+1} \tag{11.64}$$

Proceeding as before to strip all of its real-frequency axis poles and zeros yields

$$Z_{11}(s)=\cfrac{1}{s+\cfrac{1}{s+\cfrac{1}{s+1}}} \tag{11.65}$$

which can be identified as a lossless ladder terminated in a 1-Ω resistor, as indicated in Figure 11.8. Removing the 1-Ω resistor results in N, which when terminated in $Z_2(s)$ yields the input impedance $Z_1(s)$, as shown in Figure 11.9.

Alternatively, the input impedance $Z_{11}(s)$ can be computed directly from (11.50) as

$$Z_{11}(s)=Z_1(s)$$

$$-\frac{(s^4+1)(s^2+3s+2)(5s^2+2s+1)}{(6s^5+6s^4+18s^3+13s^2+9s+2)(s^9+4s^8+7s^7+9s^6+8s^5+6s^4+7s^3+9s^2+7s+2)}$$

$$=\frac{6s^7+12s^6+30s^5+37s^4+40s^3+24s^2+11s+2}{(6s^5+6s^4+18s^3+13s^2+9s+2)(s^3+s^2+2s+1)}$$

$$=\frac{s^2+s+1}{s^3+s^2+2s+1} \tag{11.66}$$

confirming (11.64).

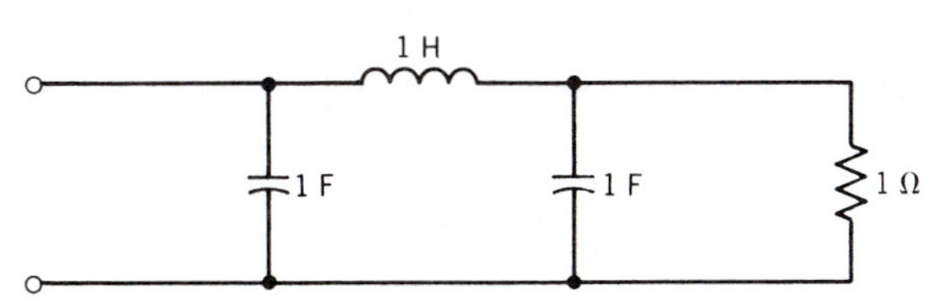

Figure 11.8

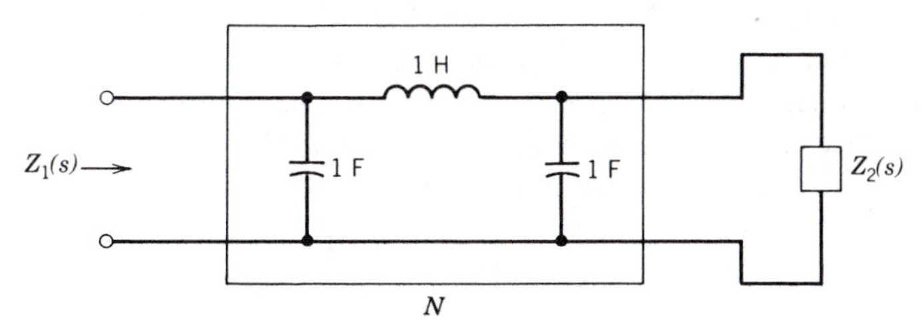

Figure 11.9

11.4 SPECIAL COMPATIBLE IMPEDANCES

As demonstrated in Chapters 4 and 10, if $Z_2(s)=R$, a positive resistance, then any rational positive-real impedance $Z_1(s)$ is compatible with $Z_2(s)$. This result also follows directly from Theorem 11.1 by observing that for $Z_2(s)=R$, all of the three conditions of the theorem are satisfied. In the following, we shall present two other special classes of compatible impedances, and determine the most general lumped, lossless, and reciprocal two-port network N associated with these two classes.

11.4.1 Equal Compatible Impedances

We consider the problem of determining the two-port configuration that, when terminated in a rational positive-real impedance $Z_2(s)=Z(s)$, yields an input impedance equal to $Z_1(s)=Z(s)$. To facilitate our discussion, two cases are distinguished. For simplicity, write $r_1(s)=r_2(s)=r(s)$.

Case 1: $r(s)$ possesses an open RHS zero. Under this situation, we have

$$\frac{h_2(s)}{h_2(-s)}=\frac{k_1(s)n_1(-s)n_2(-s)}{k_1(-s)n_1(s)n_2(s)} \tag{11.67}$$

and the only possible choice of the all-pass function $\hat{\eta}(s)$ in (11.19) of Theorem 11.1 is one, and from (11.21) we have

$$\eta^2(s)=\frac{h_2(s)k_1(-s)}{h_2(-s)k_1(s)} \tag{11.68}$$

But this is the same condition as given in (11.52) where $\mathbf{Y}_{a1}(s)$ is singular. Hence, (11.55) applies and $c=1$ and $Z_{LC}(s)=0$. The two-port network N of Figure 11.5 degenerates into a pair of wires connecting the ports, a trivial solution.

Case 2: $r(s)$ is devoid of any open RHS zeros. Under this stipulation, $k_1(s)=h_2(s)$ and (11.16c) becomes

$$y_{a22}(s)=\frac{1-\eta^2(s)}{2r(s)}+\frac{\eta^2(s)}{Z(s)+1} \tag{11.69}$$

Since $1-\eta^2(s)$ is analytic in the open RHS and since $r(s)$ is devoid of poles on the $j\omega$-axis, from Problem 11.11 we conclude that, for $[1-\eta^2(s)]/2r(s)$ to be positive real, $r(s)$ must be analytic in the closed RHS. But the poles of $r(s)$, being an even function, possess quadrantal symmetry. It follows that $r(s)$ must be analytic in the entire s-plane, infinity included. Consequently, $Z(s)$ must be a nonnegative resistance R. Without loss of generality, we may set $R=1$. From (11.12), the scattering matrix of N, normalizing to the 1-Ω resistance, is given by

$$\mathbf{S}(s)=\begin{bmatrix} 0 & \eta(s) \\ \eta(s) & 0 \end{bmatrix} \tag{11.70}$$

which can be realized by a constant-resistance all-pass structure. The conclusion is that this is the only situation under which two equal compatible impedances will produce a nontrivial two-port network N.

11.4.2 Constant Impedances

Assume that the impedance $Z_1(s)$ is a constant. Without loss of generality, we may set $Z_1(s)=1$. Thus, $r_1(s)=1$ and condition (i) of Theorem 11.1 requires that $r_2(s)=R$, a nonnegative constant. This shows that the only admissible $Z_2(s)$ is of the form

$$Z_2(s)=R+Z_{LC}(s) \tag{11.71}$$

$Z_{LC}(s)$ being a reactance function. If $Z_2(s)$ is restricted to be minimum reactance, $Z_{LC}(s)=0$ and the only impedance that is compatible to a constant $Z_1(s)$ is another resistance R. The corresponding two-port network N is obviously a transformer. As will be demonstrated in Section 11.6, we can arrive at the same conclusion without making the assumption that $Z_2(s)$ be minimum reactance. In other words, minimum reactance $Z_2(s)$ is a necessary condition for its compatibility with a constant $Z_1(s)$.

11.5 PROOF OF THE COMPATIBILITY THEOREM

In the present section, we prove the assertion that the three conditions of Theorem 11.1 are both necessary and sufficient for two rational positive-real impedances $Z_1(s)$ and $Z_2(s)$, which are not reactance functions and with $Z_2(s)$ being devoid of any real-frequency axis poles, to be compatible.

To prove the theorem, we must show that the scattering matrix $\mathbf{S}(s)$, whose elements are defined in (11.5), (11.10), and (11.11), indeed represents a linear, lumped, time-invariant, lossless, and reciprocal two-port network N normalizing to the reference impedances 1 and $Z_2(s)$. For this, we appeal to a result that characterizes the scattering matrix of a linear lossless, reciprocal two-port network.†

Theorem 11.2 The necessary and sufficient conditions for a 2×2 matrix $\mathbf{S}(s)$ to be the scattering matrix of a linear, lumped, time-invariant, lossless, and reciprocal two-port network, normalizing to the two minimum reactance functions, are that

(i) $\mathbf{S}(s)$ be rational, symmetric, and para-unitary whenever it is defined;

(ii) the matrix

$$\mathbf{Y}_{a1}(s)=\tfrac{1}{2}\mathbf{h}^{-1}(s)[\mathbf{d}(s)-\mathbf{S}(s)]\mathbf{h}^{-1}(s) \tag{11.72}$$

be analytic in the open RHS;

(iii) $\mathbf{Y}_{a1}(s)$ have at most simple poles on the real-frequency axis and the residue matrix at each of these poles be nonnegative definite;

†W. K. Chen 1976, *Theory and Design of Broadband Matching Networks*, Chapter 2. New York: Pergamon Press.

where $\mathbf{d}(s)=\mathbf{h}(s)\mathbf{h}^{-1}(-s)$, and $\mathbf{h}(s)\mathbf{h}(-s)$ is the para-hermitian part of the reference impedance matrix $\mathbf{z}(s)$ corresponding to the two minimum reactance functions.

Now we proceed to show that the scattering matrix we constructed in Section 11.1 possesses the above three attributes.

Since, by construction, $\mathbf{S}(s)$ is symmetric, and unitary whenever it is defined, condition (i) of the above theorem is satisfied. From (11.14) and (11.16), we see that since

$$y_{a11}(s)=\frac{1}{Z_1(s)+1} \tag{11.73}$$

is analytic in the closed RHS (Problem 11.13), the second condition is satisfied if

$$y_{a12}(s)=-\frac{k_1(s)}{h_2(s)}\cdot\frac{\eta(s)}{Z_1(s)+1} \tag{11.74}$$

and

$$y_{a22}(s)=\frac{1}{2h_2^2(s)}\left[\frac{h_2(s)}{h_2(-s)}-\eta^2(s)\frac{k_1(s)}{k_1(-s)}\cdot\frac{1-Z_1(-s)}{1+Z_1(s)}\right] \tag{11.75}$$

are analytic in the open RHS. To satisfy the third condition, we must demonstrate that $\mathbf{Y}_{a1}(s)$ has at most simple poles on the $j\omega$-axis and that the residue matrix evaluated at each of these poles is nonnegative definite.† To complete our proof, we show that the above two constraints are satisfied if and only if the three conditions of Theorem 11.1 hold.

Necessity Assume that the above two constraints are satisfied. Let $s=j\omega_0$ be a simple pole of $\mathbf{Y}_{a1}(s)$. Then since $y_{a11}(s)$ is analytic on the $j\omega$-axis, the residue matrix evaluated at this pole must be of the form

$$\mathbf{K}=\begin{bmatrix} 0 & k_{12} \\ k_{12} & k_{22} \end{bmatrix} \tag{11.76}$$

For $\mathbf{K}$ to be nonnegative definite, it is necessary that $k_{12}=0$ and $k_{22}\geqq 0$. But k_{22} cannot be zero; for, otherwise, $s=j\omega_0$ would not be a pole of $\mathbf{Y}_{a1}(s)$. Thus, condition (iii) of Theorem 11.1 is necessary. In addition, $y_{a12}(s)$ must be analytic on the entire real-frequency axis, infinity included. From (11.74), we see that this is possible only if $k_1(s)$ includes all the $j\omega$-axis zeros of $h_2(s)$ to at least the same multiplicity, since $\eta(s)$ is devoid of zeros on the $j\omega$-axis. In other words, the even part $r_1(s)$ of $Z_1(s)$ must contain all the $j\omega$-axis zeros of the even part $r_2(s)$ of $Z_2(s)$ to at least the same multiplicity.

Now we assert that the function

$$f_1(s)=\frac{k_1(s)}{k_1(-s)}\cdot\frac{1-Z_1(-s)}{1+Z_1(s)} \tag{11.77}$$

is analytic in the open RHS except at the points s_1 where $r_1(s_1)=0$, and the order of

†A second-order hermitian matrix $\mathbf{A}=[a_{ij}]$ is *nonnegative definite* if and only if $a_{11}\geqq 0$, $a_{22}\geqq 0$, and $a_{11}a_{22}-a_{12}a_{21}\geqq 0$.

these poles s_1 is equal to the order of the zeros of $r_1(s)$. To see this, we observe that, by construction, the poles of $k_1(-s)$ contain all the open RHS poles of $Z_1(-s)$ to the same multiplicity, meaning that the open RHS poles of $Z_1(-s)$ cannot be the poles of $f_1(s)$. The open RHS zeros of $k_1(-s)$ are also the open RHS zeros of $r_1(s)$. To complete our justification, it remains to be shown that no cancellations exist. For this, we show that, at an open RHS zero s_1 of $r_1(s)$, $1-Z_1(-s_1)\neq 0$. That this is the case follows directly from the fact that

$$2r_1(s_1)=Z_1(s_1)+Z_1(-s_1)=0 \tag{11.78}$$

showing that if $Z_1(-s_1)=1$ then $Z_1(s_1)=-1$, which is not possible for a positive-real $Z_1(s)$.

Also, observe that the all-pass function $h_2(s)/h_2(-s)$ is analytic in the closed RHS and contains the open RHS zeros of $h_2(s)$ to exactly the same multiplicity.

Let s_0 be an open RHS zero of $r_2(s)$ of multiplicity $m_2>0$. For simplicity, we say that s_0 is a zero of $r_1(s)$ of multiplicity $m_1\geqq 0$, and of $\eta(s)$ of multiplicity $m\geqq 0$. For $m_1=m=0$, $r_1(s_0)\neq 0$ and $\eta(s_0)\neq 0$. Write

$$h_2(s)=(s-s_0)^{m_2}h_2''(s) \tag{11.79}$$

$$f_1(s)=(s-s_0)^{-m_1}f_1''(s) \tag{11.80a}$$

$$\eta(s)=(s-s_0)^{m}\eta''(s) \tag{11.80b}$$

Substituting these in (11.75) yields

$$y_{a22}(s)=\frac{1}{2(s-s_0)^{m_2}h_2''^2(s)}\left[\frac{h_2''(s)}{h_2(-s)}-(s-s_0)^{2m-m_1-m_2}\eta''^2(s)f_1''(s)\right] \tag{11.81}$$

where $h_2''(s)$, $h_2(-s)$, $\eta''(s)$, and $f_1''(s)$ are finite nonzero at $s=s_0$. Thus, for $y_{a22}(s)$ to be analytic in the open RHS, it is necessary that

$$\frac{h_2''(s)}{h_2(-s)}-(s-s_0)^{2m-m_1-m_2}\eta''^2(s)f_1''(s) \tag{11.82}$$

possess a zero at $s=s_0$ to at least the multiplicity m_2. Clearly, this is possible only if

$$2m-m_1-m_2=0 \tag{11.83}$$

Observe also from (11.74) that for $y_{a12}(s)$ to be analytic in the open RHS, $\eta(s)$ must contain all the open RHS zeros of $h_2(s)$ to at least the same multiplicity, since $k_1(s)$ is devoid of zeros in the open RHS. Thus, we can write

$$m=m_2+k \tag{11.84}$$

k being a nonnegative integer. Combining (11.83) and (11.84) gives

$$m_1=m_2+2k \tag{11.85}$$

showing that the open RHS zeros of $r_1(s)$ must contain all the open RHS zeros of $r_2(s)$ to at least the same multiplicity. Since the zeros of $r_1(s)$ and $r_2(s)$ possess quadrantal symmetry, we conclude that the zeros of $r_1(s)$ must contain all the zeros of $r_2(s)$ to at least the same multiplicity. In addition, as shown in (11.83), the order of the zeros of $r_1(s)$, plus the order of the zeros of $r_2(s)$, must be an even integer at the open RHS zeros of $r_2(s)$. This in conjunction with a similar result obtained earlier for the $j\omega$-axis zeros of $r_2(s)$ shows that condition (i) of Theorem 11.1 is necessary.

From the above discussion, we see that the all-pass function $\eta(s)$ can be expressed as in (11.21) such that $\hat{\eta}(s)$ is not zero at the open RHS zeros of $r_2(s)$.

Finally, a simple analysis indicates that (11.75) can be written as

$$y_{a22}(s)=\frac{1}{2h_2^2(s)}\left[\frac{h_2(s)}{h_2(-s)}-\eta^2(s)\frac{k_1(s)}{k_1(-s)}\right]+\frac{\eta^2(s)k_1^2(s)}{h_2^2(s)[Z_1(s)+1]} \tag{11.86}$$

Under the restrictions imposed on $\eta(s)$, as given in (11.21), and the fact that the open RHS zeros of $r_1(s)$ must contain all the open RHS zeros of $r_2(s)$ to at least the same multiplicity, $\eta(s)$ contains all the open RHS zeros of $h_2(s)$ to at least the same multiplicity. Thus, the second term in (11.86) is analytic in the open RHS. Using this fact in conjunction with (11.21), we see that, for $y_{a22}(s)$ to be analytic in the open RHS, it is necessary that $f(s)$, as defined in (11.19), must be analytic in the open RHS, showing that condition (ii) of Theorem 11.1 is necessary.

Sufficiency Assume that the three constraints of Theorem 11.1 are satisfied. We show that $y_{a12}(s)$ and $y_{a22}(s)$ are analytic in the open RHS, and $\mathbf{Y}_{a1}(s)$ has at most simple poles on the real-frequency axis and the associated residue matrices are nonnegative definite.

From (11.74) and (11.86) in conjunction with (11.21), we see that conditions (i) and (ii) of Theorem 11.1 imply the analyticity of $y_{a12}(s)$ and $y_{a22}(s)$ in the open RHS.

Let $j\omega_0$ be a real-frequency axis pole of $\mathbf{Y}_{a1}(s)$. Since $y_{a12}(s)$ and $y_{a11}(s)$ are known to be analytic on the entire real-frequency axis, infinity included, the residue matrix of $\mathbf{Y}_{a1}(s)$ evaluated at this pole $j\omega_0$ must be of the form

$$\mathbf{K}=\begin{bmatrix}0 & 0\\ 0 & k_{22}\end{bmatrix} \tag{11.87}$$

Thus, the $j\omega$-axis poles of $\mathbf{Y}_{a1}(s)$ are also the $j\omega$-axis poles of $y_{a22}(s)$. But from condition (iii) of Theorem 11.1, these poles must be simple, if they exist, with positive residues, meaning that $k_{22}>0$ or $\mathbf{K}$ is nonnegative definite. This completes the proof of the theorem.

11.6 EXTENSION OF THE COMPATIBILITY THEOREM

The result stated in Theorem 11.1 can be extended to also include the situation where $Z_2(s)$ is not necessarily a minimum reactance function. According to Wohlers,† the theorem can be extended by inserting two additional constraints for the real-frequency axis poles of $Z_2(s)$, as follows:

(i) if $s=j\omega_0$ is a pole of $Z_2(s)$, but is not a pole of $Z_1(s)$, and if $j\omega_0$ is a zero of $r_2(s)$,

†M. R. Wohlers, "Complex normalization of scattering matrices and the problem of compatible impedances," *IEEE Trans. Circuit Theory*, vol. CT-12, 1965, pp. 528–535.

then $j\omega_0$ must be a zero of $r_1(s)$ to the multiplicity at least two greater than that of $r_2(s)$;

(ii) if $s=j\omega_0$ is a pole of $Z_2(s)$, then

$$\lim_{s\to j\omega_0} Z_2(s)y_{a22}(s) \tag{11.88}$$

exists, and is a real number not greater than unity.

A proof of this extension will not be presented here. The reader is referred to Wohlers' original work.†

As an illustration, consider the constant impedance $Z_1(s)$, as discussed in Section 11.4.2, in which we showed that $Z_2(s)$ must be of the form

$$Z_2(s)=R+Z_{LC}(s) \tag{11.89}$$

as given in (11.71). For $Z_1(s)=1$, (11.75) reduces to

$$y_{a22}(s)=\frac{1}{2h_2(s)h_2(-s)}=\frac{1}{Z_2(s)+Z_2(-s)}$$

$$=\frac{1}{2R} \tag{11.90}$$

But according to (11.88), if $Z_{LC}(s)\neq 0$, then

$$\lim_{s\to j\omega_0} Z_2(s)y_{a22}(s)=\frac{1}{2}+\frac{1}{2R}\left[\lim_{s\to j\omega_0} Z_{LC}(s)\right] \tag{11.91}$$

$j\omega_0$ being a pole of $Z_{LC}(s)$, must be a real number not greater than unity, which is clearly violated since $Z_{LC}(s)$ is unbounded at $j\omega_0$. Hence, $Z_{LC}(s)=0$ and the only impedance that is compatible to a constant $Z_1(s)$ is another constant, and the coupling network is simply a transformer.

11.7 SUMMARY AND SUGGESTED READINGS

We began the chapter by indicating that the general problem of compatible impedances can best be approached in terms of the *scattering matrix* with complex normalization, which leads to a complete solution. Using an existence theorem for a normalized scattering matrix, necessary and sufficient conditions were derived for the compatibility of two positive-real impedances. A careful examination of these conditions reveals that the problem reduces essentially to the existence of an all-pass function that together with its derivatives interpolates to prescribed values at desired points in the closed RHS. Using the techniques outlined in Chapters 4 and 10, we demonstrated the synthesis of a lossless coupling two-port network for two compatible impedances.

It was shown that the only situation under which two equal compatible imped-

†*Loc. cit*

ances will produce a nontrivial two-port network is that of a constant-resistance all-pass structure, and that the only impedance that is compatible to a constant input impedance is another constant, the corresponding coupling network being simply a transformer.

For a complete treatment on the subject of complex normalization of a scattering matrix of an n-port network, the reader is referred to Chen [2]. For synthesis of active compatible impedances, see Ho and Balabanian [3]. An advanced treatment on passive compatible impedances and a discussion on other aspects of the compatible impedance problem can be found in Wohlers [6 and 7].

REFERENCES

1. W. K. Chen, "The scattering matrix and the passivity condition," *Matrix Tensor Quart.*, vol. 24, 1973, pp. 30–32 and 74–75.
2. W. K. Chen 1976, *Theory and Design of Broadband Matching Networks*, Chapter 2. New York: Pergamon Press.
3. C. W. Ho and N. Balabanian, "Synthesis of active and passive compatible impedances," *IEEE Trans. Circuit Theory*, vol. CT-14, 1967, pp. 118–128.
4. J. D. Schoeffler, "Impedance transformation using lossless networks," *IRE Trans. Circuit Theory*, vol. CT-8, 1961, pp. 131–137.
5. C. Satyanarayana and W. K. Chen, "Theory of broadband matching and the problem of compatible impedances," *J. Franklin Inst.*, vol. 309, 1980, pp. 267–280.
6. M. R. Wohlers, "Complex normalization of scattering matrices and the problem of compatible impedances," *IEEE Trans. Circuit Theory*, vol. CT-12, 1965, pp. 528–535.
7. M. R. Wohlers 1969, *Lumped and Distributed Passive Networks: A Generalized and Advanced Viewpoint*, Chapter 3. New York: Academic Press.
8. D. C. Youla, "Physical realizability criteria," *IRE Trans. Circuit Theory* (Special Supplement), vol. CT-7, 1960, pp. 50–68.
9. D. C. Youla, "An extension of the concept of scattering matrix," *IEEE Trans. Circuit Theory*, vol. CT-11, 1964, pp. 310–312.
10. D. C. Youla, L. J. Castriota, and H. J. Carlin, "Bounded real scattering matrices and the foundations of linear passive network theory," *IRE Trans. Circuit Theory*, vol. CT-6, 1959, pp. 102–124.

PROBLEMS

11.1 From (11.13), derive the impedance matrix $\mathbf{Z}(s)$ of (11.17).

11.2 Show that the scattering matrix (11.35) is para-unitary whenever it is defined.

11.3 Given the positive-real impedances

$$Z_1(s)=\frac{17s^3+18s^2+306s+1}{17s^3+289s^2+17s+289} \tag{11.92a}$$

$$Z_2(s)=\frac{3s+1}{3s+9} \tag{11.92b}$$

determine whether or not they are compatible. If the answer is affirmative, realize a lossless reciprocal coupling network.

11.4 Using (11.58), compute the scattering matrix of the two-port network N of Figure 11.7 normalizing to the 1-Ω resistance.

11.5 Using cascade synthesis technique, realize the impedance (11.59) as the input impedance of a lossless reciprocal two-port network terminated in a 1-Ω resistor. Replacing the 1-Ω resistance by the impedance (11.26b), compute the input impedance of the resulting network. Compare your result with (11.26a).

11.6 Using (11.63), compute the scattering matrix of the two-port network N of Figure 11.9 normalizing to the 1-Ω resistance.

11.7 Confirm that the scattering matrix (11.47a) is para-unitary whenever it is defined.

11.8 Determine the compatibility of the positive-real impedances

$$Z_1(s)=\frac{20s^3+39s^2+64s+12}{16s^3+48s^2+44s+64} \tag{11.93a}$$

$$Z_2(s)=\frac{2s+3}{4s+8} \tag{11.93b}$$

If they are compatible, realize a lossless reciprocal coupling network.

11.9 Justify the result given in (11.54). [*Hint:* $r_1(s)$ and $r_2(s)$ are even functions.]

11.10 Determine the compatibility of the positive-real impedances

$$Z_1(s)=\frac{2s^2+s+1}{s^2+s+2} \tag{11.94a}$$

$$Z_2(s)=\frac{s+1}{s+2} \tag{11.94b}$$

11.11 Referring to (11.69), demonstrate that

$$\operatorname{Re}\frac{1-\eta^2(s)}{2r(s)}=\operatorname{Re} Z(s) \tag{11.95}$$

Using this result, show that $y_{a22}(s)$ is positive real if and only if $[1-\eta^2(s)]\ r(s)$ is positive real. [*Hint:* Invoke the stipulated constraints.]

11.12 Given the positive-real impedances

$$Z_1(s)=\frac{s+1}{s+20} \tag{11.96a}$$

$$Z_2(s)=\frac{3s+1}{3s+9} \tag{11.96b}$$

determine whether or not they are compatible.

11.13 Show that $y_{a11}(s)$, as given in (11.73), is analytic on the entire real-frequency axis if $Z_1(s)$ is positive real.

11.14 Prove that the following two positive-real impedances are compatible:

$$Z_1(s)=\frac{48s^4+42s^3+36s^2+9s+3}{48s^5+42s^4+60s^3+30s^2+18s+2} \tag{11.97a}$$

$$Z_2(s) = \frac{6s^2 + 5s + 3}{s^2 + 3s + 2} \tag{11.97b}$$

Realize a lossless reciprocal coupling two-port network.

11.15 Determine the compatibility of the following two positive-real impedances:

$$Z_1(s) = \frac{s^3 + 7s^2 + 8s + 6}{s^2 + 6s + 4} \tag{11.98a}$$

$$Z_2(s) = \frac{s + 3}{s + 4} \tag{11.98b}$$

If they are compatible, obtain a lossless reciprocal two-port network.

11.16 Derive formula (11.49).

11.17 Determine the compatibility of the positive-real impedances

$$Z_1(s) = \frac{8s^2 + 12s + 7}{3(2s + 3)} \tag{11.99a}$$

$$Z_2(s) = \frac{4}{s + 4} \tag{11.99b}$$

If they are compatible, realize a lossless reciprocal coupling network.

SYMBOL INDEX

A′ matrix transpose, 28
Im imaginary part of, 33
LHS left half of *s*-plane, 55, 154
[*n*] largest integer not greater than *n*, 180
Re real part of, 22
RHS right half of *s*-plane, 34, 55, 154
$S^F_{x_1,x_2,\ldots,x_q} = S^F_{x_1} = S^F_{x_2} = \ldots = S^F_{x_q}$, 243
S^H_x sensitivity function of *H* with respect to *x*, 241
$\hat{S}^{s_j}_x$ root sensitivity of s_j with respect to *x*, 253
$S^{\alpha(\omega)}_x$ gain sensitivity of $\alpha(\omega)$ with respect to *x*, 249
$\bar{Z}(s)$ complex conjugate of $Z(s)$, 22, 81, 107

AUTHOR/ SUBJECT INDEX